green agriculture

NIPA® GENX ELECTRONIC RESOURCES & SOLUTIONS P. LTD.
New Delhi-110 034

www.nipaers.com
Browse, Search, Read & Buy...
Print Books
eChapters
eBooks
Publishers
Forthcoming
Subject Catalogues
New Titles
Online Resources on:
✓ Current Affairs
✓ Reasoning, Logic & Aptitude
✓ Competitive Examinations
✓ English Language Lab
✓ Writing & Pronunciation Tools
✓ Personality Development
✓ Online Programmes for Professionals
✓ Interview Preparation
Pay Using
PhonePe
Google Pay
paytm
PayPal
UPI
CC Avenue

green agriculture
newer technologies

Editor

Dr. Kambaska Kumar Behera
(Ph.D, Post Doc.)
Dept. of Bio-Science and Biotechnology,
Banasthali University, Banasthali,
Rajasthan-304022, India

NIPA® GENX ELECTRONIC RESOURCES & SOLUTIONS P. LTD.
New Delhi-110 034

NIPA® GENX ELECTRONIC
RESOURCES & SOLUTIONS P. LTD.

101,103, Vikas Surya Plaza, CU Block
L.S.C. Market, Pitam Pura, New Delhi-110 034
Ph : +91 11 27341616, 27341717, 27341718
E-mail: newindiapublishingagency@gmail.com
web: www.nipabooks.com

For customer assistance, please contact
Phone: + 91-11-27 34 17 17
Fax: + 91-11- 27 34 16 16
E-Mail: feedbacks@nipabooks.com

ISBN: 978-81-19215-34-8

Composed and Designed by NIPA®.

DEDICATED

TO

MY BELOVED PARENT

AND

FAMILY MEMBERS

FOREWORD

Green agriculture is a sort of system which carries out agricultural production with "green technology". It's basic content is based on biological diversity; keeping harmony between nature and economy during the course of agricultural development, by producing agricultural products in a pollution-free and nuisance-free environment. While mankind has made enormous progress in improving material welfare over the past two centuries, this progress has come at the lasting cost of degradation of our natural environment. About half of the forests that covered the earth are gone, ground water resources are being depleted and contaminated, enormous reductions in biodiversity have already taken place and, through increased burning of fossil fuels, the stability of the planet's climate is being threatened by global warming. In order to achieve a decent living standard for the populations in developing countries, especially the billions who currently still live in conditions of abject poverty, and the additional 2 billions who would have been added to the world's population by the mid-century, a much greater economic progress will be needed.

Climate change is raising the stakes for agricultural technology as the world population grows and the amount of arable land shrinks. According to the Environmental Protection Agency, farmers will have to deal with "increased potential" for extreme events like droughts, floods and heat waves," and "enduring changes in climate, water supply and soil moisture could make it less feasible to continue crop production in certain regions." More mouths to feed, plus less arable land and changing rainfall patterns, means growing demand for technology that lets farmers do more with less.

Modern biotechnology especially in agriculture has helped to do things that people could not do before. To give some examples, the technology has produced first generation of Genetically Modified (GM) crops such as herbicide-tolerant and insect-resistant crops. The examples of second-generation plants are the nutrient content like vitamin A-enriched rice and oils that have improved lipid profile. The third generation plants are being developed to provide specific health benefits. Although we find complaints about the health and environmental problems from GM crops, it has not yet been proved. Instead, the benefit of technology is much higher since it can contribute to increasing GDP, protecting biodiversity from excessive expansion of agricultural land and safeguarding human and animal health by reducing the use of agrochemicals. Further the revolution in Information Technology for precision farming, applied research in understanding ecological systems as production ecology and gene revolution for advancement in biotechnology have brought about major technological changes in agriculture and have lot of potentials. The major challenge is to determine the most suited and affordable technology by developing suitable strategy to make them competitive and sustainable.

This book is edited by Dr. Kambaska Kumar Behera, Banasthali University, Rajasthan-304022, India, a distinguished academician and scientist who is very active in Agri-biotechnology. The contributors of this book are mainly from academic institutions. This book shall be regarded as one of the best information sources on the present status of eco-friendly agro-technology. This book is unique in providing practical knowledge and ideas in employing to Indian agriculture as a safe guard for preventing hunger and environmental pollution and shall be kept as a text for instructing beginners and for sharing ideas with researchers and policy makers. I trust that this book "Green Agriculture : Newer Technologies" will contribute to the development of ideas on balancing sustainability and productivity in agriculture and the environment for the current.

DR. DURGA PRASAD MOHAPATRA
NIT, ROURKELA

PREFACE

Promoting sustainable agriculture development for the eradication of poverty by guaranteeing environmental sustainability through agro-based eco-friendly technology is termed as Green Technology. The technology behind the first "Green Revolution" in agriculture, during the 1960s and '70s, focused on boosting crop yields, to help feed growing populations and spur economic growth in Latin America and Asia. But that revolution wasn't all that green in the present sense of the term, relying heavily on irrigation. Many technocrats like Bill Gates, Microsoft founder and uber philanthropist, wants to help accelerate a second green revolution in agriculture, again boosting yields, but this time paying more mind to the environment and turning to some technologies that could help deliver a truly sustainable movement. At the World Food Prize in Des Moines, Iowa (2009) in his speech Gates advocated the farming techniques that are both environmentally responsible and highly productive.

Green agriculture encompasses a continuously evolving group of methods or materials, from newer techniques for generating energy to non-toxic cleaning products. It is that innovation which reduces waste by changing patterns of production and consumption. It is also defined as environmental healing technology, which reduces environmental damages by the products and technologies for peoples' conveniences. The book "Green Agriculture : Newer Technologies" carries 18 chapters and covers most of the on farm adopted technology developed by our distinguished scientist mainly focusing, how to save the planet earth during agricultural activities through modern technology. The attempt is to highlight the recent agro-based development through newer technologies to make Indian agriculture productive and eco-friendly.

ACKNOWLEDGEMENTS

I wish to acknowledge gratefully the assistance, encouragement, excitement and efforts of the contribution in organizing and publishing the book in a timely manner. In particular, I appreciate individuals who made this book more philosophical than technology-driven. The excellent contents and uniqueness of the individual chapters will make this book a text book on the use of biotechnology and its advancement for a moderately long term, which usually does not occur with books associated with this topic. I thank all the contributors for their understanding and patience. There are several people who deserve special thanks but few of them are "Prof. Aditya Shastri, Vice chancellor, Banasthali University, Prof. (Dr.) Durga Prasad Mohapatra, NIT, Rourkela, Odisha who has blessed and inspired a lot to complete this task. Words are not enough to thank's, Saudamini Muddali, Baji Rout, Sikha Niketan, High School, Sector-3, Rourkela-769002, Odisha and all my friends and colleagues also inspired a lot to end up the praise worthy task.

Last but not the least; I would like to assert my intricate apprehension of gratitude, best regards and complete surrenderness to my loving parents, family members and GOD for their incessant agitated patronize and enhearten to sparkle the supreme divinity within me. Really, I am no where in this world without my parents, family members and GOD.

Dr. Kambaska Kumar Behera
(Ph.D., Post Doc.)

CONTENTS

Green Agriculture : Newer Technologies, 2012

New India Publishing Agency, New Delhi (India)
e-mail : info@nipabooks.com; website : www.nipabooks.com

Chapter-1

Micro-irrigation for Sustainable Water Management in Agriculture

D.K. Singh[1], R.M. Singh[2], S.N.S. Chaurasia[1], R.N. Prasad[1] and R.B. Yadava[1]
1. Indian Institute of Vegetable Research, Shahanshapura, P.O., Jakheni, Varanasi-22305, (U.P.) India
2. Department of Farm Engineering, Institute of Agricultural Sciences, BHU-22105, U.P. India
E-mail : dharmendradksingh@rediffmail.com

SUMMARY

Agriculture is the largest user of fresh water for irrigation to fulfill the increasing demand of food, feed, fiber, fuel and other need of ever-increasing population. Water availability to agriculture is characterized by declining in share with time, over exploitation, improper mismanagement, deterioration in quality as well as environment. This could be addressed using sustainable management of irrigation water. As per the Brundtland Commission report of United Nations, sustainable development is the development that meets the needs of the present without compromising the ability of future generations to meet their own needs. Sustainable water management aims to improve water productivity, availability and quality over an area. Improved and efficient irrigation techniques and practices enhance water application efficiencies, and thereby reduction in associated damage to environment, which include water logging, soil salinity, runoff and leaching of nutrients and other chemicals polluting water sources, and over exploitation of ground water. Micro irrigation systems have been characterized to be the most efficient irrigation techniques. These have efficiencies even more than 90%, doubled water productivity with improved quality of produce, reduced runoff and leaching of nutrients and chemicals to water sources. Application of fertilizer through drip irrigation called fertigation could be adopted for efficient use and saving of fertilizer. Fertilizer savings of 20-60% and 8-41% increase in yields of horticulture and vegetable crops has been realized through fertigation. Micro irrigation is getting popularity having large potential due to its being economic feasible, widely accepted in society and environment friendly. Therefore, micro irrigation systems could be used for sustainable water management.

Introduction

National Water Policy of India (2002) emphasizes that water is a prime natural resource, a basic human need and a national asset. The planning and management of this resource and its optimal, economical and equitable use has become a matter of the utmost urgency. The fresh water is essential for sustaining all forms of life. Agriculture is the largest user of fresh water for irrigation. Water is being over exploited to fulfill the increasing demand of food, feed, fiber, fuel and other need of ever-increasing population. Also, conventional water management has been characterized to pose problems of water logging, increased salinity and alkalinity of soil, depletion of ground water table, salt water intrusion in ground water due to poor drainage and irrigation water management.

Improper use and mismanagement of available water, climate change, urbanizations and increasing population are considered as the main causes of increasing water scarcity. The increasing population and urbanization force the greater need of water for domestic consumption. Access to water is so crucial to human well-being and development that has now become a main concern for the global community. To emphasize the need for immediate action, year 2003 was designated as the International Year of Freshwater by the United Nations to provide an opportunity to raise awareness, motivate people, and mobilizes resources in order to manage water in a sustainable way. This paper presents prospects of micro irrigation for sustainable management of water.

1. Water Associated Crisis

The four main factors of population growth, urbanization, increased consumption and climate change are responsible for aggravated water scarcity (IPCC, 2007). Water scarcity is defined as per capita supplies less than 1700 m^3/year (IPCC 2007). It has been assessed that one in three people are already facing water shortages. About 1.2 billion people which is around one-fifth of the world's population live in areas of physical scarcity, while another one quarter of the world's population which is around 1.6 billion people live in countries lacking the infrastructure to take underground and rivers water, thus facing an economic water shortage. The Water use in the world (2005) presented in Table 1 indicates two-third of total water is utilized in agriculture (Comprehensive Assessment of Water Management in Agriculture, 2007).

Table 1 : Water use in the world (2005)

Activities	**Water utilization, %**
Agriculture	67
Household	9
Water supply	8
Electricity and gas	7
Manufacturing	4
Mining	2
Others	3

The nature of rainfall in India is characterized as erratic with uneven distribution over space and time. Only about 62 mha, which is around 44% of the cropped area is irrigated. There is a need to bring more cropped area under assured irrigation to increase agriculture productivity and production. The ultimate irrigation potential of the country has been estimated to be about 140 mha out of which about 76 mha could be from surface water and about 64 million hectare from ground water sources. The per capita water availability, in terms of annual average utilizable water resource of India, which was 3450 m^3 in the year 1951, was 1250 m^3 in 2005 and would reduce to 760 m^3 in the year 2050, placing it in the category of water stressed country. The water demand will increase in future (Figure 1).

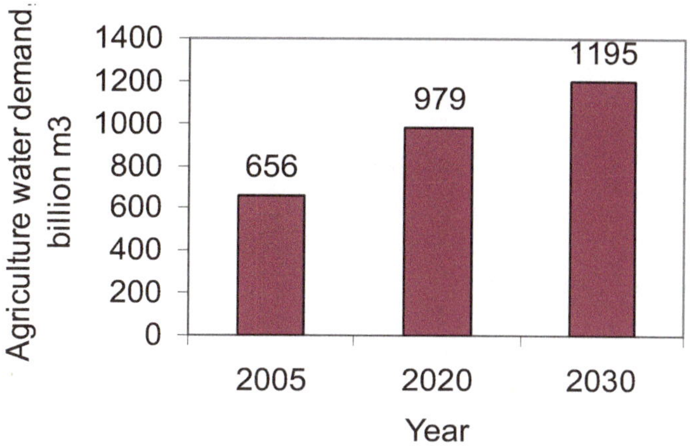

Figure 1 : The annual agriculture water demand in India is projected to increase at rate of 2.4% during 2005 to 2030. The water demand will be double by 2030.

The National Commission for Integrated Water Resources Development (NCIWRD) has assessed that about 83% of water is used for irrigation and remaining for domestic, industrial and other purposes. It has assessed the projected demand as 1180 billion cubic meters for the high demand scenario for the year 2050. The share of irrigation water in the overall demand has been estimated to reduce from the present level of about 83% to about 69% by the year 2050.

The run off from farms, untreated municipal sewage, domestic wastewater, agricultural run off, industrial effluents, fertilizer runoff, manure form livestock, all introduce pesticides, sewage and nutrient in the water. Also, some chemicals such as fluoride, arsenic, lead, nitrate, chlorinated solvents, petrochemicals contribute to water contamination. The use of contaminated water for drinking purpose leads to several diseases. The poor quality water causes water borne diseases such as, typhoid, cholera, and dysentery due to bacterial infections. These cause heavy economic loss to the individual and nation as a whole.

2. Sustainable Water Management

More and more water is required to fulfill the needs of food, feed, fiber, and energy through agriculture for increasing population. Its availability to agriculture is declining day by day. Also, over exploitation, improper use and mismanagement have led to deterioration of quantity and quality of water as well as environment. It is causing many problems to the society and responsible to loss of man, money and quality nature. This issue could be addressed with proper understanding of the problem and posing the best solution and planning through sustainable management to prevent irrigation water crises for present and future generations.

2.1. Sustainability

Sustainability is the capacity to endure. Sustainability interfaces with economics through the social and ecological consequences of economic activity. It is the potential for long-term maintenance of well being, which has environmental, economic, and social dimensions. Sustainability was one of eight goals in Millennium Development Goals agreed at the United Nations Millennium Summit in September 2000.

2.2. Sustainable development

The use of resources in a way to meet human needs for the present as well as future generations while preserving the environment may be referred as the Sustainable development. The Brundtland Commission report of United Nations in 1987, defines sustainable development as the development that meets the needs of the present without compromising the ability of future generations to meet their own needs. It is environmentally friendly forms of economic growth activities (agriculture, logging, manufacturing, etc.) that allow the continued production of a commodity without damage to the ecosystem (soil, water supplies, biodiversity or other surrounding resources) (www.ecokids.ca/pub/eco_info/glossary/ index. cfm). Sustainable develop-ment requires balance between the economy, ecology and society for present and future generations. It could be broken into three main parts viz, i) environmental sustainability, ii) economic sustainability, and iii) social sustainability.

2.3. Sustainable development models

In concentric circles model of sustainable development, three constituent parts of economy, society and environment could be represented as concentric circles (Ott, 2003). If social and environment development are considered then it is only bearable development. It is equitable if together social and economic development is taken care of. However, it is viable if environment is considered along with economic development. Sustainable development is achieved if all three developments are in balance, emphasizing sustainable development: at the union of three constituent parts (Adams 2006 and IUCN 2006).

The economic development, social development, and environmental protection are interdependent and mutually reinforcing pillars of sustainable development. Any imbalance between pillars will lead to collapse of the sustainable development. Strong foundation is required to balance three pillars for stability of sustainable development. Ethics and strong commitments should be treated as the foundation to the pillars of the sustainable development (Singh 2011).

3. Sustainable Water Management Model

The sustainable water management can be achieved by using appropriate techniques of water management to improve water productivity, and availability over an area for irrigation. A conceptualized model for sustainable water management could be visualized with its three constituent parts, the users, available water and environment all represented by circles and interacting each other (Singh et al, 2009).

The constituent parts Interact in such a way that users are affected and affecting the available water in a closed system of environment which is affecting both the users and available water and vice versa. The circle represented for user is expanding because of increasing population and the other circles remain same. The circle represented for available water remains same if assumed that quantity of available fresh water remains same in hydrologic cycle over a long period of duration. The circle represented for environment remains same under sustainable condition.

4. Strategies for Sustainable Water Management

The available water is utilized by the users, the ever increasing population. However, the water resources of country have not been harnessed adequately. On the one hand, most of the rainwater flows into sea without being harnessed and on the other hand, ground water is depleting due to its over-extraction. The following points need to be taken care of:

(i) Reduction of water losses involved during the course of water use or consumption to enhance the availability and productivity of water. It could be achieved through:
   a) Use of improved irrigation technology/ systems.
   b) Improving irrigation efficiencies, water use efficiencis and water foot print.

c) Optimum utilization of water
d) Adoption of precision agriculture using GIS, RS and GPS.
e) Adequate drainage to improve soil health and soil water plant interaction.

(ii) Enhancing the availability of both surface stored water and the ground water storage through the process of water harvesting, and conservation.

(iii) Practicing the artificial ground water recharge: it will enhance the availability of water to present and future generations without the compromising their ability to meet their own water demand.

5. Irrigation Methods

Several methods of irrigation from traditional surface flooding to modern drip irrigation systems have been evolved (Figure 2). Each irrigation method has its own advantages, disadvantages and suitability to various soil, crop, climatic and socio-economic situation.

The traditional surface irrigation poses numerous problems such as; soil salinity; seepage, conveyance and evaporative losses; higher energy cost; faster soil erosion; more wastage of fertilizer and other nutrients. Surface irrigation methods also result in higher weed population; increased operational difficulties and cost; uncontrolled, unmeasured and uneven water supply.

Surface irrigation methods are supply driven rather than crop demand driven and cause mismatch between the need of the crop and the quantity supplied. These are also associated to increased disease and increased cost of cultivation due to low production.

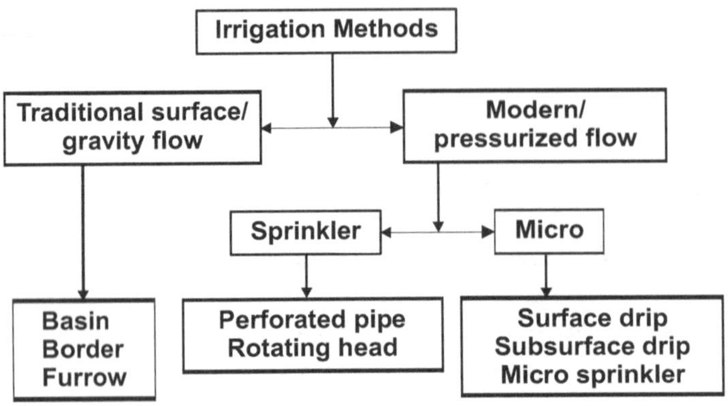

Figure 2 : Methods of irrigation are classified as traditional which flow under gravity and modern, called pressurised irrigation, that consists of sprinkler and micro irrigation.

6. Micro irrigation system

Micro irrigation system refers to low-pressure irrigation system that spray, mist, sprinkle or drip. It is the precise, slow and frequent application of water to the plants in the form of discrete drops, continuous drops or tiny streams, through the devices called the drippers or emitters or applicators located at selected points along water delivery lines called as laterals. It provides irrigation with high frequency application of water in and around the root zone of plants. It consists of network of pipes along with appropriate emitting device, control system, fertigation and filtration system and other fittings.

Emitting devices are to dissipate pressure and discharge water in micro irrigation system. These allow a small, uniform flow of water at a constant rate. Drip irrigation uses drippers as water emitting device. These are drippers, micro sprinkler, sprayers and mist according to requirement of crops. The water is also applied below the soil surface if emitting devices are buried in soil, called subsurface drip. The drip irrigation is also termed as trickle irrigation. Use of micro sprinklers is also getting popularity. However, drip irrigation; lateral placed at soil surface consists of the maximum area under micro irrigation

6.1. Suitability of Drip irrigation

Drip irrigation is adopted extensively in areas of acute water scarcity and especially for crops such as Coconut, Grape, Banana, Ber, Citrus, Sugarcane, Cotton, Maize, Tomato, Brinjal and plantation crops. It is suitable to almost 80 crops in different soil and agro climatic conditions.

6.2. Advantage of drip irrigation

Drip irrigation offers several advantages over traditional irrigation methods. These include as given below and listed in Table 3.6.:

1. Less weed growth: restricted weed growth to wetted areas only
2. Uniform and controlled water distribution closer to plant roots
3. Partial soil wetting: only to crop root zone
4. Moisture always at or near field capacity in the root zone
5. Less fertilizer/ nutrient loss due to localized application
6. High water distribution efficiency
7. Permits use of poor quality water
8. Recurrent irrigation is easily accomplished and has the effect of keeping soil moisture between field capacity and saturation
9. Soil factor plays less important role in frequency of irrigation
10. Levelling of the field not necessary
11. No soil erosion
12. Cultural operations are possible even during irrigation
13. Low labour cost
14. Possibility of regulating variation in water supply
15. Simultaneous of application water and fertilizer i.e. fertigation
16. low water delivery rate, low water pressure, precise placement of water
17. minimum application, field runoff and deep percolation losses

6.3. The limitations of drip irrigation

The drip irrigation has the following limitation:

1. Higher initial cost can be more than overhead systems
2. Durability of the components
3. Plant Performance: Studies indicate that most plants grow better when leaves are wetted as well
4. Poor system and crop performance if system not designed properly
5. Skilled persons are required to operate and maintain the drip system
6. Regular maintenance is required
7. Power source is required for operation of the system and automation except for the systems operated with gravity
8. Susceptible to crop damage in case of system failure for more than 3-5 days
9. Micro climate can not be controlled.

6.4. Moisture availability to crop

More than 80% of worlds irrigated land is under surface irrigation methods, yet its field level application efficiency is often only 40-50%. In contrast micro irrigation may achieve field level application efficiency of 80-90%, as surface runoff and deep percolation losses are minimized (Heerman *et al.*, 1990 and Postel, 2000). The moisture availability to crop under different irrigation methods has been depicted in Figure 3.

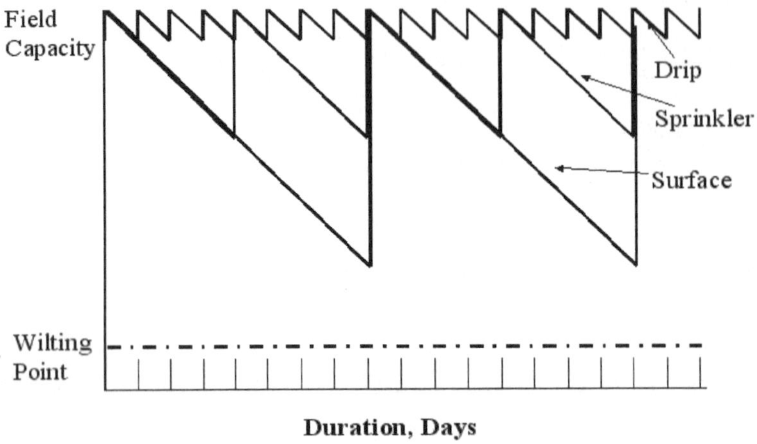

Figure 3 : Moisture availability to crop with different irrigation methods. Moisture availability to the crop through drip irrigation remains near to the field capacity and nearly uniform as compared with sprinkler and surface irrigation methods, which imposes less stress to the plants and results higher crop yield.

6.5. Enhanced productivity and water use efficiency

Using micro irrigation, water can be applied to crops efficiently with enhanced water use efficiency. Thus the micro irrigation may allow more crop cultivation per unit applied water, and allow crop cultivation in an area where available water is insufficient to irrigate through other irrigation methods. Fertilizers and chemicals could also be applied efficiently using this irrigation method. It has potential to save water and fertilizer with enhanced crop production.

Micro irrigation could utilize the marginal quality water also. The relative advantage of using MIS over conventional method of irrigation for some crops (INCID, 1994) has been presented in Table 2.

Table 2 : Advantage of micro over conventional irrigation of crops

Crop	Yield, t/ha		% Yield increase	% Water saving	Increase in water use efficiency, %
	Conventional	Drip			
Banana	57.5	87.5	52	45	176
Chilly	4.2	6.1	44	63	219
Grapes	26.4	32.5	23	48	136
Sweet Lime	100	150	50	61	289
Pomegranate	55	109	98	45	167
Tomato	32	48	50	31	119
Water Melon	24	45	88	36	195
Sugarcane	128	170	33	56	204

6.6. Potential of drip irrigation

Use of efficient pressurized irrigation systems such as drip and sprinklers irrigation can be used to improve water use efficiency including water saving and increased yield. In India, more than two million-hectare land under vegetables and high value crops is being irrigated through pressurized irrigation system with efficiencies of 60-95%. Pressurized irrigation includes sprinkler as well as micro irrigation consisting of micro sprinklers, subsurface drip and drip irrigation. Micro irrigation is growing fast in India. About one million-hectare land under vegetables and high value crops is being irrigated through drip irrigation in India during 2008 (Table 3).

Table 3 : Area under Micro irrigation (ha)

State	*Drip*	*Sprinkler*	*Total*
Rajasthan	15248	684748	699996
Maharashtra	462240	207205	669445
Haryana	6243	512657	518900
Andhra Pradesh	317935	182260	500195
Karnataka	169795	216978	386773
Gujarat	158727	128942	287669
Tamil Nadu	124951	26739	151689
West Bengal	123	150020	150143

Contd. ...

Madhya Pradesh	12518	104049	116567
Chattishgarh	2627	44763	47391
Orissa	3361	23187	26548
Uttar Pradesh	10577	10555	21132
Punjab	10427	10276	20702
Kerala	14119	2516	16635
Sikkim	80	10030	10110
Nagaland	0	3962	3962
Goa	762	332	1094
Himachal Pradesh	116	581	696
Arunachal Pradesh	613	0	613
Jharkhand	133	365	498
Bihar	107	180	287
Grand Total	1310956	2320586	3631542

(*Source :* NCPAH, New Delhi)

Government has planned to cover 14 M ha area under micro irrigation during XI Plan. The drip irrigation has vast potential in the country as depicted in Table 4. Net potential area for drip irrigation is estimated to be 21.27 mha for the country (Narayanamoorthy 2004).

Table 4 : Potential of drip and Sprinkler irrigation (area in m ha)

State	**Drip**	**Sprinkler**	**Total**
Utter Pradesh	3.35	10.53	13.88
Maharashtra	4.11	3.30	7.41
Madhya Pradesh	1.19	5.98	7.17
Rajasthan	1.86	4.01	5.87
Punjab	0.65	3.57	4.22
Others	0.84	3.04	3.88
Haryana	1.03	2.75	3.78
Karnataka	2.02	1.47	3.49
Bihar	0.84	2.44	3.28
Andhra Pradesh	2.16	0.99	3.15
Gujarat	2.04	0.89	2.93
West Bengal	1.12	0.99	2.11
Kerala	1.69	0.25	1.94
Tamilnadu	1.37	0.21	1.58
Orissa	0.84	0.42	1.26
Others	1.89	1.67	3.56
Total	27.00	42.51	69.51

To increasing the efficiency of water use and enhanced crop yield the area under micro irrigation is further likely to increase. The drip irrigation can be made more applicable for irrigating a wide range of agronomic, horticultural and fruit crops by installing the laterals below the soil surface, called subsurface drip irrigation.

6.7. Fertigation

Application of fertilizer through the drip irrigation system is called fertigation. It is the most advance and efficient practice of fertilization. Fertigation combines the two main factors in plant growth and development, water and nutrients. Fertigation is the most efficient method of fertilizer application, as it ensures application of the fertilizers directly to the plant roots (Patel & Rajput, 2001a). In fertigation, fertilizer application is made in small and frequent doses that fit within scheduled irrigation intervals matching the plant water use to avoid leaching. Fertilizer use efficiency upto 95 % can be achieved through drip fertigation (Table 5).

Table 5 : Fertilizer use efficiency

Nutrient	Fertilizer use efficiency, %		
	Soil application	**Drip**	**Drip and fertigation**
N	30-50	65	95
P	20	30	45
K	50	60	80

6.7.1. Advantages of fertigation

Fertigation has the following advantages:

- Less labour, equipment and energy needed for receiving, storing and fertilizer application
- Reduced soil compaction
- Prevents damage to crop during delivery
- No restrictions or limitation on application timing
- Accurate and uniform distribution for superior efficiency
- Application restricted to most active root zone which reduces waste
- Adaptability of nutrients supply to the growth curve resulting in better crop response
- Split applications for better control of run-off and leaching into groundwater
- Extremely efficient method of accurately delivering uniform, minute quantities of minor elements
- Complete adaptability to automation
- Can be used for other purposes, i.e. pestigation, soil amendments, maintenance
- Can overcome negative effects of saline/waste water.

6.7.2. Response of Fertigation to crops

Significant savings in the use of fertilizers and increase in yield of various crops have been reported by researchers (Table 6) (Anonymous, 2001) (Patel and Rajput, 2001a, 2001b, and 2002). The research conducted at various places on fertigation indicated that crop yields were substantially increased from 26 - 40 % in pomegranate, 11 - 41% in straw berry and 8 - 31% in grape

Table 6 : Response of crops under fertigation

Sl. No.	Crop	Saving in fertilizer, %	Increase in yield,%
1.	Okra	40	18
2.	Onion	40	16
3.	Banana	20	11
4.	Castor	60	32
5.	Cotton	30	20
6.	Potato	40	30
7.	Tomato	40	33
8.	Sugarcane	50	40

6.7.3. Potential of fertigation

Studies revealed significant fertilizer savings of 20-60% and 8-41% increase in yields of horticulture and vegetable crops. Fertigation could be adopted for increased yield and fertilizer saving. It is the essence of drip irrigation. Fertigation is a must in order to realize the full potential and benefits of drip irrigation system. With increasing area under drip irrigation in India, there is tremendous opportunity of using fertilizer through fertigation system for many crops.

7. Conclusions

The sustainable water management approach can be used to address the issues of irrigation water crises for present and future generations. It can be achieved by using appropriate techniques of water management to improve water productivity, and availability over an area for irrigation. Improved and efficient irrigation practices may be adopted to enhance water application efficiencies, and thereby reduction in associated damage to environment due to water logging, soil salinity, runoff and leaching of nutrients and other chemicals polluting to water sources, over exploitation of ground water. Micro irrigation systems have been characterized to be the most efficient irrigation techniques. These have efficiencies even more than 90%, doubled water productivity with improved quality of produce, reduced runoff and leaching of nutrients and chemicals to water sources. Fertigation could be adopted for efficient use and saving of fertilizer. Fertilizer savings of 20-60%

and 8-41% increase in yields of horticulture and vegetable crops has been realized through fertigation. Micro irrigation is getting popularity with large potential area because of being economic feasible, widely accepted in society and environment friendly. Therefore, micro irrigation systems could be used for sustainable water management.

References

Adams, W.M. (2006). The Future of Sustainability: Re-thinking Environment and Development in the Twenty-first Century. *Report of the IUCN Renowned Thinkers Meeting*, 29–31 January 2006.

Comprehensive Assessment of Water Management in Agriculture. (2007). Water for Food, Water for Life: A Comprehensive Assessment of Water Management in Agriculture. London: Earthscan, and Colombo: International Water Management Institute.

Heermann, D.F., Wallender, W.W. and Bos, M.G. (1990). Irrigation efficiency and uniformity. In: Management of farm irrigation system. Hoffman G.S., Howell, T.A. and Soloman, K.H. (Eds.), ASAE, St. Joseph, M.I.,:125-149.

IPCC Fourth Assessment Report. (2007). Intergovernmental Panel on Climate Change. http://www.ipcc.ch/ipccreports/ar4-wg1.htm.

IUCN. (2006). The Future of Sustainability: Re-thinking Environment and Development in the Twenty-first Century. Report of the IUCN Renowned Thinkers Meeting, 29–31 January 2006 http://cmsdata.iucn.org/downloads/ iucn_future_of_ sustanability.pdf.

INCID (1994). Drip irrigation in India. Indian National Committee on Irrigation and Drainage. New Delhi.

Millennium Development Goals. (2000). The United Nations Millennium Summit in September 2000.

Millennium Ecosystem Assessment. (2003). Ecosystems and Human Well-Being: A Framework for Assessment (Island Press, Washington, DC). http://en.wikipedia. org/wiki/ Millennium_Ecosystem_Assessment.

Narayanamoorthy, A. (2004). Drip irrigation in India: can it solve water scarcity. *Water Policy*. 6(2): 117-130.

NCPAH, (2001). National committee on plasticulture applications in Horticulture, progress report 2001, Department of Agriculture and Co-operation, MoA, GoI, New Delhi:44.

NCPAH. (2008). Area under Micro Irrigation. National Committee on Plasticulture Application in Agriculture, New Delhi.

National Commission for Integrated Water Resources Development. (1999), Report of the National Commission for Integrated Water Resources Development Plan. Vol. I, Ministry of Water Resources, Government of India, New Delhi.

National Water Policy. (2002). Ministry of Water Resources, Government of India, New Delhi.

Ott, K. (2003). The case for strong sustainability. In: Greifswald's Environmental Ethics. Ott, K. and P. Thapa (eds.). Greifswald: Steinbecker Verlag Ulrich Rose.

Patel, Nelam and Rajput,T.B.S. (2001,a). Effect of fertigation on growth and yield of onion. CBI Publication. 282: 451-454.

Patel, Neelam and Rajput, T.B.S. (2001, b). Fertigation of okra using commercially available granular fertilizers. Proceedings of International symposium on importance of potassium in nutrient management for sustainable crop production in India held at New Delhi from Dec. 3-5: 270-273.

Postel, S. (2000). Redesigning irrigated agriculture. In: State of the world 2000. Stark, L. (Eds). W. W. Norton and Co., New York: 39-58.

Rajput, T.B.S. and Patel, Neelam (2002). Fertigation: theory and practice. Indian Agricultural Research Institute Publication No. IARI/WTC/2002/2.

Singh, D.K., Chandola, V.K. and Singh, R.M., (2009). Sustainable management of irrigation water, An approach to address impact of climate change in agriculture. Flair Book Publication, New Delhi.

Singh, D.K. Singh, R.M. and Chandola, V.K. (2011). Sustainable water management in agriculture: issues and strategies to address. Proceeding of National Seminar on Sustainable Management of Water Resources. Banaras Hindu University, Varanasi, 14-15 January, 2011: 59-74.

United Nations. (1987). Report of the World Commission on Environment and Development. General Assembly Resolution 42/187, 11 December 1987.

□□□

Green Agriculture : Newer Technologies, 2012

New India Publishing Agency, New Delhi (India)
e-mail : info@nipabooks.com; website : www.nipabooks.com

Chapter-2

Prospective of Green Manuring in Organic Farming

Kambaska Kumar Behera
Dept. of Bio-Science and Biotechnology,
Banasthali University,Banasthali, Rajasthan – 304022, India
E-mail : kambaska@yahoo.co.in

SUMMARY

Green manuring is the practice of growing lush plants on the site into which organic matter directly incorporate to the soil while it is still fresh. Green manures, often known as cover crops, are plants which are grown to improve the structure and nutrient content of the soil. They are a cheap alternative to artificial fertilizers and an eco-friendly input for organic farming and can be used to complement animal manures. Growing a green manure is not the same as simply growing a legume crop, such as beans, in a rotation. Green manures are usually dug into the soil when the plants are still young, before they produce any crop and often before they flower. They are grown for their green leafy material which is high in nutrients and protects the soil. It is basically a holistic management system, which promotes and improves the health of the agro-ecosystem related to biodiversity, nutrient biocycles, soil microbial and biochemical activities.

Introduction

Indian agriculture has gone through major changes during the last century. It has developed from more or less subsistence farming to a highly intensive agricultural production particularly after green revolution. It relies heavily on the inputs of energy and other resources and production has become dependent on chemical pesticides and commercial fertilizers. With an increase pressure on farmland country wide, farmers are facing global problem such as soil erosion, wide spread deficiencies of macro and micro plant nutrients, salination, increasing pest problems. Besides these, damages to human beings, animals and nature from synthetic chemical pesticides have become more apparent in the different states particularly in Punjab and Haryana. Therefore, a change from high-input and chemically intensive agriculture to a more sustainable form of agriculture like organic farming is not only desirable but urgently needed to rectify the ill effects of chemically intensive agricultural production (Tandon, 1995; Stockdale et al. 2000) Organic farming is a new agricultural production system trying to work with the nature as much as possible. It involves us of locally and naturally available organic materials or agro-inputs to meet out the need of the production system without endangering our precious natural resources. In organic farming, the nutritional demands of crops are meet out mainly through on farm organic wastes, biofertilizers and green manure crops. To produce a healthy crop an organic farmer needs to manage the soil well. This involves considering soil life, soil nutrients and soil structure. Artificial fertilisers provide only short term nutrient supply to crops. They encourage plants to grow quickly but with soft growth which is less able to withstand drought, pests and disease. Artificial fertilisers do not feed soil life and do not add organic matter to the soil. This means that they do not help to build good soil structure, improve the soils water holding capacity or drainage. The soil is a living system. As well as the particles that make up the soil, it contains millions of different creatures. These creatures are very important for recycling nutrients. Feeding the soil with manure or compost feeds the whole variety of life in the soil which then turns this material into food for plant growth. This also adds nutrients and organic matter to the soil. Green manures also provide nutrients and organic matter. These are plants with high nitrogen content that are sown as part of a rotation and are dug into the soil when young.

In the present article an endeavor has been made to compile the available literature on important green manures crops of India, their potential in terms of biomass and nutrients, techniques for harnessing the maximum benefits from green manuring and their complementary effects on crops yields, physico-chemical properties of soil and quality of agricultural produce.

What is green manuring?

Green manuring is the practice of growing lush plants on the site into which you want to incorporate organic matter, then turning (tilling, ploughing and soading) into the soil while it is still fresh. The plant material used in this way is called a green manure (GM). The Criteria for the selection of green manure crops are as follows :

- Fast growing, Short duration and high nutrient accumulation ability
- Produces abundant and succulent shoots (foliages)
- Tolerance to shade; flood, drought and adverse temperatures
- Well adapted to the local condition
- Wide ecological adaptability and Photoperiod insensitivity
- Early onset of biological nitrogen fixation

- Have a high rate of nitrogen accumulation
- Timely release of nutrients
- Have high water-use efficiency when used in drier regions. The legume should use as little water as possible while still producing substantial quantities of top-growth,
- High seed production and High seed viability
- Ability to cross-inoculate or responsive to inoculation
- Pest and disease resistant
- High N sink in underground plant parts.

Generally the practice of green manuring is adopted in two ways:

(a)*In-situ* green manuring : In this system the short duration legume crops are grown and buried in the same site when they attain the age of 60-80 days after sowing. This system of on-site nutrient resource generation is most prevalent in northern and southern parts of India where rice is the major crop in the existing cropping systems. The most common green manure crops which are grown for *in-situ* green manuring are listed in Table 1.

Table 1 : Common legume crops for *in-situ* green manuring

S. No.	Common Name	Botanical Name	Growing season
1	Dhaincha	*Sesbania aculeta*	Zaid / Kharif
2	New Dhaincha	*Sesbania rostreta*	Zaid / Kharif
3	Sunnhemp	*Crotolaria juncia*	Zaid / Kharif
4	Mung	*Vigna radiata*	Zaid / Kharif
5	Cowpea	*Vigna unguiculata*	Kharif
6	Guar	*Cyamopsis tetragonoloba*	Kharif
7	Senji	*Melilotus alba*	Rabi
8	Berseem	*Trifolium alexandrium*	Rabi
9	Khesari	*Lathyrus sativus*	Rabi

(Singh et al.1992)

Besides these, some weeds growing in the fields can be buried in the soil while they are still fresh at the time of field prepration. The weeds which can be used ofr green manuring along with the common legumes to cater the nutritional requirement of crops under organic farming are presented in Table 2.

Table 2 : Mineral composition of certain weeds on dry weight basis

S. No.	Weed	Nutrient content (%)		
		N	P_2O_5	K_2O
1	*Amaranthus viridis*	3.16	0.06	4.51
2	*Cassia occidentalis*	3.08	1.56	2.31

3	*Chenopodium album*	2.59	0.37	4.34
4	*Cleome viscosa*	1.96	1.53	5.81
5	*Dactyloctenium aegyptium*	2.78	0.24	1.65
6	*Digitaria sanguinallis*	2.00	3.36	3.48
7	*Echinochloa crusgalli*	2.98	0.40	2.96
8	*Portulaca quadrifida*	2.40	0.09	5.57
9	*Solanum xanthocarpum*	2.56	1.63	2.12
10	*Trianthema partulacastrum*	2.34	0.30	1.15
11	*Eupatorium* spp.	2.93	0.49	1.47
12	*Parthenium hysterophorus*	2.66	0.88	1.29
13	*Eichhornia crassipes*	2.83	0.90	1.79

(Gupta, 2000)

(b) Green leaf manuring: Green leaves and tender plant parts of the plants are collected from shrubs and tress growing on bunds, degraded lands or nearby forest and they are turned down or mixed into the soil 15-30 days before sowing of the crops depending on the tenderness of the foliage or plant parts. The most common shrubs/trees used for green leaf manuring are given below :-

S.No.	Common Name	Botanical Name
1	Subabool	*Leucaena leucocephala*
2	Gliricidia	*Gliricidia maculata*
3	Wild daincha	*Sesbania speciosa*
4	Kranj	*Pongamia pinnata*
5	Neem	*Azadirachta indica*
6	Gulmohur	*Delonix regia*
7	Nirgundi(Begunia)	*Vitex negundu*
8	Baigaba(Jahaji)	*Jatropha curcas*

Status of GM crops in India

At present only 6.7 million hectares area is green manured which accounts for 4.5 per cent of net sown area (142 million ha) of the country (Agril Ststistics, 2005). The practice of green manuring is most common in rice growing states like A.P., U.P., Karnataka, Pumjab and Odisha which contribute 41.0, 16.0,11.0,6.0 and 5.0 per cent to the total area under green manuring in India respectively. Whereas, the share of Gujrat (3.0%), M.P. (3.0%), Himanchal Pradesh (2.0%) and Haryana (1.7%) is not very encouraging and concentrated effects are to be made out at all levels to bring more area under green manuring that too in irrigated areas if nutritional need of organic farming is to be made.

Biomass potential of GM crops

The benefits deriving from green manure crops are directly related to the amount of biomass and nutrients added in soil. Biomass production of green manure crops varies widely according to the species of the legumes, environmental conditions, soil fertility and crop management practices and age of green manure crops (Palaniappan, 1992). According to the estimates of Sigh et al. (1992), the *Sesbania aculeta* and *Crotolaria juncia* have higher rate of biomass production and both can produce dry matter to the tune of 16.0 to 19.0 t/ha within a short period of 45-60 days and on an average about 5.0t/ha dry matter can easily be produced which is sufficient for meeting out the nutritional demand of a crop grown either in *Kharif* or *rabi* season. Beside these, some weeds particularly *Eichhronia crassipes* have maximum rate of biomass production and one can get about 70 q/ha dry matter with in a period of 46-60 days and could be used for *ex-situ* green manuring (Duke 1981; Pieters 2004). Weeds like *Trianthema portulacastrum* and *Parthenium hysterophorus* are also found abundantly in different habitat with better nutrient content and dry matter production to cater the need of the organic agriculture (Table 3 and 4).

Table 3 : Nutrient compositions of common green manure crops and weeds on dry basis

Sl. No.	Crop/weed	Nutrient content						
		Major nutrients (%)*			Total micro nutrients (mg/kg)**			
		N	P_2O_5	K_2O	Zn	Fe	Cu	Mn
1	*Sesbania rostrata*	2.62	0.37	1.25	40	1968	36	210
2	*Sesbania speciosa*	3.98	0.24	1.30	50	480	44	110
3	*Crotolaria juncia*	2.86	0.34	1.27	30	1190	24	110
4	*Eichhornia crassipes*	2.83	0.90	1.79	50	470	19	420
5	*Trianthema spp.*	2.34	0.30	1.15	30	1992	19	200
6	*Parthenium hysterophorus*	2.66	0.8	1.29	70	470	19	160
7	*Gliricidia maculata*	3.49	0.22	1.30	30	550	19	150
8	*Cowpea residue*	1.70	0.28	1.25	-	-	-	-
9	*Mungbean residue*	2.21	0.26	1.26	-	-	-	-

[*Singh et al.* (1992) and Gupta (2000); Savitri et al (1999)]

Table 4 : Biomass and nutrient potentials of different green manures and weeds

S. No.	Crop	Dry matter in 45-60 DAS (q/ha)	Nutrient accumulation						
			Major nutrients (kg)			Total micro nutrients (g)			
			N	P_2O_5	K_2O	Zn	Fe	Cu	Mn
1	*Sesbania rostrata*	50.0	131.0	18.5	62.5	200	9840	180	1050
2	*Sesbania speciosa*	30.0	119.4	07.2	39.0	150	1440	132	330
3	*Crotolaria juncia*	52.5	150.2	47.3	93.9	262	2467	100	2205
4	*Eichhornia crassipes*	70.0	198.1	63.0	125.3	350	3290	133	2940
5	*Trianthema spp.*	25.0	58.5	07.5	28.7	75	4980	47	500
6	*Parthenium hysterophorus*	40.0	106.4	35.2	51.6	280	1880	76	640
7	*Gliricidia maculata*	36.0	125.6	7.9	46.8	108	1980	68	540

[Palaniappan (1992); Singh et al. (1992)]

Nutrient potential of GM crops

Almost all GM crops which are used for *in-situ* or *ex-situ* green manuring contains all the plant nutrients which are essential for completing the life cycle of any palnt grown in community. Among the different GM crops, dhainch (*Sesbania aculeta*) and Sunhemp (*Crotolaria juncia*) have higher accumulation of major and micro nutrients on account of more biomass production and better nutrient composition compared to food legumes which are inferior due to low contents of nutrients coupled with less dry matter production (Table 3, 4). Water hyacinth has great nutrient potentials and it could contribute 198 kg N, 63.0 kg P_2O_5, 125.3 kg K_2O and 350 g Zn, 3290 g Fe, 133 g Cu and 2940 g Mn when about 70 q/ha dry matter is added in the soil and could serve as better source of plant nutrients through *ex-situ* green manuring.

Table 5 : Biomass production and N accumulation of few green manure crops

Sl. No	Vernacular Name	Crop	Age (Days)	Dry matter (t/ha)	N accumulated in (kg)
1	Dhaincha	*Sesbania aculeata*	60	23.2	133
2	New Dhaincha	*Sesbania rostrata*	50	5.0	96
3	Sunnhemp	*Crotalaria juncea*	60	30.6	134
4	Cow pea	*Vigna unguiculata*	60	23.2	74
5	Cluster bean Guar	*Cyamopsis tetragonolobus*	50	3.2	91
6	*Pillipesara*	*Phaseolus Trilobus*	60	25.0	102

Techniques for harvesting the benefits of green manuring

The maximum benefit from green manuring can be obtained through better knowledge of suitable sowing time of GM crops, age or stage of GM crop for burial and time interval between burial and sowing of next crop.

(a) Sowing time of GM crops

Sowing time of GM crops for in-situ green manuring varies according to local conditions and farming situations. The green manures can be grown as catch crop during summer season particularly in irrigated agro-ecosystem. The quick growing crops like sunhemp and dhaincha etc. are sown in month of April-May and buries in the field in the month of June and July before the planting of main *kharif* crop. In rainfed areas as well as in intensive farming areas, the dhaincha is intercropped with paddy in 4:1 row proportion, whereas sunhemp and cowpea are intercropped in between the rows of widely spaced crops like cotton, maize and sugarcane and buried in the soil when these crops attain the age of 30-45 days with the help of hoes or mould board plough. In *kharif* fallow areas sunhemp, guar, cowpea and dhaincha are sown in June, July and buried in the soil during Ausgut-Spetember. This practice is most common in Punjab, U.P., Rajasthan, Bihar, Odisha and Madhya Pradesh.

(b) Stage of GM crop at burial

Knowledge of time of burial of GM crops is of utmost importance for deriving maximum benefit from green manuring. The chemical composition of most plants changes identically during growing season. During early period of crop growth its content of N, protein and water soluble constituents are maximum, while the amount of fiber, cellulose, hemicelluloses, lignin and the C:N ratio are also less. Therefore, tissues of immature plants usually decompose more rapidly as compared to those of matured plants. Singh et al (1992) reported while reviewing the nutrient transformations in soils amended with green manures that the green manure crops are to be buried in the soil when they are 2 months old and two weeks delay in the incorporation reduced their N content and increased the C:N ratio, cellulose, hemicelluloses and lignin contents.

(c) Time interval between burial of GM and sowing of next crop

Knowledge of time interval between burial of GM crops and sowing of next food crop for just to facilitate the complete decomposition of the turned in green matter is essential. Ghose et al. (1960) reported from their studies conducted at CRRI, (Cuttack) that the time interval was not so important when succulent green manure crop of eight weeks age was buried because transplanting of paddy immediately after burying of green manure crop was as good as any other treatment. But it was necessary to give the time interval of 4 – 8 weeks before planting paddy when the GM was 12 weeks of age.

Effect of green manuring on crop yields

In most of the studies conducted in different parts of the world, the crop yields under organic management are somewhat lower than conventional systems. In developing countries, organic farming methods provided similar outputs and income per labour day to that of high-input systems using inorganic fertilizers (Andrew and Hidka, 1998) In Sambalpur district of Orissa, the studies of Patra *et al.* (2000) revealed that there was reduction in the yield of rice to the tune of 15-23 per cent due to alone green manuring as compared to 100% recommended dose of NPK through fertilizers which produced the maximum yield (42.97 q/ha) of rice. In Samastipur (Bihar), Thakur et al. (1999) assessed the impact of green manuring on yield of rice-wheat system and reported that green manuring with dhaincha significantly improved the productivity of rice over other sources, whereas residual effect of green manuirng on succeeding wheat was marginal. Nair and Gupta (1999) also recorded 25% more yield of rice over no-green manuring which produced only 34.94q/ha of rice at Pantnagar, Uttranchal. Similar views were also endorsed by Hemalatha et al. (2000) at Madurai in Tamil Nadu

Effect of green manuring on soil productivity

The physico-chemical properties of soils are affected significantly due to addition of organic matter in the form of green manures particularly in plots receiving green manuring through *Sesbania rostrata* and *Crotolaria juncia*. Consequently, marked improvement in soil structures, infiltration rate, bulk density and water holding capacity of soil. It is evident from the results of studies of Badanur et al. (1990) that incorporation of subabool, sunhemp and crop residues were equally effective in increasing infiltration rate of soil while the water use efficiency of sorghum was increased significantly with the green leaf manuring of sunhemp, subabool and fertilizer application over crop residues. Aggregate stability and porosity increased identically with the addition of organic inputs particularly green manures and consequently it improves the soil aeration and water holding capacity of the soils under organic management (Droogers et al. 1996). Lower rates of runoff and soil erosion have also been reported under organic systems (Logsdon et al.1993; Reganold et al.1987). On the contrary in several studies no change in the physical properties of soil have been observed when managed organically or conventionally (Niggli et al. 1995)

The changes in chemical properties of soil could be predicted well through the nutrient budget. The results of several studies showed that nitrogen, phosphorus and potassium, organic carbon etc. ranged from deficit to surplus in organic farming systems (Fagerberg et al.1996; Nguyen et al. 1995 and Hemalatha et al. 2000). Green manuring under submerged conditions markedly increased Fe and Mn concentration and partial pressure of CO_2 and decreased pH, Eh and Zn in soil solution. The increase in Fe and Mn concentration attributed to the formation of complexes of Fe^{2+} and Mn^{2+} with organic acids produced during anaerobic decomposition of green manure and also due to sharp decrease in Eh and pH and increase in partial pressure of CO_2 (Sadana and Chahal 1995; Sadana and Nayyar 2000).

Effect of green manuring on product quality

There is no clear-cut scientific evidence with some studies showing increases in vitamin C, minerals and proteins (Lampkin 1990) because these are controlled by a complex of interaction in added manures and fertilizers. Therefore, it is difficult to distinguish the effects of the environment and farming systems on quality of crop products. Studies of Starling and Richards (1993) showed that organic wheat had lower protein levels compared to conventionally grown crops. Results of studies conducted at Madhurai in Tamil Nadu have also indicated that incorporation of Sesbania 12t/ha increased the optimum cooking time, total amylose content, crude protein in rice and reduced the gruel loss (%) of grain (Hemalatha et al. 2000).

Advantages of green manuring

- Improves the soil fertility by the increasing the humus content of the soil
- Adds nutrients and organic matters
- Improves the soil structure
- Improves soil aeration
- Increases the water holding capacity of the soil
- Increases the nutrient holding capacity of the soil
- Helps to control the soil born insects/ mite (Neem, Vitex, etc), nematodes (Marigold) and diseases (Crucifers)
- Used as a soil binder in the sloppy areas
- Conserves and recycles plant nutrients
- Protects the soil from erosion.
- Promotes habitat for natural enemies
- Serves as a good food for the earthworms
- Increases soil biodiversity of beneficial microbes by stimulating their growth.

References

Agricultural Statistics (2005). Agricultural statistics at a glance. Ministry of Agricultural, New Delhi.

Andew, D.A. and Hidaka, K. (1998). Yield loss in conventional and natural rice farming system. *Agric Ecosystem Environ*. 70 : 151-158

Badanur, V.P., Poleshi, C.M. and Naik, B.K. (1990). Effect of organic matter on crop yield and physical and chemical properties of vertisol. *J. Indian Soc. Soil Sci*., 38 (3) : 426-429

Droogerss, P., Fermont, A. and Bouma, J. (1996). Effects of ecological soil management on workabiltiy and trafficiability of a loamy soil in Netherlands. *Geoderma* 73 :131-145.

Duke, A.J.(1981) Handbook of Legumes of World Economic Importance Plenum Press, 233 Spring Street, New York, USA.

Fagerberg, B., Salomon, E. and Jonsson, S. (1996). Comparisons between conventional and ecological farming systems at Ojebyn : nutrient flows and balances. *Swedish J. Agric. Res.,* 26:169-180.

Ghosh, R.L.M., Ghate, M.B. and Subrahmanyam, T. (1960) Rice in India. I.C.A.R. New Delhi-12

Gupta, O.P. (1984). Scientific weed management in tropic and subtropics. *Today and Tomorrow Printers and Publishers*, New Delhi.

Hemalatha, M., Thriumurugan, V. and Balasubramaniam, R. (2000). Effect of organic sources of nitrogen on productivity, quality of rice and soil fertility in single crop wetlands. *Indian J. Agron.* 45 (3) : 564-567

Lampkin, N. (1990). Organic farming. Farming Press, *IPSWICH, U.K.,* pp 715

Logsdon, S.D., Radke, J.K. and Karlen, D.L. (1993). Comparison of alternative farming systems. I. Infiltration techniques *Am. J. Alt. Agric*. 8 : 15-20.

Nair, A.K. and Gupta, P.C. (1999) Effect of green manuring and nitrogen levels as nutrient uptake by rice and wheat under rice wheat sequence. *Indian J. Agron.*, 44 (4) : 659-663

Nguyen, M.L., Haynes, R.J. and Goh, K.M. (1995) Nutrient budgets and status in three pairs of conventional and alternative mixed cropping farms in Canterbury, Newzealan. *Agric. Ecosystems Environ.,* 52 : 149-162.

Niggli, U., Alfoldi, T., Mader, P., Pffiffner, L., Speciss, E. and Besson, J.M. (1995). DOK-Versuch Vergleichende Longzeit-Untersuchungen in den drei Anbausystemen biologish- dynamisch, Organish-biologish und Konventionell VI. *Synthese 1und 2 Fruchffloge periode. Schweiz Landw Fo. Snderheft DOK* 4 : 1-34.

Palaniappan, S.P. (1992) Green manuring : nutrient potential and management. (In) fertilizers, development and recyclable waste and biofertilziers. *Ed.* Tandon, H.L.S. Development and consultation Organization, New Delhi

Patra, A.K., Nayak, B.C. and Mishra, M.M. (2000). Integrated nutrient management in rice-wheat cropping system. *Indian J. Agron*., 45 (3): 453-457.

Pieters, A. J.(2004). Green Manuring: Principles and Practice/ Agrobios, Reprint. Jodhpur 356 p.ISBN 81-7754-188-9.

Reganold, J.P., Elliot, L.F and Unger, Y.L. (1987). Long term effects of organic and conventional farming on soil erosion. *Nature* 330: 370-372.

Sadana, U.S. and Nayyar, N.K. (2000). Amelioration of iron deficiency in rice and transformations of sol iorn in course textural soils of Punjab. Indian J. Pl. Nutr. 23 (11-12): 2061 – 2069.

Sadana, V.S. and Chahal, D.S. (1995). Iron availability, electrochemical changes and nutrient content of rice as influenced by green manuring in a submerged soil (In) : Iron nutrition in soil and plants (Netherlands) *Ed.* J. Abadia pp 105-109.

Savithri, P., Poongothai, S., Joseph, B and Vennila, R.K. (1999). Non-conventional sources of micronutrients for sustainable soil health and yield of rice-green gram cropping system. *Oryza*, 36 (3) : 219-222.

Singh, V., Singh, B. and Khind, C.S. (1992). Nutrient transformations in soil amended with green manures *Adv. Soil. Sci*., 20 : 238-298

Starling, W. and Richards, M.C. (1993). Quality of commercial samples of organically grown wheat. *Aspects Appl. Boil*., 36 : 205 – 209.

Stockdale, E.A., Lampkin, N.H., Hovi, M., Keatinge, R., Lennartsson, E.K.M., Macdonald, D.W., Padel, S., Tattersall, F.H., Wolfe, M.S. and Watson, C.A. (2000). Agronomic and environmental implications of organic farming systems. *Adv. Agron.* 70 : 261-327.

Tandon, H.L.S. (1995) Sustainable agriculture for productivity and resource quality- A model for India. *Fert. News,* 40 (10) ; 39-45

Thakur, R.B., Choudhary, S.K. and Jha, G. (1999). Effect of combined use of green manure crop and nitrogen on productivity of rice-wheat under lowland rice. *Indian J. Agron.*, 44 (4): 664-668.

□□□

Green Agriculture : Newer Technologies, 2012
© *Kambaska Kumar Behera (ed.), pp. 27-71*
New India Publishing Agency, New Delhi (India)
e-mail : info@nipabooks.com; website : www.nipabooks.com

Chapter-3

Citrus Rootstocks in India : Problems and Prospects

Sandeep Singh, H.S. Rattanpal, P.S. Aulakh,
D.R. Sharma, A.K. Sangwan, Anita Arora and Sarabjit Kaur
Department of Horticulture, Punjab Agricultural University,
Ludhiana 141 004 Punjab, India
E-mail : sandeep_pau.1974@yahoo.com

SUMMARY

Citrus is third most important fruit crop in India after banana and mango. Citrus crops have occupied an area of 8.43 lakh ha with total production of 75.7 lakh tons and the productivity of 9.9 tons/ha in India. Such a quantum of citrus production contributes 13 per cent of total fruit production demonstrating the significance of citrus in horticultural research and development in India. Rootstocks are known to have profound effect on the vigour, precocity, productivity, internal quality, and longevity of the scion varieties grafted on them. They are also known to influence the susceptibility of trees to various diseases and insects. Furthermore, it is recognized that optimum performances depends on the proper selection of rootstocks for a given set of growing conditions. The most commonly and widely used citrus rootstocks in India are Rough lemon, Cleopatra and Pectinifera. Rough lemon (Jatti Khatti) is the most important and widely used rootstock in the country for most of the scions followed by Kharna Khatta being more popular in Uttar Pradesh. While the problem of rootstock selection is important in all major fruit crops, it is of even greater importance in citrus due to tristeza virus disease now present in many areas of the world. It is also possible that citrus decline in India may be due partly to stionic incompatibilities. This chapter deals with problems and prospects of rootstocks in citrus in India, particularly related to horticultural aspects, insect-pests, diseases, nematodes, physiological disorders, post harvest quality, aspects related to tissue culture and rootstock breeding.

1. Introduction

Citrus is considered as an important cash crop fruit in world trade of horticulture. Over the years, citrus has come off age to establish as a vibrant industry through many blockbuster technologies. Citrus is third most important fruit crop in India after banana and mango. The most important commercial citrus groups in India are the mandarin (*Citrus reticulata* Blanco), followed by sweet orange (*Citrus sinensis* Osbeck) and acid lime (*Citrus aurantifolia* Swingle). Commercially, Kinnow mandarin in grown in Punjab, Haryana, Himachal, western parts of Rajasthan and Uttar Pradesh; Khasi mandarin in north-eastern regions comprising states of Assam, Mizoram, Meghalaya, Manipur, Nagaland, Arunachal Pradesh, Tripura, Sikkim; Darjeeling mandarin in Darjeeling; Coorg mandarin in Coorg area; Acid lime in Kheda district of Gujarat, Akola in Maharashtra and Periyakulam in Tamilnadu; Nagpur mandarin in Vidarbha region of Maharashtra and adjoining areas of Madhya Pradesh; Mosambi in Marathwada region of Maharashtra and Sathgudi in Andhra Pradesh (Tirupati, Anantpur) (Singh and Aulakh, 2010). Citrus has occupied an area of 8.43 lakh ha with total production of 75.7 lakh tons and the productivity of 9.9 tons/ha in India. Such a quantum of citrus production contributes 13 per cent of total fruit production demonstrating the significance of citrus in horticultural research and development in India (Singh, 2010). In Punjab, the farmers took the lead and brought back the lost glory in citriculture. Now Kinnow has emerged number one fruit of the state, both from area and production point of view. The total area under Citrus in Punjab is 41821 ha out of which Kinnow occupies major share of 38837 ha. Kinnow alone produces 8.7 lakh tones. The districts of Ferozepur, Hoshiarpur, Muktsar and Bathinda occupy 87 per cent of the area under Kinnow in the state. Sweet orange is grown only in the arid-irrigated region of Punjab i.e., the districts of Ferozepur, Bathinda and Muktsar. Punjab govt. has established five Citrus Estates in Bhunga and Hoshiarpur in District Hoshiarpur; Abohar and Tahliwalan Jattan in District Ferozepur and Badal in District Muktsar (Brar, 2010).

In Punjab, the need of suitable rootstocks was felt during the early sixties when large number of citrus plantations declined due to exocortis viroid. Since then, systematic work undertaken by the scientists of PAU on stionic relationships in citrus has been a continuous process to solve the burning problems of decline and diseases. Now-a-days, *Phytophthora* foot and root rot, adaptability to various soil types, alkalinity, salinity and excessive $Caco_3$ have been recognized as area specific problems for the citrus industry of northern India especially Punjab. Therefore, it is an appropriate time to review the general citrus rootstock situation as prevailing at this time and to consider the rootstocks available to the citrus growers in north-western India with the change of time.

1.1 Rootstock studies

Rootstocks are known to have profound effect on the vigour, precocity, productivity, internal quality, and longevity of the scion varieties grafted on them.

They are also known to influence the susceptibility of trees to various diseases and insects. Furthermore, it is recognized that optimum performances depends on the proper selection of rootstocks for a given set of growing conditions. While the problem of rootstock selection is important in all major fruit crops, it is of even greater importance in citrus due to tristeza virus disease now present in many areas of the world. It is also possible that citrus decline in India may be due partly to stionic incompatabilities (Chadha, 1970).

1.2 Prevalent rootstocks in India

The most commonly and widely used citrus rootstocks in North-western India are Rough lemon, Cleopatra and Pectinifera. Rough lemon (Jatti Khatti) is the most important and widely used rootstock in the country for most of the scions followed by Kharna Khatta being more popular in Uttar Pradesh. Rough lemon is vigorous and it has been well established and is generally considered adequate for the problems faced by the citrus industry in North-western India. Rough lemon and Kharna Khatta for Grapefruit are usually employed as rootstocks in the Punjab. Cleopatra has been recommended for Blood Red and Pectinifera is recommended for Mosambi. In Uttar Pradesh, Rough lemon and Kharna Khatta are the principal rootstocks. But Sweet lime is recommended for Mosambi in the wet subtropics with a high temperature. In Haryana and Rajasthan, mainly Rough lemon is used for mandarins and Grape fruits.

2. Citrus Genetic Resources

A total of 33 genera and 224 species of citrus and wild relatives (Aurantoides subfamily) are globally reported which can be used for citrus improvement. Citrus and its relatives of the subfamily Aurantioideae are considered native of Southeast Asia. North-east India, south China and north Myanmar, acclaimed to be the primitive centres of origin of contemporary *Citrus* species. Among different systems of classifications adopted by taxonomists, Swingle and Tanaka System of classification are widely accepted and acknowledged by majority of scientists' world over. Many critics often claim that Swingle put too many types under one species, like all the mandarins are kept under *C. reticulata*, whereas Tanaka was so liberal in allotment of species, name even different hybrids were given a species name. Swingle reported 16 species under citrus, whereas Tanaka reported 145-162 species. Hodgson revised the taxonomic system of Swingle and Tanaka and with his own observations suggested the system having 36 species, whereas others proposed different system of classification separating *Citrus* into 6-42 species. As per classification, subgenus Papeda comprises maximum number of truly wild species of citrus (Table 1). These species are mostly reported to grow wild in forest in India, China, Myanmar and Malaysia. The true citrus species are grouped into 6 genera (*Citrus, Eremocitrus, Fortunella, Clymenia, Microcitrus* and *Poncirus*). The genus *Eremocitrus* and *Microcitrus* are found in Australia, *Clymenia* in New

Guinea, *Poncirus* and *Fortunella* distributed in China, and genus *Citrus* distributed in India, China, Myanmar and Malaysia (Singh and Aulakh 2010).

Table 1 : List of true wild Citrus species and their distribution in the world

S. No.	Common name	Species	Natural Habitat
1	Khasi Papeda	*Citrus latipes*	Shillong area of Meghalaya (India) and Northern Myanmar
2	Honghe Papeda	*C. hongoensis*	Yunnan of China
3	Small flowered Papeda	*C. micrantha*	Phillipines
4	Celebes Papeda	*C. celebica*	Northeastern Celebes and Phillipines
5	Melanesian Papeda	*C. macroptera*	Shella and Dawki (Meghalaya); Jampui area of Tripura and Manipur; S.W. China, Thailand, Srilanka, Northern Myanmar, Malaysia, New Guinea and Phillipines
6	Maurities Papeda	*C. hystrix*	S.W.China, Srilanka, North eastern India, Northern Myanmar, Malaysia and Phillipines
7	Ichang Papeda	*C. ichangensis*	Nagaland (India), Central and southwestern China
8	Citrus *halimi*	*C. halimi*	A jungle tree species in Malaysia and Southern Thailand
9	Tachibana	*C. tachibana*	Southwestern China and Japan
10	Ban-Nimboo	*Glycosmis pentaphylla*	Nainital area of Uttaranchal (India) and Hainan, Yunnan provinces in China
11	Indian wild orange	*C. indica*	Turu area of Meghalaya, Nagaland and Assam (India)
12	Fuming evergreen trifoliate orange	*Poncirus polyandra*	Yunnan province of China
13	Daoxian wild tangerine	*C. daoxianensis*	Southwestern China
14	Mangshan wild mandarin	*C. mangshanensis*	Southwestern China
15	Athanni	*C. rugulosa*	Nainital hills of Uttaranchal India)
16	Ada jamir	*C. assamensis*	Meghalaya and Assam (India)
17	Sour Pummelo	*C. megaloxycarpa*	NEH region of India, maximum concentration in Mamit area of Mizoram

(*Source :* Singh and Aulakh (2010))

In India, number of commercial citrus cultivars has shown their worth and presented in table 2.

Table 2 : Commercial Citrus fruits in India

Species	**Scientific name**	**Cultivars**	**Distribution**
Mandarin	*Citrus reticulata* Blanco	Nagpur mandarin, Khasi mandarin, Coorg mandarin, Kamala orange, Darjeeling mandarin and Kinnow	Maharashtra, Karnataka, Tamil Nadu, Madhya Pradesh, Punjab, Rajasthan, North Eastern region, West Bengal
Sweet orange	*C. sinensis* Osbeck	Mosambi, Sathgudi, Blood Red Malta, Hamlin Sweet, Pineapple, Valencia and Jaffa	Deccan Plateau, Punjab, Andhra Pradesh, Tamil Nadu
Acid lime	*C. aurantifolia*	Kagzi lime, Vikram, Pramalini, Jai Devi, Sai Sharbati, Baramasi Kagzi, Seedless lime (*C. latifolia* Tan.)	Andhra Pradesh, Maharashtra, Gujarat, Bihar, Karnataka, Tamil Nadu
Lemon	*C. limon* Burm.f.	Baramasia, Assam lemon, Pant lemon, Gandhraj, Seville, Nepalli Oblong, Nepalli lemon, Italian, Eureka	All over the country in homestead gardens
Grapefruit	*C. paradisi* Macf.	Duncan, Marsh Seedless, Ruby Red, Sharanpur Special	Limited in cultivation (NE region, foothills of Uttarachal and HP, Karnataka)
Pummelo	-	Chakotra, Gagar, local selections (Red fleshed and white fleshed)	In home gardens
Gajanimnna	*C. pennivesiculata* Tan.	Gajanimma, Baduvapuli	Home gardens in south India
Belladikithuli	*C. maderaspatana* Tan.	Valadipudi, Kichli, Belladikithuli	Commercial orchards in Guntur (Andhra Pradesh)

(*Source :* Singh and Aulakh (2010))

In India, there are number of *Citrus* germplasm collection centres and presented in table 3.

Table 3 : Citrus Germplasm collection in India

S. No	Station/University
1	National Research Centre for Citrus (NRCC), Nagpur
2	ICAR Regional Station, Umiam, Meghalaya
3	Regional Station, Abohar, Punjab
4	Dr. PDKV, Akola, Maharashtra
5	Regional Research Station, Chethalli, Karnataka
6	Indian Institute of Horticultural Research, Bangalore
7	RFRS (Dr. PDKV), Katol, Maharashtra
8	Horticultural Research Station (TNAU), Periyakulam, Tamil Nadu
9	Mahatma Phule Krishi, Vidyapeeth, Rahuri, Maharashtra
10	HRRS (YSPUHF), Dhaulakuan, Sirmour, HP
11	Citrus Research Station (AAU), Tinsukia, Assam
12	S.V. College of Agriculture, Tirupati, AP
13	PAU, Ludhiana

(*Source :* Singh and Aulakh (2010))

Citrus rootstock species of India has been compiled in table 4.

Table 4 : Citrus rootstock species of India

Species	Variability
Rough lemon (*C. jambhiri* Lush)	Mithi, Mitha-tulia, Jambhiri Kotagiri, Renuka lemon, Florida (C. jambhiri Lush) Rough, jatti Khatta, Jullandhuri, Jatti Khatti, Pathankot, Jambhiri Pedagaon, Jambhiri Kodur, Italian-76, Soh-myndong, Khatta Patiala, Jambhiri Brown, Rough lemon Wynad, Rough lemon Chethalli, Jambhiri Poona, Jambhiri Local Angul, Jambhiri Nagpur, Khatta, Jallandar Khatti, Khata Jamir, Moogunimbe, Brazlian Rough lemon, Chase Rough, Eastes Rough, Florida Rough-8748, Limoneria, Rough lemon 58-III-4, Rough lemon Tharssa, Rough lemon South Africa, Tulia Renga
Rangpur lime (*C. limonia* Osbeck)	Brazil orange, Brazlian Rangpur, Florida Rangpur-8748, (Citrus limonia Osbeck) Khasi lime, Pookling Ming, L-2 Rangpur lime, Lime Cravo Brazil, Marmalade Orange, Philippine Red lime, Sauranto, Sindhuri, Nemutenga, Sylhet lime, Pink Fleshed lime

Small fruited mandarins (*C. reshni* Tan.)	Billikichilli, Kodakithuli, Citrus China, Cleopatra mandarin, (C. reshni Tan.) Morocco, CM Australia Hallikithuli, Seiz-in-com, Soh-seim, Karpura Tenga
Sour orange (*C. aurantium* L.)	Karun Jamir, Molepuli, Seville orange, Willow leaf sour orange, Herale, Apple Bee Smooth, Godh-hunta Chinnato
Sweet lime (*C. limettioides* Tan.)	Aabbu, Mithanembu, Sweet lime, Mitha, Sarbati lebu, Chinni
Trifoliate orange (*P. trifoliata* L).	Christensen, English Dwarf, English Large, English Small, (Poncirus trifoliata L). Flying Dragon, Gotha , Large Flowered, Pomeroy, Ponciroy, Rubidoux Trifoliate Argentina, Trifoliate Florida, Trifoliate William
Trifoliate hybrids	Citrangequat, Citrange Morton, Rusk citrange, Citrumello, Swingle citrumello, Savage, Troyer citrange, Carrizo citrange, Eccaten citrange
Karna Khatta (*C. karna* Raf.)	Karna Khatta, Karna, Soh-sarkar, Karna Nimboo

(*Source :* Chadha and Singh (1996))

2.1 Problems and prospects of existing rootstocks

(a) Fruit characters

i) Physical characters: Fruit weight, fruit length, fruit yield, fruit diameter, rind thickness, rind weight, rind percentage.

ii) Quality: Number of segments, number of seeds, seed weight, juice volume, juice weight, juice percentage and Rag.

iii) Chemical characteristics: Total soluble solids, acidity, ascorbic acid, reducing sugars, nonreducing sugars, total sugars.

(b) Plant characters : Incompatability, vigour, precocity, productivity, stock:scion girth, number of fruits/plant, fruit quality, canopy volume, flowering, Shoot length and no. of leaves/shoot, longevity of scion varieties, tree health, tree survival, maturity, dwarfing behaviour, bud take, uptake of nutrients like N, K, Mg, Fe, Mn, P, Ca, Zn and Cu, adaptability to various soil types, excessive $Caco_3$, salinity, pH, reaction to insect-pests, diseases (Exocortis, greening, Citrus tristeza virus, Citrus yellow mosaic virus and Citrus yellow vein clearing virus, citrus canker, fruit drop Indian citrus ringspot virus, Yellow corky vein disease, hop stunt viroid, xyloporosis viruses), chlorosis and nematode (*Tylenchulus semipenetrans*),

susceptibility/tolerance to cold, frost, drought, salinity, alkalinity, granulation, bud union (smooth/creased/bottle neck),
(c) Seed character like germination.

(d) Post harvest qualities like fruit physiological weight loss (PWL) and shelf life

(e) Tolerance to physiological disorders like sunscald

2.2 Rootstocks trials in Punjab

PAU, Ludhiana made comprehensive efforts to overcome all these problems by the use of resistant and most suitable rootstocks. In this direction almost all the available rootstocks were tested in more than 17 rootstock trials at different research stations of PAU with all probable scion varieties to determine the suitability of different rootstocks for the varieties adapted to local conditions.

The earliest known rootstock trial in India was conducted by Brown (1920) at Peshawar (now in West Punjab) using two scion varieties, viz., Malta (*C. sinensis* Osbeck) and Sangtra (*C. reticulata* Blanco) and four rootstocks varieties, viz., sour orange, galgal citron (*C. medica* Linn.), Rough lemon (*C. limon* Burm. f.) and sweet lime. It was found that Malta made vigorous growth on Rough lemon, medium growth on sweet lime, and poor growth on sour orange and citron. Sangtra made best growth on sweet lime and fairly satisfactory growth on sour orange but was markedly dwarfed and unsatisfactory on Rough lemon and citron. These results showed that there was a remarkable influence of the rootstock on the scion. In 1962, an observational rootstock trial was laid out at the Regional Fruit Research Station, Abohar, with a view of developing preliminary information on the performance of different rootstocks for sweet orange (Anonymous 1966). In this trial, 17 stocks were used for Blood Red scion but only 10 survived. Of the surviving trees, only those on Jatti Khatti and Iran lemon performed satisfactorily. Plants on other rootstocks including Savage and Troyer citrange, calamondin, Sacaton, *Severinia buxifolia* (Poir.) Ten., *C. pectinifera,* kharna khatta and *C. taiwanica* Tanaka and Shimada did not make the desired growth (Chadha 1970).

2.3 Performance of different Citrus station combinations evaluated at Punjab Agricultural University, Ludhiana

Elaborate rootstocks studies were conducted at main Campus and different Research Stations of Punjab Agricultural University (PAU), Ludhiana starting from 1968. Salient achievements of these scientific investigations are discussed below (Table 5).

Table 5 : Citrus varieties evaluated on different rootstocks at PAU, Ludhiana

S. No.	Variety	Rootstock
1.	Kinnow mandarin	Carrizo Cleopatra Jatti Khatti Kharna Khata Nasvaran Rangpur Lime Rubidoux Troyer
2.	Jaffa sweet orange	Carrizo Citrumelo (Swingle) Cleopatra Jatti Khatti Kharna Khata Nasvaran Pectinifera Pineapple Rangpur Lime Rubidoux Troyer
3.	Mosambi sweet orange	Carrizo Citrumelo (Swingle) Cleopatra Jambhiri Jatti Khatti Kharna Khata Nasvaran Pectinifera Rangpur Lime Rubidoux Troyer
4.	Blood Red sweet orange	Carrizo Citrumelo (Swingle) Cleopatra Jambhiri Jatti Khatti Kharna Khata Nasvaran Pectinifera Rangpur Lime Rubidoux Troyer

Contd. ...

5.	Pineapple sweet orange	Carrizo Jatti Khatti Kharna Khata Troyer
6.	Valencia Late sweet orange	Carrizo Citrumelo (Swingle) Cleopatra Jatti Khatti Kharna Khata Nasvaran Pectinifera Rangpur Lime Rubidoux Troyer
7.	Hamlin sweet orange	Carrizo Jatti Khatti Kharna Khata Troyer Sacaton Savage
8.	Campbell Valencia	Carrizo Jatti Khatti Kharna Khata Troyer
9.	Dancy tangerine	Carrizo Citrumelo (Swingle) Cleopatra Jatti Khatti Kharna Khata Nasvaran Nucellar Blood Red Pectinifera Rangpur Lime Rubidoux Troyer
10.	Wilking mandarin	Carrizo Jatti Khatti Kharna Khata Troyer
11.	Pearl tangelo	Carrizo Jatti Khatti Kharna Khata Troyer

12.	Marsh Seedless Grapefruit	Carrizo Citrumelo (Swingle) Cleopatra Jatti Khatti Nasvaran Rangpur Lime Troyer

Source : Chohan *et al.* (1986) ; Chohan *et al.* ; (1990) ; Vij. and Kumar (1990) ; Dhatt and Singh (1993); Bal and Chohan (1982); Mehrotra *et al.*, (1984); Anonymous (1974 -2003).

1. Kinnow : Trial was planted using Jatti Khatti, Kharna Khata, Troyer, Carrizo, Rangpur Lime, Nasnaran, Cleopatra and Rubidoux rootstocks. Kinnow on Jatti Khatti and Kharna Khatta, performed well in growth. But the trees on Jatti Khatti rootstocks bore more number of fruits and proved to be better having comparatively good tree health and fruit quality as compared to other rootstock.

2. Sweet oranges

(i) Jaffa : Rootstocks namely Jatti Khatti, Kharna Khatta, Carrizo, Troyer, Pectinifera, Citrumelo, Pineapple, Rangpur Lime, Nasvaran, Cleopatra and Rubidoux were tried for jaffa sweet orange. The trees on Jatti Khatti rootstock produced significantly more tree stock/scion girths, tree volume/growth as compare to other followed by Nasnaran. Minimum tree volume found in Kharna Khatta and fruit yield is also maximum in case of Jatti Khatti.

(ii) Mosambi : Rootstock namely Jatti Khatti, Kharna Khatta, Carrizo, Troyer, Pectinifera, Citrumelo, Rangpur Lime, Jambhiri, Nasnaran, Cleopatra and Rubidoux were tried for Mosambi sweet orange. The tree survival, fruit yield and excellent fruit quality was achieved on the trees raised on Pectinifera rootstock. The Mosambi trees on pectinifera had smooth bud union as compared to Jatti Khatti which exhibited creasing at bud union with declined symptoms. Mosambi on citranges declined due to incompatibility. The worst peformance of Mosambi scion was observed on Troyer and Carrizo severe yellowing at veins and acute chlorosis of leaves observed.

Blood Red : Rootstocks namely Kharna Khatta, Jatti Khatti, Carrizo, Troyer, Pectinifera, Citrumelo, Rangpur Lime,Cleopatra, Nasnaran and Rubidoux were tried for blood red sweet orange. The tree survival, growth, fruit yield and quality were found better on the trees raised on Cleopatra rootstock. Blood Red on Kharna Khatta declined due to exocortis.

Pineapple : Four roostocks namely Jatti Khatti, Kharna Khatta, Carrizo andTroyer were tried for their performance. Trees budded on Jatti Khatti roostock prouced significantly more tree volume and yield than other rootstocks.

Valencia Late : Rootstock viz. Jatti Khatti, Kharna Khatta, Troyer, Carrizo, Nasvaran, Rubidoux, Pectinifera, Rangpur lime, Citrumelo andCleopatra were tried. The tree of Valencia Late raised on Jatti Khatti, Carrizo, Cleopatra and Troyer rootstock produced significantly more tree volume than those on Rangpur lime and Citrumelo. The yield was recorded more on Carrizo, Kharna Khatta and Jatti Khatti. Minimum volume was found in Nasaaran.

Hamlin : The seven rootstock viz. Jatti, Kharna Khattra, Troyer, Carrizo, Cleopatra, Sacaton and Savage were tried for Hamlin sweet orange. The results revealed that treed budded on Jatti Khatti produced significantly more tree growth and fruit yield as compared to other roostock.

Campbell Valencia : Four roostocks namely Jatti Khatti, Kharna Khatta, Troyer and Carrizo were tried for Campbell Valencia. The trees on Jatti Khatti rootstock performed better in respect of tree health, fruit yield and quality.

Dancy : Following roostock Jatti Khatti, Kharna Khattra, Troyer Carrizo, Rangpur Lime, Nasvaran, Cleopatra, Rubidoux, Citrumelo, Rectinifera and Nuceller Blood red were used. In this high incidence of granulation was observed on all the roostock/scion combination. So, it was dropped from the experiment.

Marsh Seedless : Following roostocks Jatti Khatti, Kharna Khatta, Carrizo, Troyer, Cleopatra, Rangpur Lime and Nasvaran were used. Marsh Seedless also being vigorous scion performed equally good on all the stocks with respect to tree growth and fruit yields, highest tree canopy volume and fruit yields were however, recorded under Troyer. Fruit quality parameters were not affected much by different stocks; however, Cleopatra recorded higher percent TSS than other.

Wilking : Four roostock Jatti Khatti, Kharna Khatta, Troyer and Carrizo were used and its maximum tree volume was observed on Kharna Khatta followed by Jatti Khatti and minimum on Carrizo. Maximum fruit was also observed on Kharna Khatta. Among Kinnow, wilking, Pearl and Campbell Valencia scion varieties wilking showed least tree volume.

Pearl (Tangelo) : Four rootstock viz. Jatti Khatti, Kharna Khatta, Troyer and Carrizo were used maximum stock/scion girth, height and spread was observed on Kharna Khatta which is at par with Jatti Khatti.

Grapefruit : Jatti Khatti, Kharna Khatta, Troyer and Carrizo rootstocks were tried for Grapefruit cultivar Marsh Seedless. The trees of March seedless on Carrizo and

Troyer Citranges rootstocks were better in tree volume, trunk girth, fruit yield and quality than Jatti Khatti and Kharna Khatta.

In all rootstock trials, Jatti Khatti (Rough Lemon) proved the best for Kinnow mandarin, Pineapple, Jaffa, Valencia oranges, Hamlin, Campbell Valencia and Pearl. It is resistant to citrus tristeza, exocortis and high pH (up to 8.5). However, Blood Red and Mosambi sweet orange budded on Rough lemon formed bud union crease and started declining. So it was rejected for Mosambi and Blood Red (Jawanda and Mehrotra, 1974; Chohan and Kumar, 1983). Cleopatra mandarin was found most suitable for Blood Red as it was resistant to exocortis, citrus tristeza and tolerant to foot rot and was also free from bud union creasing disorder. It also imparted dwarfness to the scion. Similarly, for Mosambi, Pectinifera proved most suitable as it was slow growing initially, tolerant to foot rot and resistant to exocortis and tristeza (Vij and Kumar, 1990, Jawanda and Mehrotra, 1974, Chohan et. al. 1986). Kharna khatta for Mosambi, Jaffa, Pineapple and Blood Red oranges proved to be a total failure due to decline caused by exocortis and other virus diseases (Chohan *et al.* 1990), however performance of Wilking mandarin was best for this rootstock.

3. Factors Affecting Citrus Rootstocks

3.1 Salinity

Effect of soil salinity on growth of citrus rootstocks in Punjab was studied by Patil and Bhambota (1978) in pot experiments with six citrus species on sandy loam soil, the fresh weight of whole plants, roots and tops, root length and leaf chlorophyll content decreased with increasing salinity. At salinity concentration of 30 meq/l and above this adverse effect was most pronounced in Carrizo citrange (*P. trifoliata* x *C. sinensis*) and Kharna katta (*Citrus karna*). Jullunduri khatti, Jatti Khatti and Cleopatra (*C. reshni*) rootstocks were relatively resistant to salinity, while Carrizo citrange (*P. trifoliata* x *C. sinensis*) and Kharna khatta (*C. karna*) were the most sensitive ones. Although the indices of Carrizo and Kharna khatta for 20, 50 and 80 per cent growth depressions were the same, the salinity index based on the growth depression of more than 80 per cent was greater in Kharna khatta. Rangpur lime was intermediate although its tolerance index based on 80 per cent growth-depression seemed to be the greatest.

3.2 Granulation

Incidence of granulation was studied by Jawanda *et al.* (1978) in 21 sweet orange cvs, 13 mandarin orange cvs, 4 tangelo cvs and 4 grapefruit cvs; in 4 sweet orange cvs at several different places; in several cv./rootstock combinations. None of the cultivars were free from granulation but its incidence was low in Valencia and Blood Red sweet oranges, Kinnow and Kara mandarins, Minneola and Orlando tangelos and Duncan grapefruit. Trees on sweet lime rootstock and less vigorous trees were not severely affected.

3.2.1 Sweet orange varieties

Pineapple orange on 9 different rootstocks was evaluated by Mann and Naurial (1978) with respect to growth, cropping and fruit characteristics of 8-year-old trees of. *C. jambhiri* was the most vigorous rootstock, followed by Jullundari Khatti (probably a Rough lemon hybrid) and grapefruit. Trifoliate orange (*P. trifoliata*) and *C. pseudolimon* had a dwarfing effect. Juice and TSS contents were highest in fruits grown on sweet orange, and vitamin C content was highest on sour orange.

Sweet orange cvs Pineapple, Jaffa, Musambi and Valencia Late, each budded on Jatti Khatti, Rough lemon, Kharna Khatti, Troyer and Carrizo citranges and Rangpur lime were evaluated by Singh et al. (1980). The girth of the stock and scion was greatest on Jatti Khatti, followed by Kharna Khatta and least on Rangpur lime. Valencia Late and Jaffa showed greater growth vigour than Musambi and Pineapple. In some combinations on Troyer and Carrizo, tree mortality occurred. Tree volume was significantly greater on Jatti Khatti and Kharna Khatta than on Troyer, Carrizo or Rangpur lime. Symptoms resembling bark-shelling, associated with exocortis virus, were noticed on some of the combinations on Rangpur lime. All cultivars had a smooth graft union on Jatti Khatti, Kharna Khatta and Rangpur lime, whereas on Troyer and Carrizo a few trees had a bottleneck union. The stock:scion girth ratio was not a certain indication of compatibility. Bud-union crease was confined primarily to cultivars on Jatti Khatti and Kharna Khatta rootstocks. All cvs on Troyer, Carrizo and Rangpur lime were free from bud-union crease.

Old line Blood Red and old line Musambi and nucellar Blood Red and nucellar Musambi sweet oranges budded on Rough lemon or Troyer citrange rootstocks in various combinations by Dhillon et al. (1981). Combinations on Rough lemon were generally significantly more vigorous than those on Troyer citrange. Trees raised from nucellar sources (showing no decline) had greater growth and vigour than those from old line sources (showing decline). Similarly, at Abohar, Chohan et al. (1982) assessed trees of Blood Red sweet orange grafted on 8 different rootstocks for tree survival, growth, yield and fruit quality. Trees on Rough lemon and Cleopatra mandarin did best in all cases except for fruit quality, which was best on citrumelo and Carrizo citrange. Mean fruit yield (1975-1980) for trees planted in 1968 was 185 fruits/ tree on Rough lemon, and 141.7/tree on Cleopatra mandarin. Also at Abohar, Hamlin sweet orange was evaluated on 7 rootstocks. Trees budded on Jatti Khatti were the most vigorous and productive. Cleopatra (*C. reshni*) was the next best rootstock. No significant differences were observed with regard to physico-chemical parameters of fruits on trees on different rootstocks. Trees on Cleopatra produced fewer granulated fruits than those on Jatti Khatti. Sandhu and Singh (1989) carried out studies at Bathinda, Punjab on 10-year-old trees on 6 different rootstocks and the fruits were size-graded (as an average of length and breadth) as very small, small, medium, large and extra large. Fruits of trees on Jatti Khatti and Kharna Khatta showed a high incidence and degree of granulation. Minimum granulation occurred on Rangpur lime followed by Cleopatra mandarin. Extra large (> 8 cm) and very small (< 5.9 cm) fruits showed a high incidence of granulation. The incidence was lowest in small and medium-sized fruits. Fruit quality decreased with increase in granulation. Chohan et al. (1991) reported that of 8 rootstocks tested from 1968 onwards, Pectinifera gave the most smooth unions and conferred other favourable features on Mosambi trees at RFRS, Abohar, Punjab. Likewise, Kumar et al. (1994) reported that the sweet orange cv. Pineapple was established on the rootstocks Jatti Khatti, Kharna Khatta and Troyer and Carrizo citranges during 1969. Assessment during 1992 revealed that Jatti Khatti was the most vigorous rootstock, with a tree volume of 61.1 m^3. Survival on all rootstocks was more or equal to 92%. Differences in yield in terms of number of fruits/tree, weight of fruits/tree, and fruit quality in terms of granulation index, juice percentage, acidity and vitamin C content, were not significant between the cultivars. Troyer and Carrizo citranges produced the lowest peel percentage (24.7 and 24.8%, respectively) and the highest TSS (10.1 and 9.9%, respectively). In Punjab at Abohar, Chohan et al. (2000) reported that a rootstock trial using Jatti Khatti, Kharna khatta, Troyer, Carrizo, Rangpur lime, Pectinifera, Cleopatra and Citrumelo rootstocks for Blood Red cultivar of sweet orange was planted in 1968. Results obtained upto 1990 revealed that Jatti Khatti performed better for the first 12-13 years age of the trees due to high vigour and better fruit yield. After 1983, trees of Blood Red raised on Jatti Khatti started dying and by 1990, only 59% trees survived; whereas on Cleopatra all the trees survived. Tree volume was 36.1 m^3 on Jatti Khatti against 64.1 m^3 on Cleopatra rootstock during 1990. The trees on Jatti Khatti rootstock showed only 21% healthy trees with lower mean yield (181 fruits per tree) compared to Cleopatra which showed cent per cent healthy trees with more mean yield (265 fruits per tree) upto 1990. The trees of Blood Red budded on Cleopatra rootstock had better fruit quality compared to those on Jatti Khatti. Also from Abohar, Mehrotra et al. (2000) reported use of 4 rootstocks viz., Jatti Khatti, Karna Khatta, Troyer and Carrizo citranges (*P. trifoliata* x *C. sinensis*) to determine the most suitable rootstock(s) for cv. Pineapple. Pineapple budded on all the survival (100%), healthy trees (75%), and tree volume (61 m3). Trees on Kharna Khatta had the highest pooled mean fruit yield (271 fruits/tree) during 1974-96. Pooled analysis (during 1994-95 and 1996-97) of fruit quality showed trees on Kharna Khatta had the highest fruit weight (195.9 g), fruit size (7.1 x 7.3 cm), peel thickness (0.62 cm), and peel (28.0%). Troyer had the highest juice (56.1%), total soluble solids (8.9%), acidity (0.659%) and ascorbic acid (27.3 mg/100 ml juice). Mehrotra et al. (2000) also reported that a rootstock trial of Jaffa cultivar of sweet orange using eight rootstocks (Jatti Khatti, Karna Khatta, Troyer citrange, Carrizo citrange, Pectinifera (*C. pectinifera*), Pineapple orange, Rangpur lime and citrumelo was laid out in 1969 at Abohar, Punjab. Results obtained till February, 1997 revealed that Jaffa trees budded on Jatti Khatti showed highest survival (83.5%) with healthy trees; whereas no tree was found healthy on Karna Khatta and Troyer rootstocks. All the trees on citrumelo, Carrizo, and Rangpur lime died up to 1991. Jatti Khatti rootstock produced significantly more stock (108 cm) and scion (92.7 cm) girth, and tree volume (70.1 m^3) compared with Karna Khatta (65 and 61.3, and

11.3 m3, respectively) and Pectinifera (75.7 and 71.8 cm, and 39.3 m^3, respectively). Jatti Khatti also produced significantly better fruit yield (391 fruits per tree) and weight per fruit (178 g) compared with other rootstocks. Peel content of the fruits was significantly less on Jatti Khatti (26.9%) and Pectinifera (25.4%) rootstocks. However, fruits on Pectinifera rootstock had significantly more total soluble solid content (9.6%) compared with other rootstocks. Mosambi sweet orange was evaluated on eight rootstocks viz., Jatti Khatti, Kharna Khatta, Troyer, Carrizo, Rangpur Lime, Jambhiri, Pectinifera and Citrumelo Mehrotra et al. (2002). Troyer & Carrizo citranges and Citrumelo were found to be incompatible rootstocks for Mosambi. Upto February 1981, the trees budded on Rangpur Lime produced maximum volume (32.2 m^3) and higher fruit yield followed by Jatti Khatti. However, Pectinifera excelled all the rootstocks producing significantly more tree volume in February 1991, and this trend is still continuing. Mosambi trees budded on this rootstock showed significantly more survival (100%), stock (94.0 cm) and scion (96.5 cm) girth than that of Kharna Khatta, Jambhiri and Rangpur Lime rootstocks. Average fruit yield (1975 to 1996) of Mosambi cultivar was significantly more (252 to 312 fruits/tree) on Pectinifera than the rest of the rootstocks tried. Pectinifera produced significantly higher TSS (9.8%) and less peel (25.2%) content in Mosambi fruits as compared to that of all other rootstocks. On the basis of these results, it may be concluded that upto the age of 12 years Rangpur Lime, and afterwards upto 28 years age of the trees, Pectinifera rootstock is suitable for commercial cultivation of Mosambi sweet orange under arid-irrigated conditions of Punjab. Similarly, Sharma *et al.* (2002) reported that an experiment was initiated in 1968 in Abohar, Punjab, to determine the effect of Campbell Valencia sweet orange on 4 rootstocks, namely Jatti Khatti, Kharna Khatta, Troyer and Carrizo citranges (*P. trifoliata*). Tree survival on Carrizo rootstock combination remained cent per cent, 75% on Troyer, while on Jatti Khatti only 25% of the plants survived after 31 years of age. The average fruit yield was maximum on Jatti Khatti and minimum on Kharna Khatta. Fruit weight was maximum on Troyer and minimum on Jatti Khatti rootstock. Maximum total soluble solid (TSS) was recorded in fruits on Troyer and Carrizo and minimum on Kharna Khatta. The TSS:acid ratio was maximum in fruits sampled from Carrizo and Troyer rootstocks. On the basis of 20 season data with respect to tree survival, yield and fruit quality attributes, Carrizo, Troyer and Jatti Khatti are the most suitable rootstocks for Campbell Valencia sweet orange.

Rootstocks evaluation trials were conducted at Abohar, Punjab by Kumar et al. (2002) to study the performance of Blood Red, Mosambi and Pineapple cultivars of sweet orange on different rootstocks, viz., Jatti Khatti, Cleopatra, Sadaphal (*Citrus semperflorens*), sour orange (*C. aurantium*), Mithi (*C. jambhiri cv. Ducis*) and Pectinifera (*C. depressa*). Results revealed that Blood Red trees budded on Jatti Khatti, Cleopatra, Sadaphal and sour orange showed significantly more survival than those on Mithi rootstock. Significantly more fruit yield was produced on sour orange followed by Jatti Khatti rootstock than that recorded on Cleopatra and Mithi rootstocks. Mosambi trees budded on Sadaphal rootstock produced significantly more stock and scion girth, tree volume and higher fruit yield per tree than noted on Pectinifera and Mithi rootstocks. Pineapple trees budded on Jatti Khatti showed significantly more stock girth and tree volume than those found on Pectinifera, Sadaphal and sour orange. Pineapple trees on Jatti Khatti out yielded all other rootstocks. Aulakh (2008) conducted a rootstock trial on Valencia Late cultivar of Sweet orange using seven rootstocks at RFRS, Abohar. The results showed that the survival of trees on Jatti Khatti was cent percent in healthy condition and maximum tree growth (96.3m^3 volume/tree) followed by Carrizo (88.2m^3 volume/tree). Mean yield was in the range of 182-185 fruits/tree on Jatti Khatti, Kharna Khatta and Carrizo rootstocks. There was no mortality of Valencia Late trees budded on Troyer, while on Citrumelo, all trees died and proved as incompatible rootstock. Maximum fruit weight (220 g/fruit) and good quality fruits were found on Cleopatra rootstock. However, maximum TSS (9.2 %) was observed in fruit juice of Carrizo.

Rootstock studies conducted at Abohar were summarized by Aulakh and Baidwan (2004). They reported that studies conducted in 1982, 1987, 1992 and 1997 in Abohar, Punjab to investigate the performance of orange cv. Hamlin on different rootstocks: Jatti Khatti, Kharna Khatta, Troyer, Carrizo, Savage and Sacaton. Trees budded on Kharna Khatta and Cleopatra rootstocks showed 100 per cent survival while those on Jatti Khatti showed 75 per cent survival. The mean fruit yield/tree from 1983-84 to 1996-97 was higher on Jatti Khatti than the other rootstocks, except Cleopatra. Jatti Khatti produced a higher number of granulated fruits than Kharna Khatta and Cleopatra.

3.2.2 Mandarins

Kinnow mandarin grafted onto 5 rootstocks (*C. jambhiri* [Rough lemon] cv. Jatti Khatti, Cleopatra mandarin, *P. trifoliata* cv. Rubidoux, *C. karna* cv. Kharna Khatta and Troyer citrange) was investigated by Dhillon et al. (1993). The highest scion:rootstock ratio was obtained on Kharna Khatti and the lowest on Rubidoux (0.90 and 0.74-0.76, respectively). The bud union on Rubidoux was bottle-necked in shape, whereas with the other rootstocks the union was smooth. Jatti Khatti rootstock produced the greatest canopy volume (24.0-26.4 m^3) and Rubidoux the smallest (14.8-15.4 m^3). With the rootstocks Jatti Khatti and Cleopatra mandarin, there was more sugar in the bark below the bud union than above, whereas the opposite occurred with the other rootstocks. There was very little difference in carbohydrate or phenolic content above and below the bud union. Sharma *et al.* (2002) investigated the effects of rootstocks on the production and quality of Kinnow mandarins at Abohar. The rootstocks used were: *C. jambhiri, C. aurantium, C. tachibana, C. depressa,* and *C. reshni.* A 100 per cent tree survival was observed in scions on *C. jambhiri* rootstocks, and the highest fruit yield was also obtained on this rootstock. Josan and Thatai (2008) evaluated the performance of Kinnow mandarin grafted on the following rootstocks in Abohar, Punjab: Jatti Khatti, Karun Jamir (*C. aurantium*), Shekwasha (*C. depressa*), Jambhiri, Pectinifera (*C. depressa*), Estes Rough lemon (*C. jambhiri*) and Cleopatra (*C. reshni*). Tree volume and percentage of survival of healthy trees at 18 years after planting (in 1988) were greatest on Jatti Khatti (96.24 m^3 and 83.33%, respectively). This rootstock registered the highest mean fruit yield (796 fruits per tree, which was equivalent to 144 kg) and lowest percentage of declined trees on the 18th year (16.67%). The rootstock had no significant effects on scion/stock ratio and crop quality. Similarly, Josan et al. (2010) reported from a rootstock trial on Kinnow mandarin started during 1988 at Abohar, Punjab on seven rootstocks viz., Jatti Khatti, Karun Jamir, Jambhiri, Shekwasha, Pectinifera, Estes Rough lemon and Cleopatra. Jatti Khatti was observed to be the most vogoruous rootstock. Tree survival was maximum on Jatti Khatti (83.33 %) after 19 years of the tree age, while in some rootstocks, declining starts at the age of twelve years. Average of fruit yield of five seasons was the maximum (748 fruits/tree) on Jatti Khatti whereas it was minimum (470 fruits/tree) on Cleopatra rootstock. No significant variation was recorded in the physic-chemical characters of Kinnow fruits on various rootstocks. On the basis of five season data (15th to 19th years of tree age), they concluded that Jatti Khatti is the most suitable rootstock fro Kinnow mandarin under the north Indian condtions. Also at Abohar, Kumar *et al.* (2001) conducted a field experiment (started in 1991) to evaluate the performance of Lahore Local and Nagpur mandarins (*C. reticulata*) budded on various rootstocks, i.e. Jatti Khatti, Karun Jamir (*C. aurantium*), Pectinefera (*C. depressa*), Cleopatra (*C. reshni*), Troyer (*P. trifoliata x C. sinensis*), and Kharna Khatta (*C. karna*) for Lahore Local and Nagpur mandarins, and Citrumelo (*P. trifoliata* x *C. paradisi*) for Nagpur mandarin only. Trees of Lahore Local mandarin budded on Jatti Khatti, Kharna Katta, Pectinifera, and Karun Jamir rootstocks showed higher survival (83.3-100.0%) than trees on Cleopatra and Troyer (50.0%). Trees on Kharna Khatta rootstocks had greater volume (10.2 m^3 per tree) than trees on the other rootstocks. Scions on Jhatti Khatti, Kharna Khatta, Troyer, Karun Jamir, and Pectinifera showed greater survival rate (66.7-100.0%) than those on Cleopatra and Citrumelo (50.0%). The percentage of healthy trees on all rootstocks was 66.7-100.0%. Nagpur mandarin scions budded on Jatti Khatti, Kharna Khatta, and Karun Jamir had greater stock (33.5-41.7 cm) and scion girths (33.5-42.6 cm) than those on Troyer (25.2 and 22.1 cm, respectively). The scion:stock ratio of all rootstock combinations revealed incompatibility among scions and rootstocks, as trees budded on all rootstocks had smooth bud unions lacking union crease formation.

3.2.3Grapefruit

Many worker studied the effect of rootstocks on grapefruit cultivars in Punjab.Chohan et al. (1990) reported that Marsh Seedless scions were budded on Jatti Khatti, Kharna Khatta and Troyer and Carrizo citranges (*C. sinensis* x *P. trifoliata*) in 1976. The maximum trunk girth increase, greatest canopy volume, the highest yields and the best fruit quality and tree health were obtained by Carrizo citrange, closely followed by Troyer citrange. Similarly, Mehrotra et al. (1999) conducted a rootstock trial at Abohar, Punjab on Marsh Seedless cultivar of grapefruit, using Jatti Khatti, Kharna Khatta, Troyer and Carrizo as rootstocks. The results of 20-year trial revealed that the trees budded on Troyer and Carrizo (citranges) rootstocks produced 100% survival with all the trees in healthy condition; whereas Jatti Khatti and Kharna Khatta showed only 50 and 17% survival of the trees, respectively. Both the citranges also showed significantly more stock or scion girth and tree volume than Jatti Khatti rootstock. Significantly more mean fruit yield from 1986-87 to 1996-97 was observed on Troyer and Carrizo rootstocks than on Jatti Khatti. Mean results of fruit-quality attributes showed that TSS percentage was more on Troyer than on Jatti Khatti rootstock. Heavier fruits with less peel percentage were also recorded on Troyer and Carrizo citranges than on Jatti Khatti rootstock.

3.2.4 New Mandarin and Sweet orange studies

Exotic citrus varieties imported by PAGREXCO from USA and these were planted under submountane and arid irrigated region of Punjab. Varieties comprised ten sweet orange varieties i.e Early Gold, Itaborai, Hamlin, Ruby Nucellar, Westin, Vernia, Rhode Red, Trovita, Midnight Valencia, Olinda Valencia and two mandarins/tangerines i.e Clemenules and Marisol budded on six different rootstocks i.e. Carrizo, Swingle, C-32, C-35, Bentone and Volkamariana.

Based on two years preliminary studies revealed that Itaborai, Westin and Hamlin on Bentone rootstocks, Westin, Itaborai on Volkamariana and Ruby Nucellar and Hamlin also on C-35 rootstocks are performing better in respect of tree volume, scion girth and stock girth as compared to other scion varieties budded on other three different rootstocks i.e. Carrizo, Swingle and C-32 Aulakh et al. (2008b). Similarly, Aulakh et al. (2008a) observed that by the year 2007 Hamlin on Carrizo and Swingle, Ruby Nucellar and Clemenules on Carrizo are performing better in respect of growth and tree survival whereas Early gold on 852, Clemenules on C-35 and Midnight Valencia on Carrizo were not good combinations. Plant growth was more in Clemenules on Carrizo (50.55) followed by Ruby Nucellar on Carrizo (49.44), Clemenules on C-35 (36.0), Marisol (32.60) and Hamlin on Carrizo (31.75) as compared to other stock-scion combinations. With regards to quality parameters, Clemenules on Carrizo & C-35, Marisol on Carrizo bore granulated fruits with 15.63 to 18.88 per cent juice content, whereas Itaborai, Ruby Nucellar and Early Gold Sweet Oranges on Carrizo rootstock produced fruits with more juice content (52.13 to 53.13 %). Similarly, these Sweet Oranges and Mandarins bore fruits with TSS content (10.2 to 10.4%) and acidity (0.67 to 0.80%). In the same stionic combinations, Singh et al. (2008) in submountain region of Punjab observed that the maximum rootstock and scion girth was noted in Clemenules/C-35, whereas, minimum in Carrizo/Marisol and Olinda/Rubidoux Trifoliate, respectively. Plant height in varios combinations ranged from 2.05 m to 3.27 m, being the highest in Hamlin/Carrizo and lowest in Midnight Valencia/Swingle. Maximum tree spread (average of N-S and E-W) was recorded in Clemenules/C-35 and minimum in Midnight Velencia/Swingle. Scion rootstock ratio also ranged from 0.72 in Midnight Valencia/Swingle and 0.92 in Marisol/Carrizo. On the basis of tree volume, it was observed that various scion varieties grafted on rootstocks i.e. Swingle and Rubiduox Trifoliate showed dwarfing effect, whereas Carrizo rootstock resulted in vigorous tree canopy.

Kahlon et al. (2010) has reported that new strains of Rough lemon, Rangpur lime and hybrids of *P. trifoliata* were introduced at PAU, Ludhiana from different different research stations and NRCC, Nagpur and have established in the nursery (Table 6).

Table 6 : New rootstocks introduced in recent years at PAU, Ludhiana

S. No.	Variety	S. No.	Variety
1	Benton	9	Rich 16-6
2	C-32	10	Rubidoux trifoliate
3	C-35	11	Sun Chu Sha
4	Carrizo	12	Swingle
5	Goutou	13	Troyer
6	Kin Koji	14	US 852
7	Huharska	15	Volkamarina
8	Rangpur lime		

(*Source :* Kahlon *et al.* (2010))

3.3 Research in other parts of India

The information about rootstocks trials conducted in various parts of India was reviewed by Sonkar et al. (2002) which varied greatly with scion variety and agro-climatic conditions. Commercial citrus rootstock cultivars are highly polyembryonic but there can be 1-40 zygotic seedlings in the seed bed. Inspite of susceptibility to *Phytophthora* disease and restricted longevity, Rough lemon (*C. jambhiri*) has been the most time tested and widely used rootstock in India. However, attempts to develop better substitute rootstock have shown Cleopatra mandarin, Rangpur lime and pectinifera. Some of the citranges and Alemow (*C. macrophylla*) are found to be promising rootstocks in certain parts of the country.

3.3.1 Sweet orange

Sathgudi itself is the only recommended rootstock for Sathgudi scions (Rao *et al.* 1971). At Kodur, Andhra Pradesh, Gajanimma (*Citrus pennivesiculata*) gave the highest yields but was most susceptible to disease and produced fruit of poor quality. Other promising rootstocks not recommended because of disease susceptibility were jamberi (*C. jambhiri*), acid lime (*C. aurantifolia*) and kichili (*C. maderaspatana*). Woodapple (*Feronia limonia*) rootstock had a marked dwarfing effect and the fruit was of good quality. The frost tolerance of 1-year-old trees of 6 sweet orange cvs on 6 rootstocks was evaluated by Gupta et al. (1978). Of the cultivars Hamlin and Mosambi, and of the rootstocks Jatti Khatti, Cleopatra mandarin and Troyer citrange were the least susceptible. Blood Red and Carrizo citrange rootstock suffered the most damage. Vijayakumari et al. (1990) reported the effect of rootstock on yield of mandarin, *C. reticulata* in Shevroy Hills of Yercaud, Tamil Nadu. Seedlings of 6 different rootstocks were budded with scions of Coorg mandarin in 1976 and the trees were assessed for number of fruits and fruit weight between 1980 and 1989. The highest cumulative yield was obtained on Rangpur lime followed by Rough lemon, and these 2 rootstocks are recommended for Shevroy hills. Likewise, Mustafa and Reddy (1990) carried out a field study at Hessaraghatta, Bangalore with sweet orange cv. Mosambi on 9 different rootstocks: Rough lemon, Rangpur lime, Cleopatra mandarin (*C. reshni*), citrumelo (*P. trifoliata* x *C. paradisi*), Carrizo and Troyer citranges (*P. trifoliata* x *C. sinensis*), Pomeroy and Rubidoux trifoliate oranges (*P. trifoliata*), and Kodaikithuli (*C. reshni*), planted in 1979. Highest tree height (312 cm) and tree volume (20.41 m^3) in 1986 were obtained on Rough lemon followed by Rangpur lime (295 cm and 19.03 m^3, respectively). *P. trifoliata* and its hybrids induced dwarfing. Highest mean fruit weight was obtained on Rough lemon. Highest per cent juice, number of fruits/tree and yield/tree were obtained on Rangpur lime. Highest TSS was obtained with Pomeroy trifoliate orange, highest % acidity with Kodaikithuli and highest ascorbic acid content with Rough lemon. Kumar and Ganapathy (1992) evaluated sweet orange cv. Mosambi on Troyer citrange, Carrizo citrange, Poncirus trifoliata cultivars Rubidoux and Pomeroy, Rough lemon, Kodakithuli [mandarin], Cleopatra mandarin and Rangpur lime. Taking into account all the parameters recorded, the best rootstocks was Carrizo citrange followed by Rough lemon and then Troyer citrange. Rao et al. (1996) tested the performance of sweet orange cv. Sathgudi in 10 citrus rootstocks on sandy loamy soils. Graft unions were smooth on sweet orange, Rangpur lime, Cleopatra mandarin, Jatti Khatti and Khatta Jamir (both *C. jambhiri*) while grafts with Troyer and Carrizo citranges, Trifoliate orange (*P. trifoliata*), Gajanimma (*C. pennivesiculata*) and Lime Karna (*C. karna*) were incompatible. Tree canopy volumes were larger on Lime Karna, Rangpur lime, Gajanimma and Jatti Khatti than on trifoliate orange and citranges. Fruit yields in terms of fruit weight or number/tree were significantly higher on Rangpur lime followed by Gajanimma, Cleopatra and Jatti Khatti than on the other rootstocks. Fruit quality (higher TSS/acid ratio) was greater on trifoliate orange, Troyer citrange, Rangpur lime and Cleopatra mandarin. Singh and Daulta (1995) investigated the extent of scald on fruits of 6 orange cultivars grafted on 6 rootstocks. A higher incidence of scald was observed for fruits facing south, followed by those facing east, north and west. Jatti Khatti was the most susceptible and Rangpur lime the least. The lowest incidence of scald (5.73%) was observed in Campbell Valencia on a pectinifera rootstock. The highest incidence of scald (54.11%) was observed in Mosambi on a Jatti Khatti rootstock.

Kodur sathgudi budded on Troyer or Carrizo citrange, Robidoux or Pomeroy trifoliate orange, Rough lemon, Kodakithuli or Cleopatra mandarin or

Rangpur lime rootstocks was evaluated by Ganpathy (1998). Budding on Carrizo citrange produced the most vigorous growth and highest fruit yields. Troyer citrange and Rough lemon were the next best rootstocks, respectively. Kodakithuli rootstock performance was poorest as the plants died at the age of 3 years. Similarly, Grace et al. (2005) conducted studies in Tirupati, Andhra Pradesh to evaluate the yield and fruit quality of Sathgudi sweet oranges budded on different rootstocks (Sathgudi, Rangpur lime, Troyer citrange, Cleopatra mandarin and Trifoliate orange). The highest numerical yield was recorded on Troyer citrange (254.79 fruits per tree), followed by Rangpur lime (251.76 fruits/tree). However, the highest yield in terms of weight per tree was recorded on Rangpur lime (40.39 kg/tree), followed by Troyer citrange (31.64 kg/tree). Significant rootstock effects were observed on different fruit physical (fruit weight, fruit length, fruit diameter, rind thickness, rind weight, rind percentage), quality (number of segments, number of seeds, seed weight, juice volume, juice weight, juice percentage and Rag) and chemical characteristics (total soluble solids, acidity, ascorbic acid, reducing sugars, nonreducing sugars, total sugars).

Nutrient uptake in citrus is influenced by rootstocks. Srivastav et al. (2005) carried out an investigation during 2002-04 to assess the performance of Mosambi sweet orange on different rootstocks, Jambhiri, Karna Khatta, Cleopatra Mandarin, Soh Sarkar and on its own roots, at main orchard of IARI, New Delhi. The elevated levels of different essential nutrients in the leaf tissue suggest the better performance of Karna Khatta as rootstock in terms of vegetative growth, number of fruits per plant and juice recovery of Mosambi sweet orange. The rootstock Cleopatra Mandarin performed well for total soluble solids and ascorbic acid contents. However, the acidity of juice of Mosambi fruits was minimum where Mosambi was budded on its own roots. Similarly, Marathe et al. (2006) studies the effect of various rootstock genotypes of Rough lemon, Rangpur lime, Cleopatra mandarin, sour orange and trifoliate orange on leaf nutrient composition of 3-year-old Nagpur mandarin grown on black soil. Almost all the nutrients except Mn have shown significant difference in their concentration in the Nagpur mandarin scion due to rootstock influence. Rough lemon (Tirupati) rootstock recorded the highest uptake of P, Cu and Fe with highest canopy volume and intermediate fruit yield. Similarly, Rangpur lime (Gonnicoppal) recorded highest uptake of N and K with highest fruit yield and higher plant vigour. Cleopatra mandarin (Morrocco) recorded the highest uptake of Mg, Mn and Zn with lower canopy volume and fruit yield. The nutrient uptake pattern of N, K and Cu showed strong positive correlation with canopy volume as well as fruit yield of the plant. On the other hand, Mn uptake showed negative significant correlation with fruit yield. Bauri et al. (2006) evaluated the performance of Darjeeling mandarin at RRSS, Pedong, west Bengal under six different rootstocks viz., Rough lemon, Rangpur lime, Cleopatra, Jatti Khatti, Trifoliate orange and Darjeeling mandarin seedling. The data revealed that the overall performance of Rough lemon was observed to be better among the six rootstocks tried. Tolerance against gummosis and twig blight diseases was more in Jatti Khatti, Rough lemon and Trifoliate orange as stock. Darjeeling mandarin as stock was most susceptible to the diseases tested, though the growth and yield were comparatively higher. During sixth year, yield varied from 91.7 to 136 fruits/ plant on different rootstocks. The TSS varied from 13.2 to 14.9 °Brix, while acidity varied from 0.51 to 0.82 per cent. Shinde *et al.* (2007) studied the performance of orange (cv. Nucellar) budded on various rootstocks in Parbhani, Maharashtra, India, during 1977-80. The rootstocks consisted of Jambhiri local (*C. jambhiri*), Rough lemon Chettali (*C. jambhiri*), Nemutenga (*C. limonia*), Carrizo citrange (*C. sinensis x P. trifoliata*), Cleopatra mandarin (*C. reticulata*), *C. macrophylla*, Sochmyndong (*C. jambhiri*), Woodapple (*Feronia limonia* [*Limonia acidissima*]), calamondin (*C. madurensis*), L-19 Rangpur lime (*C. limonia*), L-2 Rangpur lime (*C. limonia*) and Kichili (*C. maderaspatana*). Bud take percentage was highest on Rough lemon Chettali (88.88%), *C. macrophylla* (100.0%) and calamondin (88.33%). Plant height and spread at 480 days after bud take were greatest on Jambhiri local. The rootstocks were classified based on their effects on scion growth parameters. Jambhiri local, Sohmyndong, Rough lemon Chettali and Nemutenga resulted in greater scion height and spread; thus, these rootstocks were considered vigorous. *C. macrophylla*, Woodapple, L-19 Rangpur lime, L-2 Rangpur lime and Cleopatra mandarin resulted in intermediate scion height and spread; thus, these rootstocks were considered intermediate in vigour. Carrizo citrange, Kichili and calamondin were considered dwarfing rootstocks as they resulted in the least scion height and spread. Musmade et al. (2007) studied the performance of mosambi budded on 4 rootstocks (Cleopatra mandarin Moracco, Rough lemon 58-III-IV, Marmalade orange and Rough lemon Chethalli) in Rahuri, Maharashtra. Pooled data revealed that tree volume (89.50 m^3) and scion:stock ratio (0.90) were highest on Cleopatra mandarin Moracco. Fruit yield did not significantly vary among the rootstocks. Marmalade Orange also recorded the lowest percent decline in 2001-02 (18.50%), as well as the lowest fruit acidity (0.41%) and highest total soluble solid content (9.36%). Ghosh and Tarai (2007) laid out a rootstock trial on sweet orange cultivar 'Mosambi' budded on five rootstocks viz., Jambhiri, Karna Khatta, Kichili, Rangpur lime and Sour orange. Tree growth was maximum on Jambhiri and minimum on Rangpur lime. Fruit yield (both in number and weight) was highest on Karna Khatta, rootstock followed by Rangpur lime while, fruit size and juice content were maximum on Rangpur lime. Total soluble solids and ascorbic acid content were highest in Karna Khatta, while TSS to acid ratio was maximum in Rangpur lime. Foliar nitrogen content was highest in Karna Khatta followed by Rangpur lime. On the basis of four season's data in respect of yield and fruit quality, Karna Khatta and Rangpur lime were the observed as suitable rootstocks for 'Mosambi' sweet orange grown on laterite soil of West Bengal. Root distribution studies in 21 years old trees of different stionic combinations of sweet orange were carried out at CCSHAU, Hisar by Singh et al. (2008). Rootstock Pectinifera (*C. depressa* Hayata) had significantly more length of fibrous roots (having diameters < 0.15 cm) as compared to Jatti Khatti in majority of root zones studied thereby suggesting Pectinifera is a superior rootstock in the long run. Rootstocks exerted major influence on the fibrous root system whereas influence of scions on these fine roots was minimum. Jaffa as scion induced significantly more length of fibrous roots as compared to other two scions Blood red and Pineapple in eight root zones out of twelve zones studied. This shows more length of fine roots induced by Jaffa as compared to other scions. On the basis of overall evaluation of root system of different stionic combinations, Pineapple budded on pectitnifera was the best combination followed by Jaffa on Pectinifera. Jaffa on Jatti Khatti, Blood red on Jatti Khatti and Pineapple on Jatti Khatti. The feeder roots were more at the nearest radial distance and these decreased as the radial distance from the tree trunk increased. Maximum feeder roots were found near the syrface i.e. 0-20 and 21-40 cm soil depths.

3.3.2Mandarin

At New Delhi, Singh et al. (1978) observed Troyer citrange to be the most precocious rootstock and trees on Karna Khatta, Sacaton citrumelo and D. Spineless (*Citrus megaloxicarpa*) gave the higher yields. Trees on Troyer citrange had the lowest mortality rate. Mandge and Chakrawar (1981) reported that in trials with 17 citrus rootstocks at Parbhani, bud take in both species was highest on Rangpur lime (95%) followed by *P. trifoliata* (90%) and lemon and Rough lemon (88%) and lowest on Cleopatra tangerine (40%). Effect of different rootstocks on foliar nutritional status was investigated by Kunwar and Singh (1983) of Srinagar trees, grafted on 10 different rootstocks, were sampled for N, P, K, Ca and Mg at Srinagar, Garhwal, UP. Leaf N, P, K, Ca and Mg contents were generally highest in trees on Hill lemon, Troyer citrange, Malta sweet orange, Baduvapuli (*C. pennivesiculata*) and Hill lemon, respectively. Likewise, Kinnow, Nagpur, Khasi and Coorg mandarin cultivars and satsuma were grown on 4 rootstocks, viz. Rough lemon, *P. trifoliata*, Troyer citrange and Cleopatra mandarin (Devadas et al. 1988). Coorg mandarin on Rough lemon rootstock was the most vigorous combination in respect of all the biometric characters recorded. Rough lemon was the most vigorous rootstock, whereas *P. trifoliata* induced dwarfness. Preliminary results for cropping (1984-86) indicated that Coorg on Rough lemon was the heaviest cropping combination and produced good quality fruits. Sulikeri et al. (1990) assessed Coorg mandarin trees, grafted on 6 different rootstocks for height, rootstock and scion girth, rootstock:scion growth ratio, crown spread, number of trees that flowered in the 1976-79 period, and number of fruits/tree in 1979. The preliminary trials indicated the superiority of Rangpur lime as a rootstock for Coorg mandarin. Root characteristics for 18-month-old plants of Nagpur mandarin were determined on rootstocks of Troyer citrange, Mudkhed orange, Cleopatra [mandarin], Karna Khatta, Coorg [mandarin], Jamberi [Rough lemon] or Rangpur lime at Katol, Maharshtra by Allurwar and Parihar (1992). Coorg was rated as the best rootstock in terms of its greater taproot length, greater number and length of lateral roots and higher DW of tap and feeder roots. It was closely followed by Cleopatra. Both Jamberi and Rangpur lime, which are commonly used for oranges in Nagpur, had a less spreading root system than Coorg or Cleopatra. The latter rootstocks could prove good substitutes in times of water scarcity because of their deeper root penetration.

The effect of rootstock on production of Nagpur and Kinnow mandarins at Jhargram, West Bengal was studied by Ghosh and Chattopadhyay (1993). Rough lemon rootstock was the most vigorous, while Rangpur lime supported a higher fruit yield and quality. Pal and Lal (1993) conducted experiments at Srinagar, Uttar Pradesh to determine the most suitable rootstock for Srinagar Santra (*C. reticulata*). Ten rootstocks were evaluated. Based on tree vigour, yield and fruit quality (fruit size, juice, total soluble solids, sugars, acidity and ascorbic acid content), hill lemon and Jambhiri were the most suitable rootstocks for propagation of Srinagar Santra. Singh and Dass (1999) studied the effect of rootstock on pre-bearing performance of Nagpur mandarin. One-year-old seedlings of 10 exotic and 2 indigenous rootstocks were budded during February 1992 with Nagpur mandarin and planted in September 1992 at the experimental farm of the NRCC, Nagpur. The smallest trees were produced on Flying Dragon Infoliate Orange and Rich 16-6 Trifoliate Orange (both *P. trifoliata*). Smooth bud union was observed in all rootstocks, with stock:scion girth ratios ranging from 1.09 to 1.29, except Flying Dragon, Swingle Citrumelo and Rich 16-6. Incidence and degree of granulation in Kinnow mandarin at IARI, New Delhi was determine on Troyer citrange, Karna Khatta and Sohsarkar by Sharma and Saxena (2004). The incidence of granulation was influenced by different rootstocks, being very high for fruit from trees on Sohsarkar (38.3%) and very low for fruit from trees on Troyer citrange (5.9%).

The performance of Nagpur mandarin on to five strains of Rough lemon and eleven strains of Rangpur lime rootstocks during 2004-2007 on an alkaline Inceptisol under sub-humid climate of central India has been studied by Sonkar *et al.* (2008). They reported that smooth bud union was observed on all the strains of Rough lemon and Rangpur lime indicating congenial relationship with scion. Canopy volume was observed directly proportional to fruit yield with respect to Rangpur lime Saurathan (48.76 m^3 and 59.17 kg/tree), Rangpur lime Phillipine (45.71 $m^3$and 53.83 kg/tree) and at par with Rangpur lime USA. The average fruit weight was found to be higher on Rough lemon (South Africa) and Rangpur lime (Phillipine) as 152.6 g and 152.3 g/fruit, respectively. Rangpur lime (8747) produced more juicy fruits (40.03 %) as compared to other strains. TSS/acid ratio was observed maximum on Rough lemon Chethalli (16.36 %). Rough lemon (Srirampur) registered highest concentration of leaf N (2.14 %), P with Rough lemon South Africa (0.091 %) and K with Rough lemon Chethalli (1.03 %). Amongst Rangpur lime strains, Rangpur lime (Srirampur) accumulated maximum concentration of leaf N (2.33 %), P with Rangpur lime 8747 (0.162 %) and K with Rangpur lime Pookling Minj (1.27 %). In case of micronutrients, maximum uptake among Rough lemon with Fe was observed with Rough lemon Assam (147.4 ppm), Mn (52.95 ppm) and Zn (19.15 ppm) with Rough lemon Chethalli strain. Rangpur lime (8748) displayed minimum gummosis incidence (0.87 %) but found at par with Rangpur lime (Texas), Ragpur lime (7247), Rangpur lime (USA) and Rough lemon (Jullandhari khatti). The Rangpur lime (pookling Minj) showed maximum gummosis intensity (2.87 %). Cent per cent tree survival was recorded with

Rangpur lime strains 8747 whereas in Rough lemon Chethalli, only 37.5 per cent plant survived. The results indicated that Rangpur lime strains of Saurathan, Philippine and USA may prove to be promising rootstock for Nagpur mandarin. Similarly, Ram et al. (2008) tested Nagpur mandarin using five and eleven strains of Rough lemon and Rangpur lime, respectively. The compatible stionic combination was recorded in the scion of Nagpur mandarin on all the rootstock strains. Pookling minj strains of Rangpur lime had mmaximum plant height (98.13 cm), stock:scion ratio (1.33) and canopy volume (2.23 m^3). While Jallundhari khatti strains of Rangpur lime produced highest girth of stock (8.75 cm) and scion (8.63 cm). However, minimum plant height (70 cm) and stock girth (6.88 cm) was observed in plants on Rangpur lime. Similarly, minimum increase in scion girth (7.0 cm), stock:scion ratio (1.08) and canopy volume (1.44 m^3) exhibited by plants raised on Rangpur lime (Poona shrirampur), Rough lemon (Chethalli) and Rangpur lime (Poona shrirampur), respectively.

3.3.3 Lime and Lemons

Acid lime (*C. aurantiifolia*) was budded on 5 different rootstocks, including acid lime by Jawaharlal et al. (1987). Tree height and mean 4-year yield were highest in trees on Rough lemon rootstocks, followed by those on Troyer citrange. Similarly, Iyengar and Keshava Murthy (1988) determined the influence of seven citrus rootstocks on the absorption of phosphorus fertilizer by the scion cultivars seedless lime (*Citrus latifolia* Tanaka) and Italian lemon (*C. limon* Birm) using 32P-labelled superphosphate. The greatest uptake of fertilizer P for Italian lemon was with Carrizo citrange and citrumelo hybrid rootstocks and with Rough lemon and trifoliate orange for seedless lime. Italian lemon took up more P from the fertilizer than seedless lime. Misra et al. (1989) reported that in 3-year trials, cv. Pant Lemon-1 budded on 9 different rootstocks and the effect on leaf N, P, K, Ca, Mg, Zn, Fe, Mn and Cu contents was assessed. The results showed marked annual variations between rootstock treatments in leaf N, P, K, Zn, Fe, Mn and Cu contents but not in Ca and Mg. In furthere studies, Misra and Singh (1989) reported that rootstocks Karna Khatta, Rough lemon, Bhadri lemon, Sour orange, trifoliate orange (*P. trifoliata*), Kodakithuli, Troyer citrange, Rangpur lime and Cleopatra mandarin rootstocks were budded with Pant lemon-1 and plants of Pant lemon-1 on their own roots. Average fruit weight, length and diameter, number of seeds/fruit, fruit-shape index and percentage juice and waste were not affected by rootstock. Rind thickness was least (1.09 mm) in fruits from trees on Rangpur lime and greatest (1.70 mm) in those from trees on Jambheri. Highest juice TSS (7.34%) and ascorbic acid contents (40.18 mg/100 ml juice) were obtained on Cleopatra mandarin. Highest total sugar content (0.482%) was obtained on sour orange and highest total acidity (5.31%) on Bhadri lemon. Fruits from Pant lemon-1 on its own roots had the lowest juice TSS, total sugars, acidity and ascorbic acid contents. With the same stionic combinations Misra and Singh (1991) found that shoot length was significantly affected by rootstock with highest shoot length obtained for trees on their own roots. Jawaharlal et al. (1991) used five species viz. acid lime, trifoliate orange (*P. trifoliata*), Troyer citrange, Rangpur lime and Rough lemon as rootstocks for acid lime scions at Periyakulam, Tamil Nadu. Acid lime seedlings served as controls. The fruit physiological weight loss (PWL) was highest for trees on Troyer citrange (14.3%) and lowest on Rangpur lime (6.6%). Fruit shelf life was longest on Rangpur lime (6 days) and shortest on trifoliate orange and Troyer citrange (3 days in each case). Ascorbic acid content ranged from 23.7 mg/100 g for fruits on acid lime to 29.17 mg/100 g for those on trifoliate orange; the latter rootstock also resulted in the highest fruit TSS content. Similarly, Tayde et al. (1995) evaluated Jambheri, Kharna khatta, Marmalade orange (*C. limonia*), Nasnaran (*C. japonica* [*Fortunella japonica*]), Nemutenga (*C. limon* [lemon]), Rangpur lime, sweet lime (*C. limettioides*), trifoliate orange (*P. trifoliata*) and Troyer citrange as rootstocks for Kagzi lime. Sweet lime rootstock gave the most vigorous and productive trees, followed by Jambheri and Marmalade orange. Fruit quality did not differ significantly between rootstocks. Also, Ram et al. (1999) conducted a study during 1994-96 on the pre-bearing performance of acid lime in relation to bud union, stock:scion circumference ratio, canopy volume, vegetative flush distribution, flowering and leaf nutrient uptake pattern on 20 rootstock strains at Nagpur, maharashtra. The lowest stock:scion ratio was in plants grafted on Cleopatra mandarin (Coorg and Tirupati strains). Stock:scion ratios for Troyer citrange (Gonicoppal and Chethalli strains), Carrizo citrange (Chethalli strain), C 32 and C 35 were high indicating uncongenial bud union, whereas the remaining 12 rootstock strains showed smooth bud union and exhibited congenial stionic combination with acid lime. Acid lime, Alemow (*C. macrophylla*) and Rangpur lime (Brazilian) rootstocks produced the largest canopies. Canopies were smallest on Cleopatra mandarin (Coorg and Tirupati). Although all the rootstocks showed a cyclic flush and ambia (Nov.-Dec.), mrig (June-July) and hasta (September) flushes were induced in 1995-96. However in 1996-97, flush induction varied between rootstock scion combinations. The induction of generative flushes (spring flowering) of acid lime was observed for all the rootstock strains in 3rd year after planting except for C 32 (only 25% flowered), acid lime seedling and Chethalli strain of Troyer citrange (only 50% flowered) and Chase Rough lemon and Coorg and Gonicoppal strains of Cleopatra mandarin (only 75% flowered).

3.3 Rootstock characters

About 90 per cent of the citrus plants are being propagated on Rough lemon due to its high vigour and good yield, but it is susceptible to *Phytophthora* and citrus nematode, even the fruit quality on Rough lemon is inferior. There are many rootstocks now available which produce higher yield with good quality fruits also has resistance/tolerance to *Phytophthora* and citrus nematode.Singh et al. (1997) collected rootstocks from exotic and indigenous sources. Thirteen improved rootstocks from exotic sources and four rootstocks from indigenous sources were planted in the field gene bank at NRCC, Nagpur in 1991. The results revealed that plant height was maximum in Swingle citrumelo (346.8 cm), Schaub Rough lemon (344.2 cm), C-35 citrange (335.8 cm) and minimum in Flying Dragon and trifoliate orange 9155.8 and 190 cm), respectively). Stem girth was more in Rough lemon (39.33 cm), Schaub Rough lemon (37.67 cm) and Rangpur lime (37.50 cm) and low in Flying Dragon and trifoliate orange (10.05 and 12.83 cm, respectively). Larger canopy volume was recorded in Rangpur lime (16.8 m^3), Schaub Rough lemon (9.57 m^3) and Alemow (9.42 m^3) whereas canopy volume was low in Flying Dragon (0.18 m^3) and trifoliate orange (0.60 m^3). Alemow, Schaub Rough lemon, Rangpur lime and Rough lemon were found the most vigorous rootstocks. Marathe et al. (2000) reported that in studies involving 20 rootstocks, there was a significant variation in uptake of N, K, Mg, Fe and Mn while P, Ca, Zn and Cu showed no significant variation. The nutrient uptake was not found related to growth habit but it was as per the inherent nutrient absorption characteristics of a particular rootstock. Sun chu sha rootstock showed better performance in vegetative growth with balance amount of nutrient absorption. Trifoliate, Rangpur lime and Cleopatra mandarin rootstocks absorbed less amount of Mn and may be suited for acidic soils having Mn toxicity. Large variation in different nutrient uptake was observed in Trifoliate, Cleopatra mandarin and *C. macrophylla* and hence nutrient has to be supplied to these roostocks carefully.Singh and Singh (2001) reported that a large amount of genetic variation exists within a citrus species and each species has a number of cultivars which are collected and maintained at different research centres in germplasm collections. The highest level of variation was recorded for type of flowers which ranged from 100% perfect in the California line to 65 per cent in Texas and Local lines and seeds/fruit which ranged from 14.65 in Texas to 9.25 in Brazilian. Petiole length was greatest (11.5 mm) in Abohar and least (9.85 mm) in Brazilian. Leaf length was greatest (86.4 mm) in 8784 and least (77.65 mm) in Local. Leaf width was greatest (43.65 mm) in Texas and least (38.45 mm) in Brazilian, whereas leaf area was greatest (25.99 mm^2) in 8784 and least (20.16 mm^2) in Local strain. Flower characters such as bud length and width, number of petals and stamens, and length of anthers and pistils were variable among the lines. The percentage of perfect flowers was higher in solitary flowers as compared to inflorescence and different lines showed variation for this character. These studies have indicated similarity in Brazilian and California strains and should prove helpful for characterization of rootstock germplasm. Low soil pH is a major constraint in citrus production on acidic soils due to toxic levels of aluminium that hinders rootstock growth. Very limited information is available on aluminium tolerance in citrus. Therefore, an attempt was made by Singh and Mannivannan (2008) in Arunachal Pradesh, to assess the variability for aluminium tolerance in citrus rootstocks. To study the response of low pH and aluminium toxicity in hydroponic culture, fifteen citrus rootstocks were grown in nutrient solution. The solutions contained 5 levels of aluminium ranging from 10 to 100 ppm aluminium (a.i.). The nutrient solution pH was maintained at 4.5 with 25 ± 1° C temperature. Root and shoot growth were suppressed in all the rootstocks at high ai concentrations. Seedlings of some rootstocks had yellow, mottled and withered appearance. Considerable variability was observed in tolerance to aluminium toxicity among the rootstocks which may be useful in selection for aluminium tolerance. Singh et al. (2009) conducted an experiment to observe the performance of different rootstocks using Carrizo (*C. sinensis*) (L.) Osbek. x (*C. trifoliata* L.), Rangpur lime (*C. limonia* Usbek.), Rough lemon and Sour orange. Rough lemon responded quickest germination followed by Rangpur lime, Carrizo and Sour orange. Irrespective of the species, germination commenced within 23 days and it continued for next 17 days in Rangpur lime and Rough lemon, 32 days in Sour orange and for 99 days in Carrizo. The maximum number of node/plant was found in the Carrizo plants which show a promise for use as dwarfing rootstock. However, vigor of plant was maximum in Rough lemon.

3.4. Nursery characters of citrus rootstocks

Germination and polyembryony of 27 citrus rootstock seedlings were studies by Shinde et al. (2007) at the nursery stage in Parbhani, Madhya Pradesh, India. High seed germination (54-81%) was noted in L-19 Rangpur lime, Iambheti local, Rough lemon chettali, Malta lemon, L-12 Epreka lemon, Rangpur lime local, L-2 Rangpur lime, Sohmyndong, Citrus macrophylla, Nemu-tenga, Narangi cborg, Calamondin and lemon galgal. Poor germination (31 to 45) was recorded in Troyer citrange, Canizo citrange, Cleopatra mandarin (Grabstan), Troyer citrange (Punjab), Citrange A.P., Mannalade orange and savage citrange. The remaining rootstocks were intermediate (46 to 53%) in seed germination. Germination percentage, in general, showed positive relationship with the vigour of rootstock seedlings. Polyembryony was highest in lambheri local, Sohmyndong, Narangi coorg, *Citrus macrophylla*, Kumquat, Lemon galgal, Kichili and Marmalade orange, while it was lowest in Malta lemon, Savage citrange and Bengal citrange. Singh and Viswakarma (2008a) evaluated the growth performance of mandarin grafted/budded on different rootstocks namely Pummelo, Rangpur lime, Trifoliate, Rough lemon, Cleopatra, *Bael*, wood apple and grapefruit. After one year of growth, maximum success percentage of 65.7 per cent was recorded in mandarin budded on pommel followed by grafting in wood apple (60.4 %). Lowest success percentage (21.2 %) was recorded with budding on pummel and minimum height (7.32 cm) with budding on trifoliate. Maximum diameter of rootstock (1.234 cm) was recorded with Cleopatra closely followed by Rough lemon with diameter of (1.231 cm). The maximum scion diameter (0.937 cm) was recorded with budding on Rough lemon, closely followed by budding on pummel with scion diameter of 0.919 cm. The lowest scion diameter (0.256 cm) was recorded under budding on trifoliate rootstock. Hadli and Raijadhav (2005) reported that five rootstock Rangpur lime strains (Shrirampur, Florida 8747, L-2, Brazil and Marmalade orange) were budded with cv. Mosambi under greenhouse conditions. The highest bud take was in Shrirampur followed by Florida 8747 and Brazil. The maximum scion shoot length was in Brazil and L-2, while stem diameter and scion shoot diameter were maximum in Shrirampur. Scion:stock ratio was maximum in Shrirampur, L-2 and Brazil, with vigorous stock. The wide scion ratio indicated slow growth as in the case of Florida 8747 and Marmalade orange. The total dry matter of scion was related to the increased stem diameter and leaf size. Singh et al. (2008) compared the performance of different rootstocks namely Carrizo, Rangpur lime, Rough lemon and sour orange in Rajasthan. Rough lemon responded quickest germination followed by Rangpur lime, Carrizo and Sour orange. Irrespective of the species, germination commenced within 23 days and it continued on for 17 days in Rangpur lime and Rough lemon, 32 days in sour orange and for 99 days in Carrizo. The maximum number of nodes/plant were found in the Carrizo plants which showed a promise for use as dwarfing rootstock. However, maximum plant vigour was observed in Rough lemon. Singh and Viswakarma (2008b) evaluated the growth performance of different rootstocks under poly house. After two years, maximum plant height (363.83 cm) was recorded with variety smooth flat Seville closely followed by Latipes with plant height of 360.42 cm. Maximum stem girth (3.622 cm) was recorded with variety Rough lemon followed by Latipes (3.345 cm). The maximum leaf length (15.13 cm) and leaf width (8.76 cm) was recorded with variety Seminlo tangelo whicle it was lowest in case of variety Nelsp (6.39 cm) with 3.16 cm length and 1.31 cm width, respectively.

3.5 Rootstocks and diseases resistance

Trees budded on Cleopatra are tolerant to foot rot and young tree decline, and to tristeza and xyloporosis viruses by Dhillon and Sharma (1977). Also the incidence of tree decline and bud-union creasing amounted to 69 and 100 per cent, respectively, on Jatti Khatti, compared with 12 per cent and nil, respectively, on Cleopatra. The combination of nucellar scion on *C. limonia* and *C. amblycarpa* was highly tolerant of citrus tristeza and psorosis viruses, and the least affected by fovoid [xyloporosis], vein enation and greening. *Citrus limonia* was highly tolerant of and *C. amblycarpa* weakly susceptible to *Phytophthora*. Stocks of some Rough lemons and Rangpur lime had some horticultural defects but were proposed for further trials. Certain cultivars from 11 Citrus spp. were unsuitable as rootstocks for *C. sinensis* (Choudhari and Mali 1978). Similarly, Tayde *et al.* (1988) evaluated 9 rootstocks under Akola conditions. The tallest trees were obtained on *C. limonia* (Rangpur lime). Trees on Cleopatra (*C. reshni*) and *P. trifoliata* were smallest. Kharna Khatta (*C. karna*) was the most susceptible rootstock to foot rot disease [*Phytophthora*]. Kinnow trees on Rangpur lime and Jambhiri (*C. jambhiri*) exhibited moderate growth. Trees on Rangpur lime showed a spreading habit. Prasad et al. (1990) evaluated strains of *C. limonia*, *C. jambhiri* and *P. trifoliata* for resistance to *Phytophthora* root rot. There was variation among different strains of the species and also among *P. trifoliata* strains in resistance. The most suitable pollen parent was a strain of *P. trifoliata* combining resistance with moderate growth vigour, Argentine Trifoliate. Dhillon et al. (1994) reported that the sweet orange cultivars Blood Red and Musambi, budded on nucellar Rough lemon seedlings, were shown to decline as a result of a corky layer at the bud union that developed into a crease and then into separation of the scion and rootstock bark at

PAU, Ludhiana, Punjab. In 10-year-old orchards of both cultivars, commercial stocks of scion produced greater creasing of the bud union and corresponding tree decline than virus-free stocks: with commercial and virus-free stocks, tree decline amounted to 83.7 per cent and 51.5 per cent, respectively, for Blood Red, and 90.5 per cent and 63.7 per cent, respectively, for Musambi. Kale et al. (1996) reported that acid lime budded on trifoliate orange (*P. trifoliata* was more tolerant of canker (*Xanthomonas campestris* pv. *citri* [*X. axonopodis* pv. *citri*]) than acid lime on Kharna Khatta and Rough lemon at Akola, Maharashtra.

Kale and Peshney (1997) reported that Rough lemon (Akola) and Rangpur lime (Akola) strains are resistant to *Phytophthora* root rot prevalent in Vidarbha region and very well recommended as rootstock for preparation of grafts of Nagpur mandarin and/or Sweet orange. Influence of rootstock on fruit drop in Kinnow mandarin was investigated by Sharma et al. (1999) and they observed that plants on Troyer citrange showed the lowest fruit drop in April-May and the highest pre-harvest fruit drop compared with the other rootstocks. Overall fruit drop was highest in Sohsarkar (86.41%) and lowest in Troyer citrange (69.79%). Gade (2009) collected soil samples from nurseries of Vidarbha region in India. Almost all samples were found associated with *P. parasitica* (28-46 cfu/g soil) when tested on PARPH medium. Rangpur lime was found tolerant rootstock, whereas *C. jambhiri* was found susceptible to the disease.

3.5.1 Shoot-tip grafting in citrus rootstocks

Shoot tip grafting (STG) consists of grafting under aseptic conditions, with a small shoot tip or < 0.2 mm to young seedling rootstock growing *in vitro*. Now, this technique is widely used in citrus, peach, apple, cherry, avocado, etc. for recovery of plants free of virus and virus like diseases. Singh et al. (2003) initiated work to study the graft uptake of mandarin (*C. reticulata*) cv. Desi on four different rootstocks species (*C. jambhiri, C. limonia, C. aurantium* and *C. karna*) through pretreatments given to scion meristem and light conditions during incubation. The in vitro graft uptake varied with different rootstock species and the highest success (43.9%) was noted on *C. jambhiri*, when the shoot tips were pretreated with 100 micro M kinetin and cultured in the dark for 36 h followed by 16/8 h light regime (30 micro mol m^{-2} s^{-1}) post-grafting. Similarly, Hoa et al. (2004) standardized a protocol to develop virus-free nucleus planting material of Kinnow mandarin andd Mosambi sweet orange through shoot tip grafting (STG) at IARI, New Delhi. Among the rootstocks tested, Rough lemon was found more suitable for Kinnow, but Mosambi grew better on its own root. Shoot tip grafts with apical meristem having two leaf primordia were successful up to 20.83% and were free from viruses. Similarly, Roy and Ramachandran (2008) developed viroid-free rootstock plant for citrus through micropropagation. Viroid-free micropropagated plants were indexed by NASH test using radiolabelled cDNA probes to two groups of viroid, citrus exocortis viroid and hop stunt viroid. Kumar et al. (2010) carried out studies in Sriganganagar, Rajasthan on Kinnow mandarin to know the better rootstock and its age for shoot tip grafting. They used three rootstocks namely *C. jambhiri, C. carrizo* and *C. reshni* for micro grafting when the seedling attained 3-5 cm length and 1.6 to 1.8 mm diameter. They observed that 56.33 per cent was achieved in *C. carrizo* when *in vitro* generated shoot tip comprising of apical meristem and two leaf primordial were used as scion. The success rate in case of *C. jambhiri* and *C. reshni* was 50.2 and 40.2 per cent, respectively with 16 day old seedling.

3.5.2 Micropropagation and biotechnological studies

Micropropagation protocol for citrus rootstock pectinifera (*C. depressa*) was standardized by Gill and Gosal (2002). The MS medium supplemented with benzyl amino purine (1 mg/l) was found to be optimum for shoot bud induction on epicotyl segments. Average number of shoots/explant on the liquid medium after 90 days of culturing were higher as compared to the solid medium. Further increase in BAP concentration upto 10 mg/l exhibited frequent drop in percent bud induction and subsequent shoot regeneration/explant. The *in vitro* regenerated shoots rooted on this medium also showed the highest plantlet survival upon transfer to soil. Sharma et al. (2009) standardized a protocol for *in vitro* propagation of citrus rootstocks viz. Rough lemon, Cleopatra mandarin Pectinifera and Troyer citrange. maximum number of shoots per explant was obtained through the callus in Pectinifera, Rough lemon and Cleopatra mandarin in MS basal media +BAP 1 mg/l. Maximum rooting of shoots (1.11%) was noted in rootstock Rough lemon followed by Cleopatra mandarin for the 1/2 MS media supplemented with 10 mg/l IBA. The potting media consisting of soil, sand and FYM in the ratio of 1:1:1 by volume was better with maximum survival rate of hardened plants six weeks after transferring to the pots under greenhouse for Rough lemon followed by Pectinifera and Cleopatra mandarin rootstock. Likewise, micropropagation protocol for 'Troyer' citrange from axillary buds was standardized by Sen and Dhawan (2009a). Multiple shoots can be obtained from single nodes of field grown trees cultured on Murashige and Skoog (1962) medium containing Benzylamino purine (BA; 1.11 micro M), Kinetin (Kn; 1.1625 micro M), and 3% sucrose. A shoot multiplication fold of 3.86 every four weeks was achieved. Proliferated shoots were rooted on half strength MS medium with Naphthalene acetic acid (NAA; 0.5 micro M) and plantlets were transferred to a mixture of soil and agropeat. 100% survival was observed during hardening of the rooted plantlets. Again Sen and Dhawan (2009b) developed optimal media for Rangpur lime, C 35 citrange, Troyer citrange, Swingle citrumelo and Alemow macrophylla. Cultures from adult field-grown trees were established from nodal stem segments for axillary proliferation. Among the five rootstocks tested, shoot proliferation rate was highest in 'Rangpur' lime and lowest in 'Alemow' macrophylla. The results demonstrate that genotypic differences played a significant role in the induction and growth of axillary shoots. Similarly, in trifoliate orange (*P. trifoliata*) Murkute et al. (2009) developed an *in vitro* regeneration protocol for, the important *Phytophthora* collar rot tolerant rootstock genotype of citrus. Among different plant growth regulators used, 1 mg/l BAP, 0.5 mg/l NAA, and 40 mg/l adenine sulphate significantly improved culture response including percentage forming shoot buds, days required for shoot initiation, shoot length and number of resultant shoots per explant. Medium containing 0.5 mg/l IBA+0.5 mg/l NAA gave the maximum root initiation response. The hardening medium consisted of cocopeat and soilrite (2:1) in which the survival rate of hardened plants was found to be 82.2% after one month.

Higher number of buddable Kinnow plants in nursery on Rough lemon rootstock were obtained by Kaur and Aulakh (2008) by the use of gibberellic acid (25 and 50 ppm), urea (0.5 & 1.0 %) and potassium nitrate (1.0 and 1.5 %) for enhancing the seedling growth at Abohar, Punjab. The results revealed that there was an increase in percentage of buddable seedlings by all the treatments, but the maximum increase (6.98 %) was attained with urea (1.0 %) which was closely followed by KNO_3 (1.5 %) which generated an increase of 5.94 per cent buddable seedlings as compared to control.

3.6 Studies on rootstock breeding

Three hybrids viz., CRH-12, CRH-47 and CRH-57 were highly resistant to both *Phytophthora* root rot and citrus nematodes were developed by Prasad et al. (1997). All the three hybrids were moderately tolerant to salinity while CRH-57 and CRH-12 were tolerant to drought. These hybrids were developed by intergeneric hybridization, involving Rangpur lime, Rough lemon and Cleopatra mandarin as seed parents and Trifoliate orange as pollen parent. Dass et al. (1998) crossed rangpur lime and Rough lemon with Troyer citrange and trifoliate orange. Hybrid seedlings were separated on the basis of dominant trifoliate leaf marker. Leaf morphology varied from completely unifoliate to completely trifoliate, but many intermediate forms were also observed, indicating non-uniform expression of trifoliate leaf character in the hybrid progeny. Some sort of incompatibility was observed between Rough lemon and trifoliate orange. In Punjab, Sidhu et al. (2010) made attempts to develop *Phytophthora* tolerance in Rough lemon by selection of mutated cells and callus cultures tolerant to pathogen culture filtrate. Increase in mortality of the callus was recorded with increase in concentration of pathogen culture filtrate (PCF) and the age of the cultures from which the PCF was derived. With the increase in selection pressure of PCf from 5 to 10 percent on mutagen treated (0.3 and 0.4 % EMS and MMS) calli, the survival decreased. Out of all the survivals with various selection pressures, 10 regenerants were obtained.

3.7 Effect of rootstocks on Insect pests

Batra et al. (1992) conducted field studies in the Indian Punjab to screen 134 citrus species/cultivars for resistance to *Phyllocnistis citrella*. The varieties Carrizo, Sacaton, Savage, Troyer, Yama Citrange, Citrumelo (*P. trifoliata* x grapefruit), Cambell Valencia, Pomary and Rubidoux, and *Murraya koenigii* were resistant to *P. citrella* on the basis of leaf infestation. Nineteen species/cultivars were fairly resistant, 27 slightly susceptible, 53 moderately susceptible and 25 highly susceptible. Cleopatra, a promising rootstock for sweet orange, was slightly susceptible whereas the commercial rootstock Jatti Khatti was highly susceptible. Of 26 hybrids studied, *P. trifoliata* x sweet orange, Rangpur Lime x Troyer and Kinnow x Mosambi were the least susceptible.

Shevale and Pokharkar (1992) screened eleven Citrus rootstock varieties for susceptibility to leaf miner, *P. citrella* in the laboratory and in field studies conducted in Maharashtra, India. In field studies, Trifoliate orange and the Cleopatra mandarin variety Morocco were the least susceptible to attack. Rough lemon M.P., marmalade orange and Rough lemon Chethalli were highly susceptible. In laboratory studies, Cleopatra mandarin Morocco and Trifoliate orange were the least preferred varieties; the remaining varieties were moderately or highly susceptible to *P. citrella*.

Batra et al. (1998) carried out population studies of the leafminer *P. citrella* in Ludhiana, Punjab. The results showed two peaks in April and September, coinciding with flushes on all commercial citrus rootstocks in the nursery from the seedling to the budding stage. There was no significant correlation between the population and weather factors irrespective of the rootstock. Rough lemon strain Jatti Khatti was more susceptible to leafminer injury.

Sharma et al. (2002) studied the effects of weather factors (relative humidity, temperature, total rainfall, total number of rainy days, sunshine hours and wind velocity) on the population of citrus leaf roller, *Psorosticha zizyphi* on newly transplanted Rough lemon (cv. Jatti Khatti), *C. reshni* (cv. Cleopatra) and *C. depressa* (cv. Pectinifera) rootstocks under nursery conditions. *P. zizyphi* was active from June to October in Rough lemons and *C. depressa*, and from July to October in *C. reshni,* with mean *P. zizyphi* populations being lowest in *C. reshni*. The population of *P. zizyphi* was positively correlated with temperature, relative humidity, total rainfall and total number of rainy days on all 3 rootstocks and non-significantly correlated with temperature on *C. reshni* rootstock. A 95 per cent variation in *P. zizyphi* population on Rough lemons was attributed to the combined effects of weather factors and rootstocks, whereas the 95.5 and 91 peer cent variation in *P. zizyphi* population in *C. depressa* and *C. reshni,* respectively, was attributed to weather factors alone. A partial effect of relative humidity on *P. zizyphi* population on Rough lemons (69.9%), *C. depressa* (66.9%) and *C. reshni* (96%) was recorded. The partial effect of the total rainfall on *P. zizyphi* populations was non-significant regardless of the rootstocks.

Kumar et al. (2003) conducted field studies in Ludhiana, Punjab during 2000 and 2001, to determine the association of the development and spread of the sooty mould complex (*Capnodium citri* [*Microxyphium citri*], *Alternaria* sp., *Botryodiplodia* sp. *Cladosporium* sp. and *Colletotrichum* sp. among others) with the citrus whitefly (*Dialeurodes citri*) infestations on different Citrus rootstocks. There was a direct and positive relation between sooty mould development and varying levels of *D. citri* nymphal infestation. Among the different Citrus rootstocks, sweet orange and chakotra (*C. grandis* [*C. maxima*]) were highly susceptible to sooty mould and required less *D. citri* nymph population (9.87±0.37 and 5.07±0.39) compared to other rootstock cultivars such as Rough lemon, Cleopatra and Carrizo (*P. trifoliata x C. sinensis*), the later requiring comparatively more *D. citri* populations for the development of the same level of sooty mould incidence (14.03±0.47, 14.16±0.40 and 14.25±0.40), respectively.

Singh et al. (2010) evaluated different exotic citrus varieties comprising of Sweet Oranges, mandarins/tangerines and rootstocks imported by Punjab Agri-Export Corporation Limited (PAGREXCO) from USA received from Pepsi Foods Development Pvt. Ltd. and planted at PAU Fruit Research Station, Gangian (District Hoshiarpur), Govt. Garden and Nursery, Attari (District Amritsar), Farm of Council for Citrus and Agri-Juicing, Jallowal (Bhogpur, District Jalandhar), Govt. Potato Farm, Khanaura (District Hoshiarpur) and PAU Regional Station, Abohar (District Ferozepur), against insect and mite pests during 2007-2008. At Gangian, leaf miner infestation was low in all the varieties, ranging from 3.6 to 12.0 per cent leaf infestation. Population of aphid was nil. Thrips population/25 leaves ranged from 1.7 to 3.33 nymphs. Damage of thrips on fruits was low on Itaborai/Carrizo, Westin/Swingle, Ruby Nucellar/Carrizo while it was severe on Hamlin/Swingle, Olinda Valencia/Carrizo, Olinda Valencia/Rubidoux Trifoliata, Trovita/Rubidoux Trifoliata, Clemenules/Carrizo and Clemenules/C 35. Mite population was very high on all the varieties on leaves, while on fruits, damage was low on Itaborai/Carrizo, Westin/Swingle, Ruby Nucellar/Carrizo while it was severe on Hamlin/Swingle, Olinda Valencia/Carrizo, Olinda Valencia/Rubidoux Trifoliata, Trovita/Rubidoux Trifoliata, Clemenules/Carrizo and Clemenules/C 35. Damage of scales was low on Itaborai/Carrizo, Westin/Swingle, Ruby Nucellar/Carrizo while it was severe on Hamlin/Swingle, Olinda Valencia/Carrizo, Olinda Valencia/Rubidoux Trifoliata, Trovita/Rubidoux Trifoliata, Clemenules/Carrizo and Clemenules/C 35. Mealy bug damage was low on all varieties. Low level of damage of fruit flies, *B. dorsalis* and *B. zonata* was observed only on Clemenules/C 35. Population of Citrus caterpillar ranged from 0.3 - 1.4 larvae/25 leaves. Damage of Katydid on fruits was low on Itaborai/Carrizo, Westin/Swingle, Ruby Nucellar/Carrizo while it was high on Hamlin/Swingle, Olinda Valencia/Carrizo, Olinda Valencia/Rubidoux Trifoliata, Trovita/Rubidoux Trifoliata, Clemenules/Carrizo and Clemenules/C 35. Grey weevil damage was low on all varieties. At Attari, leaf miner infestation was nil in all the varieties. Aphid population was low in all the varieties. Thrips population/25 leaves was nil on all varieties except on Ruby Nucellar/Carrizo (3.0 nymphs). Damage of thrips on fruits was nil on all varieties, except Ruby Nucellar/Carrizo where it was severe. Mite population on leaves was low on Clemenules/C 35, Marisol/Carrizo and Clemenules/Carrizo, while it was severe on all other varieties. On fruits, damage of mite was low on Clemenules/C 35, Marisol/Carrizo and Clemenules/Carrizo while it was severe on all other varieties except Midknight Valencia/Swingle and Early Gold/852 (No fruits were present on these varieties). Damage of scale insect was nil. Mealy bug damage was low on all the varieties except Olinda Valencia/Rubidoux Trifoliata where it was severe. Low level of damage of fruit flies, *B. dorsalis* and *B. zonata* was observed only on Olinda Valencia/Carrizo, Hamlin/Swingle and Hamlin/Carrizo. Population of Citrus caterpillar was nil. Damage of Katydid on fruits was low on all varieties where fruits were present except on Marisol/Carrizo where it was not recorded. Grey weevil damage was low on all varieties. At Jallowal, leaf miner infestation was low in all the varieties, ranging from 5.4 to 13.0 per cent leaf infestation. Aphid population was nil. Thrips population/25 leaves ranged from 0 to 11.5 nymphs. Fruits were present only on Trovita/Rubidoux Trifoliata. Damage of thrips was severe on fruits of Trovita/Rubidoux Trifoliata. Population on mite, scales, fruit fly, and Katydid was nil as no fruits were present on other varieties. Population of Citrus caterpillar ranged from 0.64 to 1.8 larvae/25 leaves. Grey weevil damage was low on all varieties. At Khanaura, damage of leaf miner, aphid, scale and mealy bug was not observed. Thrips population/25 leaves ranged from 1.44 to 4.0 nymphs. Damage of thrips on fruits was severe on all varieties, except Midknight Valencia/Swingle and Midknight Valencia/Rubidoux Trifoliata where fruits were not present. Mite population on leaves was severe on all varieties. On fruits, damage of mite was severe on all varieties, except on Midknight Valencia/Swingle and Midknight Valencia/Rubidoux Trifoliata where fruits were not present. Damage of scale and mealy bug was nil. Low level of damage of fruit flies, *B. dorsalis* and *B. zonata* was observed only on Clemenules/C 35. Population of Citrus caterpillar ranged from 0 to 0.66. Damage of Katydid on fruits was low on all varieties, except Midknight Valencia/Swingle and Midknight Valencia/Rubidoux Trifoliata where fruits were not present. Grey weevil damage was low on all varieties. At Abohar, leaf miner infestation was low in all the varieties, ranging from 0 to 15.1 per cent leaf infestation. Aphid, scale and mealy bug were not observed. Thrips population/25 leaves ranged from 2.1 to 6.0 nymphs. Damage of thrips on fruits was severe on all varieties. Mite population on leaves was severe on all varieties. On fruits, damage of mite was severe on all varieties, where fruits were present. Low level of damage of fruit flies, *B. dorsalis* and *B. zonata* was observed only on Clemenules/Carrizo and Clemenules/C 35. Population of Citrus caterpillar ranged from 1.0 to 1.3 larvae/25 leaves. Though the population of Citrus caterpillar was low, but freshly laid eggs were numerous, even in November. Damage of Katydid on fruits was low on all varieties having fruits. Grey weevil damage was low on all varieties.

3.8 Effect of rootstocks on nematodes

Misra and Edward (1977) examined Karna citrus, currently used in Allahabad as a rootstock for most of the commercial species of citrus, and 4 other potential citrus rootstocks for their relative resistance to *Tylenchulus semipenetrans* and for histopathological changes in the root tissues. All 5 species of citrus were susceptible to citrus nematode but in the following decreasing order; Hill lemon [*C. pseudolimon*] > Jambheri > Karna > sweet lime [*C. limettioides*] > Rangpur lime [*C. limonia*].

Chandel and Sharma (1989) reported that out of the 11 Citrus rootstocks tested for their reaction to *T. semipenetrans*, trifoliate orange (*P. trifoliata*) was the most resistant. Troyer citrange (*P. trifoliata* x *C. sinensis*), carrizo (*P. trifoliata* x *C. sinensis*) and scalo citrumelo (*P. trifoliata* x *C. paradisi*) were also quite resistant and can be used in citrus cultivation. Rangpur lime (*C. limonia*), grapefruit (*C. paradisi*) and *C. pectinfera* were moderately resistant.

Singh and Singh (1997) reported that citrus nematode, *T. semipenetrans* is a serious pest of citrus and is widely distributed. Studies on resistance in rootstocks against this nematode could be of practical value in its control. Seventeen rootstocks (5 trifoliate, 2 hybrids, 8 Rangpur lime and 2 Rough lemon) were screened under pot conditions at NRCC, Nagpur. One year after inoculation, the Reproduction Factor (Rf) was calculated on the basis of final and initial population. The females per gram root were also counted. Based on Rf values, all the trifoliate oranges and one hybrid CRH-41 was found resistant. All other rootstocks were found susceptible. Rf values of hybrid Carrizo citrange was 10.2. There was much variation in the Rf values of different rootstocks. The Rf value of Rough lemon (local) and Mithi Tulia was 17.09 and 21.96, respectively. The maximum Rf value (36.77) was in Phillipine red lime followed by 26.16 in Rangpur lime (Brazalian).

Bamel and Singh (2010) screened 14 rootstocks to find out resistant rootstock against *T. semipenetrans* in pots under screen house conditions at NRCC, Nagpur. Two weeks after transplanting, each seedling was inoculated with 5000 active juveniles of *T. semipenetrans*. One year after inoculation, the reproduction factor (Rf) of the nematode on each rootstock was calculated by dividing the final population with initial population. Based on reproduction factor of the nematode, plant reaction was classified into three groups: highly resistant (Rf =0-0.09), resistant (Rf = 0.10-1.00) and susceptible (Rf > 1.00). They observed a wide variation in the reaction of rootstocks to nematode. In all rootstocks, Rf value was more than one (maximum Rf = 22.5 in *C. karna*) its susceptibility to nematode. There was much variation in the Rf values of different rootstocks. This may be because of host specificity, excessive infectivity or nutritional requirements of the namatodes.

4. Conclusions

- The fast expansion of the citrus industry in India, particularly in the Punjab state is mainly due to the evaluation of new germplasm and release of new scion and rootstock varieties suitable and well adapted to the local agro-climatic conditions besides the generation of effective production, post-harvest and processing technologies.
- The use of rootstocks in citrus is an important horticultural practice and is being exploited since the beginning of the 19th century.
- The optimum performance of scion varieties depends on the proper selection of rootstocks under a given set of growing conditions.

- Moreover, the rootstocks are known to have a profound effect on vigour, precocity, productivity, quality, tolerance to cold, drought, salinity, alkalinity and longevity of scion varieties grafted on them. They also influence the susceptibility of trees to various diseases and insect pests.
- The increased demand for juice concentrates all over the world and best quality fruit in general are yet other factors associated with the rootstock-scion relationships.
- Due to the aforesaid reasons rootstock development in citriculture has assumed a great significance in every citrus growing country.
- The rootstocks development is a continuous process as the rootstocks found suitable at one time may entirely fail in the future, hence the search continues.

5. Future Strategies

Rootstock is an important unit of the orchard enterprise which occupies the soil and is responsible for the supply of nutrients to the scion variety. Rootstock development is an open unended process because success depends on the interaction of genetic potential with soil, climate, diseases, insect pests and cultural practices. Here the view of Bitters appears to be right when he says "the rootstocks of yesterday are not the rootstocks of today and the rootstocks of today will not be the rootstocks of tomorrow". So the available information on the subject needs to be interpreted in the phase of new emerging problems for fast attention and also new germplasm need to be tested:

1. Use standard Rough lemon (Jatti Khatti) seeds for healthy nursery production and for future healthy orchard with high productivity.
2. Area specific approach needs to be undertaken:
   (a) For salt tolerance, Rangpur lime can be exploited in arid-irrigated zones.
   (b) Similarly, for *Phytophthora* tolerance in addition to testing of new hybrid rootstocks; Carrizo, Troyer and Citrumelo can be used for Kinnow in solis with less than 8 pH. These rootstocks in combination with Kinnow are not only resistant to foot rot but also yield the fruits with high TSS in the juice. Volkameriana may also be used for citrus industry of Punjab after proper testing as it performs better than standard Rough lemon in Pakistan.
   (c) Cleopatra for Blood red, Pectinifera for Mosambi, Carrizo, Troyer and Citrumelo for Kinnow are less vigorous and consequently give low yield per plant leading to low productivity. So plant density of trees on these rootstocks should be increased to improve the productivity. For this purpose rigorous testing and time consuming spacing trials can be avoided by interpreting the already available data of canopy volume, assuming canopy and number of plants per acre of the trees budded on Rough lemon as standard.

(d) Citrus genetic diversity should be fully characterized ad conserved. *In situ* gene sanctuary for citrus in Garo hills of Meghalaya, India having huge unexploited citrus diversity is an example of protecting erosion of citrus genetic diversity.
(e) Introduction and evaluation of exotic germplasm is of prime importance.

References

Allurwar MW and Parihar SK (1992) Comparative study of root systems of common rootstocks of orange. *J Soils Crops* 2(1): 100-101.

Anonymous (1966) Annual Report of the Department of Horticulture, Punjab Agricultural University, Ludhiana for the year 1965-1966.

Anonymous (1997) Annual Research Report of Horticulturist. AICFI project, Dr. PDKV, Akola.

Arora RK, Pal V and Yamdagni R (1988) Extent of decline and disease incidence in various stionic combinations of sweet orange. *South Indian Hort* 36(5): 264-265.

Aulakh PS (2008) Effect of different rootstocks on vigour, yield and quality of Valencia Late Sweet Orange. P. 69. Souvenir and Abstract. National Symposium on Citriculture: Emerging Trends, National Research Centre for Citrus, Nagpur (July 24-26, 2008).

Aulakh PS and Baidwan RPS (2004) Performance of sweet orange (*Citrus sinensis* (L.) Osbeck) cv. Hamlin on different rootstocks. *Prog Hort*, 36(1): 8-11.

Aulakh PS, Thind SK and Kumar A (2008a) Performance of exotic citrus cultivars imported by PAGREXCO from USA under arid irrigated region of Punjab. P. 18-19. Souvenir and Abstract. National Symposium on Citriculture: Emerging Trends, National Research Centre for Citrus, Nagpur (July 24-26, 2008).

Aulakh PS, Thind SK and Kumar A (2008b) Pre bearing performance of some exotic citrus varieties under arid irrigated conditions of Punjab. P.19. Souvenir and Abstract. National Symposium on Citriculture: Emerging Trends, National Research Centre for Citrus, Nagpur (July 24-26, 2008).

Bal, J.S. and Chohan G.S. (1982) Studies on fruit quality at maturity and ripening of Kinnow Mandarin on different rootstocks. *Indian J. Hort.* 39(1): 45-51.

Bamel V and Singh IP (2010) Screening of citrus rootstocks against citrus nematode, *Tylenchulus semipenetrans* Cobb. In: souvenir and abstracts. National Seminar on Citrus Biodiversity for Livelihood and Nutritional Security. 4-5 October. National Research Centre for Citrus, Nagpur. TS6-P7: 384.

Batra RC, Sharma DR and Chanana YR (1992) Screening of citrus germplasm for their resistance against citrus leaf miner, *Phyllocnistis citrella* Stainton. *J.Insect Sci* 5(2): 150-152.

Batra RC, Sharma N and Arora PK (1998) Population studies of *Phyllocnistis citrella* Stainton on some commercial rootstocks of citrus under nursery conditions. *Pest Mgmt Hort Eco*, 4(2): 61-64.

Bauri FK, Sarkar SK, Misra DK and Bandyopadhyay B (2006) Effect of rootstock on growth, yield and disease tolerance of mandarin orange in Darjeeling. P. 26. Abstract and Souvenir. National Seminar on Integrated production and Post-harvest Management of Tropical Fruits. AICRP on Tropical Fruits, BCKV, Mohanpur, West Bengal. 11th to 12th April, 2006.

Brar LS (2010) Diversifying towards high value horticulture. pp. 44-48. Souvenir and Abstract. National Seminar on Impact of Climate Change on Fruit Crops. 6th to 10th Oct., 2010. PAU, Ludhiana.

Brown WR (1920) The orange: a trial of stocks at Peshawar. *Pusa Agric Res Inst Bull.* 93, 7 pp.

Chadha KL (1970) Rootstocks. pp. 9-25. In: KL Chadha, NS Randhawa, OS Bindra, JS Chohan and LC Knorr (eds): *Citrus Decline in India.* A Joint Publication of PAU, OSU and USAID.

Chadha KL and Singh HP (1996) Description, Classification and Cataloguing of Genetic Resources of Citrus in India. Consultancy Report, IPGRI-APO, Singapore.

Chandel YS and Sharma NK (1989) Reaction of different citrus (Citrus species) rootstock to citrus nematode, *Tylenchulus semipenetrans* *Indian J Agric Sci* 59(9): 608-609.

Chohan GS, Kumar H and Vij VK (1982) Effect of different rootstocks on vigour, yield and fruit quality of Blood Red orange (*Citrus sinensis* Osbeck). *J Res*, PAU, 19: 2, 107-112.

Chohan GS, Kumar H and Vij V-K (1988) Effect of rootstocks on tree vigour, health, yield and fruit quality in Hamlin cv. of sweet orange. *Indian J Hort*, 45(3-4): 208-211.

Chohan GS, Kumar H and Vij VK (1991) Pectinifera - a new rootstock for Mosambi cultivar of sweet orange. *J Res* PAU 28(3): 359-362.

Chohan GS, Kumar H and Vij VK (2000) Effect of rootstocks on tree survival, health, vigour, yield and fruit quality of sweet orange (*Citrus sinensis* Osbeck). *Indian J Hort*, 57(1): 54-58.

Chohan GS, Vij VK and Kumar H (1988) Effect of rootstocks on tree vigour, health, yield and fruit quality of grapefruit (*Citrus paradisi* Macf.) cv. Marsh Seedless. *Punjab Hort J*, 28(1-2): 27-29.

Chohan GS, Vij VK and Kumar H (1990) Effect of rootstocks on tree vigour, health, yield and fruit quality of grapefruit (*Citrus paradisi* Macf.) cultivar Marsh Seedless. *Indian J Hort*, 47(3): 297-300.

Chohan GS, and Kumar H (1983) Performance of Musambi cultivar of sweet orange (*Citrus sinensis* Osbeck.) as influenced by different rootstocks. *J. Res*. PAU, 20 (3):275-280.

Chohan GS, Kumar H and Vij VK (1986) Performance of Jaffa and Valencia late cultivars of sweet orange (Citrus senensis) (L) Osbeck) on different rootstocks. *Indian J Hort* 43: 29-34.

Chohan GS, Sharma JN and Thatai SK (1978) Effect of Four rootstocks on vigour, yield and quality of Kinnow mandarin. *Indian J.Hort.* 212-215.

Chohan,GS, Vij VK and Kumar H (1990) Effect of Rootstocks on three Vigour, Health, Yield and fruit quality of grapefruit (*Citrus paradisi* Macf.) cultivar Marsh Seedless. *Indian J. Hort.* 47 (3): 297-300.

Choudhari KG and Mali VR (1978) Reaction of mosambi (*Citrus sinensis* L. Osbeck) on different rootstock to virus and other disorders under field conditions. *J Maharashtra Agric Univ* 3: 1, 40-44.

Dass HC, Singh A and Vijayakumari N (1998) Rootstock breeding - variations for leaf morphology in citrus rootstock hybrid progeny. *Indian J Hort*, 55(1): 16-19.

Devadas VS, Kuriakose JM and Kannan K (1988) Early performance of mandarin oranges (*Citrus reticulata* Blanco.) on different rootstocks in the submontane region of Wynad in Kerala. *Agric Res J Kerala*, 26(2): 183-197.

Dhatt A.S. and Singh Z (1993) Propagation and rootstocks of citrus. p. 523-549 In: Advances in Horticulture Vol. 2. Fruit crops (Chadha, K.L., ed.). Malhotra Publishing House, New Delhi.

Dhillon PS, Bindra AS, Singh R and Dhillon DS (1993) Studies on stionic ratio, canopy volume and metabolites distribution in kinnow. *Punjab Hort J*, 33(1/4): 17-20.

Dhillon RS and Sharma KK (1977) Cleopatra mandarin - a promising rootstock for citrus in the Punjab. *Punjab Hort J.* 17: 3-4, 104-108.

Dhillon RS, Kapur SP and Cheema SS (1981) Role of stionic combinations in sweet orange decline -- effect on growth and vigour in Blood Red and Musambi. *J Res* PAU, 18: 2, 144-148.

Dhillon RS, Kapur SP, Cheema SS and Singh SN (1994) Decline of sweet orange in relation to source of budwood and bud-union crease. *J Res PAU* 31(1): 32-34.

Dubey AK and Singh AK (2003) Evaluation of rootstocks of different mandarins (*Citrus reticulata*) under foot-hills conditions of Arunachal Pradesh. *Indian J Agric Sci*, 73(10): 527-529.

Gade RM (2009) Biological and chemical management of *Phytophthora* root rot/collar rot in citrus nursery. *Association Francaise de Protection des Plantes*, France, Dec. 8-9, 2009. 245-253.

Ganpathy R (1998) Performance of Kodur sathgudi (*Citrus sinensis* (L) Osbeck) on different root stocks under rainfed conditions of Bihar plateau. *Orissa J Hort*, 26(1): 25-27.

Ghosh SN and Chattopadhyay N (1993) Effect of rootstocks on tree vigour, yield and fruit quality of mandarin orange under non-irrigated condition in red and laterite soils of West Bengal. *Hort J*, 6(2): 79-82.

Ghosh SN and Tarai RK (2007) Performance of mosambi sweet orange on different rootstocks grown in laterite soil in West Bengal. *J Hort Sci* 2(2): 153-155.

Gill MIS and Gosal SS (2002) Micropropagation of pectinifera (*Citrus depressa* Hayata) - a potential citrus rootstock for sweet orange. *Indian J Citriculture.* 1(1): 32-37.

Grace JK, Ranganayakulu C and Seshadri KV (2005) Effect of rootstocks on the fruit quality of Sathgudi sweet orange grown on different rootstocks. *Indian J Hort,* 62(3): 300-302.

Gupta OP, Arora RK and Chundawat BS (1978) Variations in the extent of winter injury in different cultivars of sweet orange on various rootstocks. *Haryana J Hort Sci* 7: 3-4, 112-115.

Hadli J and Raijadhav SB (2005) Rootstock influence on the growth of scion cv. Mosambi of Rangpur lime strains. *Karnataka J Agric Sci*, 18(4): 1031-1033.

Hoa NV, Ahlawat YS and Pant RP (2004) Production of virus-free Kinnow mandarin and Mosambi sweet orange nucleus planting material through shoot tip grafting. *Indian Phytopathol*, 57(4): 482-487.

Iyengar BRV and Keshava Murthy SV (1988) Absorption of fertilizer phosphorus by Italian lemon and Seedless lime as influenced by Citrus rootstocks. *J Nuclear Agric Biol*, 17(4): 191-194.

Jawaharlal M, Durairaj P, Subburamu K and Irulappan I (1987) Performance of acid lime on different rootstocks. *South Indian Hort*, 35(3): 236-239.

Jawaharlal M, Thangaraj T and Irulappan I (1991) Influence of rootstocks on the post harvest quality of acidlime (*Citrus aurantifolia* L.) fruit. *South Indian Hort* 39(3): 151-152.

Jawanda JS, Uppal DK, Singh R and Arora JS (1978) Studies on granulation in citrus fruits. *Punjab Hort J* 18: 3-4, 180-188.

Jawanda JS and Mehrotra NK (1974) Pre-bearing performance of five sweet orange cultivars on different rootstocks. *Indian J. Agric. Sci.* 44(2): 119-126.

Josan JS and Thatai SK (2008) Studies on the evaluation of rootstocks for Kinnow mandarin under north Indian conditions. *Indian J Hort* 65(3): 332-334.

Josan JS, Kumar A and Thind SK (2010) Evaluation of rootstocks for Kinnow mandarin. pp. 87-88. Souvenir and Abstract. National Seminar on Impact of Climate Change on Fruit Crops. 6th to 10th Oct., 2010. PAU, Ludhiana.

Kahlon GS, Aulakh PS and Rattanpal HS (2010) Research highlights of the Department of Horticulture, PAU, Ludhiana. pp. 22-24. Souvenir and Abstract. National Seminar on Impact of Climate Change on Fruit Crops. 6th to 10th Oct., 2010. PAU, Ludhiana.

Kale KB and Peshney NL (1997) Evaluation of seedling resistance in Rangpur lime and rough lemon strains to *Phytophthora* disease. pp. 62-64. Proceedings of National Symposium on Citriculture, NRCC, Nagpur, 17th to 19th November, 1997.

Kale KB, Raut JG and Dudhe YH (1996) Influence of root stocks on the incidence of bacterial canker of acid lime. *PKV Res J*, 20(1): 105-106.

Kaur N and Aulakh PS (2008) Studies on the effect of growth promoters on the seedling growth of Jatti Khatti rootstock. P.69. Souvenir and Abstract. National Symposium on Citriculture: Emerging Trends, National Research Centre for Citrus, Nagpur (July 24-26, 2008).

Kumar H, Chohan GS and Vij VK (1994) Studies on tree survival, growth, yield and fruit quality of pineapple cv. of sweet orange on different rootstocks. *J Res PAU*, 31(1): 27-31.

Kumar H, Vij VK, Aulakh PS and Baidwan RPS (2002) Performance of blood red, mosambi and pineapple cultivars of sweet orange (*Citrus sinensis* Osbeck) on different rootstocks. *Indian J Citriculture,* 1(2): 135-139.

Kumar H, Vij VK, Mehrotra NK and Aulakh PS (2001) Pre-bearing performance of Lahore Local and Nagpur mandarins (*Citrus reticulata* Blanco) on different rootstocks under arid-irrigated conditions of Punjab. *J Res PAU,* 38(3/4): 178-180.

Kumar R and Ganapathy MM (1992) Performance of Mosambi on different rootstocks. *Indian J. Hort.* 49(3): 222-226.

Kumar R, Batra RC and Sharma DR (2003) Association of sooty mould development with citrus whitefly, *Dialeurodes citri* Asmead, infestation on different citrus rootstocks. *Integrated-plant-disease-management-Challenging-problems-in-horticultural-and-forest-pathology,-Solan,-India,-14-to-15-November-2003,* 71-74.

Kumar R, Kaul ML, Saxena SN and Bhargav S (2010) Effect of age and type of rootstock on success of *in vitro* shoot tip grafting in Kinnow mandarin. P. 33-36. Proceedings. National Symposium on Citriculture: Emerging Trends, National Research Centre for Citrus, Nagpur (July 24-26, 2008).

Kunwar R and Singh R (1983) The influence of different rootstocks on mineral composition of Srinagar mandarin (*Citrus reticulata* Blanco) leaves. *J Plant Nutrition*. 6: 5, 405-412.

Mandge AS and Chakrawar VR (1981) Effect of different rootstocks on bud-take and growth of budling of Nagpur mandarin and nucellar Mosambi sweet orange. *J Maharashtra Agric Univ* 6: 1, 22-24.

Mann SS and Naurial JP (1978) Performance of Pineapple sweet orange (*Citrus sinensis* Osb.) on different rootstocks. *Prog Hort* 10: 1, 37-42.

Marathe RA, Singh S, Ram L and Sonkar RK (2000) Rootstock behaviour in relation to leaf nutrients composition of acid lime (*Citrus aurantifolia* Swingle). *Indian J Hort*, 57(2): 95-101.

Marathe RA, Sonkar RK, Ram L and Singh S (2006) Nutrient uptake, canopy volume and yield of Nagpur mandarin as affected by different rootstocks. *Indian J Hort*, 63(4): 372-375.

Mehrotra NK, Kumar H, Vij VK and Aulakh PS (2000) Performance of Jaffa cultivar of sweet orange (*Citrus sinensis* (L.) Osbeck) on different rootstocks. *J Res PAU,* 37(1/2): 56-60.

Mehrotra NK, Kumar H, Vij VK and Aulakh PS (2000) Performance of Pineapple cultivar of sweet orange (*Citrus sinensis* Osbeck) on different rootstocks. *J Res PAU*, 37(3/4): 198-202.

Mehrotra NK, Kumar H, Vij VK and Aulakh PS (2002) Rootstock studies on Mosambi cultivar of sweet orange under arid-irrigated conditions of Punjab. *J Res PAU,* 39(1): 50-55.

Mehrotra NK, Vij VK, Kumar H, Aulakh PS and Singh R (1999) Performance of Marsh seedless cultivar of grapefruit (*Citrus paradisi* Macf.) on different rootstocks. *Indian J Hort*, 56(2): 141-143.

Mehrotra, NK., Jawanda, JS. and Vij, VK. (1984). Evaluation of rootstocks for Jaffa orange (*Citrus sinensis* Osbeck) under arid-irrigated conditions of Punjab. *Indian J. Hort.* (41): 29-36.

Misra KK and Singh R (1989) Effect of rootstocks on the fruit quality of lemon (*Citrus limon* Burm.) cv. Pant lemon-1. *Annals Agric Res*, 10(4): 410-414.

Misra KK and Singh R (1991) Note on the influence of rootstocks and seasons on the shoot growth of lemon. *Indian J Hort*, 48(1): 16-18.

Misra KK, Singh R and Tiwari JP (1989) Effect of rootstocks on leaf nutrient status of lemon (*Citrus limon* Burm.). *Indian J Hort*, 46(3): 357-363.

Misra SL and Edward JC (1977) Observations on histopathology of some of the more important rootstocks for commercial citrus affected by citrus nematode (*Tylenchulus semipenetrans*, Cobb, 1914). *Allahabad Farmer* 48: 1, 7-18.

Murkute AA, Sharma S and Singh SK (2009) Micropropagation of trifoliate orange (*Poncirus trifoliata*). *Indian J. Plant Physiol* 14(2): 190-193.

Murkute AA, Sharma S, Singh SK and Patel VB (2009) Response of mycorrhizal citrus rootstock plantlets to salt stress. *Indian J Hort* 66(4): 456-460.

Musmade AM, Jagtap DD and Pujari CV (2007) 'Marmalade Orange' a promising root stocks for mosambi (*Citrus sinensis* Osbeck). *Asian J Hort* 2(1): 50-53.

Mustafa MM and Reddy BMC (1990) Effect of rootstocks on growth, yield and quality of Mosambi sweet orange. *Crop Res*, Hisar. 3(2): 234-240.

Pal RS and Lal SD (1993) Studies on the rootstock for Srinagar Santra (*Citrus reticulata* balance) in the valley areas of U.P. hills. *Prog Hort*, 31(1/2): 58-63.

Patil VK and Bhambota JR (1978) Effect of soil salinity on growth of citrus rootstocks. *Res Bull, MAU*, 2: 7, 94-96.

Patil VK and Bhambota JR (1978) Relationship between relative growth of different rootstock seedlings of citrus and levels of salinity in soil. *Res Bull*, MAU, 2: 6, 71-74.

Prasad MBNV, Agarwal PK, Sawant SD and Rekha A (1991) Role of germplasm in Citrus rootstock improvement. Horticulture: New technologies and applications. Proceedings-of-the-International-Seminar-on-New-Frontiers-in-Horticulture, Indo-American-Hybrid-Seeds,-Bangalore,-India,-November-25-28,-1990.73-77. Dordrecht, Netherlands: Kluwer Academic Publishers.

Prasad MBNV, Sawant SD, Parvatha Reddy P, Palaniappan R, Srinivasa Rao NK and Rekha A (1997) Citrus rootstock hybrids resistant to *Phytophthora* and citrus nematodes and also tolerant to salinity and drought. pp. 59-62. Proceedings of National Symposium on Citriculture, NRCC, Nagpur, 17th to 19th November, 1997.

Ram, L., Singh, S., and Marathe, R.A. (1999) Performance of pre-bearing acid lime (*Citrus aurantifolia*) on various rootstock strains. *Indian J Agric Sci*, 69(3): 193-197.

Ram L, Singh S and Yadav RP (2008) Pre-bearing canopy growth, flush emergence and nutrient status of Nagpur mandarin (*C. reticulata* Blanco) plants on various citrus rootstock strains. P. 96. Souvenir and Abstract. National Symposium on Citriculture: Emerging Trends, National Research Centre for Citrus, Nagpur (July 24-26, 2008).

Rao DVR, Madhavachari S, Reddy MLN, Murti VD and Reddy GS (1996) Evaluation of certain citrus rootstocks for sweet orange in Andhra Pradesh. *Adv Hort Forestry,* 5: 33-42.

Rao SN, Rao BVR, Swamy GS and RamaRao BV (1971) Rootstock investigations on Sathgudi orange (*Citrus sinensis* Osbeck). *Indian J.Hort*, 28: 3, 189-195.

Roy and Ramachandran (2008) Development of viroid free nucleus material for rootstock of citrus: an approach for management of yellow corky vein disease. *Indian Phytopathol* 61(2): 259-263.

Sandhu AS and Singh K (1989) Effect of rootstocks and fruit size on granulation and fruit quality of Jaffa sweet orange. *Haryana J Hort Sci*, 18(3-4): 157-160.

Sen S and Dhawan V (2009) Genotypic differences in shoot multiplication among five Citrus rootstocks in vitro. *Acta Hort* (839): 51-56.

Sen S and Dhawan V (2009) Micropropagation of 'Troyer' citrange [*Poncirus trifoliata* (L.) Rat. x *C. sinensis* (L.) Osbeck]. *Acta Hort* (839): 63-70.

Sharma JN, Josan JS and Thatai SK (2002) Effect of rootstocks on tree health, yield and quality of Kinnow mandarin under arid-irrigated region of Punjab. *Indian J Hort*, 59(4): 373-377.

Sharma JN, Thatai SK and Josan JS (2002) Effect of different rootstock on tree vigour, yield and fruit quality of Campbell Valencia sweet orange. *Indian J Hort*, 59(2): 135-139.

Sharma RR and Saxena SK (2004) Rootstocks influence granulation in Kinnow mandarin (*Citrus nobilis* x *C. deliciosa*). *Scientia Hort*, 101(3): 235-242.

Sharma RR, Goswami AM, Saxena SK and Shukla A (1999) Influence of rootstocks on fruit drop in Kinnow mandarin under dense planting. *J App Hort*, Lucknow, 1(2): 133-134.

Sharma S, Batra RC and Sharma DR (2002) Role of weather factors on the activity of *Psorosticha zizyphi* on different rootstocks. *Annals Plant Prot Sci*, 10(2): 378-380.

Sharma S, Parkash A and Tele A (2009) In vitro propagation of citrus rootstocks. *Notulae Botanicae, Horti Agrobotanici Cluj Napoca*. 37(1): 84-88.

Shevale BS and Pokharkar RN (1992) Relative susceptibility of citrus on rootstocks to citrus leaf miner, *Phyllocnistis citrella* Stainton. *Indian J Ent*, 54(1): 54-61.

Shinde BN, Patil VK and Kalalbandi BM (2007) Studies on behaviour of sweet orange (*Citrus sinensis* Osbeck) variety nucellar in relation to bud-take, height and spread on different rootstocks. *Asian J Hort* 2(1): 231-233

Shinde BN, Patil VK and Kalalbandi BM (2007) Studies on germination and polyembryony of different citrus rootstocks seedling at nursery stage. *Asian J Hort* 2(1): 180-183.

Sidhu GS, Rattanpal HS, Gill MIS and Gosal SS (2010) Chemically induced mutagenesis in rough lemon (*Citrus jambhiri* Lush.) for *Phytophthora* tolerance. P 130. Souvenir and Abstract. National Seminar on Impact of Climate Change on Fruit Crops. 6th to 10th Oct., 2010. PAU, Ludhiana.

Singh A and Dass HC (1999) Pre-bearing performance of Nagpur mandarin (Citrus reticulata) on indigenous and exotic rootstocks. *Indian J Agric Sci*, 69(1): 52-54.

Singh A and Singh S (2001) Variability studies in Rangpur lime (*Citrus limonia* Osbeck) strains. *Annals Agri Bio Res*, 6(1): 85-89.

Singh A, Dass HC and Vijayakumari N (1997) Pre-bearing performance of different rootstocks collected from exotic and indigenous sources. pp. 99-104. Proceedings of National Symposium on Citriculture, NRCC, Nagpur, 17th to 19th November, 1997.

Singh B and Singh A (1997) Screening of the citrus rootstocks against citrus nematode, *Tylenchulus semipenetrans* Cobb. pp. 113-114. Proceedings of National Symposium on Citriculture, NRCC, Nagpur, 17th to 19th November, 1997.

Singh D and Daulta BS (1995) Studies on sun burning of fruits on different stionic combinations in sweet orange (*Citrus sinensis* Osbeck) cultivars. *Agric Sci Digest*, 15(1/2): 15-16.

Singh D and Mannivannan S (2008) Growth of citrus rootstocks under aluminium stress environments in hydroponics. P. 94. Souvenir and Abstract. National Symposium on Citriculture: Emerging Trends, National Research Centre for Citrus, Nagpur (July 24-26, 2008).

Singh IP and Aulakh PS (2010) Citrus genetic diversity: Characterization and conservation. Lead Lecture. pp. 1-13. Proceedings. National Symposium on Citriculture: Emerging Trends, National Research Centre for Citrus, Nagpur (July 24-26, 2008).

Singh IP and Singh S (2006) Exploration, collection and characterization of citrus germplasm - a review. *Agric Rev*, 27(2): 79-90.

Singh J, Bhatnagar P, Jain MC, Dashora LK and Jakhar RP (2008) Growth performance of different rootstocks of citrus. P.64-65. Souvenir and Abstract. National Symposium on Citriculture: Emerging Trends, National Research Centre for Citrus, Nagpur (July 24-26, 2008).

Singh J, Bhatnagar P, Jain MC, Dashora LK and Jakhar RP (2009) Growth performance of different rootstocks of citrus. *Env Ecol* 27(2): 536-538.

Singh MP, Dhillon RS, Sharma KK and Singh SN (1980) Prebearing performance of sweet orange on different rootstocks. *J Res, PAU* 17: 2, 127-134.

Singh NP, Gill PPS and Randhawa JS (2008) Pre-bearing performance of different citrus exotic varieties under submontane zone of Punjab. P.23. Souvenir and Abstract. National Symposium on Citriculture: Emerging Trends, National Research Centre for Citrus, Nagpur (July 24-26, 2008).

Singh R and Viswakarma AK (2008a) Growth perfromance of budded mandarin on different rootstocks. P.79-80. Souvenir and Abstract. National Symposium on Citriculture: Emerging Trends, National Research Centre for Citrus, Nagpur (July 24-26, 2008).

Singh R and Viswakarma AK (2008b) Performance of different citrus rootstocks varieties under poly house. P.80. Souvenir and Abstract. National Symposium on Citriculture: Emerging Trends, National Research Centre for Citrus, Nagpur (July 24-26, 2008).

Singh R, Saxena SK and Singh A (1978) Rootstock trial in Kinnow [mandarin]. *Punjab Hort J*, 18: 3-4, 143-148.

Singh S (2010) Foreword. Proceedings. National Symposium on Citriculture: Emerging Trends. National Research Centre for Citrus, Nagpur (July 24-26, 2008).

Singh S, Chharia AS and Singh S (2008) Studies on total root length of different stionic combinations of sweet orange (*Citrus sinensis* Osbeck). P.105-106. Souvenir and Abstract. National Symposium on Citriculture: Emerging Trends, National Research Centre for Citrus, Nagpur (July 24-26, 2008).

Singh Sandeep, Rattanpal HS and Sangwan AK (2010) Evaluation of exotic citrus varieties on different rootstocks against insect-pests in different agro-climatic zones of Punjab. In: Souvenir and abstracts. National Seminar on Citrus Biodiversity for Livelihood and Nutritional Security. 4-5 October. National Research Centre for Citrus, Nagpur. TS6-P9: 385-387.

Singh SK, Khawale RN and Singh SP (2003) A comparative studies on success of in vitro shoot tip grafting in *Citrus reticulata* Blanco on different rootstocks. *Haryana J Hort Sci*, 32(3/4): 171-173.

Sonkar RK, Huchche AD, Ram L and Singh S (2002) Citrus rootstocks scenario with special reference to India - a review. *Agric Rev*, 23(2): 93-109.

Sonkar RK, Srivastava AK, Gupta SG and Das AK (2008) Screening of certain strains of rough lemon and Rangpur lime for Nagpur mandarin in black clay soils of Nagpur. P. 97. Souvenir and Abstract. National Symposium on Citriculture: Emerging Trends, National Research Centre for Citrus, Nagpur (July 24-26, 2008).

Srivastav M, Dubey AK and Sharma RR (2005) Effect of rootstocks on leaf nutrients, tree growth, yield and fruit quality of 'Mosambi' sweet orange (*Citrus sinensis*) under Delhi conditions. *Indian J Agric Sci,* 75(6): 333-335.

Sulikeri GS, Upadya VR, Shekarappa GK and Yaraguntaiah RC (1990) Pre-bearing performance of Coorg-mandarin on different rootstocks under Mudigere conditions. *Curr Res*, UAS, Bangalore,

19(1): 1-2.

Tayde GS, Kulwal LV and Deshmukh PP (1988) Rootstock trial on Kinnow mandarin. I. Prebearing performance. *PKV Res J* 12(1): 40-44.

Tayde GS, Kulwal LV, Rakhonde BM and Deshmukh PP (1995) Rootstock trial on Kagzi lime. *PKV Res J*.19(2): 198-199.

Uppal DK, Grewal SS and Chanana YR (1990) Inherent problems in Kinnow in comparison to other citrus varieties of north-western India. pp. 109-115. *In:* Citriculture in North-western India. Punjab Agricultural University, Ludhiana.

Vij, V.K and Kumar, H (1990). Citrus rootstocks- Selection, raising and performance.pp. 123-130 *In:* Citriculture in North-western India. Punjab Agricultural University, Ludhiana.

Vijayakumari RM, Mohideen MK, Thamburaj S, Seemanthini R and Mariappan S (1990) Effect of rootstocks on yield of mandarin orange (*Citrus reticulata* Blanco) in Shevroy hills. *South Indian Hort,* 38(1): 38-39.

□□□

Green Agriculture : Newer Technologies, 2012
© Kambaska Kumar Behera (ed.), pp. 73-87
New India Publishing Agency, New Delhi (India)
e-mail : info@nipabooks.com; website : www.nipabooks.com

Chapter-4

Fly Ash : A Potential Soil Ameliorant for Sustainable Crop Productivity

R.B. Yadava, D.K. Singh, S.N.S. Chaurasia and R.N. Prasad
Indian Institute of Vegetable Research,
Varanasi- 221305, Uttar Pradesh
E-mail: raj_yadava@rediffmail.com

SUMMARY

Fly ash, a waste product generated by coal based thermal power plants. The annual production of fly ash in the country has increased from 40 million tones in 1994 to 110 million tones in 2004 and it is expected that by 2012 the annual production of fly would reach the mark of 170 million tones. Owing to its physico-chemical characteristics, it has a vast potential for use in agriculture, forestry and wasteland reclamation. Application of fly ash into soil has been reported to increase the water holding capacity, hydraulic conductivity, porosity and decrease bulk density, modulus of rupture and surface encrustation. The studies conducted by various organizations on fly ash use in different crops have revealed significant increase in the crop yields such as wheat, maize, paddy, soybean, sunflower and vegetables like brinjal, tomato, potato, cabbage. Application of fly ash has been found to increase the level of heavy metals viz., Co, Pb, Mo, As and Se as well as the activity of radio-nuclides i.e., ^{40}K, ^{226}Ra, ^{228}Ac in the soil- plant system; however, the levels of all these heavy metals and radio-nuclides were found to be well below the permissible limits.

Introduction

Fly ash, a waste product generated by coal based thermal power plants, poses serious environmental problems all over the world. In India, nearly 73% of the total installed power generation capacity is thermal, of which 90% is coal- based. Nearly

15- 30% of the total amount of residue generated during burning of coal in thermal power plants constitutes fly ash. The annual production of fly ash in the country has increased from 40 million tones in 1994 to 110 million tones in 2004 and it is expected that by 2012 the annual production of fly would reach the mark of 170 million tones. Despite several alternate usages like making of bricks, manufacturing of cement, construction of roads and embankments and mines/ landfills, lot of fly ash still remains unutilized (Fig. 1, 2 & 3). Disposal of such a huge quantity of fly ash poses challenging problems of land usage, environmental pollution and human and animal health hazards. Fly-ash particles, when dumped, cause a serious problem to human and animal health (Page et al. 1979; Borm 1997) as well as to the normal functioning of higher plants. The usage of fly-ash for agriculture as a potential fertilizer or soil amendment has been a global problem related to the concept of so-called zero emission or ecological recycling, whereas its utilization in many other industrial purposes has been established (EPD, 1993). However, the major part of fly-ash is disposed off in unmanaged landfills or lagoons, which serve as a major source of environmental pollutant in the area through fly-ash erosion and leachate generation. To overcome the problems of safe disposal of fly ash, some alternatives such as use of fly ash in brick making, manufacturing of asbestos and cement and conversion into zeolite have been developed.

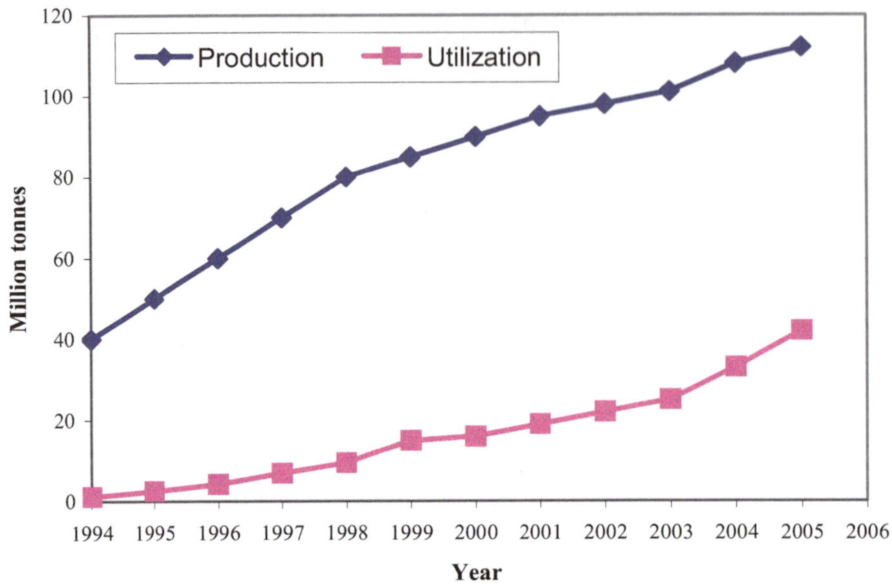

Fig. 1: Fly ash production and utilization pattern in India

Owing to its physico-chemical characteristics, fly ash has a vast potential for use in agriculture, forestry and wasteland reclamation. Application of fly ash into soil has been reported to increase the water holding capacity, hydraulic conductivity, porosity and decrease bulk density, modulus of rupture and surface encrustation (Chang et al. 1977; Adriano et al. 1980; Singh et al. 2000; Yadava et al. 2005). It increases the electrical conductivity of soil mixture. Depending upon the source it has variable pH, thus it may be used to reclaim both alkaline as well as acidic soils. Fly ash also consists of a number of macro and micronutrients and other heavy metals. Some of these are essential to plants and animals whereas some others are toxic and hazardous to them. Use of fly ash alone or in combination with various materials such as farm yard manure, sewage sludge, water hyacinth lime, gypsum and microbial cultures has been found to improve the growth, yield and nutrients uptake of various agricultural crops (Sikka and Kansal 1993; Suresh, et al. 2005). An attempt has been made to present in brief the physico-chemical properties of fly ash and its implications on soil properties and crop yields in this chapter.

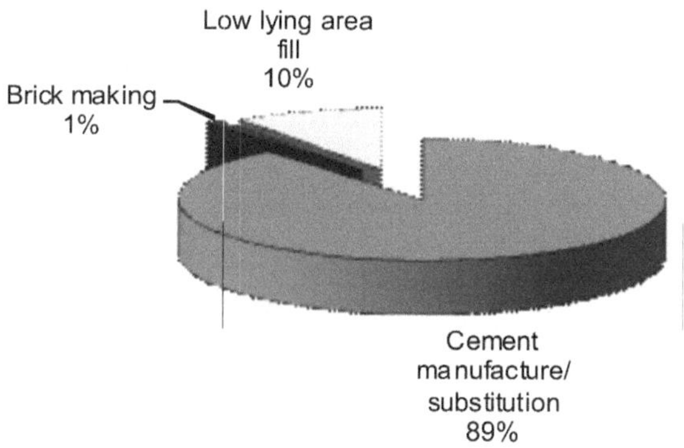

Fig. 2 : Fly ash Utilization – 1994
(Total utilization ~ 1 million tones/year)

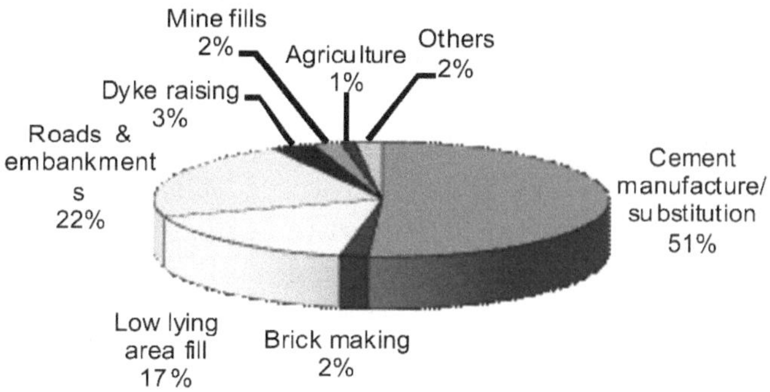

Fig. 3 : Fly ash Utilization – 2005
(Total Utilization ~ 42 million tones/year)

1. Properties of Fly Ash

The components and characteristics of fly ash are largely dependent upon the composition and property of the parent coal and the conditions of combustion and final handling. The concentration of various elements also varies according to particle size (Davison et al. 1974; Page et al. 1979; Khan et al. 1996).

1.1 Physical properties

Fly ash, the residue of combustion of coal, originally a mixture of vegetation, clay and rocks, comprises a wide range of inorganic matters. Physically, fly ash occurs as very fine spherical particles, having average diameter in the range from few μ to 100μ with sandy silt to silt loam texture, low to medium bulk density, high surface area, high porosity and water holding capacity (Salter et al. 1971; Petruzzelli et al. 1987; Kumar et al. 2005).

Table 1 : Physical properties of fly ash

Parameters	Value
PH	6.0 - 10.0
Electrical conductivity (dS/m)	0.15 - 0.45
Specific gravity	1.66 - 2.55
Bulk density (g/cc)	0.85 - 1.2
Particle size distribution	Sandy silt to Silt loam
Porosity (%)	45 – 55
Water holding capacity (%)	45 – 60

(*Source :* Kumar, et. al., (2005))

1.2 Chemical properties

Fly-ash is a complex heterogeneous material consisting of both amorphous and crystalline phases (El-Mogazi et al. 1988; Mattigod et al. 1990). It is generally considered a ferro-alumino silicate mineral with Al, Si, Fe, Ca, K and Na as the predominant minerals (Adriano et al. 1980). Fly-ash has a high salinity and alkalinity i.e., pH 6.0- 10.0 when dissolved in water (Carlson and Adriano 1993; Singh and Yunus 2000). However, some acidic fly-ash having pH 6.0– 6.7 is derived from high sulfur coal. The fly-ash components and resulting alkalinity are largely dependent upon the composition and property of the parent coal and the conditions of combustion and final handling. The concentration of various elements also varies according to particle size (Davison et al. 1974; Page et al. 1979; Khan et al. 1996). Various elements that constitute fly-ash are Si, Ca, Mg, Na, K, Cd, Pb, Co, Cu, Fe, Mn, Mo, Ni, Zn, B, F, Ca and Al. This means that fly-ash contains most of the elements required for plants growth and metabolism, with the exception of nitrogen and available phosphorus, as well as most toxic non-essential elements for plants (Plank and Martens 1973, 1974; El-Mogazi et al. 1988; Singh and Yunus 2000). The average chemical composition, concentration of micronutrients and trace and heavy metals as well as the activity of radio-nuclides have been presented in tables 2- 5.

Table 2 : Chemical properties of Indian fly ash

Parameters	%
SiO_2	38-63
Al_2O_3	27-44
TiO_2	0.4 -1.8
Fe_2O_3	3.3-6.4
MnO	0.1-0.5
MgO	0.01-0.5
CaO	0.2-8.0
K_2O	0.04-0.9
Na_2O	0.07-0.43

(*Source :* Kumar et al. (2005))

Table 3 : Total and available micronutrients content in fly ash

Micronutrients	Total (ppm)	Available (ppm)
Cu	40-80	0.5-1.6
Zn	50-150	0.4-1.8
Mn	500-750	0.9-1.5
Fe	33000-64000	10-15
B	17–38	0.5-0.8
Mo	2.2 – 6.7	0.1-0.6

(*Source :* Kumar et. al. (2005))

Table 4 : Total and available trace / heavy metal in fly ash

Trace/ Heavy metals	Total (ppm)	Available (ppm)
Se	0.6 – 2.6	0.1-0.4
Cr	50-150	0.3-0.6
Pb	10-70	BDL
Co	10-50	0.05-0.15
Ni	50-150	0.15-0.25
Cd	5-10	0.03-0.07
As	1.0 – 4.0	BDL
Hg	BDL*	BDL

* BDL= Below detectable limit

(*Source :* Kumar et al., (2005))

Table 5 : Radioactivity levels in fly ash

Radionuclide	Radioactivity level (Bq kg^{-1})
^{226}Ra	30-110
^{228}Ac	30-110
^{40}K	180-500

(*Source :* Kumar et al. (2005))

2. Effect of Fly Ash Application on Soil Properties

The addition of fly ash alters the soil environment by influencing certain physical, chemical and biological characteristics.

2.1 Physical soil properties

The effect has been evaluated on important soil groups of the country (alluvial, black and red & lateritic) at varied application rates. Fly-ash application to sandy soil could permanently alter soil texture, increase microporosity and improve the water-holding capacity (Ghodrati et al. 1995; Page et al. 1979). Fly-ash addition @ 70 t/ ha has been reported to alter the texture of sandy and clayey soil to loamy (Capp, 1978). Addition of fly-ash @ 200 t/ acre improved the physical properties of soil and shifted the textural class from sandy loam to silt loam (Buck et al., 1990). The particle size range of fly ash is similar to silt and changes the bulk density of soil. Chang et al. (1977) observed that among five soil types, Reyes silty clay showed an increase in bulk density from 0.89 to 1.01 g/ cc and a marked decrease in soils having bulk density varying between 1.25 and 1.60 g/ cc when the corresponding rates of fly ash amendment increased from 0 to 100%. Application of fly-ash at 0, 5, 10 and 15% by weight in clay soil significantly reduced the bulk density and improved the soil structure, which in turn improved porosity, workability, root penetration and moisture-retention capacity of the soil (Kene et al., 1991). Prabhakar et al., (2004) observed that addition of fly ash up to 46% reduced the density of the soil in the order of 15-20%. A gradual increase in fly ash concentration in the soil (0, 10, 20 up to 100% v/v) was reported to increase the porosity and water holding capacity (Khan and Khan, 1996). Amendment with fly ash up to 40% also increased soil porosity from 43 to 53% and water holding capacity from 39 to 55% (Singh et al., 1997; Yadava et al. 2005). Fly ash has been shown to increase the amount of plant available water in sandy soils (Taylor and Schumann, 1988; Yadava et al. 2005). The Ca in fly ash readily replaces Na at clay exchange sites and thereby enhances flocculation of soil clay particles, keeps the soils friable, enhances water penetration and allows roots to penetrate compact soil layers (Jala and Goyal 2006). Water holding capacities of fly ashes from different thermal power plants in Eastern India were compared and the effect of size fractionation on the water holding capacity was determined in an investigation by Sarkar and Rano (2007).

The fly ash, generally, reduced bulk density and increased water holding capacity and porosity of soils. Fly ash itself is not effective in retaining water, but it significantly increases water holding capacity of the soil mixture (Chang et al. 1977). Hydraulic conductivity of soils can be improved by the application of limited amounts of fly-ash, but it deteriorates rapidly when fly-ash input exceeds 20% (v/v) in calcareous soils and 10% in acidic soils. Fly-ash is an inorganic substrate that is rich in electrolytes and does not contain significant organic matter; therefore, its addition to soil results in increased hydraulic activity limited at lower application doses. The impedance of water flow is related to the pozzolonic reaction of fly-ash, which reacts with water to cement soil particles under wet conditions (Adriano et al. 1980).

Fly-ash also reduces the modulus of rupture (cohesiveness of soil particles) in soils (Adriano et al. 1980). Yadava et al. (2005) reported that application of fly ash @ 50- 100 t/ha in red and black soils of Bundelkhand region reduced the crust strength significantly.

2.2 Chemical properties

Calcium in fly ash readily reacts with acidic components in soil and releases nutrients such as S, B and Mo in the form and amount beneficial to crop plants. Fly ash improves the nutrient status of soil (Rautray et al. 2003; Yadava et al. 2005). The fly ash has been used for correction of sulphur and boron deficiency in acid soils (Chang et al. 1977). Application of fly ash for increasing the pH of acidic soils (Phung et al. 1979). Most of the fly-ash produced in India is alkaline in nature; hence, its application to agricultural soils could increase the soil pH and thereby neutralize acidic soils (Phung et al. 1978). The hydroxide and carbonate salts give fly ash one of its principal beneficial chemical characteristics i.e., the ability to neutralize acidity in soils (Cetin and Pehlivan, 2007). Flyash has been shown to act as a liming material to neutralize soil acidity and provide plant-available nutrients (Taylor and Schumann 1988). Researches have shown that the use of fly-ash as liming agent in acid soils may improve soil properties and increase crop yields (Matsi and Keramidas 1999).

The electrical conductivity of soil increases with fly ash application and so does the metal content. Metals like Fe, Zn, Cu, Mn, Ni and Cd have been shown to be available at higher concentrations in DTPA extracts of fly ash (Gupta et al. 2007). Sarangi et al. (2001) observed a gradual increase in soil pH, conductivity, available phosphorus, and organic carbon with increased application rate of fly ash. Fly ash is considered to be a rich source of Si and application of fly ash in Si-deficient soils has been demonstrated to improve the Si content of rice plants as well as their growth (Lee et al. 2006).

The addition of fly ash increases the availability of nutrients significantly in some cases. The increases are also noticed of certain heavy metals (Se, As and Pb) and radio-nuclides(^{226}Ra, ^{228}Ac and ^{40}K) at some places. However, their contents were reduced to some extent with the application of organic manures. Field experiments have revealed that the application of fly ash with other solid wastes e.g. sewage sludge to soils serves as both a means of disposal and as a medium for plant growth. Organic amendments improve soil conditions by increasing the cation-exchange capacity and organic matter content of the soil, thereby resulting in increased immobilization of toxic elements, higher fertility and enhanced microbial activity. Certain inhibitory effects to soil microbes by toxic components of fly-ash may, furthermore, be attenuated by the application of organic materials (Chaney and Giordano 1977).

2.3 Biological properties

Use of fly ash has been found to modify soil biological properties. Studies have revealed that application of fly ash did not affect the microbial population adversely even though there was an increase in the soil pH, but this effect was negated by improvement in physical conditions and due to supply of some essential trace elements for growth of microorganisms. Lal et al. (1996) reported that the cumulative CO_2 evolution increased only up to 8 per cent fly ash level. However, application of fly ash along with recommended dose of NPK fertilizers increased the population of bacteria significantly over control in all the treatments without NPK fertilizers. Such an increase might be attributed to complementary effect of fly ash and NPK fertilizer. Patil and Sahu (1990) also observed intensive respiration of fly ash amended soil, which reflected an increased activity of soil bacteria. Vallini et al. (1999) reported an increase in the bacterial and actinomycetes count due to application.

The enzymatic activity of soil is also an important factor for measuring soil biological properties after fly ash application in soil. The high pH and electrical conductivity of fly ash have been suggested to be important elements limiting microbial activity (Elliott et al. 1982). Sarangi et al. (2001) reported that invertase, amylase, dehydrogenase and protease activity increased with increasing application of fly ash up to 15 t/ ha/ y, but decreased with higher levels of fly ash application (Table 6).

Pati and Sahu (2004) studied the CO_2 evolution and enzyme activities of dehydrogenase, protease and amylase. They found little or no inhibition of soil respiration and enzyme activities up to 2.5% fly ash amendment. However, with further addition of fly ash, all the above activities were significantly decreased. On the other hand, significant stimulation of soil respiration and microbial activities were observed up to 5% fly ash amendment when the soils contained earthworms. This may be due to increased microbial activity induced by substrates that are produced by the earthworms.

Lal et al. (1996) reported that fly ash added to soil at 16% (w/w) increased enzyme activities i.e., urease and cellulase. However, acid phosphatase activity was depressed and with fly ash application. Mix application of FYM and fly ash has been proved to be beneficial in augmenting proliferation and activity of microorganisms in soils (Das et al. 2005). Fly ash composted with wheat straw and

2% rock phosphate (w/w) for 90 days enhanced chemical and microbiological properties of the compost and fly ash up to 40-60% and did not exert any detrimental effect on either C:N ratio or microbial population (Gaind and Gaur, 2004).

Table 6 : Effect of fly ash on the activity (μ g/ g soil) of soil enzymes

Enzymes	Days after amendment	Fly ash dose (t/ha)					
		0	**10**	**12.5**	**15.0**	**17.5**	**20.0**
Protease	35	30.90	34.50	37.20	49.90	42.60	43.70
	110	26.20	36.40	39.30	39.10	38.90	36.00
Amylase	35	53.90	58.20	62.60	78.30	35.30	40.30
	110	44.20	48.30	50.60	63.00	20.80	23.50
Invertase	35	553.40	363.60	382.00	495.80	511.80	635.80
	110	65.00	75.00	89.10	128.70	80.60	62.90
Dehydrogenase	35	6.64	9.12	9.30	12.70	8.00	10.85
	110	0.34	0.31	0.32	0.36	0.33	0.31

3. Effect of Fly Ash on Crop Yields

Fly ash may either have a positive or negative effect on plant growth and yield if not used in optimum doses. The effect is determined primarily by chemical composition and the ash dose applied. In a study by Kalara et al .(2003), application of 5 to 12 tonnes/ ha/ yr modified the soil physico-chemical properties viz., reduced the bulk density, increased the water holding capacity, improved the exchangeable calcium and magnesium status of the soil which in turn enhanced the wheat yield. On the other hand, Khan and Khan (1996) in their studies on the effect of increasing concentration of ash in the soil (from 10 to 100 % of the volume) on the growth of tomatoes observed a positive effect of ash on plants.

The application of fly ash has been reported to increase the yields of a large number of crops on major soil groups with varied application rates (Table 7). The yields increased by 6-18 % on alluvial soils of Uttar Pradesh and Delhi (@ 10-20 t/ha), 32% on alluvial soil of Haryana (@20%w/w), 29% on alluvial soil of West Bengal (@200 t/ha/3 years), 10-21% on black soils of Maharashtra (@10 t/ha), 28-45% on red and black soils of Karnataka (@80 & 40 t/ha), 14-25% on red lateritic soils of TamilNadu (@40 t/ha), 12 % on lateritic soils of West Bengal (@ 10 t/ha) and 31 % on red soil of West Bengal (@200 t/ha/3 years).

Changes in soil properties caused by fly-ash may directly or indirectly affect microbial activity and the root growth of plants. Fly-ash increases water-holding capacity of soil mixtures, but this capacity does not appear to significantly increase the available water to plants (Chang et al. 1977). Field and greenhouse studies both indicate that many chemical constituents of fly-ash benefit plant growth and can improve the agronomic properties of the soils (Chang et al. 1977). A lower application of fly-ash (5–10%) in soils enhances seed germination as well as seedling growth, although higher application (20–30%) either delays or drastically inhibits plant growth, development and other specific parameters (Singh et al. 1997). When *Beta vulgaris* roots were grown in fly-ash-amended soil it was found that low fly-ash application of up to 2% (kg/m^2 plot) was stimulatory for sugar production whereas higher doses (4 and 8%) were inhibitory (Singh et al. 1994; Singh and Yunus 2000). Detrimental effects of higher applications of fly-ash on plants are primarily due to a shift in the chemical equilibrium of the soil (Singh and Yunus 2000).

Table 7 : Effect of fly ash application on crop yields

Location	**Application rate**	**Crop**	**Percent yield increase**
Dadri (U.P.) & IARI (Delhi)	10- 20 t/ha	Wheat, Mustard, Rice, Maize	6- 18
Hissar (Haryana)	20% (w/w)	Pearlmillet, Wheat	32
Murshidabad (W.B.)	200 t/ha/3yrs (one time application)	Wheat, Rice	29
Vidarbha region (Maharshtra)	10-15 t/ha	Seed cotton, Sorghum, Gram, Soybean, Summer Groundnut, Wheat	10- 46
Raichur (Karnataka)	30- 60 t/ ha/ 3 yrs. (one time application)	Sunflower, Groundnut	10- 26
Coimbatore and Vridhichalam (Tamil Nadu)	40 t/ha	Rice, Groundnut	14- 25
Kharagpur (W.B.)	10 t/ha	Kharif rice, Mustard	12
Birbhum (W.B.)	200 t/ha/ 3 yrs. (one time application)	Kharif and Boro Paddy, Potato	31
Rihand Nagar (U.P.)	25% in 1st year + 5% in 2nd year + 5% in 4th year + 5% in 5th year + 5% in 6th year (Total 45% = 1170 t/ha)	Brinjal Potato Pea Tomato Cabbage Maize Wheat Sunflower Paddy	50 45 45.6 28 29.6 36.6 23.6 22.7 15.1
Angul (Orissa)	280- 560 t/ha)	Maize Paddy Onion Sunflower	16- 31.8 5- 19.9 4.7- 23.9 3.7- 23.2

Contd. ...

Contd. ...

Angul (Orissa)	140- 280 t/ha	Wheat Paddy Groundnut Sunflower	9.8 19.0 8.6 8.5
Damanjodi (Orissa)	260 t/ha	Tomato Cabbage Sunflower Bottle gourd	50 35.3 38.4 11.4
Sarni (M.P.)	100- 300 t/ha	Wheat Maize Soybean Sunflower	7.75- 24.9 6.8- 28.8 17.8- 40.6 16.0- 29.4
Jhansi (U.P.)	50- 100 t/ha	Wheat Paddy Groundnut	22.9 21.2 14.3
Jhansi (U.P.)	50- 100 t/ha	Sorghum Cowpea Oats Berseem	17.8- 28.0 18.5- 23.3 11- 34 20.5- 34
Panki (Kanpur), U.P.	50- 100 t/ha	Wheat Paddy Maize	31.1 19.0 15.0

(*Source :* Samra et al. (2003); Saxena et al. (2005); Suresh et al. (2005))

4. Effect on Cost of Cultivation

Studies have revealed that use of fly ash alone or in combination with other organic, inorganic and/ or biofertilizers reduces the cost of cultivation and increased the benefit cost ratio. Use of fly ash along with chemical fertilizers and organic materials in an integrated way can save chemical fertilizer as well as increase the fertilizer use efficiency (FUE). According to Mittra et al. (2003) integrated use of fly-ash, organic and inorganic fertilizers saved N, P and K fertilizers to the range of 45.8, 33.5 and 69.6%, respectively and gave higher FUE than chemical fertilizers alone or combined use of organic and chemical fertilizers in a rice-groundnut cropping system.

5. Conclusions

The potential of fly ash as a resource material for use in agriculture and related areas is now well-established and `users' are getting convinced with its utility as soil ameliorant. The major attributes, which make fly ash suitable soil amendment, are its texture and the content of almost all the essential plant nutrients except organic carbon and nitrogen. Although fly ash cannot substitute the need of chemical fertilizers or organic manure it can be used in combination with these to get additional benefits in terms of improvement in soil physico-chemical characteristics and increased crop yields. As in the case with fertilizers and any other agriculture input, the amount and method of fly ash application would vary with the type of soil, the crop to be grown, the prevailing agro-climatic conditions and also the type of fly ash available. Although, fly ash has many benefits as an input material for its use in agriculture, presence of some amounts of heavy and toxic elements call for some cautions and hence, repeated application of fly ash in the same field should not be done at least for 5- 7 years.

References

Adriano, D.C., Page, A.L., Elseewi, A.A., Chang, A.C. and Straughan, I. (1980). Utilization and disposal of fly-ash and other coal residues in terrestrial ecosystems: a review. *J. Environ. Qual.* 9: 333–334.

Aitking, R.L., D.J. Campbell and L.C. Bell (1984). Properties of Australian fly ash, relevant to their agronomic utilization. *Aust. J. Soil Res.* 22 : 443-453.

Borm, P.J. (1997). Toxicity and occupational health hazards of coal flyash (CFA). A review of data and comparison to coal mine dust. *Ann. Occup. Hyg.* 41: 659–676.

Buck, J.K., Honston, R.J. and Beimborn,W.A. (1990). Direct seedling of anthracite refuge using coal flyash as a major soil amendment. Proceedings of the Mining and Reclamation Conference and Exhibition, April 23-26, Charleston, West Virginia, pp: 213-219.

Capp, J.P. (1978). Power plant fly ash utilization for land reclamation in the eastern United States. In: Reclamation of Drastically Disturbed Lands, Schaller, F.W. and Sutton, P.(eds.).Madison,WI.,ASA.,pp:339-353.

Carlson, C.L. and Adriano, D.C. (1993). Environmental impact of coal combustion residues. *J. Environ. Qual.* 22:227–247.

Chaney, R.L. and Giordano, P.M. (1977). Microelements as related to plant deficiencies and toxicities. In: Elliott LF, Stevenson FJ (eds) Soils for management of organic wastes and waste water. American Society of Agronomy, Madison, pp 235–279.

Chang, A.C., Lund, L.J., Page, A.L. and Warneke, J.E. (1977). Physical properties of fly-ash amended soils. *J. Environ. Qual.* 6:267–270.

Clemens, S., Kim, E.J., Neumann, D.and Schroeder, J.I. (1999). Tolerance to toxic metals by a gene family of phytochelatin synthase from plants and yeast. *EMBO J.* 19:3325–3333

Cerevelli, S., G. Petruzzelli, A. Perna and R. Menicagli, 1986. Soil nitrogen and fly ash utilization: A laboratory investigation. *Agrochemica*, 30: 27-35.

Cetin, S. and Pehlivan, E. (2007). The use of flyash as a low cost, environmentally friendly alternative to activated carbon for the removal of heavy metals from aqueous solutions. *Colloids Surfaces A: Physicochem. Eng. Aspects*, 298: 83-87.

Das, S.K., Yadava, R.B., Pathak, P.S., Bhatt, R.K., Suresh, G., Mojumdar, A.B. and Kareemulla, K. (2005). Influence of fly ash application on dehydrogenase activity in soils. In: P*roceedings International Congress on Fly Ash Utilization.* Dec. 4-7, 2005, New Delhi. Vol. V, pp. XII 26.1-XII 26.5

Davison, R.L., Natusch, D.F.S., Wallace, J.R.and Evans, C.A. Jr. (1974). Trace elements in fly-ash: dependence of concentration on particle size. *Environ. Sci. Technol.* 8:1107–1113.

El-mogazi, D., Lisk, D.J. and Weinstin, L.H. (1988). A review of physical, chemical and biological properties of fly-ash and effects on agricultural ecosystems. *Sci. Total Environ.* 74:1–37.

Elliott, L.F., Tittemore, D., Papendick,R.L., Cochran, V.L. and Bezidicek,D.F. (1982). The effect of Mount St. Helen's ash on soil microbial respiration and numbers. *J. Environ.Qual.*11:164-166.

EPD (Environmental Protection Department) (1993). Environmental Hong Kong 1993. Government Printer, Hong Kong.

Gaind, S. and Gaur, A.C. (2004). Evaluation of fly ash as a carrier for diazotrophs and phosphobacteria. *Bioresour.Technol.*95:187-190.

Ghodrati, M., Sims, J.T., and Vasilas, B.S. (1995). Evaluation of flyash as a soil amendment for the Atlantic coastal plain. I. Soil hydraulic properties and elemental leaching. *J.Water Soil Air Pollut.* 81:349-361.

Gupta, A.K., Dwivedi, S., Sinha, S., Tripathi, R.D., Rai, U.N. and Singh, S.N. (2007). Metal accumulation and growth performance of *Phaseolus vulgaris* grown in fly ash amended soil. *Bioresour. Technol.* 98: 3404-3407.

Gupta, D.K. (2002). Ecophysiological studies on *Cicer arietinum* L. growing under fly-ash stress condition. PhD thesis, Lucknow University, Lucknow, India

Jala, S. and Goyal, D. (2006). Flyash as a soil ameliorant for improving crop production-a review. *Bioresour. Technol.* 97: 1136-1147.

Kalra, N., Jain, M. C., Choudhary, R., Hari, R. C., Vatsa, B. K., Sharma, S. K. and Kumar, V. (2003). Soil properties and crop productivity as influenced by fly ash in corporation in soil. *Environment Monitoring Assessment* 87: 93-109.

Kene, D.R., Lanjewar, S.A., Ingole, B.M. and Chaphale, S.D. (1991). Effect of application of fly ash on physico- chemical properties of soil. *J. Soils & Crops* 1 (1): 11-18.

Kene, D.R.; Lanjewar, S.A.; Darange, O.G. and Deotale, Y.D. (1991). Physical-chemical properties of coal fly ash from thermal power station, Koradi. *J. Soil & Crop* 1 (2): 120-123.

Khan, M.R. and W.N. Singh, 2001. Effects of soil application of flyash on the fusarial wilt of tomato cultivars. *Int. J. Pest Manage.*, 47: 293-297.

Khan, R.K. and Khan, M.W. (1996). The effect of fly ash on plant growth and yield of tomato. *Environ. Pollut.* 92: 105-111.

Khan, S., Bagum, T. and Singh, J. (1996). Effect of fly-ash on physico-chemical properties and nutrient status of soil. *Indian J. Environ.* Health 38:41– 46.

Khan, M.R., Khan, M.W. and K. Singh, 1997. Management of root-knot disease of tomato by the application of fly ash in soil. *Plant Pathol.*, 46: 33-43.

Kumar, A., Vajpayee, P., Ali, M.B., Tripathi, R.D., Singh, N., Rai, U.N.and Singh, S.N. (2002). Biochemical responses of *Cassia siamea* Lamk. grown on coal combustion residue (fly-ash). *Bull Environ. Contam. Toxicol.* 68:675– 683.

Kumar, K.V., Singh, N., Behl, H.M. and Srivastava,S. (2008). Influence of plant growth promoting bacteria and its mutant on heavy metal toxicity in *Brassica juncea* grown in fly ash amended soil. *Chemosphere* 72: 678-683.

Kumar, V., Goswami G. and Zacharia, A.K. (2001). Fly-ash use in agriculture: issues and concerns. Technology demonstration projects commissioned by fly-ash mission under technology information forecasting assessment council (TIFAC) News and Views, pp. 1–6.

Kumar, V., Singh, G. and Rai, R. (2005). Fly ash: A material for another green revolution. In: P*roceedings International Congress on Fly Ash Utilization.* Dec. 4-7, 2005, New Delhi. pp. XII 2.1- XII 2.16

Lai, K.M., Ye, D.Y. and Wong, J.W.C. (1999). Enzyme activity in sandy soils amended with sewage biosolid and coal fly ash. *Water, Air and Soil Pollution* 113:261-272.

Lal, J. K., Mishra, B., Sarkar, A. K. and Lal, S. (1996). Effect of fly ash on growth and nutrition of soybean. *Journal of the Indian Society of Soil Science* 44 : 310-313.

Lee, H., Ha,H.S., Lee,C.S., Lee Y.B. and Kim, P.J. (2006). Fly ash effect on improving soil properties and rice productivity in Korean paddy soil. *Bioresour. Technol.* 97: 1490-1497.

Matsi, T. and Keramidas, V.Z. (1999). Flyash application on two acid soils and its effect on soil salinity, pH, B, P and on ryegrass growth and composition. *Environ. Pollut.* 104: 107-112.

Mattigod, S.V., Rai, D., Eary, L.E. and Ainsworth, C.C. (1990). Geochemical factors controlling the mobilization of inorganic constituents from fossil fuel combustion residues. I: Review of the major elements. *J. Environ. Qual.* 19:188–201.

Mittra, B.N., Karmakar,S., Swain, D.K. and Ghosh, B.C. (2003). Fly ash-a potential source of soil amendment and a component of integrated plant nutrient supply system. http://www.flyash.info/2003/28mit.pdf.

Page, A.L., Elseewi, A.A. and Straughan, I, (1979), Physical and chemical properties of fly-ash from coal fired power plants with reference to environmental impacts. *Residue Rev.* 71:83–120.

Pati, S. S. and Sahu, S. K. (2004). CO2 evaluation and enzyme activities (dehydrogenase, protease and amylase) of fly ash amended soil in presence and absence of earthworms (Under laboratory condition). *Geoderma,* 118: 289-301.

Petruzzelli, G., Labrano, L. and Cervelli, S. (1987). Heavy metal uptake by wheat seedlings grown in fly-ash amended soils. *Water Air Soil Pollut.* 32:389–395.

Phung, H.T., Lund, L.J. and Page, A.L. (1978). Potential use of fly ash as a liming material. In: Environmental Chemistry and Cycling Processes, Adriano, D.C. and Brisbin, I.L. (Eds.). US Department of Commerce, Springfield, VA, pp: 504-515.

Phung, H.T., Lam, H.V., Lund, H.V. and Page, A.L. (1979). The practice of leaching boron and salts from fly ash amended soils. *Water, Air Soil Pollut.* 12: 247-254.

Plank, C.O. and Martens,D.C. (1974). Boron availability as influenced by application of fly ash to soil. *Soil Sci. Soc. Am. Proc.* 38: 974-977.

Prabakar, J., Dendorkar, N. and Morchhale, R.K. (2004). Influence of fly ash on strength behavior of typical soils. *Constr. Build. Mater.* 18: 263-267.

Rautaray, S.K., Ghosh, B.C. and Mittra, B.N. (2003). Effect of fly ash, organic wastes and chemical fertilizers on yield, nutrient uptake, heavy metal content and residual fertility in a rice-mustard cropping sequence under acid lateritic soils. *Bioresour. Technol.*, 90: 275-283.

Salter, P.J., Webb, D.S. and Williams, J.B. (1971). Effects of pulverized fuel ash on the moisture characteristics of coarse textured soil and on crop yields. *J. Agric. Sci. Cambr.* 77: 53–60.

Samra, J. S., Sharma, P. D.and Kumar, Vimal. (2003). The use of fly ash in agriculture in India. CBIP –3rd International Conference- Fly Ash Utilization & Disposal 19-21 Feb.,2003, New Delhi, India.

Saxena, Mohini, Murali, S., Asokan, P. and Chaudhari A. K. (2005). Technology demonstration on bulk utilisation of pond ash in agriculture. In: *Proceedings National Seminar cum Business Meet on Use of Fly Ash in Agriculture*, September 16, 2005, New Delhi. pp 24-36.

Sarangi, P.K., Mahakur, D. and Mishra,P.C. (2001). Soil biochemical activity and growth response of rice *Oryza sativa* in fly ash amended soil. *Bioresour. Technol.* 76: 199-205.

Sarkar, A. and Rano,R. (2007). Water holding capacities of fly ashes: Effect of size fractionation. Energy Sources, Part A: *Recovery, Utilization Environ. Effects* 29: 471-482.

Schutter, M.E. and J.J. Fuhrmann, 2001. Soil microbial community responses to fly ash amendment as revealed by analyses of whole soils and bacterial isolates. *Soil Biol. Biochem.*, 33: 1947-1958.

Sharma, M.P., U. Tanu and A. Adholeya, 2001. Growth and yield of *Cymbopogon martini* as influenced by flyash, AM fungi inoculation and farmyard manure application. *Proceedings of the 7th International Symposium on Soil and Plant Analysis*, July 21-27, Edmonton, AB, Canada, pp: 43-43.

Sikka, R. and Kansal, B. D. (1995). Effect of fly ash application on yield and nutrient composition of rice, wheat and on pH and available nutrient status in soils. *Bioresource Technology* 51: 199-203.

Sikka, R. and Kansal, B.D. (1993). Impact of fly ash on soils and plants. *National Seminar on Development in Soil Science.* pp 113.

Sikka, R. and Kansal,B.D. (1994). Characterization of thermal power plant fly ash for agronomic purposes and to identify pollution hazards. *Bioresour. Technol.* 50: 269-273.

Sims, J.T., Vasilas,B.L. and Ghodrati, M. (1995). Development and evaluation of management strategies for the use of coal fly ash as a soil amendments. Proceedings of the 11th International *Symposium of the American Coal Ash Association*, (ISACAA'95),

Singh, N. and Yunus, M. (2000). Environmental impacts of fly-ash. In: Iqbal, M., Orlando, Florida, pp: 8.1-8.18.

Singh, N., Singh, S.N., Yunus, M. and Ahmad, K.J. (1994). Growth response and element accumulation in *Beta vulgaris* L. raised in fly-ash-amended soils. *Ecotoxicology* 3:287–298.

Singh, S.N., Kulshreshtha, K. and Ahmad, K.J. (1997). Impacts of fly-ash soil amendment on seed germination, seedling growth and metal composition of *Vicia faba* L. *Ecol. Eng.* 9:203–208.

Singh, V.K., Behal, K.K. and Rai, U.N. (2000). Comparative study on the growth of mulberry (*Morus alba*) plant at different levels of fly-ash amended soil. *Biol. Mem.* 26(1):1–5.

Singh, L.P. and Z.A. Siddiqui, 2003. Effects of flyash and *Helminthosporium oryzae* on growth and yield of three cultivars of rice. *Bioresour. Technol.*, 86: 73-78.

Singh, S.N., K. Kulshreshtha and K.J. Ahmad, 1997. Impact of fly ash soil amendment on seed germination, seedling growth and metal composition of *Vicia faba* L. *Ecol. Eng.*, 9: 203-208.

Stoewsand, G.S., W.H. Gutenmann and D.J. Lisk, 1978. Wheat grown on fly ash: high selenium uptake and response when fed to Japanese quail. *J. Agric. Food Chem.*, 26: 757-759.

Straughan, I., A.A. Elseewi and A.L. Page, 1978. Mobilization of Selected Trace Elements in Residues from Coal Combustion with Special Reference to Fly Ash. In: Trace Substances in Environmental Health-XII, Hemphill, D.D. (Ed.). University of Missouri, Columbia, pp: 389-402.

Suresh, G., Yadava, R.B., Pathak, P.S., Das, S.K., Bhatt, R.K., Mojumdar, A.B. and Kareemulla, K. (2005). Effect of fly ash application on growth and yield of forage crops. In: Proceedings International Congress on Fly Ash Unitization held at New Delhi, Dec.4-7, 2005, p XII 25.1-25.5

Taylor, E.M. and Schumann, G.E. (1988). Fly ash and lime amendment of acidic coal soil to aid revegetation. *J. Environ. Qual.* 17: 120-124.

Vallini, G., Vaccari, F., Pera, A., Angolucci, M., Scatena, S. and Varallo, G. (1999). Evaluation of composted coal fly ash on dynamics of microbial populations and heavy metal uptake. *Compost Science and Utilization* 7 : 81-90.

Wong MH, Cheung KC, Lan CY (1992) Factors related the diversity and distribution of soil fauna on Gin Drinkers B landfill. Hong Kong. *Waste Manage. Res.* 10:423–434.

Wong MH, Wong JWC (1986) Effects of fly-ash on soil microbial activity. *Environ Pollut Ser. A.* 40:127–144.

Wong, J.W.C. and Wong, M.H. (1990). Effects of fly-ash on yield and elemental composition of two vegetables, *Brassica parachinensis* and *B. chinensis*. *Agric. Ecosys. Environ.* 30:251–264.

Yadava, R.B., Das, S.K., Pathak, P.S., Bhatt, R.K., Suresh, G., Mojumdar, A.B. and Kareemulla, K. (2005). Effect of fly ash application on soil physico-chemical properties. In: *Proceedings International Congress on Fly Ash Unitization* held at New Delhi, Dec.4-7, 2005, pp. XII 24.1-24.

□□□

Green Agriculture : Newer Technologies, 2012
© Kambaska Kumar Behera (ed.), pp. 89-108
New India Publishing Agency, New Delhi (India)
e-mail : info@nipabooks.com; website : www.nipabooks.com

Chapter-5

Sustainable Soil Management for Agricultural Development

Ashok Kumar Rajani
Deptt. of Agril. Chem. & Soil Science
J.A.U., Junagadh-362 001, Gujarat
E-mail : avrajani@jau.in

SUMMARY

Sustainable agricultural development can only be possible, when all agricultural inputs used judicially. Soil management practices, like nutrient management, time schedule of fertilizer application, irrigation management etc. are should be manage in such away that resulted in best remuneration and sustain overall soil health. More over that, soil physical properties like, tilth, aeration and drainage are also play important role in soil productivity, so tillage operation should be carry out at proper level of soil moisture condition to maintain good soil tilth which result in better soil environment for high crop yield. The most important factor in sustainable soil management for agricultural development is nutrient management and it's only possible through integration. So nutrient management through integrated approaches, like use of chemical fertilizer is decreases by integrating with organic manures, biofertilizer, green manure, neem cake, caster cake etc which resulted in sustainable crop production and good soil health on long-term basis.

1. Introduction

Sustainable soil management is the successful management of soil for agricultural production to satisfy the human needs while maintaining or enhancing the quality of soil and conserving it. The sustainable soil management system can be considered sustainable if it ensures that "today's development is not at expense of tomorrow's development prospects".

Following are a few important aspects of sustainable soil management:
(i) Soil fertility management,
(ii) Conservation of soil and water,
(iii) Soil productivity as an integrated index of both above factors.

Any system of soil management which affect adversely on soil quality is unsustainable. The sustainability has a time dimension and could not be evaluated over a short period.

2. Soil Fertility Management for Sustainable Agriculture

Soil fertility refers to the inherent capacity of soil to supply essential nutrients to plants in adequate amount and in right proportion for their optimum growth. Management of the fertility of Indian soils demands its build up and sustenance at high level to produce adequate food for the ever increasing population. Fertilisers, therefore, posses a great importance and integral part of the prime inputs for deriving high productivity. The Green Revolution in India would not have been possible without proper increase in fertilizer input which could deteriorate the potential of fertilizer responsive high yielding varieties in the presence of adequate water and pest management.

Nutrient use efficiency is the most important concern in the sustainable management of soil fertility. Balance nutrition is a pivotal point in plant and soil fertility management (Nambiar and Ghosh, 1984; Nambiar and Abrol, 1989; Sarkar et al., 1997). During the early stage, the increase in food grain production was mainly due to increase in area of cultivation but after 1970-71, it has been largely depending on the productivity. Following aspects require considering:-

- Optimization of fertilizer rates
- Balance use of fertilizer nutrients
- Potential use of fertilizer
- Nutrient management through integration.

2.1. Optimization of fertilizer rates

The nutrient supplying capacity of soil to crop and requirement of crop largely affect on the amount of fertilizer required to add in soil. It is also depends on soil and climatic conditions and management practices.

In Indian soils, nitrogen is the most limiting factor for crop production due to its poor organic carbon content. The response of crops to nitrogen is invariable in Indian soils, except in soils containing high organic matter. The chemistry of applied fertilizers is vary with upland crops and submerged rice culture. The low nitrogen use efficiency in rice is the most important things to concern because rice alone consumes 40% of the total fertilizer N in India. Rice, wheat and maize generally respond significantly up to 120 kg N ha^{-1} , though at some places response up to 150-180 kg N ha^{-1} have also been reported (Narang and Bhandari,

1992).Leguminous crops require a starter dose of about 10 kg N ha-1 as a bulk of the requirement is met through fixation of atmospheric N.

Deficiency of phosphorus is also observed in many regions of Indian soils. Low and medium categories of Phosphorus in soil require application of P for optimizing yields. Wheat crop is found higher responsive to P application. In general, it responds to P applied up to 60 kg P_2O_5 ha^{-1} while in mixed black and red and yellow soils, its response to P continue up to 120 kg P_2O_5 ha^{-1}. Similarly, rice also responds generally up to 60 kg P_2O_5 ha^{-1}. However, in medium black soils, very low in available P, it responds up to 90 kg P_2O_5 ha^{-1}. Studies in Indo-Gangetic plains of India have indicated that submergence of soil causes an increase in the solubility of soil phosphates and consequently, rice crop generally shows small or no response to P application even on soils testing low in available P.

Response to potassium application on crops is depends on amount of reserve K and several other factors, like availability of soil moisture, K competition with calcium and magnesium ions present in soil solution. Groundnut in monsoon responds up to 80 kg K_2O ha^{-1}, while it in summer responded up to 120 kg K_2O ha^{-1} in calcareous soils of Saurashtra (Rajani et al, 2006). Crop responses to K are large on laterites, red and yellow and mixed red and black soils. Crops grown on coarse textured alluvial soils and some shallow black soils need K fertilization for optimum crop production (Subba Rao and Srinivasa Rao, 1996).

Due to continuous use of sulphur free fertilizers and reduction in FYM application to the soils, resulted in wide spread of sulphur deficiency in different soils of India. Variable response to sulphur application of different crops has been reported. In Punjab and Haryana, 20 kg S ha^{-1} to groundnut, linseed, pigeonpea and sorghum, while in UP, 30 kg S ha^{-1} to groundnut, wheat, potato, chickpea, black gram and maize and 40 kg S ha^{-1} to green gram and lentil corrected the S-deficiency in crops. In black soils, 30 kg S ha^{-1} to potato and 20 kg S ha^{-1} to groundnut efficiently corrected S-deficiency. Rice and maize responded well up to 60 kg S ha^{-1}, whereas mustard, wheat and berseem responded up to 90 kg S ha^{-1}. The beneficial effects of sulphur application ranging from 2 to 12 q ha^{-1} in different cropping systems have been reported (Tandon, 1989).

All soils of high rainfall area, acid soils and sodic soils are inherently deficient in Ca and/or Mg and crops respond to the application of these elements.

Zinc deficiency is critical in many states of India, among the micronutrients disorders of crops. Responses to Zinc differ widely among crops in each state as well as across states. *Kharif* crops generally give a better response to Zn application than *rabi* crops; the magnitude is affected by several factors, such as crops and their varieties, soil types, deficiency status of zinc and other nutrients, soil environment, climate and cultural practices, etc. The deficiency of zinc, copper, iron and manganese are more common in soils having high pH, high calcium carbonate, low organic matter and clay contents. On the other hand, deficiency of B and Mo are more observe in acidic soils; B is deficient in highly calcareous soils also. Crop responses to micronutrient application have been reviewed over the years by several workers (Nayyar et al. 2001).

2.2. Balance use of fertilizer nutrients

Crops are removed continuously nutrients from the soils and replenishment of nutrients through fertilizers and manures is essential. The proper ratio of nutrients for potential growth depends on the crops and soils. When the added fertilizer contains nutrients in less or higher quantity than that needed by the crop, the imbalance thus caused results in poor crop growth. The application of phosphorus along with nitrogen increases the nitrogen use efficiency. Sometimes even with NP application, decreases in yield occurs because K becomes the limiting factor. Similarly, with optimum dose of NPK, the deficiency of zinc may appear. Soil test based fertilizer recommendations ensure balance use of fertilizers and increase yield and profits. Balanced fertilizers encompass, besides major nutrients, secondary and micronutrients whose deficiencies appear in different soils in agro-eco-regions due to intensive cultivation.

2.3. Potential use of fertilizer

Increasing nutrient potentiality is the key to the management of soil fertility. The proportion of the applied fertilizer actually consumed by plants is a measure of fertilizer potentiality. Soil characteristics, crop characteristics and fertilizer management techniques are the major factors that determine fertilizer potentiality.

2.3.1. Soil Characteristics

• **Status of Nutrient in Soil :** The response of any crop or cropping system to applied nutrient depends mostly upon the fertility of soil to supply that nutrient as per the demand of crop. Chemical test have long been used to estimate the nutrient availability in soil to predict the possibility of getting a well response to applied nutrients. On that bases soils are classified low, medium and high in plant nutrients and proper fertilizer amounts are recommended. In a low nutrient soil, the crop responded well to its application, while in a high nutrient soil, the crop may give minute or no response. In medium nutrient soil, crop response is moderate. Soil testing useful in calculating the amount of fertilizer and so improves the potentiality fertilizer use. By surveying the areas responding variably to various plant nutrients, correct type and amount of fertilizer can be applied to them.

At present, most of the soil testing laboratories in India follow similar critical limits for adequacy or deficiency of plant nutrients, without considering of the mineralogy or texture of soil or even crop species & varieties. However, the proportion of available nutrients depends on the amount and types of nutrient reservoir and climatic situations which always affect on losses or transformation of nutrients to less available forms.

• **Fractions of Nutrients :** All pools of nutrients are not available equally to crops. Some pools of nutrients even do not show in soil test but may supply considerable quantity of the nutrient during the crop growth period. As per example, sandy soils high in mica may supply adequate potassium to crops even though the soil tests indicate low in available potassium. Response to potassium on such soils cannot expect in a short period of time but may show on in long-term experiment.

• **Transformation and Losses of Nutrients :** Because of leaching, volatilization, de-nitrification and transformation of available forms of nutrients, whole quantity of its not available to plants. Nitrate nitrogen is not held by exchange sites in the soil and that why its leaching losses occur at greater extent. Such losses are more in light texture soils, like in sandy soils. Leaching losses of calcium, sulphate, potassium and magnesium are more common in acid soils. Volatilization of ammonia from urea fertilizer is at a greater rate, when it applies at soil surface which high in pH. Denitrification losses of nitrogen prevail under waterlogged condition in rice field, particularly under higher temperature and in the presence of easily decomposable organic materials.

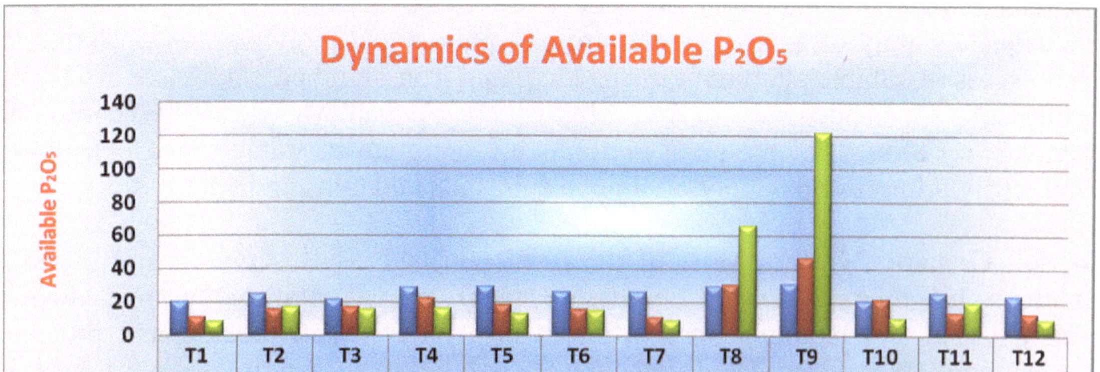

	T1	T2	T3	T4	T5	T6	T7	T8	T9	T10	T11	T12
Initial	20.03	24.78	21.10	28.84	29.73	25.78	26.19	29.69	30.80	20.48	25.55	23.21
4th year	10.41	15.38	16.83	21.85	18.22	15.21	10.46	30.30	45.99	21.21	13.57	12.63
8th year	8.17	17.07	15.36	15.93	12.89	14.71	8.63	65.67	121.5	9.94	19.29	9.21

Figure1 : Status of Available P_2O_5 (kg ha^{-1}) in soils of LTFE in 1st, $4^{th}$and 8^{th}year at Junagadh (Gujarat).

T_1- 50 % NPK of recommended doses in G'nut-Wheat sequence, T_2- 100 % N P K of recommended doses in G'nut -Wheat sequence, T_3 -150 % N P K of recommended doses in G'nut -Wheat sequence, T_4 - 100 % N P K of recommended doses in G'nut -Wheat sequence + $ZnSO_4$ @ 50 kg ha^{-1} once in three year to G'nut only (i.e. '99, 02, 05 etc), T_5 - N P K as per Soil Test, T_6 - 100 % N P of recommended doses in G'nut -Wheat sequence, T_7 - 100 % N of recommended doses in G'nut -Wheat sequence, T_8 - 50 % N P K of recommended doses + FYM @ 10 t ha^{-1} to G'nut and 100 % N P K to Wheat, T_9 - Only FYM @ 25 t ha^{-1} to G'nut only, T_{10} - 50 % N P K of recommended doses + Rhizobium + PSM to G'nut and 100 % N P K to Wheat, T_{11} - 100 % N P K of recommended doses in G'nut -

Wheat sequence (P as SSP) and T_{12} – Control.(Recommended dose: 12.5-25-0 N, P_2O_5, K_2O kg ha^{-1} to groundnut & 120-60-40 to Wheat)

In figure-1, it is clearly observe that application of of FYM increases status of available phosphorus in soil after 4th and 8th year of cropping sequence, even though application of half dose of recommended N P K to groundnut in T_8 and only FYM to groundnut only in T_9 in LTFE experiment on groundnut-wheat cropping sequence at Junagadh (Gujarat), because of transformation of unavailable form of P to available form of P due to secretion of organic acid from FYM.

In calcareous soil, applied phosphorus converted to calcium phosphate, which is not available to plants, due to this reaction, efficiency of added phosphorus is 20-30 per cent. Immobilizations of available nutrients in to micro organism's body also convert soluble forms of nutrients in to insoluble forms, but it is temporarily.

Soil reaction is the one of the soil property that affects plant growth. The detrimental effects of low soil pH are more due to secondary effects except in extreme case. The important secondary effects of low pH in a soil are the inadequate supply of available calcium, phosphorus and molybdenum on one end and excess of soluble aluminum, manganese and iron on other. Likewise, in saline-sodic soil, the inadequacy of Ca, Mg, P, Zn, Fe and Mn is mostly common. Availability of most of the plant nutrients are optimum in suitable soil pH range, that's why it improve by applying lime in acid soils and gypsum in sodic soils.

• **Soil Organic Matter:** Most of the plant nutrients are reserve in soil organic matter so it is index of soil fertility, beside that O. M. play vital role in improving of soil physical and biological properties. It also reduces soil erosion and protects surface soil losses. Overall soil organic matter is the prime criteria of the soil health. Due to rapid decomposition of organic matter at high temperature, it is very difficult to maintain it in tropics. The cultivation of soil generally adversely affects on organic carbon content due to increased rate of decomposition by the present cultivation practices. Long-term fertilizer experiment have shown that the integrated use of organic manures (FYM) and chemical fertilizers can increases productivity and sustain crop production (Fig. 2). Recent studies have indicated that a periodic addition of large amount of Farm Yard Manure to soil increases the nitrogen and organic carbon level (Fig. 3). The application of FYM, compost and cereal residues effectively maintains or increases the soil organic matter. There is a significant increase in soil organic matter due to incorporation of rice or wheat straw in to the soil instead of removing or burning it.

• **Soil Moisture:** Application of fertilizer increases root extension into deeper layer and leads to greater root proliferation in the root zone. Irrigated wheat fertilized with nitrogen used 20-38 mm more water than the unfertilized crop on loamy sand and sandy loam soils and increased dry matter production (Gajri et al. 1989). Soil moisture also affects root growth and plant nutrient absorption. The nutrient absorption is affected directly by soil moisture and indirectly by the effect of water on metabolic activities of plant, soil aeration and concentration of soil solution. If soil moisture becomes a limiting factor during critical stage of crop growth, fertilizer application may adversely affect the yield.

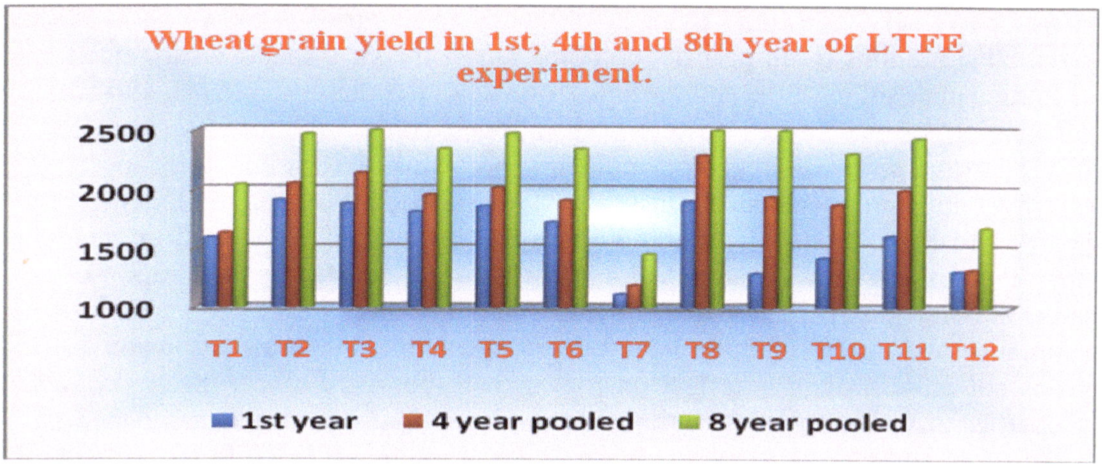

Figure 2 : Yield (kg ha^{-1}) of wheat grain in 1st, 4th and 8th year of LTFE at Junagadh (Gujarat).

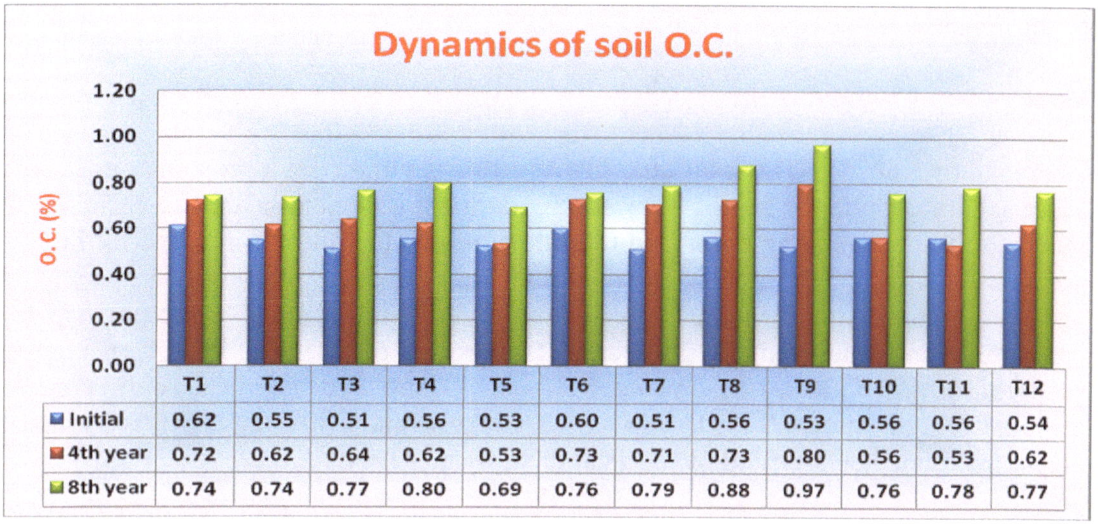

	T1	T2	T3	T4	T5	T6	T7	T8	T9	T10	T11	T12
Initial	0.62	0.55	0.51	0.56	0.53	0.60	0.51	0.56	0.53	0.56	0.56	0.54
4th year	0.72	0.62	0.64	0.62	0.53	0.73	0.71	0.73	0.80	0.56	0.53	0.62
8th year	0.74	0.74	0.77	0.80	0.69	0.76	0.79	0.88	0.97	0.76	0.78	0.77

Figure 3 : Status of O.C. (%) in soils of LTFE in 1st, 4th and 8th year at Junagadh (Gujarat).

T_1- 50 % NPK of recommended doses in G'nut-Wheat sequence, T_2- 100 % N P K of recommended doses in G'nut -Wheat sequence, T_3 -150 % N P K of recommended doses in G'nut -Wheat sequence, T_4 - 100 % N P K of recommended doses in G'nut -Wheat sequence + $ZnSO_4$ @ 50 kg ha^{-1} once in three year to G'nut only (i.e. '99, 02, 05 etc), T_5 - N P K as per Soil Test, T_6 - 100 % N P of recommended doses in G'nut -Wheat sequence, T_7 - 100 % N of recommended doses in G'nut -Wheat sequence, T_8 - 50 % N P K of recommended doses + FYM @ 10 t ha^{-1} to G'nut and 100 % N P K to Wheat, T_9 - Only FYM @ 25 t ha^{-1} to G'nut only, T_{10} - 50 % N P K of recommended doses + Rhizobium + PSM to G'nut and 100 % N P K to Wheat, T_{11} - 100 % N P K of recommended doses in G'nut - Wheat sequence (P as SSP) and T_{12} – Control.(Recommended dose: 12.5-25-0 N, P_2O_5, K_2O kg ha^{-1} to groundnut & 120-60-40 to Wheat)

• **Soil Physical Condition:** Even though sufficient nutrient supply, unfavourable physical conditions resulting from combination of the size, shape, arrangement and mineral composition of the soil particles, may lead to poor crop growth and activity of microorganisms. In fine texture soil, nitrogen content is more as compared to coarse texture soil. Optimum soil moisture and aeration, optimum soil temperature and freedom from mechanical stress are the basic need of crop growth in terms of soil physical conditions. Incorporation of organic matter, tillage irrigation, mulching and other amendments like application of gypsum to sodic soils and liming of acid soils are the major field management techniques that improving soil physical properties and make suitable for better crop growth. Tillage affects water use by crops not only through its effect on root growth but also affects the hydrological properties of soils. Mulching with residues, plastic film, etc. influences evaporation losses from soil by modifying the hydro-thermal regime of the soil and affects root growth and rooting pattern. Use of organic mulch also decreases maximum soil temperature in summer and increases minimum soil temperature in winter and helps in the conservation of soil moisture.

2.3.2. Crop Characteristics

• Uptake of Nutrient: The total quantity of nutrient uptake by a crop may not serve as an accurate guide for fertilizer recommendations; it does indicate the variation in their requirements among crops and the rate at which the nutrient reserve in the soil are being depleted. The nutrient removal may vary depending upon the crops and its varieties, nutrient level in the soil, soil type, soil and climatic conditions, plant population and management practices. It is estimated that 8 tons of rice grain remove 160 kg N 38 kg P, 224 kg K, 24 kg S and 320 g zinc as compared to removal of 125 kg N, 20 kg P, 125 kg K, 23 kg S and 280 g zinc by five tons of wheat from one hectare field (Tandon and Narayan 1990).

• **Characteristics of Root:** Principal organs of nutrient absorption in plants are roots. A proper knowledge of their characteristics aids in developing potential fertilizer practices. The absorption of nutrients depends upon the proliferation of roots in soil. The plants having shallow root system are more dependent on fertilizer. Hence, any soil manipulation which enhances deep rooting will encourage better utilization of fertilizers. It is well established that some plants are better scavengers of certain nutrients than others. This is mainly due to the preferential absorption of these nutrients by the roots of those plants. As like, legumes have a marked preference for divalent cations like Ca^{2+} whereas grasses feed better on monovalent cations like K^+.

The potentiality of applied fertilizer can be improved remarkably if the rooting habits of various plants during early growth stages are understood. This is specifically true for relatively immobile nutrients and for situations where the fixation of applied nutrients is very high. If the plant produce tap root system early, fertilizer can best be placed directly below the seed. On the other hand, if lateral root are formed early, side placement of fertilizer would be helpful.

Mycorrhizal fungi often associated with plant roots, increase the ability of plants to absorb nutrients particularly under low soil fertility. However, fertilizer addition generally reduces their presence and activity.

• **Rotation of Crop:** Potentiality of fertilizer and its requirement, mainly affected by the nature of cropping sequence. Crops are known to differ in their feeding capacities on added as well as native nutrients. Totally applied fertilizer may not be use by highly fertilized crops like maize, potato and some amount of nutrient may be left in the soil which can be utilized by the succeeding crop. Phosphorus, among the major nutrients, is worthy of consideration because only less than 20 per cent of applied phosphatic fertilizer is used by the first crop. Same things true for zinc, only 3 % of applied zinc is utilized by the first crop. The quantum of residual effect is, however, dependent on the rate and kind of fertilizer used, the cropping and management system followed and to great extent on the type of soil. Crops are habituated to luxury consumption of N and K and may not leave any residual effect unless doses in excess of the crop requirement are applied. On the other hand, if sub-optimal doses of fertilizers are applied to a crop, they may leave the soil in a much exhausted condition and the fertilizer requirement of the succeeding crop may increase. The residues of legume in the soil, reduces nitrogen requirement of succeeding crop.

2.3.3. Characteristic of Fertilizer and Management Techniques

The potentiality of fertilizer varies with in large extent on the type of fertilizer, time of application and method of application.

• **Fertilizer Type:** In fertilizer, nutrient content and its form are differing. In case of nitrogenous fertilizer, the nutrient may be present in ammonium, nitrate or amide form. Similarly in the case of phosphatic fertilizers, phosphorus may be present in water soluble, citrate soluble or citrate insoluble form. As like that, nutrient content of a fertilizer may also differ largely, *i. e.* CAN (Calcium Ammonium Nitrate) content total 25 % of nitrogen in which 12.5 % ammonium & 12.5 % nitrate form of nitrogen as compare to 46 % of urea, 20.6 % of ammonium sulphate and 82 % of anhydrous ammonia. Nitrogen may also be present as a one of the component of the complex fertilizer sources such as 18 % in diammonium phosphate and 13 % in potassium nitrate. Single super phosphate content only 16 % P_2O_5 and DAP supply 46% P_2O_5, whereas the K_2O content of potassium chloride is 60%. The fertilizer may also variable in their water solubility and their ability to get fixed in soil.

Nitrate fertilizer has high water solubility and are subjected to leaching in high rainfall areas or due to heavy irrigation, as these do not get fixed with soil exchange cite after their application to soil and thus move to greater depth. In upland soil, generally all sources of nitrogenous fertilizers are equally potential, but in submerged rice soil, fertilizers which content nitrogen in ammonical and amide forms are more potential. Water soluble form of phosphorus content in DAP and SSP fertilizers found more effective for most of the crops grown on neutral or alkaline soils as compared to citrate soluble or citrate insoluble form. Both single super phosphate (SSP) and diammonium phosphate (DAP) found equally potential on the basis of equivalent phosphate content but SSP has the benefit in soils low in available sulphur and particularly for sulphur loving crops like oilseeds and pulses. Water insoluble form like in rock phosphate has proved useful in acid soils, rice soils or for long duration legumes.

Most of the K fertilizers used in Indian agriculture is imported mostly in the form of muriate of potash (KCl) and only about one per cent as sulphate of potash (K_2SO_4). In soils where sulphate deficiency is expected (Like in acid soils) and crops like tobacco which are chloride sensitive, potassium sulphate should be selected.

$FeSO_4$, $ZnSO_4$, $MnSO_4$ and $CuSO_4$ which are carrier of micronutrients have been found potential and cheap as compared to other sources of micronutrient cations.

• **Time of Application:** Time of application of fertilizer is important particularly for those nitrogen containing fertilizers which liable to leach with irrigation or rainfall. Split application of nitrogen is popular and widely accepted practice for almost all crops. The basic concept is to apply nitrogen in two or more splits to coincide with critical period of nitrogen requirement of the crop. Split application of nitrogen more beneficial on light textured soil in increasing the potentiality of applied nitrogen. Its application in three split doses at transplanting, tillering and panicle initiation in rice, at sowing, knee high stage and tasselling stage in case of maize, is usually recommended. In general, crops grown in rainy season should receive nitrogen fertilizer in split doses so that the leaching of the nutrient during heavy rains may be checked and an optimum supply of the nutrient at critical stages is sure. On the contrary, best results are derived in crops like wheat, barley, Brassica sp, which are grown in winter season when rains are minute by applying whole dose of nitrogen as a basal in medium to heavy textured soils and in light soils in two splits at sowing and at first irrigation. Recent studies have depicted that nitrogen applied in form of urea at pre-sowing irrigation even to a loamy sand soil gives higher grain yield of wheat as compared to its application at preparatory tillage and increases both nitrogen use efficiency (7 to 34%) and water use efficiency (2 to 24%) at different nitrogen and irrigation levels (Sidhu et al. 1994). For unirrigated crops, however, split application is beneficial.

Best results found with most of the crops, in case of potassic and phosphatic fertilizers as well as for zinc and copper among micronutrients cations, when whole amount applied at sowing. However, in the case of iron and manganese where foliar spray of these nutrients is recommended, initiation of foliar spray is very important. In case of manganese, foliar spray initiated 2-3 days before the first irrigation to wheat have been found to be significantly superior to their application after the first irrigation (Takkar et al. 1986). Similarly, foliar application of iron should be initiated as soon as the symptoms of its deficiency appear on crops.

• **Application Method :** Many techniques have been developed to improve the potentiality of applied fertilizers. Since nitrates are easily leached and lost by denitrification, retardation of nitrification of ammonium containing or ammonium producing fertilizers by using nitrification inhibitors in rice fields subjected to help in increasing the potentiality nitrogen. Chemical nitrification inhibitors as well as indigenous materials like *neem* and *karanj* cakes have been successfully used. Application of urea with sulphur or soil or neem cake gave significantly higher grain yield of wheat as compared to urea alone (Anon 1984). Also, some slow release nitrogen materials like urea-form, isobutylidene diurea as well as coated fertilizers such as sulphur-coated urea, lac-coated urea, etc. have been used to reduce the nitrogen losses. Particularly in rice field of high pH soil, volatilization losses as ammonia can be minimize by root zone placement of urea supergranuales or urea briquettes. Under rainfed condition, foliar application of nitrogen as a supplemented to soil application has been found beneficial.

Fixation of water soluble phosphatic fertilizers can be reduced by its placement rather than its broadcast. However, phosphates of low solubility react slowly in soil and are usually more effective as a source of phosphorus when broadcast. When rock phosphate is used in rice field, it should be mixed with soils a fortnight or month before rice planting.

In long duration crops and light textured soil to reduce leaching losses of potassium, split application is found more effective. Benefit of split application of potassium has been observed in sugarcane and rice. Tow to six splits of potassium has been recommended for fruit crops.

Top dressing of ammonium sulphate is beneficial in wheat crop. For correcting sulphur deficiency, application of it in neutral to calcareous soil is equally potential whether broadcast, drilled or applied in split doses.

For efficient utilization of applied zinc, the best time of its application for wheat and rice is at seedling or transplanting of crop (Takkar 1996). Among various methods, broadcasting and mixing zinc is more potential than its drilling or band placing and top dressing 60 days after seedling. Application of zinc sulphate @ 25-50 kg ha^{-1} to soil is better than foliar spray of 0.5% zinc sulphate solution neutralized with lime. Foliar spray of 0.5 to 1.0% manganese sulphate solution has been found more effective and economical as compared to soil application (20 kg Mn ha^{-1}) of manganese in Mn deficient soil of Punjab. Soil application of ferrous sulphate in coarse-textured soils for rice is less efficient than foliar spray, but in upland crops, the soil application of iron has been found equal effective as foliar sprays.

2.4. Nutrient management through integration

Nutrient management through integration comprises the use of chemical fertilizers in conjunction with organic manures, legume in cropping system, biofertilizers and other locally available nutrient resources for sustaining soil health and productivity. The combined application of organic manures and chemical fertilizers produces higher crop yields than when each is applied alone. As shown in fig. 4, groundnut pod yield in LTFE at Junagadh (Gujarat) is higher in treatment which received 50% NPK + 10 tons FYM ha^{-1} (T_8) then 50% NPK alone (T_1) and 25 tons FYM ha^{-1} alone(T_9). This increase in crop productivity may be due to the combined effect of nutrient supply, synergism and improvement of soil physical and biological properties. FYM (Farmyard Manure) constitutes an important component of nutrient management through integration for sustaining soil fertility and yield stability.

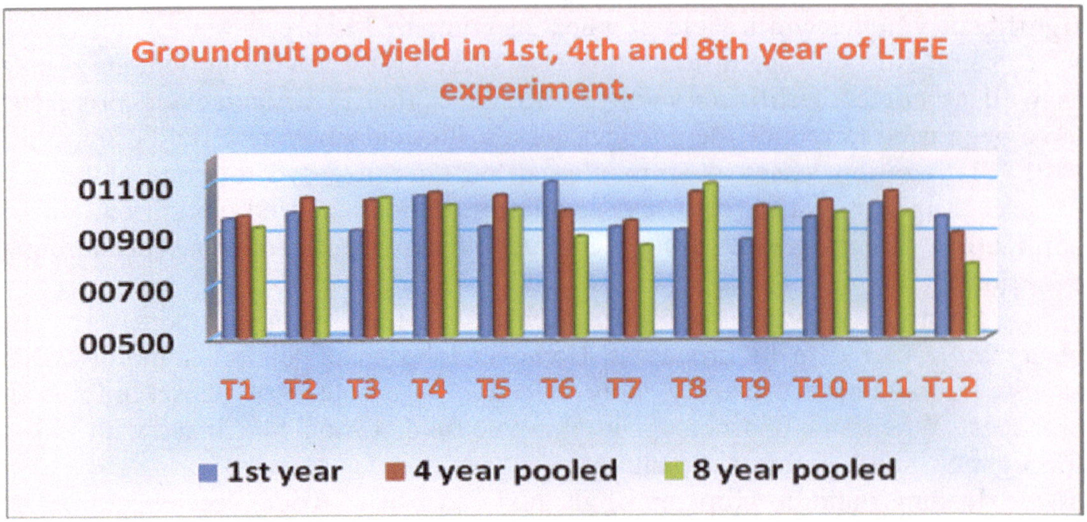

Figure 4 : Yield (kg ha^{-1}) of groundnut in 1st, 4th and 8th year of LTFE at Junagadh (Gujarat).

T_1- 50 % NPK of recommended doses in G'nut-Wheat sequence, T_2- 100 % N P K of recommended doses in G'nut -Wheat sequence, T_3 -150 % N P K of recommended doses in G'nut -Wheat sequence, T_4 - 100 % N P K of recommended doses in G'nut -Wheat sequence + $ZnSO_4$ @ 50 kg ha^{-1} once in three year to G'nut only (i.e. '99, 02, 05 etc), T_5 - N P K as per Soil Test, T_6 - 100 % N P of recommended doses in G'nut -Wheat sequence, T_7 - 100 % N of recommended doses in G'nut -Wheat sequence, T_8 - 50 % N P K of recommended doses + FYM @ 10 t ha^{-1} to G'nut and 100 % N P K to Wheat, T_9 - Only FYM @ 25 t ha^{-1} to G'nut only, T_{10} - 50 % N P K of recommended doses + Rhizobium + PSM to G'nut and 100 % N P K to Wheat, T_{11} - 100 % N P K of recommended doses in G'nut - Wheat sequence (P as SSP) and T_{12} – Control.(Recommended dose: 12.5-25-0 N, P_2O_5, K_2O kg ha^{-1} to groundnut & 120-60-40 to Wheat)

The green manures are an important potential source of N and organic matter. *Sesbania,* sunnhemp and cowpea incorporated before transplanting of rice partially meet the nutrient requirement of the crop.Growing of legumes in rotation and incorporation of legume residues after picking of pods also significantly improve the yield of cereal crop in rotation.

3. Conservation of Soil and Water

Soil and water conservation is the combination of all management and land use practices which protect soil against depletion or deterioration by natural or man-induced factors and improve the productivity of the soil as natural resources on sustained basis. It is said that nature takes centuries to produce the soil but man may lose it in a few years due to its indiscriminate exploitation for his need and greed. Soil once lost, is difficult and expensive to replace. The conservation of soil and water is not only anti-erosion and anti-runoff approach; it is a comprehensive and integrated approach for judicious use of these natural resources rather than their negligent and wasteful use.

3.1. Soil conservation measures

The following soil and water conservation measures may be adopted for controlling the soil erosion depending on the degree and length of the slope and physical configuration of the land.

3.1.1. Mechanical Measures

(i) Contour Bunds: Contour bunds are mechanical barriers built across the slop for safe diversion of excess runoff and retaining eroded soil. The land area in between the two bunds gets leveled in due course of time. Due to deposition of eroded soil along the bund, the latter takes the shape of a riser. These risers should be planted with grasses to check their erosion.

(ii) Graded Bunds: The graded bund is a small earthen bund with slight grade constructed across the slop for safe disposal of runoff. The graded bunds are recommended up to 10% slope for areas where annual rainfall exceeds 800 mm, particularly on clayey and black soils with poor drainage. However, efficacy of graded bunds gets reduced gradually beyond 4% slop. The purpose of graded bunds is to reduce the velocity of runoff water, for *insitu* conservation of rain water and minimize the erosion.

(iii) Bench Terraces: Bench terraces are flat beds constructed on hills across the slope. The height of the riser should not be more than one meter and width of bench terrace depends on the degree of slope. The bench terraces are important because they promote uniform distribution of soil moisture, irrigation water, and etc. and control soil erosion. The bench terraces may be table top (level), outward sloping or inward sloping, with or without mild longitudinal grades. On steep slopes, it is better to construct terraces on foot hills for agricultural crops when soil depth is more than one meter.

(iv) Half Moon Terraces: Half moon terraces are semi-circular beds of appropriate diameter with the shape resembling a half moon. These terraces are recommended for fruit trees or other plantation crops on steep slopes.

(v) Grassed Water Ways: Grasses are well known for their soil binding characteristics. They are most effective in moderating the flow and reducing the erosive velocity of runoff, particularly on rolling topography. The runoff water moves with high velocity down the slope, carrying with it soil and nutrients. If some suitable grasses are planted on the runoff rout or natural channels, the soil and nutrient losses can be reduced. These grassed water ways are laid on natural drainage lines in the watershed. Stilling basins or water ponds are constructed in rout at appropriate locations, with earthen and boulder pitched bunds for retention of runoff water. By reducing the velocity of runoff water, the erosion losses can be minimized.

(vi) Water Harvesting Ponds: water harvesting structure can be dug out for retaining runoff on seasonal or perennial basis. These are generally constructed down the slope. Earthen dams should be used for retaining silt loads at appropriate locations on the slope of watershed. The water thus harvested or stored can be used for pisciculture and supplemental irrigation.

(vii) Conservation Bench Terraces (CBT): These are used to stabilize the yield of rainfed crops by inter-field water harvesting. A part of the field is leveled to retain runoff originating from rest of the field.

(viii) Gully Control Structures: Gully control structures are provided to reduce the erosive velocity of runoff, to facilitate establishment of vegetation or to provide protection at points that cannot be adequately protected by other methods. Loose boulder check-dams perform well in gullies which do not carry much runoff and it also helps in silt deposition, thereby helping the stabilization of gully beds. Permanent gully control structures are constructed to control the over falls either at gully head or in gully bed. Erosion from the extending heads and sides of gully and main channel are the major sources of sediment. There is also need to construct diversion bunds to divert surplus water to water harvesting structures or to the grassed water ways.

(ix) Contour Trenches: Contour trenches are dug-out, pilling up the dug-out earth on lower side of the trench, for trapping sediment and runoff at early stage of their movement. These trenches also improve soil moisture and favour quick growth of trees and grasses.

(x) Stream Bank and Torrents Control: The vulnerable stream banks should be protected by providing spurs and retaining walls, etc. To control torrents, structure like barrages, paved channels, etc. need to be provided.

3.1.2. Agronomical measures

For preventing soil erosion on cultivated lands, proper choice of crops and cropping patterns is necessary, particularly on hill slopes. The protection through vegetative shield, forest cover, grasses, crops and mulches, etc. are some important measures to prevent soil erosion. Such a protection by absorbing the energy of rain impact prevents the loss of both water and soil. The following crop management practices can be useful in minimizing the erosion of soil and nutrients.

(i) Cropping Systems: crops with ability to develop canopy quickly provide an early protection to the soil. Interplanting of erosion-resistant crops like cowpea, soybean, etc. are also useful. Strip cropping of erosion resistant legumes along with cereals can conserve rain water and reduce the velocity of runoff.

(ii) Crop Geometry: It is essential to manipulate the crop layout in the field in a manner which may prevent soil erosion. A closer spacing of rows across the slope can help in this regard.

(iii) Contour Cultivation: Contour cultivation reduces the runoff to large extent, thereby reducing the soil and nutrient losses. Contour cultivation as well as furrows and ridges have been found useful.

(iv) Tillage: Low intensity tillage favours consolidation of soil through better structure, infiltration and pore distribution. This imparts erosion resistance. A study of conventional method of cultivation of maize with zero tillage, with or without live mulch has shown that runoff and soil loss are greatly reduced with low intensity tillage (Bhardwaj, 1998).

(v) Agroforestry: Agroforestry has become popular as a useful land use system on slopes in the recent past. Growing of trees along with agricultural crops satisfy multifarious needs of farmers. The data presented in Table 1, give an idea of the usefulness of Agroforestry system as a soil conservation measure

Table 1 : Runoff and soil losses under different Agroforestry systems

Land use	**Runoff (%)**	**Soil loss (tons ha^{-1})**
Maize	18.3	17.70
Maize + *Subabul*	8.9	5.00
Maize + Eucalyptus	3.6	0.91
Chrysopogon fulvus	1.6	0.33
Grass + *Subabul*	0.6	0.13
Subabul	0.4	0.04
Grass + Eucalyptus	0.1	0.02
Eucalyptus	0.1	0.01

[Agroforestry system as a soil conservation measure (Narain et al.*1994)].*

(vi) Grasses: grasses are perhaps the best friend of soil conservationists. Low and evenly distributed canopy and fibrous root systems with much soil binding capacity makes grasses highly effective in controlling soil erosion. The selection of grasses should be based on their production potential considering edaphic conditions and local preferences.

3.2. Water Management

Water is a prime limiting factor for crop production and scientific water management is key to development of sustainable agriculture for both irrigated and rainfed farming systems. The low water use potential, prohibitive costs of irrigation development, increasing problems of salinity, sodicity and water logging in the wake of development of irrigation and increasing competition for good quality water between agriculture, industry and urban uses necessitate assignment of high priority to the development of potential system of irrigation water management. The most impact of irrigation on productivity and sustainability of agriculture and environment can be realized only when water use is integrated with potential soil, water, crop, fertilizer and other agronomic management systems. Drainage is an essential component of the irrigation system. It is use of irrigation and lake of adoption of best soil and water management techniques, and mismatch of nutrient inputs and agronomic management of a farming system which lead to lower potentiality of irrigation water.

The modern irrigation techniques such as drip irrigation can become remunerative if these are used for high value crops, with high technology and optimum agricultural inputs. Its demand for irrigated farming in the future, especially for horticulture and vegetable farming is increasing in India. While in the developed countries this system is getting computerized, in India it is still in its infancy and need to be considered as an innovative technology for sustainable high productivity, profitable and sustainable farming systems.

For dryland farming, increased potentiality of rain water is essential and it can be achieved in the following ways:

(i) Retain precipitation *in situ* and minimize the runoff.
(ii) Use drought tolerant crops that fit in the rainfall pattern.
(iii) Reduce evaporation in relation to transpiration.
(iv) Use watershed concept for maximization of rainwater harvesting and recycling.
(v) Recycle the run-off and drainage water for high value crops adopting life saving irrigation approach.

In fact, it is the combination of all these practices that ensures the best results with minimum degradation of environment. Since the crucial factors of rainfed farming are moisture stress and nutrient-deficiency stress, an integrated technology is required for removing these stresses to ensure successful sustainable farming. The Rio de Janeiro declaration of 1992 emphasized the following principles relevant to sustainable farming:

- Human beings are at the centre of concern for sustainable development and environmental issues are best handled with the participative co-operation of all concerned people.
- The criteria of productivity, equity and environmental safety are critical for the present and future generations for development of sustainable farming.
- For sustainability it is necessary to remove all negative factors associated with unsustainability.

4. Soil Productivity

Soil productivity is the capacity of soil to produce crop yield per unit area of the soil in a specific management practices. It is the resultant effect of all factors like soil moisture, environment, sol fertility, etc. if the physical soil constraints limit the crop production, the removal of such constraints through mechanical, hydrological and chemical methods generally result in sustained high productivity, provided the operations are repeated after periodical intervals and do not damage the soil structural elements permanently.

The All India Coordinated Research Project on Improvement of Soil Physical Conditions to Increase Agricultural Production of Problematic Soils has provided adequate information on the amelioration of such soils to sustain high productivity. Compaction technology suitable for management of high permeable and low water retaining sandy and loamy sand soils, increases the productivity of rainy season crops and reduce the irrigation water needs of succeeding wheat crop. In the case of rice soil of lateritic type and of sandy loam texture, the compaction helps in reducing percolation losses of water from upland rice fields. Chiseling of dry soil up to 40 cm depth and at 50-120 cm interval has been found suitable for management of shallow-soils with highly impermeable sub soil. The broad bed and furrow system developed by ICRISAT for improving the surface drainage of Vertisols has been found effective for increasing not only yields but also cropping intensity. This system has been found effective in sustaining high productivity of sorghum, pigeonpea or maize-chickpea system and reducing soil erosion. The system could also provide opportunity for water harvesting and its reuse. Likewise at Jabalpur in a higher rainfall area of >1000 mm, raised bed and sunken bed technology on Vertisols made the rice-based cropping system sustainable. The crop residue recycling technology for the management of

Alfisols-the red sandy loam soils, which are prone to crusting, has been found to improve productivity and sustainability of the system. The best means of improving and maintaining soil quality which determines soil productivity and environmental quality is adoption of alternative agricultural practices such as crop rotation, recycling of crop residues and animal manures, green manures, biofertilizers and integrated nutrient management for encouraging balance use of fertilizers and manures, and reduced use of pesticides.

These are some of the components of a strategy for obtaining sustainable high productivity in any farming system. The relationship between the soil degradation processes and soil conservation practices as outlined by Hamick and Parr (1987) is shown in Figure 5.

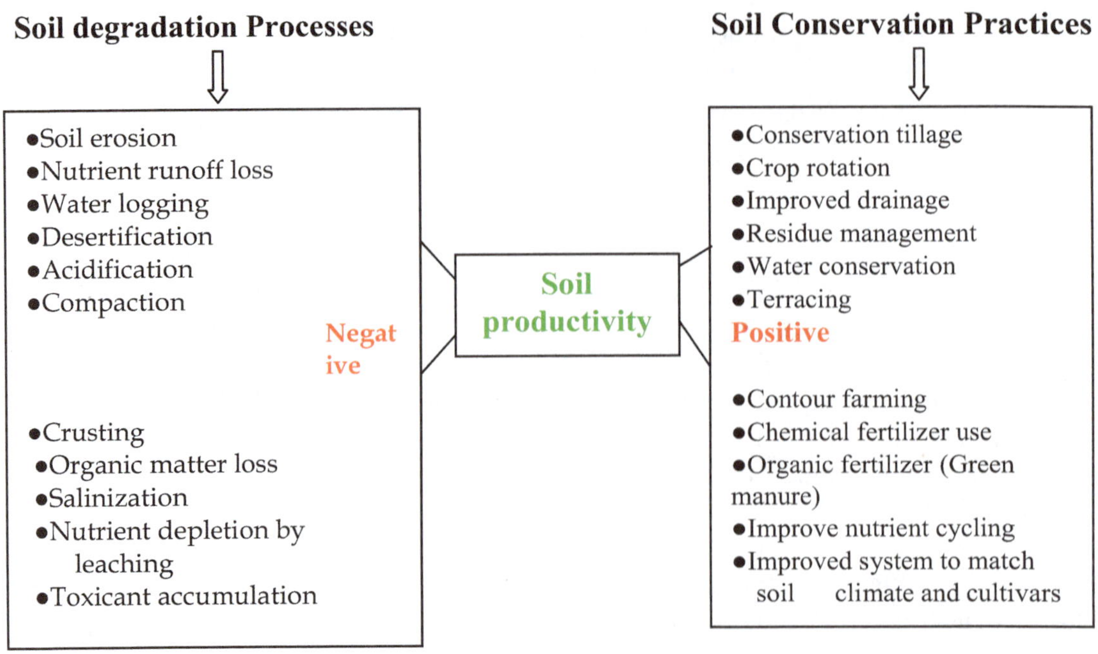

Figure 5 : The relationship between soil degradation processes and soil conservation practices

4. Sustainable Soil Management Problems

The main problem of sustainable soil management is linked up with the sustainable land management and it has many technological, socio-economic and environmental aspects. A sustainable land management includes a set of technologies, policies and activities aimed at integration of socio-economic principles with environmental concepts so as to simultaneously.

- Maintain or enhance productivity and production,
- Reduce the level of production risks,
- Achieve environmental stability, and
- Be economically viable and socially acceptable.

Sustainable land management system is more comprehensive than the sustainable farming. It is because; the latter is concerned with the onsite constraints, whereas the former deals with the offsite or watershed based issues and sustained productivity.

Agricultural sustainability may be defined as a function described by Equation (1) (Lal 1994):

$$\text{Agricultural sustainability} = d\,(P_1 \times S_p \times W_1 \times C_1)\,dt\ldots \quad (1).$$

Where, P_l is the productivity potential with input of the limiting or non-renewable resources; S_p is the measure of critical soil property, *i.e.* rooting depth, soil organic matter content, cation exchange capacity; W_l is the plant available water resources and quality; and C_l is the climatic factor with reference to evaporation and soil temperature flux of radioactive gases.

Thus, it is a product of numerous parameters affecting productivity and sustainability.

CONCLUSION

From foregoing discussing, it is concluded that for sustainable agricultural development, sustainable soil management is the essential prime factor. Sustainable soil management can only be possible through soil fertility management, conservation of soil and water and maintenance of soil productivity. Balance nutrition is a pivotal point in soil fertility management. For increasing soil productivity, optimization of fertilizer rate, balance use of fertilizer, potential sue of fertilizer and nutrient management through integration are the most important aspects. Nutrient availability in soil can be maintained by application of organic fertilizers in soil. Optimum soil moisture always interacts positively with fertilizer application. Soil physical conditions also play important role in crop production. Time of fertilizer application, influence very much on crop productivity. Soil and water conservation are the most important parameter for sustaining soil resources for future generation. Overall, nutrient management through integration has great potential to offset the growing heavy nutrient demands, to achieve maximum yields and, to sustain the crop productivity and good soil health on long-term basis.

References

Anonymous (1984) Annual report of department of Agril. Chem. and Soil Sci. G.A.U. Junagadh.

Bharadwaj, S.P. (1998) Zero tillage and weed mulch for erosion control on sloppy farmland in Doon valley. *Indian Journal of Soil Conservation* 26, 81-85.

Gajri, P.R., Prihar, S.S. and Arora, V.K. (1989) Effects of nitrogen and early irrigation on root development and water use by wheat on two soils. *Field Crop Research* 21, 103-114.

Harmick and Parr (1987) World Commission on Environment and Development (1987) *Our Common Future.* Oxford University Press, UK.

Lal, R. (1994) *Soil Conservation: Stressed Ecosystem and Sustainable Agriculture* (S.M. Virmani, J.C. Katyal, H. Eswarn and I.P. Abrol, Eds), Oxford & IBH Publishing Co., New Delhi.

Nambiar, K. K. and Abrol, I. P. (1986) Longterm fertilizer experiment in India: An over view. *Fertilizer News* 34, 11.

Nambiar, K. K. M. and Ghosh, A .B. (1984) Highlights of research on longterm fertilizer experiment in India. (1971-82). *Longterm Fertilizer Experiment Project.* pp 100.

Narain, P., Chaudhary, R.S. and Singh, R.K. (1994) Efficacy of conservation measure in north-eastern hill regions, *Indian Journal of Soil Conservation* 22, 42-62.

Narang, R.S. and Bhandari, A.L. (1992) Integrated nitrogen management in rice-wheat system. In *Rice-Wheat Cropping System* (R.K. Pandey, B.S. Dwivedi and A.K. Sharma, Eds) Project Directorate of Cropping Systems Research, Modipuram, Meerut, India. Pp 68-83.

Nayyar, V.K., Arora, C.L. and Kataki, P.K. (2001) Management of soil micronutrient deficiencies in the rice-wheat cropping system. *Journal of Crop Production* 4, 87-131.

Rajani, A.V., Shobhana, H.K. and Golakiya, B.A. (2006) Comprehensive appraisal of potassium fertilization in groundnut. In proceeding of the international symposium held at Punjab Agricultural University, Ludhiana, India, 22-25 November, 2006 on *Balance Fertilization for Sustaining Crop Production.* Pp 302-305.

Sarkar, A.K., Lal, S. and Singh, B.P. (1997) Balanced fertilizer use in red and lateritic soils. *Fertilizer News* 42, 49-55.

Subba Rao, A. and Srinivasa Rao, Ch. (1996) *Potassium status and Crop Response to Potassium on Soils of Agro-ecological Region of India. IPI Research Topics No.* 20. International potash Institute, Basal, Switzerland, pp 1-57.

Tandon, H.L.S. (1989) Long term fertilizer experiment in India- Lesson and practical expectations. *Fertilizer News* 25, 3-20.

Tandon, H.L.S. and Narayan, P. (1990) *Fertilizers in Indian Agriculture- past, present and future (1950-2000).* Fertilizer Development Consultation Organization, New Delhi.

□□□

Green Agriculture : Newer Technologies, 2012
© Kambaska Kumar Behera (ed.), pp. 109-141
New India Publishing Agency, New Delhi (India)
e-mail : info@nipabooks.com; website : www.nipabooks.com

Chapter-6

Biodrainage : An Overview

Afroz Alam, Kambaska Kumar Behera and Sharad Vats
Department of Bioscience and Biotechnology,
Banasthali University,
Rajasthan-304022
E-mail : afrozalamsafvi@gmail.com

SUMMARY

The primary requirement for sustainable agriculture is the maintenance of a satisfactory balance of water, air and salt in the root zone encouraging for plant growth. This balance can be achieved by sufficient drainage. About 22% of the World's irrigated area is affected by waterlogging and salinity. Vertical drainage by pumping can control the increase of ground water table in those areas where the ground water is of good quality. Horizontal sub-surface drainage in which the saline ground water is drained away through pipes and drains, is the other regularly used anti waterlogging methods. In bio-drainage system, the characteristic of the trees to transpire water is used to control the increase of ground water table. It is an alternative drainage method, which would be most economical and environmental congenial. But there are worries and doubts on the long-term sustainability of biodrainage. This chapter describes various aspects of biodrainage.

Introduction

Biodrainage, a biological alternative of controlling water logging which is a major problem in arid and semi-arid climates related with irrigation, and salinization, the another issue that causes land ruin and consequently yields reduction (Abrol et al. 1973). One method to fight these effects is to set up a subsurface drainage system that keeps the water table down and allows for discharge of salts from the root zone. Primary requirement for sustainable agriculture is the maintenance of a satisfactory balance of water, air and salt in the root zone encouraging for plant growth. This balance can be achieved by sufficient drainage. Drainage can be either natural or artificial. Most land has some natural surface and subsurface drainage. When natural drainage is insufficient, artificial drainage is required to increase the drainage capacity. Artificial drainage is necessary to sustain irrigated agriculture. Often, subsurface drainage is needed in irrigation schemes to control the rising water table and avoid water logging and salinization. Usual subsurface drainage systems are of two types, vertical (tubewells) and horizontal (drainpipe) systems. When properly designed, installed and maintained, these systems are competent in lowering the water table and preventing salinization of irrigated lands. Though, they have two drawbacks, explicitly, they are costly and they generate drainage effluents, which will have to be either carefully reused or safely disposed of. A choice is biodrainage, which can be less costly and more environmentally responsive. It is a combined drainage-cum-disposal system. Biodrainage relies on vegetation, rather than mechanical means, to remove surplus water. The driving force behind the biodrainage concept is the consumptive water use of plants. Biodrainage is economically attractive because it requires only an initial investment for planting the vegetation, and when established, the system could produce economic returns by means of fodder, wood or fiber harvested (Anon 2003). There is consent that biodrainage, when properly implemented, can lower the water table (Anonymous, 2009). It could solve problems related with waterlogged areas and canal seepage. The unclear point is the capability of the biodrainage system to maintain a salt-balance.

The consequences of irrigated agriculture are often that the unavoidable deep percolation losses of irrigation cause groundwater tables to go up. Natural or internal drainage can no longer survive with the concentrated human activities that constitute irrigated agriculture. This causes water logging of the root zone, which certainly leads to yield reduction. Water logging is a common problem in non-irrigated areas, where low-lying depressions serve as discharge areas, and on irrigated lands (Chhabra and Thakur 1998).

Conventional solutions to combat waterlogging and salinity are vertical and horizontal subsurface drainage systems consisting of respectively pumped tubewells or horizontal buried pipes and deep open drains. Subsurface drainage and drain water evacuation systems accomplish the following:-

1. Dewatering means inducing water flow through the soil towards the subsurface drain (tubewell, pipe or open drain).
2. Moving of the drain water through the lateral, or field drain, to collector drains and thereafter to main drains.
3. Often pump lifting the drain water to higher elevations in the migration system and/or further transportation by gravity outflow to disposal sites.
4. Final disposal at selected sites (e.g. by evaporation ponds).
5. Salinity control: dewatering of the rootzone and leaching of salts happens at the same time.

Biological systems make use of the evapotranspirative power of plants, especially of trees, to lower groundwater tables. They can perform the above functions together. Salinity control, function, is more difficult to achieve and this requires additional means in the long-term. However, biodrainage systems may delay the salinization process (Batra 1988). Biodrainage relies on vegetation, rather than engineering mechanisms to remove excess soil water through evapotranspiration. It is often considered attractive because it requires only an initial investment in site development (plantation of a "biodrainage crop") and (potentially) returns a benefit when the biocrop is harvested for fodder, wood or fibre. In addition, under some management scenarios, viz. certain cropping systems and slightly saline conditions, it might offer limited scope to achieve nutrient and/or salt balance through removal of biomass, thus alleviating the problem of the disposal of polluted drainage effluent from the biodrainage crop area by reducing volumes and improving the quality of the effluent. Presently about one-third of the world's irrigated area faces the threat of waterlogging (Shakya and Singh 2010).

Traditional methods

The conventional engineering-based techniques most commonly used to drain excess water from land are: surface drainage, horizontal subsurface drainage and vertical subsurface drainage.

Alternative methods

The limitations and shortcomings of the conventional drainage techniques call for alternative approaches to help keep agriculture sustainable over the long term. Alternative techniques must be effective, affordable, socially acceptable and environmentally friendly and not cause degradation of natural land and water resources. Biodrainage is one of these alternative options. The absence of effluent makes the system attractive. However, for biodrainage systems to be long-term sustainable, careful consideration is required of the salt-balance under the biodrainage crops. The term biodrainage is relatively new, although the use of vegetation to dry out soil profiles has been known for a long time. The term bio pumping is used to describe the use of trees for water table control. Another term relating to the "bio" aspect of soil water removal is biodisposal, which refers to the use of plants for final disposal of excess drainage water

Execution of biodrainage

In considering their long-term feasibility, biodrainage systems will have to be subjected to the same analysis as other plant-based biological systems with regards to nutrient and salt-balance, soil considerations, etc. In addition, as large-scale adoption of biodrainage crops could include considerable tracts of land, socio-economic considerations will have to be taken into account when considering the implementation of this technique (Denechke 2001).

Irrigated systems

In landscapes with undulating topography, recharge and discharge areas are often relatively easy to delineate. Recharge occurs at the higher parts of the landscape and discharge lower down the slope. In irrigation areas, with their flat topography and (often) shallow water tables, the distinction between recharge and discharge is less clearly delineated and frequently areas that are discharging groundwater by evapotranspiration between irrigation events temporarily turn into recharge areas during and immediately after irrigation (Grattan et al. 1987).

Water table control

Superficial water table levels cause a threat to agricultural crops as they often result in salinization of the plant root zone. The management of irrigation areas often aims to keep water tables below the critical depth, which is defined as the depth at which capillary salinization is negligible. Sustainability of irrigation is determined by the leaching capability of soils. To avoid salinity problems, the salts present in the irrigation water will have to be removed from the root zone by leaching them either laterally to adjoining non-irrigated areas or streams or vertically down to levels below the vegetation root zone.

Plants can remove water from the soil either

(1) Directly from the saturated zone below the water table (water table control).
(2) From the unsaturated capillary fringe above the water table (water table control).
(3) From unsaturated topsoil layers after rainfall or irrigation (recharge control).

Doctrine of preparation

The following issues should be considered in the development of biodrainage systems:

- Water balance
- Drawdown of water table
- Salt balance
- Economic aspects
- Social acceptance
- Plantation area
- Salt tolerance

Design considerations for biodrainage

- Tree water use
- Water table drawdown by plantations
- Quality of groundwater

- Water balance through biodrainage in irrigation areas
- Sustainability and combined biodrainage and conventional drainage systems
- Root depth and density
- Trees adjacent to groundwater pump

'Walking' plantations

After harvesting, trees are replaced by shallow-rooted irrigated crops, water table control could be provided by a new plantation established on an neighboring site. This second tree crop should be planted about five years before the planned harvesting date of the original plantation so that the new trees would be old as much as necessary to have an impact on the water table and protect the adjacent paddocks from shallow water table induced salinity problems. The tree plantation width would depend on its drawdown impact (Kapoor 2002).

Under this system tree plantations 'walk' through the landscape, each new planting providing water table and salinity protection to the preceding site.

- Salt balance
- Salt storage in vegetation components

Scientific basis of biodrainage

In natural environments the components of the hydrological system, i.e. rainfall, evapotranspira-tion, change in soil-water storage and drainage, are almost in balance. Periods of high rainfall might temporarily result in amplified drainage flows, a rise in the groundwater table and/or soil moisture storage, then over a period of about 5-10 years equilibrium is established. Vegetation plays a vital role in the evapo-transpiration and soil-water storage components of this balance. It enables control of water-logging and salinisation of soils. The total minerals removed annually by crop and forest biomass can match the total annual import of minerals with the irrigation water (Kapoor 2002). The Indira Gandhi Canal Project, Rajasthan, India is presented as a case study of the feasibility of biodrainage (Kapoor and Denecke 2001). Various Salinity Control and Reclamation Projects have been implemented in Pakistan to control these problems. In spite of these measures, the problems are still not being controlled (Akram, 2006). This is attributed to the huge expenses needed to run and maintain them. Removal of excess water from the root-zone is necessary to sustain irrigated agriculture. Biodrainage is suggested as an economical strategy to tackle the situation. The technique utilizes certain categories of plants that draw their main water supply directly from groundwater and are tolerant to water-logging/salinity. A study was carried out on a 4 ha area of 6 year old *Eucalyptus* plants near Bahawalnagar. During canal irrigation of cotton, rice and sugarcane the water table varied considerably. Although generally deep under the Eucalyptus, the water-table rose away from the plantation, with a maximum rise of over 30%. Salinity was maintained in the area under plantation. The annual water use by the plantation ranged from 300 to 2100 mm. By comparison on a sodic soil near Lahore the water use by *Eucalyptus camaldulensis* and *E. microtheca* was 1400 and 1050 mm per year. Also described are drylands, non-irrigated situations, where biodrainage can be used to partially restore the hydrological balance caused by inappropriate land development. The use of biodrainage in irrigated areas is explained where the system is used to protect irrigated land against the hazards of water-logging and salinization (Tomar and Kumar 1999). When natural vegetation is cleared and replaced by crops or tree plantations, the ooze losses to the groundwater table under the new land use system are either higher or lower than under the pre-clearing situation (Denecke 2001).

The increased seepage scenario prevails and the development of land for either rainfed or irrigated agriculture generally increases groundwater recharge rates; the new landscape systems "leak"! Where this increased recharge results in shallow water tables, evaporation at the soil surface causes groundwater to move upwards through the soil, and consequently salts accumulate in the root zone, limiting plant growth. For example in Australia, the area of shallow, saline water tables is expanding rapidly (Marcar et al. 1990) in response to widespread agricultural development over the past 50-100 years.

An example of increased recharge after clearing is also presented by Heuperman (2000). They showed that in a semi-arid region in southern Australia (average rainfall 250-300 mm/yr), the recharge rate beneath native Eucalyptus spp. was <0.1 mm/yr. Recharge was found to increase significantly to between 5 and 30 mm/yr following clearing and subsequent cropping. Management solutions to the problem of excessive recharge are being developed, based on improved water use efficiency of the agricultural systems (Heuperman and Kapoor 2002).The reduced seepage scenario, although less publicised, does also occur. Vertessey et al. (1996) studied the hydrology of mountain ash (*Eucalyptus regnans*) forest in a high-rainfall environment in southern Australia and analysed relationships between forest age and runoff volumes. Old-growth forest yielded up to twice as much annual runoff as younger re-growth forest. The same process was observed in a mixed-species forest in a drier catchment (www.catchment.crc.org.au). The driving mechanism behind this process is leaf area index, which was highest between the ages of 30 and 40 years. The findings have important consequences for the management of the catchments, which are used for water harvesting for urban water supplies. For these areas, long harvesting rotations are obviously preferred to obtain maximum runoff yields.

The dynamic force behind the biodrainage concept is the consumptive water use of plants. In Australia (Heuperman and Kapoor, 2002) suggested that the rates of transpiration and groundwater uptake by trees underlain by relatively shallow (5-8 m below surface) water tables, were very high, exceeding the annual evaporation from pasture (~400 mm) by a factor 3-6 (1200-2300 mm/yr). These results, coupled with a growing interest in timber production in Australia, led to the popularity of the tree-based water management strategy for agricultural areas. However, results obtained by the measurement technique used by Greenwood (ventilated chambers) have been challenged and later work suggests more modest water use figures with potential stand water use approximating standard Class A pan evaporation. *Eucalyptus camaldulensis* and *E. grandis* grown on a shallow saline water table both used approximately 300 mm per year. They also stated that the plantation's ability to transpire groundwater is reduced where the groundwater table is drawn down in soils of low hydraulic conductivity (Dagar et al. 2001).

Potential water use differences between species are also a topic of discussion. Hatton et al. (1998) recognized the need to generalize water use behaviour of Eucalypts to facilitate landscape management processes in a wide range of Australian environments. They concluded that the leaf efficiency of sympatric Eucalypt species in soil water-limited systems is similar, i.e. there was a strong linear relationship between tree leaf area and mean daily water use for a wide range of Eucalyptus species grown under similar climatic conditions. Meyers et al. (1996) arrived at the same conclusion for non-water limited situations, stating that species (including *Pinus radiata*) with unrestricted access to water have similar rates of water use at similar stages of canopy development.One of the major factors determining the sustainability of plant productivity (and thus plant water use) processes is salt balance. If the salts moving into the root zone are not either (1) taken up by the vegetation and harvested or (2) removed from the root zone by leaching, the vegetation is doomed to give way to salinity. The deep-rooting characteristics of trees make them extremely efficient users of water. While shallow-rooted grasses and crops have limited access to underlying water tables, deep-rooted trees can access water tables down to several metres. An important consideration of biodrainage issues is the definition of "water table". The water table can be defined as "the upper surface of a zone of saturation, where the body of groundwater is not confined by an overlying impermeable formation". The depth of the water table is measured in observation wells. In contrast, piezometers record pressures at a specific depth in the soil profile; they are often installed in aquifers for faster response to changes in pressure. Where the water table, as measured in observation wells, is perched, leakage through an underlying slowly permeable layer can provide a means of salt export to deeper formations in the profile. Perched groundwaters are typically shallow, small in extent and often fresh (Shakya and Singh 2010). The distinction between water tables and piezometric pressures should be kept in mind when analysing soil water movement under vegetation.

Biodrainage scenarios

Dryland / rainfed systems:

- Recharge control
- Groundwater flow interception
- Discharge enhancement

Irrigated systems:

- Water table control

- Channel seepage interception
- Biodrainage cum conventional drainage systems

Rainfed systems

A major problem with biodrainage (as opposed to conventional drainage) in rainfed conditions is that plant water requirement is generally low during cooler winter periods with high rainfall. So there is a delayed drainage response to rainfall inputs with the soil reservoir filling over winter and being depleted by vegetation water use over summer, thus creating a storage buffer to accommodate the next rainfall season (Grewal 1984).

Recharge control

The sustainability of natural environments relies on the balance between recharge and discharge or hydrological balance; water fluxes passing beneath the root zone of vegetation communities are laterally discharged through regional subsurface aquifer systems. Where vegetation is changed by agricultural 'development' (clearing) and crops with lower annual water use and/or shallower root systems are planted, recharge increases. As the conveyance capacity of the underground aquifer system is often not high enough to accommodate the increased recharge volumes, groundwater tables' rise and cause waterlogging and salinization (Grattan et al. 1987). Often the clearing of vegetation in the higher areas of the landscape results in increased recharge, followed by the formation of shallow water tables in the lower areas of the landscape. Water tables in the recharge areas are too deep to be accessed by vegetation root systems, and plants in these areas rely on rainfall for their evaporative requirements. The process to minimize deep seepage losses in the higher parts of the landscape to minimize discharge problems down-slope is often referred to as recharge control. Re-vegetation of recharge areas is a major tool in the fight against dry land salinity in Australia. Often only relatively small proportions of the landscape have to be planted to achieve the objective of reducing localized salinity discharge problems in the lower part of the landscape. However, re-vegetation of recharge areas can also have negative effects. Where the evaporative capacity of the new vegetation exceeds the pre-clearing evaporative demand, the landscape 'dries out'. This scenario is often encountered in catchments covered by newly established fast-growing plantations. It is an example of an over-designed recharge-control biodrainage system and could cause problems such as reduced river flows, the drying-up of wells and increasing groundwater salinity (Shakya and Singh 2010).

Groundwater flow interception

Break-of-slope, (where the slope 'breaks' from convex to concave) plantings have been promoted as flow interceptors for areas where groundwater flows through permeable layers overlying low-permeability strata. By tapping these layers at some point down the slope, where the quality is still relatively fresh, the trees are considered to intercept these flows and thus reduce discharge problems further down the slope. Location of the tree plantations, based on a thorough understanding of the underlying stratigraphy, is extremely important if this concept is to work. Photo 2 shows a break-of-slope planting of two-year-old blue gums (*Eucalyptus globulus*) in Northern Victoria, Australia.

Discharge enhancement

Low-lying landscape units with shallow water tables often serve as local discharge areas. Where these areas have drainage outlets and seepage flows discharge into rivers, salt balance is provided. Where the depressions are land-locked (closed basins) and percolation to deeper aquifers is inhibited, salinization of the landscape unit is inevitable. The use of biodrainage in waterlogged discharge areas is based on the concept of enhanced evapotranspiration. The long-term sustainability of biodrainage in this environment is a topic of intense debate. Smedema (1997) highlights this in his short topic paper. He suggests that biodrainage could be considered for waterlogged landscape depressions and canal seepage interception, and could be applied in 'parallel field drainage' arrangements as an alternative to conventional field drainage systems. In Australia it is now widely accepted that in discharge situations, enhanced evapotranspiration biodrainage sites will eventually succumb to salinity, unless some form of conventional drainage is installed to control salt balance to the vegetation's rootzone by removal of saline drainage effluent (Heuperman, 2000). The hills in Northern Victoria, Australia with salinity problems mainly in the lower parts of the landscape. Plants can use water both from the unsaturated part of the soil profile above the water table and from the saturated part below the water table. Plants in the latter category are called phreatophytes. They often (but not always) grow in (semi) arid climates where they tap deep water tables. Boumans et al. (1988) reports on mesquite (*Prosopis*) growing in desert washes in the southeastern United States where the groundwater is sufficiently shallow that seedlings can occasionally produce deep enough roots to reach the water table in wet years. In that same area the introduced phreatophyte *Tamarix* has lowered the water table to such a low level that other species with shallower rooting depths are being eliminated. One special application of the biodrainage concept is the amelioration of waterlogged soils during the initial reclamation or ripening phase of 'new' land development. Vegetation with a vigorous, deep and extensive root system is used to dry out waterlogged soil profiles. For example, in the Netherlands land to be reclaimed from the sea is sown with reed while a few centimetres of water remain. This accelerates the ripening process. Sustainability of irrigation is determined by the leaching capability of soils. To avoid salinity problems, the salts present in the irrigation water will have to be removed from the root zone by leaching them either laterally to adjoining non-irrigated areas or streams or vertically down to levels below the vegetation root zone.

Channel seepage interception

Channel seepage can be a major contributor to water table accessions in irrigation areas. High seepage rates will result in ground water mounds beneath channels, causing waterlogging and salinity problems in the adjoining land. Water quality in supply channels is normally good and the seepage water, if not left to evaporate and increase in salinity, can be productively used by vegetation and commercial crops. The issue of salt balance, although less critical than for more saline groundwater situations, is still a matter of long-term concern (Boonstra and de Ridder 1990).

Biodrainage cum conventional drainage systems

Biodrainage crops are no exception to the basic rule that irrigation, or for that matter plant growth, is not sustainable without some form of root zone salt balance. Where biodrainage results in salt accumula-tion, engineering assistance is needed to make the system sustainable.

Ethics of planning and design *(IPTRID/FAO, 2002)*

The aim of biodrainage is to remove excess groundwater through the process of transpiration by vegetation. This is achieved by enhancing the transpiration capacity of the landscape by introducing high-water use vegetation types in large enough areas to balance recharge/discharge processes to maintain groundwater balances below the root zone of the agriculture crops. The following issues should be considered in the development of biodrainage systems:

i. Water balance: Biodrainage plantations should be able to extract groundwater volumes equal to the net recharge. The water balance is to be maintained such that the water table is kept below the rootzone.
ii. Plantation area: The biodrainage plantation area should be kept as small as possible. Agriculture (particularly irrigated agriculture) is practised primarily to produce high-value crops. Conversion of high-value cropping land to relatively low-return forestry may be difficult. Often good quality water is in short supply while land is not a limited resource. Particularly in arid and semi-arid regions, dryland areas surrounded by irrigated land could be earmarked for tree plantations without loss of productive resources.
iii. Salt tolerance: Biodrainage crops need to be salt tolerant. Groundwater qualities can vary greatly spatially, normally they have a higher salinity than irrigation supplies. The water use capacity of trees and other crops decreases with increase in water salinity. In the case of Eucalypt species, it reduces to about one-half of potential when the water salinity increases to about 8 dS/m (Shakya and Singh 2010).
iv. Drawdown of water table: Crops, including trees, act as biopumps; they depress the water table directly underneath plantation areas and consequently lower the water table in the surrounding area. The drawdown effect under trees/crops

depends on the tree/crop's water use, the rate of recharge in the surrounding area, the hydraulic conductivity of substrata and the depth to deeper barrier layers. Biodrainage plantation should be established in blocks or strips and spaced to keep water table levels in the irrigated farmland in between the plantings below the root zone. The harvesting of the biodrainage plantations would need to be planned in such a manner that the "drainage" function is not lost (thinning regimes).

v. Salt balance: The introduction of irrigation always upsets the salt balance. Although irrigation supplies often have relatively low salinities, the large volumes of water that are introduced in the landscape increase salt imports significantly. Drainage of effluent to export these salts is therefore generally considered a necessity. To achieve salt balance without conventional drainage, the irrigated crops, along with interspersed biodrainage plantation, would have to accumulate the salts introduced by irrigation and subsequently have to be harvested and removed from the region. This is only (potentially) achievable in situations where very low-salinity water is available to the plants (Minhas 2001).

vi. Economic aspects: The growing of biodrainage trees and crops requires a different operational management approach than the growing of agricultural crops. Up-front costs associated with planting and maintenance precedes the income from harvesting by many years. Some form of contract growing, based on annual payments might have to be considered to make the system acceptable to land holders.

vii. Social acceptance: The introduction of new crops such as tree plantations affects rural social societies. New markets might have to be developed, security arrangements differ from those for normal crops (illegal pruning or cutting for fuelwood) and fires could destroy the results of many years of labour in a single day. Active participation of local communities in the development of tree plantation-based biodrainage systems is extremely important to overcome problems and ascertain the benefits of biodrainage systems are reaped to the maximum extent.

Requirements to be met by bio-drainage (Kapoor 2002)

For biodrainage to be effectively adaptable, the following requirements must be met:

(a) Water balance	:	The quantity of water removed from the ground water annually should equal the quantity of recharge.
(b) Salt balance	:	The quantity of minerals removed annually should be nearly equal to the quantity of mineral import.

(c) Area under plantation	:	Irrigation is practiced primarily to promote agriculture, horticulture, dairy etc. Therefore in term of economic returns afforestation or agro-forestry should be comparable with that from other alternative uses of land. If it is not so, afforestation may still be justified, on considerations of the environmental and drainage benefits.
(d) Water for plantations	:	Under ideal situation, trees in afforestation area on full development should be able to draw most of their requirement of water from the ground water table, so that surface irrigation water can be put to other productive uses. If, and so long it is not possible, plantation trees would need some irrigation water. They may also need some water periodically to leach down salts from the root zone, if and when the salinity levels approach threshold limits.
(e) Ground water quality	:	The quality of ground water, when the water table approaches the root-zone of trees, should be such as can be tolerated by the plant species, otherwise the trees would need to be supplied irrigation water.
(f) Effect on lowering ground water table	:	Trees can lower the ground water table directly underneath the plantation area, to a depth upto which the tree roots can extend. This can be upto 15 m from ground surface or even more. To be effective as a drainage measure, the ground water table must be lowered in the irrigated area to a minimum critical depth (say 2 m below ground level), at the farthest point from the edge of the plantation area.

Whether and to what extent the above requirements can be met, would depend upon the prevailing field conditions. Kapoor (2002) has described methods of estimating these.

Common doubts about biodrainage

Evacuation in Irrigated Agriculture

All plants and vegetation contain some minerals, notably Calcium (Ca^{++}), sodium (Na^{+}), Magnesium (Mg^{+}) as cations and sulfate (SO_4), chlorides (Cl^{-}) etc. as anions. The composition and quantity of mineral content in bio mass depends on the plant specie and characteristics of the soil where the plant grows. The plant analysis results show that the weight of Ca^{++}, Mg^{++} and Na^{+} cations in dry biomass of a plant is about 3.3 per cent of the weight of the dry biomass. When biomass is harvested and removed from the field, minerals to this extent are also evacuated from the field along with the biomass. Therefore, if the dry biomass produce (grain + foliage) be 10 tons/ha, the weight of cations of minerals evacuated along with the biomass of harvested crop would be at the rate of 0.33 tons/ha (Singh 1985).

Comparison of drainage methods

A range of issues has to be considered before an appropriate drainage technique can be selected. In Table 1 biodrainage is compared with conventional drainage techniques and various factors that should be considered in the selection of the appropriate technique are summarized (Kapoor 2003).

Comparison of drainage methods

Parameter	Drainage method		
	Conventional		Biodrainage
	Horizontal drainage	Vertical drainage	
(A) Efficiency and dependability	Well tested method Needs outfall Evaporation ponds have shown mixed results Solar evaporators may be part of drainage system	Well tested method Effluent reused Evaporation ponds have shown mixed results Solar evaporators may be part of a drainage system	Tried and tested at many locations with success Not presently adopted as large-scale drainage method on irrigation projects No outfall needed
(B) Cost	Medium cost	Medium cost	Low net cost (after returns)
(C) Advantages/ disadvantages	Reclamation of waterlogged area Water table control Local salinity control effective if disposal available	Reclamation of waterlogged area Water table control Additional water available for conjunctive use Local salinity control effective if disposal available	Reclamation of waterlogged area Water table control Could provide shelter belts against wind hazards Could provide additional wood and forest products Biodiversity Occupies (potentially valuable) land. Salinity control very limited

(D) Operation and maintenance	Periodic maintenance of pipe or open drains; Evaporation ponds require periodical removal of salts	Periodic maintenance, including pipe lines, pumps, screens, etc.; Power or fuel supply needed	Thinning, pruning, harvesting. Disease control
(E) Requirements for:			
(i) Land	Nil for sub-surface drains Open drains and evaporation ponds would require land	Only small areas needed	Relatively large areas of land needed; e.g. in IGNP India, land requirement is estimated at about 10 % of irrigated area
(ii) Good quality water	Nil	Nil	Water needed during establishment
(F) Energy requirement	Pump-lifting from drainage outlets to final disposal sites may be necessary	Energy needed for pumping groundwater and possibly further removal.	Nil
(G) Environmental impact off-farm	Adverse; the drainage can contain high concentrations of salts, chemicals and/or nutrients; Evaporation ponds can pose environmental problems	Adverse; pumped water can be of inferior quality and cause deterioration in quality of stream or canal water on disposal	Positive; dewatering, transport and disposal combined in one system; For final salt balance control disposal system required

Tree water use

Tree plantations often use water at higher rates than shorter vegetation types. This is for three reasons: (1) the high aerodynamic roughness of forests leads to greatly enhanced evaporation rates, which on an annual basis can be as much as twice that for grass; (2) this effect may be even more pronounced because of the so-called clothesline effect prevailing in rows of trees, substituting for a conventional drain pipe; (3) deep root system of trees with access to good-quality groundwater leads to high annual transpiration rates. The water use by tree plantations is not less than 1.5 times that of agriculture crops or about 1.25 times of Class A pan. A conservative tree water use figure of 1.0 Class A pan could therefore be assumed to estimate the potential biodraining capacity of tree plantations under conditions of good water quality. For design purposes this must be corrected for future estimated salinity levels (Westcot, 1988).

Water table drawdown by plantations

The water table under vegetation falls when discharge (evapotranspiration, surface runoff and groundwater outflow) exceeds recharge (infiltration and groundwater inflow) and stabilizes when they are equal. A depressed water table beneath a tree plantation induces groundwater flow from the surrounding areas (where the water table is higher) towards the plantation area, thus providing water table control to these areas. If tree plantations were planted in parallel strips, the water table profile would be similar to the profile found between parallel, open drainage ditches (Minhas, 2001) The relationship between depression of the water table, rate of recharge, hydraulic conductivity, depth to barrier layer and distance between plantations can be described using following equations:

$$L^2 = \frac{8\,KY_0h}{R} + \frac{4\,Kh^2}{R}$$

where : L = distance between parallel plantation strips (m); R = rate of recharge (m/day); Y_0 = height of water table above barrier layer under the tree plantations (m); K = hydraulic conductivity of substrata (m/day); h = head difference (m)

Root depth and density

The lateral extent of the impact of the water table depression beneath plantations on the surrounding land obviously depends on the vertical and lateral size of the trees' root system. Root systems have a remarkable ability to expand to access water and nutrients. The selection of appropriate species is important in the design of efficient biodrainage plantations. Gill (1985) measured roots up to 20 m from the trunk of individual eucalyptus trees. In the Indira Ghandi Nahar Project, Rajasthan, India, a perched water table along a seeping irrigation canal resulted in the development of pools in the borrow pits. Plantations of (mainly) *Eucalyptus camaldulensis* and *Acacia nilotica*, established around the waterlogged areas, transpired enough water to lower the water table by 15 m over a period of 6-7 years. Open pits excavated in the plantations down to a depth of 10 m, showed that tree roots were extending at least to that depth (Kapoor and Denecke, 2001).

Theiveyanathan and Benyon (2000) compared water use of Flooded Gum (*E. grandis*) and Spotted Gum (*Corymbia maculata*) on a shallow water table site in southeastern Australia. *E. grandis* used 300 mm groundwater per year while *C. maculata* used 675 mm over the same period at the same site. The researchers attribute the difference to the trees' root systems. Both species showed dense root growth in the top half metre of the soil, Spotted Gum had a lot of roots in the capillary fringe just above the water table and seemed better equipped to tap water at depths of around 3 metres.

Quality of groundwater

In semi-arid regions, the groundwater table is normally quite deep before irrigation is introduced and groundwater is commonly saline and unsuitable for irrigation. After introduction of irrigation, with good quality water brought in from outside the region, the deep percolation losses increase and two things can happen.If there is a barrier layer above the groundwater table, most of the deep percolating water may collect over the barrier layer and form a perched water body. The perched water table will rise and will eventually cause waterlogging. Since the quality of water in the perched water body is generally good, the groundwater can be pumped out, either directly by conventional drainage or via biodrainage crops (Kumar 2004).Where the percolating water infiltrates down to the saline regional groundwater table, this water table then rises and causes waterlogging and salinization problems. The poor quality of the groundwater limits its use. Subsurface drainage presents problems for disposal. Biodrainage can be practised with certain limitations. The transpiring capacity of trees reduces progressively as the groundwater salinity increases. When the groundwater salinity is about 8 dS/m, eucalyptus trees may transpire only one-half as much water as they do under non-saline conditions (Shakya and Singh 2010). However, this is true for many crops, not just trees.

Water balance through biodrainage in irrigation areas

When irrigation is introduced, the pre-existing water balance is disturbed; groundwater recharge increases and causes the water table to rise. The recharge to the groundwater occurs by way of (1) seepage losses from the water conveyance system, (2) irrigation water application and (3) rainfall events, the latter (especially) during cold seasons with limited crop growth. Seepage losses from the conveyance system depend on construction techniques and materials used. Recharge accessions directly from irrigation can be small when appropriate efficient irrigation techniques are used. Winter rainfall events can be significant contributors to groundwater recharge as crops use little water during that time of the year and non-cultivated fields are often fallow. If biodrainage is applied to obtain regional water balance, the total water use of the biodrainage plantations in a region should balance the recharge processes described above, minus the net regional subsurface drainage flows out of the region through underlying aquifer systems. The latter can be substantially different from the pre-irrigation situation when water tables were often much deeper (Khan and Yadav 1962).Water balances and recharge volumes are inherently difficult to determine, especially at a regional level, as recharge processes are extremely variable, both spatially and over time. This makes it more difficult to plan regional level biodrainage activities. Planning should focus at local level implementation such as seepage interception or break-of-slope plantings.

Sustainability and combined biodrainage and conventional drainage systems

The long-term sustainability of non-irrigated biodrainage tree plantations growing in shallow saline water table areas may be questionable. At some stage in their commercial life their growth performance could be affected by increasing root-zone salinity. After the trees are harvested, in the absence of subsurface drainage to provide salt balance, the accumulated salts in the rootzone will move to the surface by capillarity and impact on successive land use. A number of management options could be considered to minimize, delay or even avoid this problem (Shakya et al., 1995). Changing vegetation cover in a landscape is fraught with danger as it normally results in a redistribution of salts in the landscape, both vertically and/or laterally. This chapter sets out salt balance processes and their impact on the sustainability of biodrainage systems. Two salt balance mechanisms can be considered in plant systems: (1) salt balance through removal of salts from the vegetation rootzone by leaching and (2) salt applied to the plant is taken up and removed through grazing or harvesting of plant matter. The former needs little discussion, as it is well understood and covered in many textbooks on irrigation technology. The latter mechanism however needs some elaboration as it is often mentioned in biodrainage related information (Singh 1992).There appears to be a general consensus that the salt uptake by plants is negligible compared to the total salt applied in irrigation supplies. Gill et al. (1984) mentions that under most agricultural conditions where salinity is a concern, salt removal by crops can be ignored in the salt balance equation. The United States Salinity Laboratory (1954) suggests disregarding of salt removed from the soil in the harvested crop. Chhabra and Thakur (1998) of the Central Soil Salinity Research Institute in Karnal, India, mention that trees do not bio-harvest the salts and thus do not remove the salts from the soil. Heuperman (1999) mentions that the tree roots exclude salts during water uptake; the trees skim water off the top of the saturated part of the profile, causing the formation of a saltwater lens. The Nuclear Institute for Agriculture and Biology (1997) reports that when trees take up water, most of the dissolved salts remain in the soil. Although in high-salinity environments plant salt uptake might be negligible in relation to the salts present in the system, under low-salinity scenarios this might not be the case and salt balance by plant uptake and removal might be achievable. This option needs to be critically reviewed. Important aspects to be considered in the salt balance analysis are (i) mineral content in supply (irrigation or ground) water and (ii) mineral content in plant biomass (Rhoades et al. 1989).

Salt uptake/exclusion processes in plants

Water taken up by plants carries some soil solutes that, following transpiration, are eventually deposited in leaves so that salt in these leaves builds up gradually over time. For this reason older leaves generally have a higher mineral salt content than young leaves. The mineral content in plants largely depends on the species and (to a lesser extent) the mineral composition of the soil solution. The mineral content is maximum in leafy vegetables. The average dry weight content of minerals is around (1) leafy vegetables - 14 percent, (2) other vegetables - 8 percent, (3) roots and tubers - 6.5 percent, (4) pulses and legumes - 3.5 percent and (5) cereal grains - 2 percent (Shakya and Singh, 2010). Depending on prevailing climatic conditions, plants transpire from 30-70 times more water than they retain. Consequently any soil solutes not excluded by the roots, will end up in leaves in concentrations 30 - 70 times that of the water taken up at the root tips (Batra, 1988). For plants to achieve maximum growth rates, they should exclude most of the soil salts at their point of uptake. For example, if a plant is transpiring 40 times more water than it retains, it should admit only 1/40 or 2.5 percent of soil salt, and exclude the other 97.5 percent. If this was achieved, leaf salt concentration would stay comparable to soil salt concentration and the plant would survive indefinitely, provided salts remained compartment-alized. If salts were not excluded at all, shoot concentrations would soon be 40 times the external concentration (Abrol et al. 1973). All plants control Na and Cl uptake to some extent. Certain species excel and are able to maintain very low concentrations of Na and Cl ions in their sap flow from root to leaves. For example barley (a relatively salt tolerant plant) is a strong excluder, while lupin (salt sensitive legume) shows poor ability to regulate xylem salt concentration once the soil solution exceeds about 125 mM (7 250 mg/litre or EC of about 12 100 dS/m). *Atriplex* spp. (saltbush) are well known for their salt uptake capability, reports on salt uptake by halophytic *Atriplex* spp. as measured in field experiments under a range of saline conditions. Yields ranged from about 4 tonnes dry weight/ha (*A. lentiformis*) to nearly 10 tonnes dry weight/ha (*A. nummularia and A. undulata*) at a plant density of 10000 bushes/ha, after one-year establishment with fresh water irrigation and two seasons of saline irrigation. The highest yield of 10 tonnes dry weight/yr/ha was achieved with the highest applied irrigation salinity of 10,000 mg/litre. All treatments showed reduced yields in the third year of saline irrigation. Higher irrigation salinities resulted in higher Na and Cl concentrations in the leaf; K showed no trend with increased salinity and Mg^{++} and Ca^{+} leaf tissue concentrations actually decreased with increasing irrigation salinity. Most trees and shrubs are classified as non-halophytes; they show growth reduction with increased salt concentrations. Some trees and shrubs are halophytic; they commonly require some salt to achieve maximal growth. Whilst they exclude salts to a certain extent, these plants are much better adapted to managing salt accumulation in their leaves.

Salt storage in vegetation components

Halophytes survive in saline environments by absorbing salts. Most halophytes accumulate (relatively) large amounts of salt in their leaves. It is reported that *Atriplex nummularia* (Old Man Saltbush) grown near its optimum salinity of 200 mM NaCl (11 600 mg/litre or EC of about 19.5 dS/m) contains about 10 percent NaCl on a dry weight basis. The measured average yields of *A. nummularia* of 0.6 kg dry weight per plant per year across a range of applied irrigation salinities (100-10 000 mg/litre; NaCl-dominant water) over a three-year period. At planting densities of 10 000 bushes/ha this translated to 6 000 kg dry weight/ha/year. He also measured leaf ions for the range of applied water salinities for five *Atriplex* species, i.e. *A. ammenicola*, *A. cinerea*, *A. lentiformis*, *A. nummularia* and *A. undulata*. On average across the five species, leaf Na and Cl increased with increasing irrigation salinities, K showed no trend and Ca and Mg showed decreasing trends. With the average leaf component of dry weight production being 43 percent (ranging between 38 and 51 percent for the five *Atriplex* spp.), the salt export in the leaves would be between 350 and 433 kg/ha/year for irrigation salinities of 100 mg/litre and 10 000 mg/litre respectively, if all leaf matter was harvested and removed from the site. The annual irrigation application of about 10 million litres/ha applied 1 tonne/ha/yr and 100 tonne/ha/yr for respectively the low and high salinity treatment. This suggests that with the low salinity irrigation water the plants made a significant contribution to salt removal, but with the higher salinity values, salt balance control by vegetation was not possible. In this experiment only the salt uptake in the leaves was considered, probably to simulate grazing; this only marginally affects the total salt uptake as the plants' stems accumulate only minor quantities of salts (Shakya and Singh 2010).

Non-halophytic trees are salt excluders. The relatively small amounts of salts/nutrients that are taken up by trees are recycled by foliar drop. The only export of salts takes place through removal of timber during harvesting and through seepage losses where trees grow on deep water tables that cannot be accessed by roots. *Eucalyptus* species are not known for their salt-uptake capability. Marcar et al. (1990) discusses the results of a large number of bark and wood studies for Australian tree species grown in forest environ-ments. She found very high variability between species and between the woody components in the same tree (sapwood-heartwood-bark). However, the mineral ash content of the heartwoods (the major bulk of the timber removed in logging operations) was quoted as generally being less than 0.1 percent, with some species showing higher levels, e.g. *E. maculata* (4 percent). Mineral content in plants growing in the Indira Gandhi Nahar Project, Rajasthan, India, determined by the di-acid wet digestion method are presented in Table-1a and 1b. Stem-wood percentages for the tree spp. are of the same order of magnitude as mentioned by Lambert (1981).

Table 1a : Mineral (cations: Na^+, K^+, Ca^{++}, Mg^{++}) content in tree and crop biomass

Tree species	Tree component (% dry weight)						
	Stem bark	**Stem wood**	**Branch bark**	**Branch wood**	**Twigs**	**Leaves**	**Fruits**
Eucalyptus	5.8	1.8	6.6	2.4	2.7	2.2	3.2
Acacia nilotica	2.8	2.0	2.9	1.1	1.3	4.0	2.5
Ziziphus	3.8	2.1	3.5		2.6	4.3	
Prosopis cineraria	6.0	2.0	4.6		2.5	4.5	
Tecomella undulata	3.3	1.1	2.6	1.5	1.9	3.5	1.7
Delbergia sissoo	4.6	4.8	4.6	1.3	3.7	5.1	3.0
Azatirachta mica	1.5	1.7	2.3	0.9	3.3	3.6	

(Data from IGNP project, India.)

Table 1b : Crop component (% dry weight) in some plant

Crop	Crop component (% dry weight)				
	Stem	**Leaves**	**Grains**	**Twigs**	**Seeds**
Wheat	3.1	4.8	0.9	-	-
Groundnut	5.0	4.8	-	5.0	1.2
Mustard	-	-	-	-	1.5
Cow pea (guar)	-	-	-	-	7.7
Gram	-	-	1.8	-	-
Cotton	2.2	4.9	-	1.0	-

Consequences for biodrainage systems

Halophytes, with their high leaf salinities, on first sight seem to offer the best scope to achieve salt-balance through export of plant matter from sites. However, experimental field data do not support this. A yield of 1 500 kg dry matter per ha (NaCl-dominant environment) would translate to 150 kg of salt per ha if the foliage was removed from the site. If water applied to the site contained about 5 000 mg salts/litre, and the saltbush would use around 5 million litres/yr (conservative), 25 tonnes of salt would be added to the site and only 150 kg (or 0.6 %) would be exported; hardly a salt-balance scenario (Suresh and Shakya, 1993). A practical challenge would be the development of a technique to harvest the saltbush plants. Grazing would be the most convenient system although this would result in a return of most of the salts back to the site. Cut-and-carry systems that do not harm the productive capacity of the saltbush would have to be developed. The five species in the Schulz trial were harvested by cutting back at about 20 cm from ground level, using a flail mower. The bushes showed excellent re-growth during the first and second year. Yields declined in the third season and continuous heavy cutting back can be expected to result in dieback of the saltbush. For Eucalyptus plantations the prospects for salt balance in saline environments through harvesting and export are not much better. Following data from the highest recorded ash percent for the heartwood of 4 percent (for *E. rossii*), an annual growth increment of say 10 tonnes (high for saline environments) would result in an average salt removal through harvesting of 400 kg/ha per year. The data on sodium and chloride accumulation in different components of five and 22-year old *E. grandis* plantations. For five-year old plantations, 35 kg/ha NaCl was stored in total wood plus 17 kg/ha in bark; for 27-year old plantations the figures were 186 kg/ha in total wood and 170 kg/ha in bark. This means that over the 22-year period the uptake in both the wood and the bark was about 7 kg/ha/yr or a total of 14 kg/ha/yr for the combined components. These quantities are small in relation to the salt inputs in the plantations, even under some rainfed scenarios where annual inputs of sodium and chloride in coastal areas can be as high as 65 kg/ha/yr (e.g. Coffs Harbour, 1 km from the coast in eastern Australia (Thorburn and George 1999). Leaching would still have to take care of considerable salt export from under tree plantations grown at these sites. It also has to be noted that bark and leaves and twigs are normally not exported from plantation sites at harvesting.

Suitable plant, tree and shrub species

Selection of species for biodrainage purposes will depend on the environmental conditions for which they are planned. Salt tolerance will be an important criterion for (potentially) saline discharge environments, water use considerations will prevail in recharge control situations where salinity is of no concern and in channel seepage scenarios with low-salinity water supply.

***Eucalyptus* – the bio-drainer** : Tree like *Eucalyptus* can grow well under high water table conditions as compared to other forest species. Evapotranspiration from forest tree species is 2.5 to 3.0 times higher than agricultural crops. Some *Eucalyptus* tree species transpire higher than pan evaporation. Pan evaporation in arid climate is more than 2 m per year. Assuming a specific yield of 0.2 and evaporation of water from forest plantation area equivalent to pan evaporation, it works out that plantation in 10 ha area will lower the water table area by 1 m in 1000 ha annually. Appropriate drainage removes excess water from crop root zone and excess nutrient salts. A suitable soil environment is created for growth of plant roots which results in higher crop production on sustainable basis. This chapter is associated with the biodrainage systems and its implementation. In considering their long-term practicability, biodrainage systems will have to be subjected to the same analysis as other plant-based biological systems with regards to nutrient and salt-balance, soil considerations, etc. Moreover, as large-scale implementation of biodrainage crops could include sizeable tracts of land, socio-economic considerations will have to be taken into account when considering the support for this technique (Ram et al. 2007; Sharma et al. 1990).

Crop selection

For non-agricultural tree and bush species, reliable information is more difficult to obtain. Marcar et al. (1995) provide detailed information on the use of 30 tree species for use on salt affected land and less detailed summary descriptions for an additional 30 species. Ram et al (2007) provides comparisons for five saltbush species grown on a range of saline irrigation regimes and other authors have investigated water use of different tree species under a range of saline conditions (Kapoor, 2002).The water use of *Acacia nilotica*, *A. ampliceps* and *Prosopis pallida* on 3-5 year old plantation sites with contrasting soil and groundwater salinity in the Indus Valley in Pakistan. Annual water use by *A. nilotica* was 1248 mm on a severely saline site and 2225 mm on a moderately saline site. This was considerably higher than the annual rainfall, indicating that much of the water was taken up from the saline water tables underlying the sites (20 dS/m at 1-1.5 m below surface at the saline site and 1.5 dS/m at 2 m below surface at the moderately saline site). The other species used less water, this was considered to be a result of a lower planting density. The authors concluded that trees can evaporate large volumes of saline groundwater but they warn against the dangers of salt accumulation as observed under the trial sites. Rootzone salt-balance might be achievable in the long term, but at a much reduced tree water use and thus shallower water table.The Central Soil Salinity Research Institute (CSSRI) at Karnal, India, presents data on the tolerance of tree species to soil salinity as shown in Table-2 (Tomar and Gupta 1999).

Table 2 : Suitability of tree spp. for saline soils

Tolerant (EC_e 25-35 dS/m)*	*Tamarise troupii, T. artiaulata, Prosopis juliflora, Pithe cellobium dulce, Parkinsonia aculeata, Acacia farnesiana*
Moderately tolerant (EC_e 15-25 dS/m)	*Callistemon lanceolatus, Acacia nilotica, A. pennatula, A. tortilis, Casuarina glauca* 13144, *C. glauca* 13987, *C. obessa* 27, *C. glauca* (FRI), *C. equisetifolia (FRI), Eucalyptus camaldulensis, Leucaena leucocephala, Erescentia alata*
Moderately sensitive (EC_e 10-15 dS/m)	*Casuarina cunninghamiana* (FRI), *C. cunninghamiana* (Aust.), *Eucalyptus tereticornis, Acacia auriculiformis, Guazuma ulmifolia, Leucanea shannonii, Samanea saman, Albizzia caribea, Senna atomeria, Ferminalia arjuna, Pongamia pinnata*
Sensitive (EC_e 7-10 dS/m)	*Syzgium cumimi, S. fruticosum, Tamarindus indica, Salix app., Acacia deanei, Albizia quachepela, Alelia herbertsmithi, Ceaselpimia eriostachya, C. velutina, Halmatoxylon brasiletto*

* EC_e is the average root zone salinity as measured in a saturation extract

Reported salt tolerance data for the same tree species vary widely. Deep root systems make measurement of average rootzone-EC difficult and sometimes meaningless; the trees develop active roots in the least-saline part of the rootzone. For example 'moderately tolerant' is defined as EC_e 4-8 dS/m by Ram et al. (2007) but is reported as 15-25 dS/m by CSSRI Karnal.

Eucalyptus species are generally considered to be effective for biodrainage purposes. *Eucalyptus camaldulensis* is a hardy tree that grows under a wide range of climatic conditions and soil types. Some provenances of the species tolerate saline conditions quite well. They grow fast when good quality water is available. In a study in IGNP, Rajasthan, India, five-year old subirrigated plantations (channel seepage) produced dry biomass of 185 tonnes/ha. The utilizable biomass production was 29 tonnes/ha/year. *Acacia nilotica*, *Dalbergia sissoo*, *Tecomella undulata* and *Ziziphus mauritiana* are other species that have performed quite well in plantations along leaking canals in arid conditions. Species suitable for non-irrigated conditions are Acacia tortilis, Prosopis cineraria, Prosopis juliflora and Parkinsonia. Poplar and tamarix trees are also reported to perform well for biodrainage (Kapoor 2001).

Improvement of alkali soils by tree plantations

Alkali soils are characterized by high pH (8.2-10.5) and high exchangeable sodium percentage (ESP >15). High pH and ESP primarily occur as a result of the presence of measurable amounts of soluble sodium carbonates and bicarbonates in these soils. The tolerance of plants is governed by several factors such as the Na/K ratio of shoots and the capacity to take up K under strong Na competition, selective absorption or retention of cations (mainly Na in the foliage, roots and shoots), ability to excrete sodium through leaves, tolerance to oxygen stress in the rhizosphere and nutritional stresses. Available research information on the soil alkalinity tolerance of important fuelwood/timber and fruit trees as compiled by Gill et al., (1987) and Singh and Gill (1990) is presented in Table-3

Table 3 : Relative tolerance of tree species to soil alkalinity

Average soil pH in 1:2 soil water suspension	Fuelwood/timber tree species	Fruit tree species
>10.0	*Prosopis juliflora*	*Acchras japota*
	Acacia nilotica	
	Casuarina equisetifolia	
9.0-10.0	*Tamarix articulata*	*Zizyphus mauritiana*
	Terminalia arjuna	*Sapindus laurifolius*
	Albizzia lebbeck	*Emblica officinalis*
	Pongamia pinnata	*Carissa carandas*
	Sesbania sesban	*Psidium guajava*
	Eucalyptus tereticornis	*Phoenix dactylifera*
		Aegle marmelos

8.2-9.0	*Dalbergia sissoo*	*Punica granatum*
	Morus alba	*Pyrus persica*
	Grevillea robusta	*Pyrus communis*
	Azadirachta indica	*Vitis vinifera*
	Tectona grandis	*mangifera indica*
	Populus deltoides	*Syzygium cumini*

Experiments at the Central Soil Salinity Research Institute, Karnal, India, (Gill et al. 1984), have demonstrated that special tree establishment techniques in alkali soils require the planting of tree saplings in 0.15 - 0.25 m diameter, 1.2 - 1.8 m deep post-holes filled with either good non-alkali soil or alkali soil amended with 2-3 kg gypsum and 6-8 kg farm yard manure/compost per posthole. Use of gypsum is a must for achieving desired establishment (commercial survival rates) and boosting early growth of trees in the inhospitable alkali soil environments. Tree growth by itself initiates biological amelioration once a good canopy cover is developed. Significant decrease in soil pH, electrical conductivity, and increase in organic carbon and other nutrients (N, P, K, etc.) was observed to result in recycling of nutrients through litter production (Gill et al. 1987). Nitrogen-fixing trees do exceptionally well in this respect. Irrigation promotes root development and decreases soil sodicity. Pruning and lopping help tree growth. Close spacing of trees (1 x 1 or 2 x 2 m) produces less biomass yield per tree but more biomass per ha. Growing of trees has been reported to ameliorate alkali soils by improving physical, chemical and biological properties (Todd 1959).

Tree cropping systems management

Selection of species for biodrainage purposes will have to be based on currently prevailing and expected future environmental and economic conditions, rather than natural conditions as they existed before agricultural clearing and development. For example, in regions where water tables were deep before agricultural development, clearing and the introduction of irrigation have often resulted in the formation of shallow water tables. Under this scenario tree species selected for biodrainage should be able to survive and grow in shallow-water table environments, rather than in the pre-clearing deep-water table environment.

The management of forest plantations in saline waterlogged environments is described in detail by Thorburn and George (1999) and Marcar et al. (1990). Issues such as site preparation (deep ripping, liming, gypsum application, weed control), soil mounding, mulching and fertilizer application all need to be considered if tree vegetation is to be established successfully at sites with unfavourable characteristics. On sites with high soil salinities, (temporary) remedial drainage might have to be applied before biodrainage plantings can be established.On-going management issues, such as pruning and pest control are dependent on the purpose of the final product (e.g. timber or fuelwood) and local economic considerations. Generally selection of a mix of species rather than a monoculture tree plantation minimizes the risk of severe insect or disease attack.Animals and young trees do not mix well. Young seedlings might have to be protected from small native wildlife species such as rodents or rabbits. Individual tree guards might be required to avoid major losses. When animals are introduced for grazing/weed control at a later stage, close supervision is initially needed to prevent damage to bark.The planting of crops between widely spaced tree rows can be considered, especially at an early stage before competition for light becomes an issue. Crops such as wheat, mustard, lentil, berseem and gram or pasture species can be grown. However, the main objective of biodrainage plantations is the prevention of waterlogging and salinity management. The use of plantation plots for the growing of inter-crops may divert landholders' attention from the trees in favour of the crops and for this reason inter-cropping is not recommended until the water table and salinity conditions are fully and finally stabilized. In general it is recommended to assign a certain function to a certain area to avoid conflict of interests negatively affecting the biodrainage function.Monitoring of plant health and soil salinities at biodrainage sites is extremely important to be able to adjust management of perennial (tree) crops (e.g. by occasional irrigation, pruning or thinning to reduce water use) and avoid tree crop losses.

Biodiversity values

Landscape biodiversity is currently in the spotlight in many agricultural areas of the developed world. In the past, biodiversity was often a neglected issue, especially in irrigation areas. Lack of awareness and short-sightedness of previous generations of policy makers, regional resource managers and landholders has often resulted in landscapes with very little left of the original native vegetation and fauna. Conditions in agricultural environments often vary considerably from those of the original landscape. For example water table regimes are changed, there are higher water and/or nutrient inputs and intensive cropping practices impact on native flora and fauna. The "monopolizing" of the landscape by the agricultural sector alone is increasingly challenged by other landscape users, including promoters of more natural environment development. However, farming and landscaping may compatibly coexist.Adoption of biodrainage practices rather than conventional drainage can contribute to diversification in the agricultural landscape. However, often the tree crops selected for biodrainage will not be the same as the indigenous species for that locality as the agricultural practices will have changed the environmental conditions at the site.Monoculture blocks of land have an inherently poor biodiversity and if biodrainage plantings are designed for dual-purpose drainage/biodiversity purposes, they should incorporate a mixture of tree, shrub and grass species. Biodrainage plantings will often incorporate (or even be fully composed of) Eucalypt species. The ecological effects of *Eucalypts* are reviewed by Poore and Fries (1985). *Eucalypts* are often perceived as environmental 'bandits' because they are alleged to have adverse impacts on soil (nutrient status and encouraging erosion), on hydrology (drying up aquifers) and on the ecosystem (poor habitat for fauna). In their discussion on the impact of Eucalyptus on the water cycle, consider Eucalypts as another 'crop' and outside their natural ecological region, Eucalypts probably have little value as habitat providers. However, eucalyptus trees provide wildlife habitat in California and especially when, apart from trees, other salt-tolerant crops are introduced, the ecological value increases remarkably.

Production and Market

The adoption of new production systems, such as biodrainage plantations, by private landholders is to some extent linked to the market value of its end products, which in turn, depends on local conditions. In fact, the biodrainage plantation is needed to sustain high yields of the main commercial crops.In several parts of India farm forestry has shown phenomenal growth rates in the past and good economic returns. Recently, however, market prices have collapsed because of the lack of marketing facilities. Obviously market research should form the basis of any large-scale move towards the production of new crops, including trees. However, the situation might be completely different for rural communities in developing countries where other economic drivers apply and where fuelwood production can be a very profitable activity (Rao et al. 1994).Economic evaluation studies were carried out on watershed management projects in India having important forestry components. In a World Bank survey (1981-1988) near Kandi, Punjab, the internal rate of return for different components was: forestry 18.2 percent, animal husbandry 15.6 percent, soil conservation 10.8 percent, horticulture 25.3 percent and irrigation minus 1.22 percent. In another study by the Indian Agriculture Finance Corporation (1988) in the catchment of the Matatila reservoir, the benefit/cost ratios for different investment sectors were assessed as: soil conservation 1.68:1, water storage structure 1.84:1 and forest/tree plantations of *Eucalyptus*, *Acacia catechu* and *Dalbergia sissoo* of 4.48:1. Similar results were obtained on several research farms in India.Timber yields of trees grown under saline conditions can be reasonably high, e.g. *E. occidentalis* grown with 9 dS/m irrigation water on a saline site in northern Victoria, Australia, showed annual growth increments of nearly 17 m^3 during its fourth year of saline irrigation. However, because of poor tree shape, the timber was not suitable for anything except fuelwood. Breeding efforts to improve form and timber quality have traditionally focused on high yielding species in more benign environments. There is scope for improvement in this area (Todd 1959).

Biodrainage and wetlands

Wetlands are areas that are covered with water for at least part of the year to a depth of less than two metres. The values of wetlands are now widely recognized as reserves of native plants and wildlife, water quality improvers (nutrient filters) and flood-protection buffers. In their natural environments, the evapotranspiration of the vegetation of the wetland system is in balance with the through-flow and seepage fluxes. Where either the vegetation or the watering regime is changed, these finely tuned balances are disturbed and habitat changes are inevitable (Silberstein et al., 1999).Water balance changes in wetlands are mentioned by Poore and Fries (1985) who state that tree plantations, especially Eucalypts, have been used to lower water tables in swampy areas, either to dry out the soil or to control mosquitoes. Obviously this practice clashes with the management of wetlands for ecological values.

Biodrainage and urban landscapes

Urban salinity is an issue associated with rising water table levels. Water use in arid urban environments can be high and leakage from water pipes substantial. Urban areas often have relatively good surface drainage but subsurface drainage is rarely considered. Only in places where groundwater is pumped for water supplies is subsurface drainage provided. This practice is not feasible in urban landscapes overlying saline water tables. In some cases, old and leaking sewage pipe systems might have been providing some form of salt-balance by intercepting shallow saline groundwater, however, where these old systems are being upgraded and leaks fixed, serious waterlogging and/or salinity problems can be expected to develop (Tomar et al. 1994; 1998).In Australia, urban salinity has been identified as a serious problem. The city of Wagga Wagga in eastern Australia suffers from extensive damage to old brick structures (Thorburn and George 1999). The strategies involved in managing this problem include biodrainage through planting of deep-rooted vegetation species adapted to the local arid climatic conditions, rather than the commonly used imported exotic species requiring irrigation to survive.

Socio-economic considerations

Commercial tree plantations are grown to yield economic returns. Plantations developed for biodrainage potentially offer an extra protection benefit, free of cost, and improve the regional economy. However, looking at the level of the individual landholder who needs annual returns to survive, the aim should be to produce economic returns comparable at least to those from agriculture.The planting of large areas of new crops, especially tree crops, will have a significant impact on regional economies and social structures. Ram et al. (2007) discusses these issues in a working paper based on two case studies in South India and Thailand. During the 1980s, social protests developed in Asia (and elsewhere in the world) over alleged adverse effects of Eucalypts on soil nutrient status, soil water depletion, soil erosion and wildlife. The biodrainage system can have significant advantages in the rural environment of many developing countries, especially when there are new projects and the choice is between (1) biodrainage, (2) installation of a conventional subsurface drainage system using a closed horizontal pipe drainage system, by open drains, or by vertical drainage, and (3) a combined system of biodrainage and conventional drainage. Often the rural livelihood benefits because biodrainage provides for fuel wood, fruits, timber, fodder, windbreak, flora and fauna, biodiversity, pollutant removal.

Potential and limitations of bio-drainage (Kapoor 2003)

Bio-drainage, like other drainage measures, has potential and limitations. All irrigation regions are not alike. They differ in physiographic and climatic conditions. Bio-drainage may be very suitable, partly by suitable or unsuitable depending upon the prevailing conditions. The important characteristics of a region that would govern whether bio-drainage is appropriate or not are whether the region is humid or dry and what is the salinity level of the ground water. Brief description on four possible regional scenarios follows:

Humid region with ground water of good quality

Generally, humid regions have soils and ground water of good quality. The natural precipitation washes down the salts in soils which are drained away naturally. The top soils and ground water are of a reasonably good quality to enable irrigated agriculture. Natural precipitation may not occur uniformly throughout the year to match with the crop water requirements and therefore supplemental irrigation may be needed. This can be done by storing surface water and making it available for irrigation when needed or by pumping ground water and using it. In case, use of stored surface water for irrigation results in rise of ground water table and there is a threat of waterlogging the best course would be to make conjunctive use of ground and surface water. This would enable best possible use of the water resource and at the same time check the rise of the ground water table. However, if this is not possible for some reasons, like abundance of surface water or non-availability of dependable power for pumping ground water (a situation that is common in developing countries), then plantation of trees on part area can provide the needed drainage.

Humid regions with saline ground water

There are very few regions which receive good precipitation and yet have saline ground water. This condition can occur in localised pockets where the ground sub-strata have salt incrustations. The infiltrating water dissolves and carries the salts to the ground water table. In a situation like this, it should be possible to make conjunctive use of surface and ground water. Bio-drainage through tree plantations can also be a viable option.

Semi-arid and arid regions with ground water of good quality

In dry regions, water is generally scarce and irrigation is often practised by bringing surface water from outside regions. Recharge from irrigation disturbs the previously existing ground water balance and the ground water table (often perched water table) starts rising. The ground water can be pumped and put to conjunctive use along with the surface water for irrigation. This should be the best way to prevent waterlogging. However, there are many examples of excessive use of surface water for irrigation and little use of ground water inspite of its good quality. The reasons of this anomalous position are surplus availability of surface water during early operative years of a project or low cost of surface water to the farmer in comparison to the cost of ground water. The situation should be remedied by appropriate management measures (Verma et al. 1988). The second best alternative to prevent or over come the threat of waterlogging should be to plant trees on part area to provide bio-drainage (Westcot, 1988).

Semi-arid and arid region with saline ground water

This is the commonly prevailing situation in most semi-arid and arid regions. The need of irrigation water is maximum in such regions. Water is the scarce resource. Land, mostly waste land, is in abundance. Waterlogging and secondary salinisation is a common problem when lands are brought under irrigation by bringing water from outside sources. Horizontal sub-surface drainage would be the most suitable and effective measure, where saline drainage water can be safely disposed off into the sea or in evaporation tanks at affordable cost without polluting natural water bodies and/or causing environmental problems. Conjunctive use of ground water can be possible to a limited extent (Yadav and Pathak 1967). Bio-drainage can be a feasible option. Availability of land should not be a problem. Transpiration rates are quite high. The limitation would be the salinity level of the ground water. Where the salinity in ground water at shallow depths reaches 12 dS/m, the transpiration rate falls down to about one-half. If and when it reaches to a level of say 25 dS/m, the tree plantations may become ineffective. The position of potential and limitations of bio-drainage and relative suitability with other drainage measures is summarised in the following table:

Table 4 : Potential and limitations of biodrainage and relative suitability of drainage measures. (Kapoor, 2003)

Characteristics of the region	Potential and limitation of bio-drainage			Relative ranking of drainage measures	
	Feasibility	Merits		Demerits/ Limitation	Region with surplus water
Humid region with ground water of good quality	Feasible	Least cost, environment friendly	Land needed for tree plantation	1.Bio-drainage 2.Horizontal sub-surface drainage 3.Vertical drainage	1.Vertical drainage with conjunctive use 2.Bio-drainage 3. Horizontal sub-surface drainage

Humid region with ground water of poor quality	Feasible where salinity of ground water is not excessive (say, more than 12 dS/m)	Least cost, environment friendly	Land needed for tree plantation	In coastal region 1.Horizontal sub-surface drainage 2.Bio-drainage 3.Vertical	In inland regions 1.Bio-drainage 2.Vertical drainage 3.Horizontal Sub-surface drainage
				1.Vertical drainage with conjunctive use. 2.Bio-drainage 3.Horizontal sub-surface drainage.	
Semi arid and arid regions with ground water of good quality	Feasible	Least cost, environmental friendly	Hardly any	In coastal regions	In inland regions
Semi arid and arid regions with ground water of poor quality	Feasible where salinity of ground water is not excessive (>12dS/m)	Least cost, environment friendly	Natural river water (unpolluted) is used for irrigating crops	1.Horizontal sub-surface drainage 2.Bio-drainage 3.Vertical drainage	1.Bio-drainage 2.Vertical drainage 3.Horizontal sub-surface drainage

Conclusions

Although the term biodrainage is relatively new, the concept is not novel. The drainage of waterlogged land through the introduction or modification of vegetation has been practiced historically in different parts of the world. The increasing concern about the environmental impact of intensive agricultural development and associated monocultures has resulted in increased interest in the use of trees and salt-tolerant crops as an integrated part of the landscape and farming on saline land. Biodrainage combined with conventional drainage systems coalesce positive and negative aspects as stated above: they are more positive and less negative than when only making use of conventional drainage systems.

Bio-drainage can be a practicable option for regulating waterlogging and salinity in irrigated lands. Its main qualities are economy in cost and environment friendly. The restrictions are requirement of land for tree plantations, limited evacuation of salts from the system and weakness of trees to high saline conditions. The requirement of about 10% of the area of land for tree plantations should be problematic, particularly in semi-arid and arid zones. All biomass have some minerals which are evacuated, when the crops are harvested and removed from the field. When natural river water (unpolluted) is used for irrigation, the mineral evacuation by biomass may equal the net import of minerals with the irrigation water, enabling a reasonable salt balance. There are many species of trees that are salt tolerant and grow adequately upto salinity levels of 12 dS/m. There are relatively large irrigated areas that have ground water salinity of less than 12 dS/m, Biodrainage should be practicable in such areas.

References

Abrol, I. P., Dargan, K. S. and Bhumla, D. R. (1973). Reclaiming Alkali Soils, Central Soil Salinity Research Institute, Karnal, India.

Akram, S. (2006). Sensitivity analysis of bio-drainage systems on salinity using SAHYSMOD model. *Proceedings of the Fourth Technical Workshop on Drainage*, IRNCID (in Farsi)

Anon. (2003). Biodrainage: status in India and other countries, Indian National Committee on Irrigation and Drainage, New Delhi, pp. 1–47.

Anonymous (2009). Anti-waterlogging project, Department of Irrigation and Drainage, Punjab Government.

Batra, L. (1988). Performance of Casuarina species, their nodulation pattern and nitrogen fixation. *Annual Report, CSSRI*, Karnal, India, pp 122-124.

Boonstra, J. and de Ridder, N. A. (1990). Numerical modeling of groundwater basins, 2nd edn. ILRI Publication 29, Wageningen, p 226

Boumans, J. H., Van Hoorn, J. W., Kruseman, G. P. and Tanwar, B. S. (1988). Water table control, reuse and disposal of drainage water in Haryana. *Agric. Water Manage.*, 14, 537–545.

Chhabra, R and Thakur, N. P. (1998). Biodrainage using tree to control water logging and secondary salinization in canal irrigated areas. *Proceeding of the National Conference on Salinity Management in Agriculture*, Central Soil Salinity Research Institute, Karnal, India

Chhabra, R., Abrol, I.P. and Chawla, K.L. (1987). Tolerance and productivity of Sesbania species for fuel wood production in sodic soil. *Proceedings of International Symposium on Afforestation of Salt Affected Soils*, CSSRI, Karnal, India, pp 33-47.

Dagar, J.C., Singh, G. and Singh, N.T. (2001). Evaluation of forest and fruit trees used for rehabilitation of semiarid alkali soils in India. *Arid Soils and Rehabilitation*, 15: 115-131.

Denecke, H. W. (2001). Concept note on biodrainage and bio-disposal systems. Available at http://www.fao.org/iptrid/drainage/bio_drainage/concept-note-bio-drainage.htm

Gill, H.S. (1985). Studies on the evaluation of selected tree species for their tolerance to sodicity and mechanical impedence in a highly sodic soil with particular reference to root growth behaviour. Ph.D. Thesis (unpublished), Kurukshetra University, Kurukshetra, India.

Gill, H.S., Abrol, I.P. and Samra, J.S. (1987). Natural cycling through litter production in young plantations of Acacia nilotica and Eucalyptus terticornis in a highly alkaline soil. *Forest Ecology and Management*, 22: 57-69.

Gill, H.S., Grewal, S.S. and Abrol, I.P. (1984). Performance of tree species in highly sodic soils. *Proceedings of the Seminar on Soil Resources and Productivity Management*, Indian Society of Soil Science, IARI, New Delhi, India, pp 152.

Govt. of Haryana (1998). Master Plan for Management of Waterlogging and Salinity.

Grattan, S. R., Shennan, C., May, D. W., Mitchell, J. P. and Barau, R. G. (1987). Use of drainage water for irrigation of melons and tomatoes. *Clif. Agric.*, 41, 27–28.

Grewal, S.S. (1984). Studies on rain water conservation for the establishment of an agroforestry system on sodic soil. Ph.D. Thesis (unpublished), Kurukshetra University, Kurukshetra, India.

Gupta, S.K., Sharma, D.P., Tyagi, N.K. and Dubey, S.K. (2002). Lavaniya-Kshariya Mridaon ka Sudhar aur Prabandh. CSSRI, Karnal, p 153.

Heuperman, A. F. (2000). Biodrainage: an Australian overview and two Victorian case studies. Proceedings of the 8th ICID International Drainage Workshop, New Delhi, India.

Heuperman, A. F. and Kapoor, A. S. (2002). Biodrainage: Principal Experiences and Applications, IPTRID, FAO, Rome, , pp. 1–79.

Heuperman, A.F. (1999). Hydraulic gradient reversal by trees in shallow water table areas and repercussions for the sustainability of tree-growing systems. *Journal of Agricultural Water Management*, 39: 153-167.

IPTRID/FAO (2002). Bio-drainage - Principles, Experiences and Applications. Knowledge Synthesis Report No. 6, International Programme for Technology and Research in Irrigation and Drainage, May, 2002.

Kapoor, A. S. (2002). Biodrainage: A Biological Option for Controlling Waterlogging and Salinity, Tata McGraw-Hill, New Delhi, , pp. 1–332.

Kapoor, A. S. (2003). Biodrainage- Potential and Limitations. Paper No 061. Presented at the 9th International Drainage Workshop, September 10–13, 2003, Utrecht, The Netherlands

Kapoor, A. S. and Denecke, H.W. (2001). Biodrainage and biodisposal: the Rajasthan experience. In: *GRID, IPTRID's network magazine* no. 17:284

Khan, M.A.W. and Yadav, J.S.P. (1962). Characteristics and afforestation problems of saline alkali soils. Indian Forester, 88: 259-271.

Kumar, R. (2004). Groundwater Use in North- West India (eds Abrol, I. P. et al.), Centre for Advancement of Sustainable Agriculture, New Delhi, pp. 1–26.

Kumar, R. and Singh, J. (1994). Drainage system for groundwater management. *In Proceedings of the National Seminar on Reclamation and Management of Waterlogged Saline Soils*, CSSRI, Karnal, 5–8 April.

Marcar, N.E., Leppert, P.M., Humphreys, E., Muirhead, W.A. and Lelij, A.V. (1990). Salt and waterlogging tolerance of frost resistant *Eucalyptus*. *Proceedings of the symposium* held at Albury, New South Wales, Australia, September 18-20, 1989, pp 373-374.

Marcum, K.B. and Murdoch, C.L. (1994). Salinity tolerance mechanisms of six C4 Turfgrasses. *Journal Amer. Soc. Hort. Sci.* 119(4): 779-784

Minhas, P. S. and Gupta, R. K. (1993). Conjunctive use of saline and non-saline waters. 1. Response of wheat to initial salinity profiles and salinization patterns. *Agric. Water Manage.*, 23, 125–137.

Minhas, P.S. (2001). Evaluating tree plantations for control of salinity and water tables (Final Report), funded by Department of Wasteland Development, Ministry of Rural Development (Govt. of India), New Delhi, Coordinating Unit, AICRP Saline Water Use, CSSRI, Karnal, India, p 105.

Poore, M. E. D. and Fries, C. (1985). Influence on water cycle. In: The ecological effects of eucalypts. *FAO Forestry Paper* No. 59, p 26.

Ram, J., Garg, V. K., Toky, O. P., Minhas, P. S., Tomar, O. S., Dagar, J. C. and Kamra, S. K. (2007) Biodrainage potential of *Eucalyptus tereticornis* for reclamation of shallow water table areas in north-west India, *Agroforest Syst.* 69:147–165.

Rao, K. V. G. K., Sharma, S. K. and Kumbhare, P. S. (1994). Drainage requirements of alluvial soils of Haryana. In : *Proceedings of the National Seminar on Reclamation and Management of Waterlogged Saline Soils*, CSSRI, Karnal, Haryana, India, 5–8 April.

Rhoades, J. D., Bingham, P. T., Letey, J., Hoffman, G. J., Dedrick, A. R., Pinter, P. J. and Replogle, J. A. (1989). Use of saline drainage water for irrigation. Imperial valley study. *Agric. Water Manage.*, 16: 25–36.

Shakya, S. K. and Singh, J. P. (2010). New drainage technologies for salt-affected waterlogged areas of southwest Punjab, India, *Current Science*, Vol. 99, No. 2, 25 July 2010.

Shakya, S. K., Gupta, P. K. and Kumar, D. (1995). Innovative drainage techniques for waterlogged sodic soils. A bulletin published by AICRP operated in the Department of Soil and Water Engineering, PAU, Ludhiana, sponsored by ICAR, New Delhi.

Sharma, D. P., Singh, K. N., Rao, K. V. G. K. and Kumbhare, P. S. (1990). Response of wheat to the reuse of saline drainage water in sandy loam soil. *Indian J. Agric. Sci.*, 60: 448–452.

Silberstein, R.P., Vertessy, R.A., Morris, J. and Feikema, P.M. (1999). Modelling the effects of soil moisture and solute conditions on long-term tree growth and water use: a case study from the Shepparton irrigation area. *Australian Journal of Agricultural Water Management*, 39: 283-315.

Singh, G.B. and Gill, H.S. (1990). Raising trees in alkali soils. Wastelands News. 6: 15-18.

Singh, G.B., Abrol, I.P. and Cheema, S.S. (1989). Effect of gypsum application on mesquite and soil properties in an abandoned sodic soil. *Forest Ecology and Management*, 29: 1-14.

Singh, J. (1985). Ground water movement in irrigated areas and its management through hydro-chemical stratification approach. *Irrig. Power J.*, 42: 259–263.

Singh, N.T. (1992). Land degradation and remedial measures with reference to salinity, alkalinity, waterlogging and acidity. In: *Natural Resources Management for Sustainable Agriculture and Environment* (Ed. D.L. Deb), Angkor Publications, New Delhi, p 442.

Smedema, L. K. (1997). Biological drainage: myth or opportunity? GRID, magazine of the IPTRID network,Issue 11, p 3

Suresh, R. and Shakya, S. K. (1991). Steady flow towards multiple well point system in leaky confined aquifer. *J. Agric. Eng.* (Special issue), 57–60.

Theiveyanathan, V and Benyon, R. G. (2000). Planting trees for sustainable rice production. In: Onwood; research updates from Commonwealth Scientific and Industrial Research Organization (CSIRO) Forestry and Forest Products, No 31, 5 pp.

Thorburn, P. J. and George, R. J. (1999). Interim guidelines for revegetating areas with shallow, saline water tables. In: *Thorburn PJ (ed) A report on a workshop held on 28 and 29 May* (1997) in Australia as a joint venture agroforestry program on agroforestry over shallow water tables/the impact of salinity on sustainability. Water and Salinity Issues in Agroforestry No. 4. RIRDC Publication No. 99/36, pp 12–17

Todd, D. K. (1959). Ground Water Hydrology, John Wiley and Sons.

Tomar, O.S. and Kumar, S. (1999). Alternate land use systems: Evaluation of multipurpose forest tree species for firewood, forage and timber production in highly alkali soils. *Annual Report*, CSSRI, Karnal, pp 19-20.

Tomar, O.S., Gupta, Raj K. and Dagar, J.C. (1998). Afforestation techniques and evaluation of different tree species for waterlogged saline soils in semiarid tropics. *Arid Soil Research and Rehabilitation*, 12: 301-316.

Tomar, O.S., Minhas, P.S. and Gupta, Raj K. (1994). Potentialities of afforestation of waterlogged saline soils. In: Agroforestry Systems for Degraded Lands (Eds. P. Singh, P.S. Pathak and M.M. Roy). Oxford and IBH Publishing Co. Pvt. Ltd. New Delhi, pp 111-120.

Verma, R. S., Shakya, S. K. and Singh, S. R. (1988). Design of multiple well point system and cost estimation. *Indian J. Power River Valley Develop.* XXXVIII: 199–201.

Vertessy, R., Connel, L., Morris, J., Silberstein, R., Heuperman, A. F., Feikema, P., Mann, L., Komarzynski, M., Coollopy, J. and Stackpole, D (2000). Sustainable hardwood production in shallow water table areas. Joint venture agroforestry programme. *Water and Salinity Issues in Agroforestry No. 6. RIRDC Publication* No. 00/163, p 105

Westcot, D. W. (1988). Reuse and disposal of higher salinity subsurface drainage water – A review. *Agric. Water Manage.*, 14: 483–511.

Yadav, J.S.P. and Pathak, T.C. (1967). Study of saline and alkaline soils and their possible improvement by afforestation. *Journal Soil and Water Conservation*, India, 5: 24-29.

□□□

Green Agriculture : Newer Technologies, 2012
© Kambaska Kumar Behera (ed.), pp. 143-195
New India Publishing Agency, New Delhi (India)
e-mail : info@nipabooks.com; website : www.nipabooks.com

Chapter-7

Sustainable Crop Production in Rainfed Areas of Central India

S. D. Billore and O. P. Joshi
Directorate of Soybean Research, Khandwa Road, Indore-452 001
E mail- billsd@rediffmail.com

SUMMARY

At present, Indian agriculture faces a severe challenge for the future in view of growing demand for food for meeting the needs of nation's ever growing population. At the same time, resources are gradually shrinking and the productivity of some resources is already being utilized threatening the sustainability and leading to environmental degradation which necessitates judicious and sustainable management of natural resources, particularly soil and water. Of 142 m ha arable land, nearly 96 m ha (67%) is devoted to rainfed farming, which contributes about 44 % of the total production. Overexploitation of water resources, unequal access to water and adverse impacts on ecosystem services are among the major issues impacting sustainable use of water resources. Crop production in rainfed areas is generally affected by *chronic* drought, *ephemeral* drought, and *apparent* drought. In order to increase production, farmers have two options, either to use extensive systems (horigental) or intensive systems (vertical). There is a meager scope for extensive system in India and the remaining intensive system is the only option for the farmers as well as policy makers. Hence, optimizing soil water management is crucial to enhancing agronomic productivity and meeting the food needs of the growing population with rising standards of living under rainfed area. The scientifically proven technologies like *ex-* or *in-situ* moisture conservation and adoption of integrated approach for nutrient and pest management are available to cope up with the aberrant weather conditions in rainfed areas and found economically viable for farmers. Further, in future, the development of drought and insect pest resistance through transgenic breeding, farmer's participatory approach and precision farming will be helpful for increasing the productivity under rainfed area.

Introduction

In spite of decreased agricultural growth in the country during past decade, it has been sufficient to move the country from severe food crises of the 1960s to aggregate food surpluses today. At present, India's agricultural sector, however, faces severe challenges for the future in view of growing demand for food for meeting the needs of its ever growing population. Despite sizeable available national food stocks, widespread poverty and hunger still remain as the agricultural and national economic growth has not adequately benefitted disadvantaged regions and the poor. The demand for basic staples, non-food grains, and exports is consistently increasing. At the same time, resources are gradually shrinking and the productivity of some resources already being utilized threatening the sustainability and leading to environmental degradation. Growth in total factor productivity is reported to have declined slightly in major crops. Although, the returns to investment in agricultural research and rural infrastructure are reported to be comparatively high, but these investments remain low and fails to express in agricultural growth.

The growing demand for food in India, due to rising population and rapid industrialization, necessitates judicious and sustainable management of natural resources, particularly soil and water. The low water- and fertilizer-use efficiency added with their un-thoughtful use is resulting in land degradation, deterioration in soil quality and increased pollution of surface and ground waters. With increasing competition for soil and water for non-agricultural uses (e.g., urbanization, industrial use), the appropriate strategy shall be to increase production per unit input of scarce resources through agricultural intensification, and by reducing losses, adoption integrated nutrient management system and optimizing the input-use efficiency. Strategies of soil water management in dry farming regions of the country encompasses surface water conservation, water harvesting and recycling, and recycling of crop residue, and extensive use of mulch and bio-solids. Soil water, being a precious and scarce resource, must be used prudently and responsibly. Undervaluing any scarce resource can lead to its misuse, overexploitation, depletion, pollution, and contamination and eutrophication (Lal 2009).

India's arable land resources are excellent, and are considered second only to those of the United States. The country is also endowed with a range of climates that permit production of diverse crops round the year in one region or another. Renewable fresh-water resources of India are adequate to vastly expand the irrigable land area from 60 M ha now to the largest in the world (Lal 2009). Indian farmers, with a remarkable track record of quadrupling grain production across four decades (1965 to 2005), are receptive, hardworking and innovative but resource poor. With abundance of natural and human resources, the country has the potentials to feed the world. Yet, Indian agricultural production has stagnated since the 1990s, farmers are demoralized to the extent of committing suicide by some, the country's grain reserves are frequently depleted, and India has imported wheat several times during the 2000s.

India's population of slightly above 1 billion at present is expected to stabilize somewhere at 1.65 billion by 2050, making it by then the most populous country in the world. The increasing population coupled with the standard of living is leading to hike in the demand for cereals and other food products. It has been predicted that the cereal grain consumption (e.g., wheat, rice, maize, sorghum, millet) is likely to increase from 220 million tons (Mt) in 2007 to 260 to 300 Mt by 2020 (Kumar, 1998; Bhalla *et al.*, 1999), necessitating increases in yields of cereals like rice and wheat by 56 % and 62 %, respectively (Paroda et al.1994). Yet, the grain yield of rice (*Oryza sativa*) and wheat (*Triticum aestivum*) is either declining or stagnant. The yield of wheat declined and that of rice stagnated between 1999 and 2006. Grain yields of maize (*Zea mays*), sorghum (*Sorghum bicolor* L.) and millet (*Panicum miliare* L.) are low and have been stagnant since 1998–99. Yields of soybean (*Glycine max*), chickpea (*Cicer arietinum*), safflower (*Carathamus tinctorius*), rapeseed (*Brassica campestris*), and groundnut (*Arachis hypogea*) are among the lowest in the world. Indeed, yields of most crops in India have stagnated or declined during the 2000s because of a range of interactive biophysical and social/ economic factors. Important among biophysical factors are low use efficiency of water, nitrogen (N), and energy-based inputs because of depletion of soil organic carbon (SOC) pool and the overall decline in soil quality. The rice-wheat system (RWS) is the dominant production system in the Indo-Gangetic Basin (IGB) (Timsina and Connor, 2001). Decline in its productivity is occurring in conjunction with either decline in ground water or rise in water table and soil salinity (canal irrigation) because of the widespread use of improper irrigation practices (e.g., flood irrigation, pre-monsoon planting of flooded rice). Soil degradation, exacerbated by depletion of SOC pool and the nutrient reserves (N, P, K, and micronutrients), is another factor responsible for declining productivity (Bhandari et al., 2002; Singh et al. 2004). Depletion of SOC and nutrient reserves is partly due to removal of wheat straw for fodder (Samra *et al.*, 2003), burning of wheat/rice straw, and use of animal dung for cooking. Yet, sustainability of agricultural production in the IGB (northwest region of Punjab and Haryana) is important because the region contributes about 52 % of India's food production (Abrol 1999).

Overexploitation of water resources, unequal access to water and adverse impacts on ecosystem services are among the major issues impacting sustainable use of water resources throughout the developing countries (Vincent, 2003). The overall water-use efficiency of irrigated crop production in India is about 30 % (Sharma and Rao 1997). Groundwater, supporting more than 50 % of the total irrigated area in India, is being depleted due to overdraft and consequently resulting in falling water table and dry wells.

Eight hundred million people are food-insecure, and 166 million pre-school children are malnourished in the developing world. Producing enough food, and generating adequate income in the developing world to better feed the poor and reduce the number of those suffering is going to be a great challenge. This challenge is likely to further intensify, with a global population that is projected to increase to 7.8 billion in 2025, putting even greater pressure on world food security, especially in developing countries where more than 80 % of the population increase is expected. Irrigated agriculture has been an important contributor to the expansion of national and world food supplies since the 1960s, and will continue to play a major role in feeding the growing world population. However, irrigation accounts for about 72 % of global and 90 % of developing country water withdrawals, and water availability for irrigation may have to be reduced in many regions in favor of rapidly increasing non-agricultural water uses in industry and households, as well as for environmental purposes. Out of concern over increasing water scarcity for irrigation, the role of water management and investments for irrigated agriculture and food security has received substantial attention in recent years (Hofwegen and Swendsen 2000; Rosegrant 1997). However, rainfed areas currently account for 58 % of world food production. Adequate attention is warranted to exploit potential of production growth in rainfed regions, which can play a significant role in meeting future food demand.

The 'Green Revolution' in mid-1960s, though turned out to be a boon to Indian agriculture, ushers in an era of disquieting disparity between productivity of irrigated and dry land agriculture. Most of the increase in agricultural output over the years has taken place under irrigated conditions. In view of limited opportunities for continued expansion of irrigated area, Indian planners increasingly diverting their attention to rainfed, or unirrigated agriculture to help meet the rising demand for food projected over the next several decades. Despite the historic bias in favor of irrigated agriculture in terms of research and infrastructural investments, rainfed agriculture has always been an important part of the agricultural sector. This is obvious from the fact that rainfed agriculture accounts for about two-thirds of total cropped area, nearly half of the total value of agricultural output and hundreds of millions of poor rural people depend on rainfed agriculture as the primary source of their livelihoods.

Rainfed areas are highly diverse, ranging from resource-rich with good agricultural potential to resource-poor with much more restricted potential. Some resource-rich rainfed areas potentially are highly productive and already have experienced widespread adoption of improved seeds. In drier areas, on the other hand, productivity growth has lagged behind, and is characterized with widespread poverty and degradation of natural resources. Although, the rainfed agriculture is suggested to receive greater emphasis in public investments, the key issue will be to ascertain the allocation among different types of rainfed agriculture. Agricultural growth in these areas will be essential for reducing poverty and environmental problems in future.

There is a need to identify the opportunities for stimulating agricultural growth and reducing poverty and environmental degradation in rainfed areas. Likewise, there is a need to assess the opportunity costs of diverting scarce public resources from resource-rich to resource-poor areas. The tradeoffs between investing in resource-rich and resource-poor areas in terms of their productivity, poverty and environmental outcomes need to be understood in order to guide public policy decisions toward productive outcomes.

1. Rainfed Agriculture Scenario

Of 142 mha arable land, nearly 96 mha (67%) is devoted to rainfed farming. Approximately 76 mha is under irrigation. As predicted by The National Commission on Agriculture in 1976 that even if the full irrigation potential is tapped by 2013 AD, over 50 % of the arable land would continue to remain rainfed in the foreseeable future. So far much of the agricultural growth achieved in past decades occurred in irrigated areas. The potential for additional production gains in these areas may lessen with time from inherent problems. It is in the rainfed belt where cultivation of coarse cereals (91%), pulses (91%), oilseeds (80%) and cotton (65%) predominates. About 44 % of the total production is being contributed by rainfed region. Rainfed agriculture supports 40 % of country's population. The rainfed areas are increasingly being warranted to help meet the rising demand for food, pulses, oilseeds, feed, fuel, fruits, vegetables etc. Thus, the country's economy largely depends on a sustained increase in the productivity from dry lands.

Dry lands are generally defined in climatic terms as lands with limited rainfall. Dry lands are characterized by low (100-600 mm annually), erratic and highly inconsistent rainfall levels. The main characteristic of *dryness* is the negative balance between annual rainfall and evapo-transpiration rates. Rainfall is scarce, unreliable and concentrated during a short rainy season, with the remaining period tending to be relatively dry. High temperatures during the rainy season cause much of the rainfall to be lost in evaporation, and the usual intensity of storms ensures that much of the rainfall runs off in floods. Water is not only meager in absolute terms, but also scarce for natural and human uses. The traditional typology of the Food and Agriculture Organization of the United Nations (FAO) is based on agro-climatic zones according to the length of the growing period, which is defined as "the period when both water and temperature permit crop growth". As defined by FAO, dry lands are zones falling between 1-74 and 75-119 growing days, representing arid and semi-arid lands, respectively. Others have adopted a practical working definition of dry lands, namely, "anywhere that rainfall is a problem because of amount, distribution and unreliability".

FAO Classification of Dry lands based on length of growing period in days/year (FAO 1978)

Growing period (days/year)	Types of dry land
1-74	Arid
75-119	Semi-arid
120-180	Dry sub-humid

Rainfed regions in India encompass a wide range of soil and rainfall conditions.

1.1. Soil

Soils of these regions belong to alluvium (Entisols and Inceptisols), red soils (Alfisols, Oxisols and Ultisols), black soils (Vertic Inceptisols and Vertisols), submontane soils (Entisols, Inceptisols and Mollisols), and sierozemic soils (Aridisols). Among these, red (132 mha) and black (72 mha) soils cover the largest area in the country. Red soils exhibit large variation in pore size distribution, and consequently in water retention, transmission and release characteristics. Black soils possess high water retentivity and unsaturated hydraulic conductivity, which make them suitable for rainfed crops in both the seasons; however, soils with high clay remain prone to oxygen stress because of poor internal drainability through micro-pores in *kharif.* Thus, there exists a wide range of water retentivity, its conductivity and availability to the growing plants. The length of water availability or growing period and growth depends on soil related constraints inherently.

1.2. Climate

It is estimated that the total precipitation in the country is 400 mha-m (million-hectare meters) annually. Of this, around 150 mha-m enters the soil, about 180 mha-m constitutes the runoff, and about 70 mha-m is lost through evaporation. India has, by now, been able to utilize less than 20 mha-m of the 180 mha-m of runoff for all major and minor irrigation projects, thus leaving about 160 mha-m of precipitation that flows freely through rivers into the sea (Vital et al. 2003, Bhaskar 2002). Rainfall exhibits a large variation in its temporal and spatial distribution. The rainfed areas in the country can broadly be classified into three climatic regions – arid, semi-arid, and sub-humid. Among many factors contributing to suboptimal and unsustainable yield levels, availability of excessive water in spells during *kharif* (rainy) season, and water stress of varying degree and duration during *rabi* (post-rainy) season figure prominently. The crop yields, thus, largely or entirely depends on the growing season rainfall plus water stored in the soil profile. Consequently the rainfed regions are characterized by relatively low and unstable crop yields. Risk is high because rain is undependable in both timing and amount making the region prone to periodic short to long-term droughts. Dry land crop production is a function of both spatial and temporal availability of soil moisture within the field during the crop growth period. The uncertain drought is a major contributing factor to low productivity, which is estimated at 0.2, 0.6 and 1.0 t/ha against a potential of 1.0, 1.9 and 3.0 t/ha under arid, semi-arid and sub-humid regions, respectively (Vittal *et al.*, 2003). Drought recurrence is a location specific interaction between soil capacity to retain moisture and precipitation in a region. Agriculture in rainfed areas, thus, continues to be a gamble because farmers in rainfed region face many uncertainties.

1.3. Drought definition

Drought is common in high as well as low rainfall areas. Farmers term drought as deficient rainfall, lack of moisture or a dry spell resulting in low crop yields including crop failure. They realize that seasonal variations in precipitation and temperature are much more important in farming than annual averages. Regardless of variability in perspective, it is clear that drought is a normal feature of climate and its recurrence is inevitable. The American Heritage dictionary (1976) defined "drought as a long period with no rain especially during planting season". The Random House dictionary (1969) defined it "as an extended period of dry weather, especially one injurious to crops". Drought is a condition relative to some long-term average condition of balance between rainfall and evapo-transpiration in a particular area, a condition often perceived as normal. Yet average rainfall does not provide an adequate statistical measure of rainfall characteristics in a given region, especially in the drier areas. Agricultural drought is usually defined as a "period when insufficient water is available to support the normal activities of a crop over a fairly normal long period of time of a fortnight or more depending on stage of crop". Drought is distinguished from aridity and it may be expected that both very wet and very dry regions experience drought. From an agricultural standpoint, a drought indicator should record crop management on the phenological drought sensitivity. Weather technology should be collineated. Thus, the partition between weather effect on yields and technology should be diffused. Emphasis should be placed on identifying periods within a given growing season when drought related weather conditions have greatest effect like yield altering impacts and crops. An operational definition would be one that compares daily precipitation values to evapo-transpiration rates to determine the rate of soil moisture diffusion and express these relationships in terms of drought effects on plant behavior at various stages of crop development. Thus, intensity of drought is a ratio of actual evapo-transpiration to potential evapo-transpiration during the growing season. Various types of droughts are stated below.

Meteorological Drought

Meteorological drought as expressed in Meteorological Dictionary is the time period of one or more days in which no rains or little rain is recorded by meteorological station.

Agronomical Drought

Agronomical drought as defined in Meteorological Dictionary is the shortage of water in soil. The mentioned shortage is caused by previous or present occurrence of meteorological drought.

Hydrological Drought

Hydrological drought is defined for surface streams as certain number of successive days, weeks, months or years characterized by low flowage compared to long-time normal. Hydrological drought occurs generally at the end of longer lasting period of dryness during which no precipitation was received.

Physiological Drought

It occurs as a result of water shortage in term of the needs of various sorts of plants. It's presence in growth process enables transport of nutrients from soil to vegetal body and provides it for growth. To ensure the development of plants the soil humidity must be more than the point of wilt.

Socio-economical Drought : Socio-economical drought arrives when the demand for economic commodity exceeds the water supply as a result of water shortage.

1.4. Types of agricultural droughts

Drought is a climatic anomaly characterized by deficient supply of moisture in rooting zone of soil resulting either from sub-normal rainfall, erratic rainfall distribution, higher water need or a combination of all the three factors. Crop production in rainfed areas is generally affected by *chronic* (permanent) *drought* (Southwest monsoon for *kharif* region and Northeast monsoon for *rabi* region with low rainfall), *ephemeral drought* (occurring in early, June-July; mid, July-August; and terminal, September-early October periods during crop growth in Southwest monsoon season), and *apparent* (Southwest monsoon with high rainfall) *drought*. Permanent drought usually had a high frequency of occurring once in 2 to 5 years while drought in ephemeral region will be once in 5 to 10 years and once in 10 to 20 years in apparent conditions (Vittal et al. 2003).

Early season drought : Early season drought generally occurs either due to delayed onset or due to prolonged dry spell soon after the onset of monsoon. This may at times result in seedling mortality needing re-sowing or may result in poor crop(s) stand and seedling growth. Further, the duration of the water availability for crop growth gets reduced due to the delayed start and the crops suffer from acute shortage of water during reproductive stage due to early withdrawal of monsoon. Therefore, for characterization of early season drought, information on optimum sowing period for different crops/varieties, quantum of initial rainfall spell expected and its ability to wet the soil profile enough to meet the crop water requirements for better germination and establishment is essential. The effect of early season drought is less on the crop, because during this period sowing is carried out. Various operations carried out during the period are primary tillage, sowing, fertilizer application and intercultural operations. On plant emergence and establishment, the effect of drought in first 2 to 4 weeks may affect initial vigour of the plants but may not have dire consequences on yield. In case of late onset of monsoon by 3 to 4 weeks, the contingency plans should be in place.

Mid-season drought

Mid season drought occurs due to inadequate soil moisture availability between two successive rainfall events during the crop-growth period. Its effect varies with the crop growth stage and the duration and intensity of the drought spell. Stunted growth results if drought occurs at vegetative phase. The occurrence of drought at flowering or early reproductive stage may adversely affect ultimate crop yield. Management of rainwater, especially *in situ* conservation of moisture, is a vital component of dry land crop management practices. In agricultural lands with 1-3 % slopes, runoff is about 25-35 % when the rainfall is in the range of 700-800 mm (Vittal et al. 2003). Though a lot of emphasis has been laid on soil and moisture conservation, the efforts mainly concentrated on construction of various types of bunds across the slope. This helped in controlling erosion and soil loss rather than achieving uniform moisture distribution. The research results have indicated that bunding increased the crop yields by a mere 6 % while a simple inter-terrace management such as contour cultivation that helped in uniform distribution of moisture, raised crop yields by 15-20 %, the impact being more pronounced in years of scanty rainfall (Vittal et al. 2003). During mid season, plant protection, top dressing of fertilizer, inter-culture, supplemental irrigation are the usual practices, in case of severity ratooning of few crops can be tried. In case the season is good, relay cropping is practiced. In case of long dry spells, location specific contingency plans are needed.

Terminal drought

The common practices executed during the terminal drought period are plant protection, soil water conservation efforts, inter-culture, supplemental irrigation, and harvesting. Late season droughts occur as a result of early cessation of monsoon rains, which can be anticipated with greater certainty during the years of late commencement or weak monsoon activity. Terminal droughts are more critical for most of the crops as the grain yield is strongly related to water availability during the reproductive/grain filling stage. Further, these conditions are often associated with an increase in ambient temperatures leading to forced maturity

Farmers are faced with many challenges. They must be efficient in order to remain in business and their production systems must be sensitive to environmental concerns. These requirements are in addition to the usual challenges of weather, pests and uncertain markets. A "Sustainable Crop Production System" is a term often used to describe a management philosophy that will be adopted by those farmers who are going to remain as the future producers of our food, feed and fiber. This philosophy includes the implementation of crop management strategies that provide: *Adequate, high quality food, feed and fiber supplies that are produced economically, and with the added responsibility to safeguard the environment.*

2. Sources of growth in rainfed crop production

In order to increase production, farmers have two options, either to use extensive systems (which expand the area planted) or intensive systems (which increase inputs on a planted area in order to increase yields). In order to meet immediate food demands, farmers in many rainfed areas have expanded production into marginal lands. These fragile areas are susceptible to environmental degradation, particularly erosion, due to intensified farming, grazing and gathering. This problem may be especially severe in areas of harsh climate, in which the transfer from extensive to intensive systems was slower than in other regions (De Haen 1997).

Expansion of production into marginal areas can cause many environmental problems. Clearance of marginal lands for crop production may result in loss of native vegetation and insurgence of disease and pest problems due to changes in the ecosystem. Soil erosion is also often a significant problem in areas of agricultural expansion. Many of the marginal areas to which agriculture expands in the developing world include hillsides and arid areas, which make soil erosion a particular concern.

These environmental impacts can lead to additional economic and health problems, particularly for the poor individuals that generally live in marginal areas. These impacts are generally greater on the poor than on other factions of the population due to the fact that they do not have adequate assets to mitigate the impacts of environmental degradation (Scherr 2000). Environmental problems can have far-reaching implications in poor communities through decreased agriculture production potential, which may further increase poverty, leading to increased malnutrition and poor health. Increasing production by expanding the planted area into marginal areas may have additional negative impacts on the population that moves into these areas, as living conditions can be much harsher than in more productive areas.

Because of these environmental consequences of area expansion, crop yield growth is a better solution than increasing the area planted in rainfed areas. McNeely and Scherr (2001) noted that under some circumstances, increasing production on more productive lands—such as irrigated areas—can ease the pressure to use more marginal lands for cropland and help to keep those natural habitats from being destroyed. But the potential for expansion of irrigated area is limited in most of the world. Therefore, intensive cropping systems that involve increased inputs such as labor, fertilizers, pesticides, or improved varieties to increase yields will be essential for rainfed crop production. Sustainable intensification of rainfed agriculture development can increase production while limiting environmental impacts. The three primary ways to enhance rainfed agricultural production through higher crop yields are (i) to increase effective rainfall use through improved water management; (ii) to increase crop yields in rainfed areas through agricultural research; and (iii) to reform policies and increase investment in rainfed areas.

3. Strategies for Enhancing Agronomic Productivity in India

An integrated package of options was conceived and developed by a multidisciplinary team of agricultural and social scientists at ICRISAT in 1974 in response to these constraints. This was later known as Vertisol technology (El-Swaify *et al.* 1985; Kampen 1982). The package targeted Vertisol areas in regions with relatively dependable rainfall, where land was left fallow during the rainy season (Flower, 1994). Vast Vertisol areas in India, such as in Madhya Pradesh, Maharashtra, and Andhra Pradesh, are left fallow during the rainy season and sown during the post-rainy season using residual moisture (Kampen 1982). It may be feasible to use a package of approaches consisting of several clusters of improved technological options to markedly increase productivity. The prerequisites for the success of the technology are dependable on early-season rainfall for dry seeding and soils with sufficient moisture-holding capacity to grow two crops without irrigation. It was estimated that Vertisol technology may be suitable over 5- 12 million ha in India, largely covering the States of Andhra Pradesh, Gujarat, Karnataka, Madhya Pradesh, and Maharashtra (Ryan *et al.,* 1982).

The options developed by ICRISAT for the management of land and water involved the use of moderate inputs and bullock power, and accessibility to small farmers in rainfed semi-arid tropics (SAT). They effectively improved drainage, reduced soil erosion, and increased/stabilized crop productivity, and were based on the concept of a micro-watershed (3 to 25 ha) as the basic natural resource management unit. Following are the essential components of Vertisols technology identified by Ryan et al. (1982).

Optimizing soil water management is crucial to enhancing agronomic productivity and meeting the food needs of the growing population with rising standards of living. An important strategy is to adapt to climate change by (i) improving soil quality; (ii) increasing water availability; (iii) adjusting time of sowing; and (iv) growing improved varieties and GM/biotech crops in appropriate crop combination and sequences. Improving soil quality involves restoring degraded soils by (i) increasing SOC pool; (ii) enhancing soil structure; (iii) balancing and increasing nutrient reserves; and (iv) improving activity and species diversity of soil biota. Similarly, alleviating drought stress involves (i) improving water-infiltration rate; (ii) water harvesting by safe disposal and storage of excess runoff for use as supplemental irrigation; (iii) using an efficient irrigation system; and (iv) conserving soil in the root zone by minimizing losses.

Crop yields in rainfed agriculture (predominantly in southern, central, and western India) are low because of severe drought stress and high temperatures. Several experiments have shown that conserving water in the root zone, through practices that enhance water infiltration and decrease evaporation losses, can increase crop yield by 33 % to 173 % (Lal, 2009). In addition, water harvesting for supplemental irrigation can be extremely important in these drought-prone regions. Increasing recharge involves installation of check dams, percolation tanks, recharge tube-wells, and rainwater conservation.

Water harvesting and storage are key options in dry lands of peninsular India (Gunnell and Krishnamurthy, 2003). Remote sensing (Ines et al. 2006), field-water management (Bouwan and Tuong, 2001), crop planning (Reddy and Kumar, 2008), and modeling to plan watershed management (Kaur et al. 2004) are some of the modern innovations needed to improve water-use efficiency (WUE). One such supplemental irrigation, applied at the critical stage of crop growth, can increase agronomic yield by 23 % to 250 %. Effectiveness of these soil-water management techniques can be enhanced when used in conjunction with improved varieties and INM strategies (Lal 2006).

3.1 Water harvesting for rainfed agriculture

Many arid or semi-arid regions experience significant problems in securing adequate amounts of water simply from the lack of sufficient rainfall for rainfed crop production. Semi-arid regions, however, may receive enough annual rainfall to support crops but it is distributed so unevenly in time or space that rainfed agriculture is not viable (Reij et al. 1988). Rockström and Falkenmark (2000) noted that due to high rainfall variation in semi-arid regions, a decrease of one standard deviation from the mean annual rainfall often leads to the complete loss of the crop. Water loss through evaporation and runoff exacerbate water scarcity problems in these areas. Low rainfall areas that receive between 300 – 600 mm annually may be able to combat these problems using supplemental irrigation methods, but regions receiving less than 300 mm of annual rainfall must resort to other methods to secure enough water to support crop production (Oweis *et al.,* 1999).

Water scarcity is a significant problem for farmers where 80 - 90 percent of water withdrawals are used for agriculture (FAO 2000). While farmers in some high-potential regions have been able to increase yields by 4 – 5 percent in recent years, farmers in the semi-arid tropics have only increased agricultural growth by less than 1 percent (Barghouti 2001). Farmers in these arid regions may be particularly hard hit, as development requires more water for domestic and industrial uses. Potential does exist, however, to increase agricultural water use efficiency through water harvesting and conservation techniques. Bruins *et al.* (1986) estimated that an additional 3 - 5 % of arid areas could be cultivated using runoff farming. Some water harvesting methods have proven successful in practice;

trials of water harvesting have shown increased yields of 2 - 3 times those achieved in dry land farming (FAO 2000).

Water harvesting is a general term usually used to describe the collection and concentration of runoff for many purposes, including agriculture and domestic uses. One characteristic is the importance of storage to many water harvesting systems due to the intermittent water flow in the arid and semi-arid areas where water harvesting takes place. In addition, most water harvesting operations consist of a catchment area and a receiving area for the capture of runoff, and are generally small both in size and in level of investment. Water harvesting activities occur near the location where the rain falls, therefore the storing of river water in large reservoirs and groundwater mining are generally not included under the category of water harvesting.

Water harvesting for agriculture (sometimes referred to as runoff farming) involves concentrating and collecting the rainwater from a larger catchment area onto a smaller cultivated area. The runoff can either be diverted directly and spread on the fields or collected in some way to be used at a later time. Pacey and Cullis (1986) classify rainwater harvesting techniques into three broad categories: external catchment systems, micro catchments, and rooftop runoff collection. External catchment rainwater harvesting (sometimes referred to as macro catchment water harvesting) involves the collection of water from a large area that is a substantial distance from the area where crops are being grown. Types of external catchment systems include runoff farming, which involves collecting sheet or rill runoff from the hillsides into flat areas, and floodwater harvesting within a streambed using barriers to divert stream flow onto an adjacent area, thus increasing infiltration of water into the soil. This type of water harvesting can be used for any number of different crops including row crops, trees or closely growing crops (Oweis et al. 1999).

Micro catchment water harvesting methods are those in which the catchment area and the cropped area are distinct but adjacent to each other. Some specific micro catchment techniques include contour or semi- circular bunds made of earth, stone or trash, pitting, strip catchment tillage, and a meskat-type system in which the cropped area is immediately below the catchment area that has been stripped of vegetation to increase runoff. These methods are often used for medium water demanding crops such as maize, sorghum, millet and groundnuts (Habitu and Mahoo 1999).

In situ water harvesting (or water conservation) methods are also used to help increase water use efficiency. Many factors influence the usefulness of rainwater harvesting in general as well as the applicability of different methods in a particular area. Rainfall harvesting is only necessary in arid and semi-arid regions that receive low levels of rainfall or in which there is high intra or inter-seasonal rainfall variability that makes traditional rainfall agriculture infeasible. Rainfall intensity is one factor that impacts the effectiveness of the chosen water harvesting method. Li et al. (2000) pointed out that while the bare ridge and furrow method has been shown to be quite effective in semi-arid areas of India where rainfall is generally high intensity.

Other factors that influence the choice of rainwater harvesting method include topography, soil characteristics – particularly those related to water infiltration, and the choice of crop to be planted. Specific topographic characteristics that are necessary for rainwater harvesting include a landscape surface that facilitates runoff and variations in altitude such that runoff flows down the slope and collects at a flat portion of the landscape. In addition, the cropped area soil must be deep enough and of a suitable texture to induce rainfall infiltration and retention (Bruins et al. 1986). Loamy soils with a medium texture are generally the best-suited soils for water harvesting projects (Critchley and Siegert, 1991). Due to the unique soil and landscape characteristics across regions, the same rainwater harvesting technique may produce quite dissimilar results in different areas. Particular attention needs to be given to the relationship between soil management and water availability to assure the best possible results from water harvesting operations.

Soil nutrient availability is essential in enhancing the effects of water harvesting and helping to ensure increased yields. Rockström (1993) while addressing the importance of the water-nutrient equilibrium in crop production noted that although fertilizer application on fields with adequate moisture will increase yields, addition of nutrients during periods of drought may actually lead to decreases in yields. The relationship between soil nutrient levels and water harvesting is particularly important in areas where soil nutrient levels are generally very low (Rockström and Falkenmark, 2000). Tabor (1995) noted that regular application of animal manure is crucial to the success of micro-catchment water harvesting as manure increases nutrient levels and improves the physical condition of the soil. Increased nutrient availability will also help to promote root development and canopy cover growth, which will increase water uptake by the crops and help to advance biomass growth (Rockström and Falkenmark 2000).

In addition to soil nutrient requirements, the physical structure of the soil also has an impact on the effectiveness of water harvesting. The degradation of the easily erodible soils in many arid and semi-arid regions leads to specific concerns regarding water harvesting methods. The erosion of the sandy surface of these soils, often due to the removal of vegetation by overgrazing or other means, results in exposure of the clayey subsurface that forms a crusty layer with lower water infiltration rates. While these crusted surfaces are often abandoned because of their low potential for agriculture, they may prove very useful for water harvesting by inducing runoff from the more impenetrable catchment area to the cultivated area below (Tabor, 1995). The impact of raindrops on the eroded surface can also help to induce crusting (Abu-Awwad and Shatanawi 1997). In some situations, the runoff may also bring nutrient-rich litter along with it to the cultivated area, thus increasing the availability of nutrients to the crops (Nabhan 1984). In cases when a crusty layer is not formed on the catchment area, the soil surface may be treated with other materials to reduce infiltration rates and promote runoff. Some chemical treatments have been used for this purpose, including asphaltic materials and paraffin wax. While these materials have increased runoff efficiency, they are generally only effective for 2-5 years (Ojasvi et al. 1999). Other materials such as plastic sheeting, fiberglass and concrete have also been used but these items are often too expensive for farmers in arid areas to afford.

Although the catchment area benefits from a layer of soil with low infiltration rates, this characteristic can be detrimental to crop production in the cultivated area. Very low infiltration rates that result from crusty surfaces can lead to water logging in the cultivated area, rendering the area unfit for crop production. The most appropriate type of soil for the cultivated area would be a deep fertile loamy soil. The presence of organic matter improves soil structure and allows for greater water infiltration and better penetration of plant roots. Deep soils are able to hold water from a water harvesting system and may also be able to provide a more nutrients for plant growth (Critchley and Siegert 1991).

3.2 Water conservation

Water conservation methods, often referred to as *in situ* rainwater harvesting; include activities such as mulching, deep tillage, contour farming and ridging (Habitu and Mahoo 1999). The purpose behind these methods is to ensure that the rainwater is held long enough on the cropped area to ensure infiltration. These techniques are best suited to areas where rainfall and water holding capacity are sufficient to meet the crop water requirement but the amount of water infiltration is not adequate to reach the required moisture level (Habitu and Mahoo 1999). Some methods, such as mulching or the addition of organic matter may also help to enhance the physical characteristics of the soil. Water conservation is often used in tandem with water harvesting techniques in order to achieve better results. The different components of conservation agriculture (Bhale and Wanjari 2010) are given in Table 1.

Table 1 : Components of conservation agriculture

Practice	Slope (%)	Topography and Rainfall	Soil type	Purpose
A. Mechanical measurers				
Contour bunding	> 6	Flat land with scanty and erratic rainfall	Medium deep black soil	To reduce length and degree of slope
Graded bund Channel terracing	6 - 10	> 800 mm	Red soil and clay soil under less rainfall condition.	To make runoff water trickle rather than to rush out
Graded border strips	2 - 3	High and low rainfall area	Deep red soil	Formed between contour or graded bunds

Bench terracing	6 - 33	Hilly areas with high	rainfall	To convert original slope into level field
Compartmental bunding	< 1	Rainfed areas	Rainfed Vertisols	Storing initial rainfall and permitting increased infiltration rate
B. Agronomic measures				
Contour farming	< 1	Rainfed areas		Act miniature barrier to the runoff over soil surface.
Close growing / cover	< 1	Rainfed areas with rainfall of high intensity		Reduce splash erosion and protect crop soil from beating action of rainfall
Strip-cropping	0.5 - 3	Rainfed areas		Check runoff and increase absorption of water.
Vegetative barrier	2 - 3	Rainfed areas		Reduce length of slope, check runoff velocity and trap silt
BBF	< 1	Semi-arid	Deep black soil	Reduce runoff provide drainage

Source : Bhale and Wanjari (2010)

3.2.1. Summer deep tillage

Deep tillage is a water conservation technique that improves soil moisture capacity by increasing soil porosity. In addition, runoff is reduced through increased roughness at the soil surface, which increases the time available for water to infiltrate the soil. This increased infiltration will increase the availability of water in the root zone to favour plant growth. It is important to note, however, that these techniques are not suitable in all situations. Soil texture and structure as well as economic limitations that may exist if high capital inputs are needed, are the factors for consideration.

Ploughing land immediately after harvesting the post-rainy-season crop, when the soil is not too hard and still contains some moisture, induce favourable conditions for water conservation. This as well prevents weeds from setting seed, thus reducing weed survival and multiplication (Shetty et al. 1977). Often, stubble/crop residues remain in the field after harvest, providing shelter to insects during the off-season and then multiplying during cropping. Stubble of many crops such as sorghum, maize, and paddy provide shelter to the stem borer. Most of the weeds in the field act as alternate hosts to pests and diseases during the off-season. Summer cultivation exposes the hibernating egg colonies and pupae of various insects to severe solar heat, thus killing them (Patil and Jawaregowda 2000). It also improves soil fertility, fills cracks, and pulverizes the soil, thereby improving moisture absorption.

3.2.2. Contour farming

Contour farming is a technique in which tilling and weeding are done along the contours to minimize water runoff. Mulching or the addition of other organic material to the soil is a water conservation method that may both increase soil water availability by increasing soil water holding capacity and decreasing evaporation and improve the quality of the soil. The effect of different in-situ moisture conservation practices (Gupta and Bhan 1997) are presented in table 2.

Table 2 : Yield (kg/ha) of maize-mustard and additional net return (Rs/ha) as influenced by *in-situ* moisture conservation and fertilization

Treatment	**Maize yield (kg/ha)**		**Mustard yield (kg/ha)**		**Additional net return over control**
	1988	**1989**	**1988**	**1989**	
Water harvesting and moisture conservation practices					
Contour sowing (Control)	1,336	1,033	688	752	
Interceptal earth bund	1,360	1,065	943	797	298
Vegetative bund (*Leucaena leucocephala*)	1,406	1,127	923	832	787
Terracing	1,436	1,183	888	848	747
Ridge and furrow	1,540	1,305	1,013	911	1,770
Ridge and furrow + Vegetative bund	1,894	1,687	1,001	974	2,459
CD (P =0.05)	61	123	36	111	
Mulching					
Control	1,422	1,105	790	728	
Paddy straw mulch@ 3 t/ha	1,665	1,363	1,029	976	2,141
CD (P =0.05)	31	58	23	45	

(*Source* : Gupta and Bhan, (1997))

3.2.3. Conservation Furrows

Conservation furrows across the slope at 3 m interval, as a measure of moisture conservation and runoff management were evaluated in ten farmers' fields in five villages of Nalgonda district, Andhra Pradesh. Castor with pigeon pea (5:1) was the test intercrop. During the growing season, the study area received 290 mm rainfall in 28 rainy days (rainfall >2.5 mm/day^{-1}). The results showed that conservation furrows plots stored 4 to 37 % additional soil moisture compared to control throughout the growing season. This additional moisture resulted in better plant growth and 12 % higher bean and grain yields of castor and pigeon pea than the control (DARE/ICAR report 2003-04).

Studies carried out under All India Coordinated Research Project on Soybean on *in situ* moisture conservation technique revealed that for achieving higher productivity and profitability from soybean opening conservation furrow after 6 rows in central zone and after 3 rows in southern zone is effective.

3.2.4. Ridge and furrow system

In this system, planting of upland crops on ridges laid out on farm where slope is less than 1 %. Furrows serve an effective conservation of moisture under low rains and also effective surface drainage and carry excess water during high rains. Spacing of cropped rows and rainfall would be deciding factors for specification of ridges. This system has proved highly effective in medium to high rainfall areas (700-1200 mm). The planting of maize, sorghum and soybean on ridge and furrow system (Table 3) increased the seed yield to the tune of 27 to 106 % as compared to flat sowing in Madhya Pradesh (Gupta et al. 1979).

Table 3 : Effect of ridge planting and slopes on productivity of various crops grown on Vertisols

Crop	Slope (%)	Yield (kg/ha)		Percent increase over flat planting
		Flat planting	Ridge and furrow planting	
Maize	0.30	1601	3376	106
	0.60	1704	3247	88
	1.20	1601	3644	38
Sorghum	0.30	2638	3608	38
	0.60	2445	3754	54
	1.20	2717	3553	30
Soybean	0.30	1890	2900	53
	0.60	1740	2650	52
	1.20	2390	3030	27

3.2.5. Graded furrows

Under moderate rainfall conditions (< 1000 mm) , the productivity of rainfed crops can be significantly increased by providing the graded furrows of 0.2 to 0.3 % slope which can conveniently carry runoff water to drainage channels and also serves for moisture conservation. The spacing between furrows may vary from 8-10 m depending upon slope and rainfall characteristics.

3.2.6. Broad bed furrow system (BBF)

BBF system consists of a series of broad beds and furrows accommodates in 90-150 cm wide parallel running strips (Kanwar et al. 1982). This system can be developed with the help of tractor drawn or bullock drawn implements. It provides safe disposal of runoff, tends to conserve soil and water *in situ*, reduces soil nutrient losses and enhanced crop productivity and sustainability. BBF system was found to reduce runoff (13.5 %), soil loss (30.1 %), N and P loss (16.1 and 13.5 %) over the flat bed planting and enhanced soybean yield, water use and water use efficiency by 20.6, 4 and 7.9 %, respectively (Table 4 and 5). Gupta and Sharma (1990a) also reported 6 % increase in soybean yield due to BBF planting in sub-normal rainfall season.

Rainfall infiltration increases with BBF and excess water is removed with minimum erosion, thus stabilizing soil moisture conditions. It helps maintain optimum moisture supply in the effective root zone, thereby improving the physical, chemical, and biological properties of the soil, which ultimately results in higher yield (Desai et al. 2000). Sengar (1998) reported the beneficial effects of surface drainage in pigeon pea in Madhya Pradesh. Mamo et al. (1994) reported significantly higher grain yields of gram with BBF than with flat seedbeds. Farmers in some parts of India are adopting BBF for post-rainy-season groundnut in order to increase yield and save on labor (Joshi and Bantilan, 1996).

Sowing of safflower on BBF resulted in conservation of moisture in soil which was observed to be 9.61 % more as compared to traditional method of sowing. The conserved soil moisture has shown higher yield in BBF method. The observed yield of crop in BBF method was 8.81 q/ha as compared to 8.27 q/ha under traditional method. It has been reported that adoption of BBF sowing for safflower le to 6.50 % higher yield than traditional method of sowing. In addition, it has also been observed that in BBF method of sowing cost of operation was only $ 16.39 per hectare in comparison with $ 26.50/ha in case of traditional sowing, thus saving 58 % cost. Vyas et al. (2004) concluded that the broad bed furrow system significantly improves the nutrient use efficiency and enhanced the sustainability as the yield levels obtained in RDF, 75 and 50 % of RDF showed non-significant differences (Table 6).

Table 4 : Rainfall, runoff and other water balance components, water use and water use efficiency of soybean due to various land treatments

Treatment	Rainfall (mm)	Runoff (mm)	Deep percolation (mm)	Water flux to root zone (mm)	Water use (mm) by soybean	Water use efficiency (kg/ha/mm)
Flat	831	111	190	77	572	3.05
BBF	831	96	148	51	595	3.29
BBTF	852	116	161	45	650	3.07
RSB	831	66	252	69	571	3.46

BBF – Broad bed furrow; BBTF- Broad bed and tied furrow system; RSB – Raised bed sowing

Source : Sharma et al. (2005)

Table 5 : Soybean yield, soil and nutrient losses under different land management treatment

Treatment	Yield (kg/ha)	Soil loss (kg/ha)	Nutrient loss (kg/ha)			
			N	P	K	S
Flat	1218	1597	17.91	0.37	0.77	8.52
BBF	1469	1117	15.03	0.32	0.84	9.11
BBTF	1435	954	17.28	0.32	0.64	8.43
RSB	1433	471	9.26	0.28	0.38	5.48

BBF – Broad bed furrow; BBTF- Broad bed and tied furrow system; RSB – Raised bed sowing

Source : Sharma et al. (2005)

Table 6 : Effect of in situ moisture and nutrient management on soybean productivity and sustainable yield index

Treatment	Seed yield (kg/ha)					Sustainable yield index				
	Fertility levels									
	RDF	75% RDF	50% RDF	*Control*	*Mean*	RDF	75% RDF	50% RDF	Control	*Mean*
Flat sowing	1509	1438	1403	1308	1415	0.45	0.45	0.44	0.46	0.45
Single row ridge sowing	1488	1273	1220	1180	1290	0.28	0.27	0.26	0.26	0.27
Two row ridge owing	1345	1274	1259	1197	1268	0.33	0.35	0.34	0.35	0.34
BBF	1645	1468	1440	1327	1470	0.34	0.45	0.44	0.44	0.44
Mean	1497	1364	1331	1253		0.37	0.38	0.38	0.38	
CD (P=0.05) Land configuration					76					0.045
Fertility level					73					0.042
Interaction					205					0.070

3.2.7 Broad bed and tied furrow system (BBTF)

BBTF system is similar to BBF except that furrows are tied with small cross section earthen bunds at a regular interval of 10 m in mid August month to retain runoff, if any, towards the end of rainy season for the benefit of rainy season crops during reproductive growth phase. The beneficial effect this system could be visualized from the data presented in table 4 and 5. This system can be very useful for alleviating adverse effect of prolonged dry spells, which are very commonly encountered during late *kharif* season.

3.2.8. Raised and sunken bed system (RSB)

The rainfed crops grown on Vertisols and its associated soils experiences the poor drainage during continuous and intense rainfall during vegetative growth period of crops and moisture deficit during reproductive phase due to long dry spells. Raised and sunken bed system is of semi-permanent nature and consists of an array of raised and sunken beds of 6-8 m and 304 m widths, respectively, with the elevation difference of 15 to 30 cm. The system is created by mechanically shifting of soil from demarcated 3-4 m wide strips, designated as sunken bed to adjoining 6-8 m wide strips called raised beds. Sunken beds are tied with small cross section earthen bunds of about 10 cm height at 20 m distance interval to ensure uniformity in runoff retention. System ensures surface drainage, encourages in-situ water conservation and retards soil erosion and nutrient losses to a considerable extent. The runoff from raised beds arrested in sunken beds supporting to relatively water tolerant crop. Gupta et al. (1978) revealed that the 3-6 wide raised beds with elevation of 20-30 cm work successfully under Vertisols of Madhya Pradesh where water intake rate is poor (4-6 mm/hr) and average rainfall is high (>1200 mm) while for western regions where water intake is fairly high (10-12 mm/hr) and rainfall is moderate (<1000 mm), 8 m wide raised bed alternated by 4 m wide sunken bed with elevation difference of 15-20 cm have been found suitable (Gupta and Sharma, 1990a, b; Sharma and Gupta, 1990; Gupta and Sharma, 1994). This system is equally suitable for saline soils, wherein sunken beds can be used for growing salt tolerant crops. The beneficial influences of the system can be seen from the data presented in table 4 and 5.

3.2.9. Planting of cover crops

The rainy season crops which develop fast and thick canopy, intercepts rainfall and dissipates energy of falling rain drops is helpful in reducing soil erosion. Soybean, maize, sorghum, green and black gram crops are quite effective in this respect. Verma (1981) suggested cropping strategies for different land types (Table 7).

Table 7 : Cropping strategies for different rainfed land types

Type of land	**Crop plan**
Bare hill slopes	Fuel fodder and fruits purposes trees, grasses, crops and their varieties chosen according to local experience
Shallow soil with slope < 3%	Grasses and legumes including fodder shrubs
Shallow soil with slope 1- 3% or light soil	Sorhum, maize and short duration crops in rainy season
Shallow soil with slope < 1%	Sorghum, maize, soybean
Deep soil with slope > 3%	Grass plantation and stabilization work
Deep soil with slope 1- 3%	Soybean, sorghum, maize, pigeon pea and intercropping

3.2.10. Mulching with crop residues

The potential benefits of crop residues as surface mulch for soil and water conservation is well documented in literature. The moderate of residue application may be effective in enhancing the infiltration and reducing runoff. However, huge amount may be required to significantly cut down evaporative losses of profile stored moisture and weed suppression. Usually they can be applied at the rate of 5 to 8 t/ha. Straw mulches besides reducing erosion and enhancing infiltration, increases water use efficiency of crops (Sharma et al. 1985 a and b). The beneficial effects of straw mulches can be seen from the data presented in table 8.

Table 8 : Impact of soil and straw mulching on yield and water use efficiency of rainfed crops

Parameter	Crop	No mulch	Soil mulch	% increase over no mulch	Sorghum cob husk mulch (6 t/ha)	% increase over no mulch
Yield (kg/ha)	Chickpea	1430	1750	22.4	1800	25.9
	Linseed	905	968	6.9	1043	15.2
	Wheat	1730	1837	6.2	1835	6.1
	Safflower	1755	1942	10.6	1822	3.8
Water use efficiency (kg/ha. Mm)	Chickpea	6.5	8.1	24.6	9.4	44.6
	Linseed	3.9	4.3	10.3	4.6	17.9
	Wheat	6.6	8.2	24.2	7.5	13.6
	Safflower	6.8	8.1	19.1	7.6	13.8

Source : Shama et al. (1985b)

3.2.11. Soil mulch/Intercultural operations

Soil mulching *in-situ* tends to reduce evaporation as it minimizes and delays developments of shrinkage cracks and besides providing a diffusion barrier (Sharma and Gupta, 1984; Gupta and Sharma, 1990.b; Sharma and Gupta, 1990). Soil mulch can be created by hand operated implements (*Khurpi*/ hand hoe) or bullock drawn implements like *dora/kulpa* (small blade harrow) which 2-3 cm soil loose and removed the weeds. On an average 6-25 % increase in water use efficiency of post-rainy season crops can be realized due to soil mulching (Table 8).

3.2.12. Vegetative barriers/ hedges

The beneficial effects of vegetative barriers or hedges are to reduce the runoff and conserve the soil and plant nutrients. For this purpose a number of grasses have been identified and found effective. Grasses like vetiver (*Cymbopogon maritini)* have proved to be useful (Table 9) Vegetative barriers are established at 0.5 to

0.75 m vertical interval. Two rows planted 30 cm apart make a good hedge which should be regularly cut to maintain 30 cm height.

Table 9 : Runoff, soil and nutrient losses due to vegetative barriers at different slopes

Treatment	Slope (%)	Runoff (%)	Soil loss (kg/ha)	Nutrient loss (kg/ha)			
				N	P	K	S
Check	2.0	115.7	986	23.85	0.018	1.42	2.98
	1.5	85.2	918	15.31	0.002	1.13	4.37
	1.0	87.9	614	10.67	0.023	0.76	2.24
Graded bunds	2.0	91.9	633	19.04	0.014	1.46	2.99
	1.5	70.5	340	13.23	0.015	1.29	1.39
	1.0	53.9	582	8.70	0.014	0.72	3.37
Vetiver grass	2.0	94.9	662	17.40	0.015	1.14	2.03
	1.5	69.1	453	13.12	0.044	1.19	4.13
	1.0	53.8	465	8.88	0.038	0.60	2.56
Bund + Cymbopogo n	2.0	94.6	567	17.18	0.012	0.89	2.91
	1.5	69.4	509	11.48	0.024	0.67	5.32
	1.0	52.9	474	8.02	0.026	0.71	3.13

Source : Ranade et al. *(1995)*

3.2.13. Use of organic manures

The addition of organic matter is often used along with other conservation methods to help increase water infiltration. When organic material is added with conservation tillage systems instead of using traditional mould-board ploughs, the decrease in erosion is even greater (Rockström and Falkenmark, 2000). Fall and Faye (1999) noted that animal traction used in place of tractors and heavy machinery decreases soil compaction, thus increasing soil aeration and water infiltration. These benefits are further increased when organic material is added to the soil. Beet (1990) found that a decrease in organic matter in soils from 5 % to 3 % decreased soil water retention from 57 % to 37 %. While animal traction may be useful in increasing soil aeration and infiltration of water, it may also result in soil degradation in the form of increased erosion (Fall and Faye 1999).

3.2.14. Micro catchments

Micro catchment water harvesting systems consist of a distinct catchment area and cultivated area that are adjacent to each other (Habitu and Mahoo 1999). Boers and Ben- Asher (1982) additionally specified that the distance between the catchment area and the runoff receiving area of micro catchments must be less than 100 meters. Some advantages of micro catchments include the high specific runoff yield compared to larger catchments (Bruins et al. 1986) and their simplicity, inexpensiveness and easy reproducibility (Boers and Ben-Asher 1982). Some authors suggest that micro catchment water harvesting systems offer significant increased cropping potential to smallholders without access to tractors in developing countries (Suleman et al. 1995).

3.2.15. Bunding and V-Shaped Catchments

Various forms of bunding and v-shaped micro catchments have been used successfully in some arid and semi-arid regions. Contour bunds are earth, stone or trash embankments placed along the contours of the hillside in order to trap rainwater behind them and allow for greater infiltration. Micro catchments using contour bunds were found to be successful for fruit tree plantations. At least twice yearly runoff was found to be sufficient for the selected tree species. The same method was found to be uneconomical for mountainous areas as the construction costs were too expensive to offset any additional gains in yield (Oweis et al. 1999). Semi-circular bunds are generally placed in a staggered formation and allow water to collect in the hoop for greater infiltration. Excess water is displaced around the edges of the bund when the hoop area is filled with water. Contour bunds are generally used on slopes less than five %, while semicircular bunds are usually only used if the slope is less than three % (Habitu and Mahoo 1999). V-shaped micro catchments are similar to semicircular bunds except that a v shaped catchment area is used instead of a hoop shaped area.

3.2.16. Meskat-type Systems

Other types of micro-catchments use a catchment area that diverts runoff water directly onto a cultivated area at the bottom of the slope. The meskat-type system differs from those previously described in that the field is divided into a distinct catchment area that is located directly above the cropped area instead of alternating catchment and cultivated areas (Habitu and Mahoo, 1999). The cultivated basin is surrounded by a U shaped bund in order to hold the runoff. Meskat-type systems are used for most types of crops (Habitu and Mahoo, 1999). Suleman et al. (1995) conducted a study using water catchment aprons of various lengths and slope gradients to examine increases in soil moisture. In this method, a flat cultivated area is located in between two apron catchment areas. They found that soil moisture is significantly increased when aprons of 4 to 5 meters with 7 to 15 % gradients are used. Moisture was increased by 59 % in the first 15 cm of soil, by 63 % in the second 15 cm, and by 80 % in the 30 to cm depth range. A study compared three experimental fields: a control area in which the entire area was planted, a water harvesting area with a ratio of 1:1 between the catchment area and cropped area, and a second water harvesting area with a ratio of 2:1 between the catchment area and cultivated area (Rees et al., 1991). While the water storage in the 1:1 and 2:1

trials increased by 55 and 43 %, respectively over the control, the yields were not always higher than the control. Averaged over the three years of the study, the 1:1 trial achieved yields that were 95 % of the control yields. The 2:1 trial, however, obtained significantly lower yields than the control due to water logging problems.

4.2.17. Comparison of Micro-catchment Techniques and Combination Methods

Kaushik and Lal (1998) conducted an experiment comparing five different water harvesting techniques (flat bed, bed and furrow, furrows on grade with eventual cultural operations, field bunding, and inter-row water harvesting) on rainy season crops in New Delhi, India. Significant differences in yield, moisture use, and moisture use efficiency were only found in lower rainfall year of the two-year study. The highest grain yield, monetary returns, soil moisture use, and moisture use efficiency were obtained using the bed and furrow method. Inter-row harvesting and bunding were found to have slightly better results than the two remaining methods. Kaushik and Gautam (1994) found increased pearl millet yields by 73.6 % over the flat bed method when using a ridge and furrow seedbed, and an increase of 54.0 % when using a flat seedbed with straw mulch. In the same study, the ridge and furrow method was shown to obtain a higher plant height, moisture use rate and water use efficiency compared to the flat bed method. Another study in Jodhpur, India tested several water harvesting and moisture conservation methods on three tree species (Gupta, 1995). The methods tested included a control, weeding only, weeding and soil working, weeding and 1 m diameter saucers, weeding and 1.5 m diameter saucers, weeding and 1.5 m diameter saucers with mulching, bunding micro-catchments around each tree in a checkerboard design, and the ridge and furrow method. Increased height, collar circumference, crown diameter, and biomass accumulation over the control were found for all three tree species using the ridge and furrow method.

Some studies have used a micro catchment system along with a runoff enhancing or mulching technique in order to increase water use effectiveness. Li et al. (2000) used a plastic-covered ridge to complement a ridge and furrow micro catchment system used in the low- intensity rainfall area. Gravel mulch was also used in this study to hold water in contact with the soil for a longer period of time to increase infiltration and reduce evaporation. The test plots using plastic covered ridges obtained higher corn yields than the bare ridge plots. The highest yields were obtained from the plot with both plastic covered ridges and gravel mulched furrows, which produced yields 1.3 times greater than the plastic covered ridge only plot, 2.6 times greater than the bare ridge and furrow field, and 1.9 times greater than the bare flat soil control field.

Another study tested the use of various forms of waste (polyethylene bags, newspaper, stone, and marble) to line catchment areas in a shallow conical micro catchment agro forestry system in India (Ojasvi et al. 1999). These linings served to harvest water and mulch jujube trees. The largest plant height was obtained using linings of stone and marble. During the first year of the study, when the lining served only as mulch due to lack of rainfall, an increase of 33.3 % and 25.0 % in tree height over the control was found for stone and marble. Increases in tree height of 97.3 % (stone) and 108.5 % (marble) over the control were obtained in the second year when the linings served as mulch and aided in water harvesting. All types of lining were found to increase the soil moisture levels compared to the control.

3.3 Nutrient management technology

3.3.1 Residue management

The potential availability of crop residues is presented in table 10. Low irrigation and nutrient-use efficiency (NUE) are partly attributed to low SOC pool. Therefore, enhancing SOC pool through crop residue management and application of bio-solids is an important strategy for long-term improvement in soil quality. The data in table 11 and 12 show that residue retention, along with the use of recommended rates of fertilizers and farmyard manure (FYM), produced the highest yield. In addition to enhancing soil quality by recycling nutrients, crop-residue retention is also important to improving soil structure and hydrological practices.

Table 10 : Availability and resource benefits of crop residues in India

Crop	Total Crop residues (Mt)[1]	Crop residues available for utilization (Mt)[2]	Nutrient content (% on oven dry basis)[3]				Nutrient potential (Mt)[4]	
			N	P_2O_5	K_2O	Total N+P+K	Available for utilization	Fertilizer replacement value
Rice	131.70	43.90	0.61	0.18	1.38	2.858	0.953	0.476
Wheat	109.55	36.52	0.48	0.16	1.18	1.994	0.665	0.332
Sorghum	15.30	5.10	0.52	0.23	1.34	0.320	0.107	0.053
Pearl millet	16.22	5.41	0.45	0.16	1.14	0.284	0.095	0.047
Maize	21.21	7.07	0.52	0.18	1.35	0.435	0.145	0.072
Pulses	13.67	4.56	1.29	0.36	1.64	0.450	0.150	0.075
Oilseeds	49.68	49.68	0.80	0.21	0.93	0.964	0.964	0.482
Sugarcane	23.45	23.45	0.40	0.18	1.28	0.436	0.436	0.218
Potato	11.80	11.80	0.52	0.21	1.06	0.211	0.211	0.106
Total	392.58	187.48				7.951	3.724	1.862

[1] Total crop residue production calculated by dividing the economic yield by the corresponding economic yield: residue yield ratio; [2] Assuming that 2/3rd of the residue is used as feed for animal and other purposes and only 1/3rd is available for utilization except oilseeds, sugarcane and potato residue; [3]Based on residue yield at (2); [4] Based on assumption that half of the total N+P+K is mineralized in a season.

(Source : Anonymous (2008)).

3.3.2. Soil organic carbon management

The SOC pool in most cultivated soils of India is low (0.1 % to 0.3 %) and below the threshold level needed for the soil's ecosystem functions. In addition to nutrient retention and uptake, managing SOC pool is essential for judicious use and conservation of soil water. The strategy is to maintain a positive C budget in the soil by application/recycling of bio-solids (e.g., crop residues, manure, compost, sludge, animal waste). It is in this context that management of crop residues needs a careful appraisal. The data show that even with application of the recommended rates of NPK and FYM, the rate of SOC sequestration ranges from 0 to <400 kg/ha/yr. Furthermore, CO_2 emissions are lower with no-tillage (NT) used in conjunction with residue retention than with plow-till and residue removal or burning. It is important to identify and adopt agricultural practices that minimize C-footprint and farming operations.

3.3.3. Nutrient management

Adopting INM is crucial to increasing and sustaining crop yields (Singh et al. 2004; 2005). Application of FYM, in conjunction with chemical fertilizers, can increase yield of rice by 70 % and that of wheat by 170 %. There is an important synergistic effect when chemical fertilizers are used in combination with organic amendments, especially in the light-textured soil of the Indo-Gangetic Basin (IGB) in which the SOC pool has been severely depleted because of residue removal for fodder and other purposes. Long-term experimental data are needed to model crop growth under a range of fertility management options (Bhattacharyya *et al.* 2007).

Stagnation or decline in agricultural productivity in India may be attributed to degradation of soil quality and excessive/improper use of the scarce water resources. Yet, the increasing demand for food production necessitates increasing crop yields. The strategy is to improve WUE by conserving water in the root zone and minimizing losses. In dry-farming systems of peninsular India, conserving water implies water harvesting and recycling for supplemental irrigation, groundwater recharge, and minimizing losses by runoff and evaporation. In irrigated regions, using appropriate time and method of irrigation, mixing saline and fresh water, using grey or urban wastewater, and decreasing losses by evaporation are important. Identifying viable alternatives to flooded rice in the northwestern region is also important. Direct sowing of wheat after rice and use of crop-residue mulch can save time, labor, and water.

According to DARE/ICAR repot (2003-04), integrated nutrient management (INM) technology for pulse-based cropping systems has been developed by conducting about 150 field experiments on 60 farmers' fields in five target districts over a period of three years. The benefit of optimum nutrient managements, especially integrated nutrient management, in conjunction with soil moisture conservation measure for rainfed pulses was demonstrated to the farmers. In a year of drought, farmers were able to harvest about 12-25 % more chickpea and 15- 28 % more lentil through proper nutrient management compared to their own practices in Bhopal and Raisen districts of Madhya Pradesh. The best INM treatment was: 75 % recommended NPK plus 2.5 tones/ha FYM (on dry weight basis) plus soil moisture conservation measure for the *kharif* crop, and 50-75 % of recommended NPK depending on soil moisture stock for the *rabi* crops. In both seasons, seed inoculation with *Rhizobium* was considered a must. Integrated nutrient management (INM) technology has been developed for seven rainfed oilseed based cropping systems by conducting more than 300 field experiments on about 100 farmers' fields in 9 target districts of 7 states over a period of three years. Results showed that conjunctive use of different locally available organic manures along with fertilizers increased the seed yield of safflower, mustard, castor, soybean, sunflower, raya and groundnut by 24.5, 25.7, 33.2, 35.7, 40.7, 51.7 and 67.2 %, respectively over the existing nutrient management practice of farmers (which is generally 50% RDF).

One of the major problems for low adoption of integrated nutrient management in dry land conditions is poor availability of organic residues. Long-term experiments conducted at the CRIDA, Hyderabad have revealed that it is possible to generate about 4 tones/ha of nitrogen rich horse gram biomass through cover cropping by utilizing the post-seasonal rains and stored soil moisture. The incorporation of this biomass has shown a positive impact on the yield of the succeeding *kharif* crops like sorghum and sunflower and also resulted in improvement of soil health.

Table 11 : Effect of different levels of N, FYM and crop residues on yield of soybean and safflower and changes in fertility after 9 years

Treatment	Yield (kg/ha)		Sustainability yield index (SYI)		Available nutrient status (kg/ha)				
	Soybean	Saf-flower	Soy-bean	Saf-flower	OC (%)	N	P	K	S
N_0P_0	1104	748	0.30	0.19	0.30	142.5	4.40	481.0	17.50
$N_{20/40}P_{40}$	1770	1241	0.44	0.29	0.46	208.5	12.67	481.0	20.37
$N_{10/20}P_{20}$	1653	1070	0.40	0.25	0.44	198.0	12.62	481.0	18.70
FYM @ 6 t/ha	1972	1647	0.52	0.37	1.15	345.0	60.60	1295.4	20.17
FYM + $N_{10/20}$	2089	1814	0.55	0.39	1.20	355.8	59.20	1247.4	19.54
$N_{20/40}P_{40}$+ Zn	1805	1158	0.44	0.30	0.48	214.5	13.34	559.0	17.84
Residues + $N_{10/20}P_{20}$	1573	1146	0.31	0.21	0.64	238.8	15.20	624.0	18.67

Table12 : Effect of different levels of N, P, FYM and crop residues and their conjunctive use on yield, water use efficiency of soybean safflower cropping system under rainfed conditions

Treatment	**Yield (kg/ha)**		**Water use efficiency (kg/ha. Mm)**		**Available nutrient status (kg/ha)**			
	Soy-bean	**Saf-flower**	**Soy-bean**	**Saf-flower**	**N**	**P**	**K**	**S**
N_0P_0	1684	637	4.25	2.98	113.8	4.9	32.8	2.2
$N_{20}P_{13}$	2012	952	4.84	4.39	137.8	7.5	47.2	3.8
$N_{30}P_{20}$	2235	1225	5.56	6.34	151.2	8.7	58.3	4.6
$N_{40}P_{35}$	2387	1392	5.95	7.13	165.5	9.2	64.5	5.5
$N_{60}P_{35}$	2432	1460	5.91	7.40	167.5	10.1	68.4	5.9
FYM @ 6 t/ha + $N_{20}P_{13}$	2429	1701	5.83	11.21	170.0	11.1	81.1	6.7
Residues @ 5 t/ha + $N_{20}P_{13}$	2139	1329	5.13	8.03	146.1	8.4	55.6	4.4
FYM @ 6 t/ha	2294	1616	5.88	10.49	161.3	9.2	62.4	4.9
Residue @ 5 t/ha	2042	966	4.88	5.32	138.6	8.4	46.6	3.6

3.3.4 Role of biofertilizers in soil fertility and agriculture

Although biofertilizers are not mutually exclusive to chemical fertilizers in achieving maximum crop yields, these play a vital roles in amending soil fertility, crop productivity and production in agriculture and are eco-friendly. Under rainfed conditions, the use of chemical fertilizers have their own limitations due to water deficit conditions. Some of the important functions played by biofertilizers in agriculture are (i) supplementing chemical fertilizers for meeting the integrated nutrient demand of the crops, (ii) can adding 20-200 kg N/ha year under optimum soil conditions and thereby increasing 15-25 % of total crop yields, (iii) minimizing the use of chemical fertilizers not exceeding 40-50 kg N/ha under ideal agronomic and pest-free conditions, (iv) increasing the mineral and water uptake, root development, vegetative growth and nitrogen fixation, (v) stimulating the stimulate production of growth promoting substance like vitamin-B complex, Indole acetic acid (IAA) and gibberellic acids etc. *(eg, Rhizobium BGA, Azotobacter* sp), (vi) phosphate mobilization /solubilization to the extent of 30-50 kg P_2O_5/ha, by converting insoluble soil phosphate into soluble forms by secreting several organic acids and under optimum conditions and thereby increasing crop yields by 10-20 %, (vii) enhancing the uptake of P, Zn, S and water, leading to uniform crop growth and increased yield and also enhance resistance to root diseases and improve hardiness of transplant stock (VAM), (viii) librating the growth promoting substances and vitamins and help to maintain soil fertility, (ix) serving as antagonists and suppressing the incidence of soil borne plant pathogens, (x) serving as cheaper, pollution free and renewable energy sources, (xi) improving physical properties of soil, soil tilth and soil health, in general, (xii) reclaiming alkaline soils (Blue green algae like *Nostoc, Anabaena, and Scytonema*), (xiii) enhancing (Bio-inoculants containing cellulolytic and lignolytic microorganisms) the degradation/ decomposition of organic matter in soil/compost pit, (xiv) leading to nitrogen economy (BGA) of rice fields in tropical regions, (xv) promotion of seed germination and initial vigour (*Azotobacter*)of many non-leguminous crop plants by producing growth promoting substances (xvi) fixing 100-150 kg N/ha (*Azolla-Anabaena* grows profusely as a floating plant in the flooded rice fields) leading to production of approximately 40-60 tones of biomass, and (xix) playing an important role in the recycling of plant nutrients.

Table 13 : Effect of co-inoculation of PGPR with *Bradyrhizobium japonicum* on nodulation and seed yield of soybean

Treatment	Nodules/ Plant (No)	Nodule dry biomass (mg/plant)	Shoot dry biomass (g/plant)	Seed yield (kg/ha)	Per cent increase over control	Percent increase over Bj
Un inoculated (control)	8.67	141.11	4.01	1391	-	-
Bradyrhizobium japonicum (Bj)	28.67	163.33	6.00	1633	17.40	-
Bj + *Pseudomonas fluorescens isolate-1* (PF 1)	31.33	48.39	6.50	1759	26.46	7.72
Bj + *Pseudomonas fluorescens isolate IV* (PF IV)	30.11	160.22	6.80	1731	24.44	6.00
Bj + *Entrobacter* Spp (Z-2)	9.11	52.78	4.14	1885	35.44	15.37
Bj + *Proteus vulgaris* (PV)	19.22	45.56	5.31	1742	25.23	6.67
Bj + *Bacillus sphaeriticus* (Bs)	34.66	97.89	8.09	1785	28.32	9.30
Bj + *Klbiesiella planticola* (Kb)	35.33	164.44	4.11	1673	20.27	8.57
unidentified bacterial isolate (RP 7)	31.00	141.00	7.09	1836	31.99	12.43
unidentified bacterial isolate (RP 24)	19.78	33.44	6.98	1591	14.38	-
CD (P=0.05)	2.25	5.17	0.19	311	-	-

The studies conducted at Directorate of Soybean Research (Billore *et al.*, 2003 and 2006) concluded that co-inoculation with *Bradyrhizobium japonicum* and PGPR had beneficial effect on soybean symbiotic traits and productivity (Table 13).

Table 14 : Influence of AM fungi, PSB and phosphorus on nodulation, yield attributes and yield of soybean

Treatment	Nodule (No/plant)	Nodule dry weight (mg/plant)	Pods (No/ plant)	Seed Index (g)	Yield (kg/ha)
Uninoculated	12.6	55.7	29.8	10.76	1191
B. japonicum	18.8	117.7	34.2	10.46	1414
G. lamellosum	16.4	93.7	32.7	11.02	1350
P. striata	17.1	83.7	28.5	11.68	1483
B. japonicum + G. lamellosum	12.9	100.0	29.5	11.08	1452
B. japonicum +P. striata	17.4	92.0	29.3	10.76	1424
G. lamellosum +P. striata	30.0	164.3	28.7	11.56	1270
B. japonicum + 60kg P_2O_5/ha	20.2	124.7	31.1	12.30	1391
B. japonicum + 40 kgP_2O_5/ha	21.2	108.3	30.0	15.30	1260
G. lamellosum + 40 kg/ P_2O_5/ha	31.6	134.3	29.5	11.32	1727
P. striata + 40 kg/ P_2O_5/ha	22.3	127.0	26.0	10.40	1706
B. japonicum + G. lamellosum.+ 40 kgP_2O_5/ha	25.2	114.0	27.7	11.21	1685
B. japonicum + P. striata + 40 kg P_2O_5/ha	25.2	125.7	30.3	10.62	1685
G. lamellosum + P. striata + 40 kg P_2O_5/ha	20.2	117.7	30.6	10.84	1554
60 kg P_2O_5/ha	33.9	160.0	27.5	11.43	1659
40 kg P_2O_5/ha	20.8	90.7	31.7	10.92	1558
CD (P=0.05)	10.3	69.8	NS	NS	408

Vyas et al. (2005) revealed that the highest nodule number was recorded with 60 kg P_2O_5/ha closely followed by *G. lamellosum* + 40 kg P_2O_5/ha, and *G. lamellosum* + *P. straita*, while maximum nodule dry mass with *G. lamellosum* + *P. straita,* which was significantly higher over control, *P. straita* and 40 kg P_2O_5/ha. Seed inoculation of G. *lamellosum, P. straita, B.Japonicum* + *G. lamellosum, B.japonicum* + *P. straita* with 40 kg P_2O_5/ha and 60 kg P_2O_5/ha significantly improved soybean yield over uninoculated control to the tune of 39.3 to 45.0 % (Table 14).

Continuous cultivation of inoculated soybean followed by wheat for 3 years in Vertisols of Jabalpur, a significant response to *Bradyrhizobium* inoculation on soybean seed yields (increase by 10.6 %) was noted additional mineral N of 16 ppm in 0-30 cm layer was recorded after soybean. In case of wheat, additional yield due to *Azotobacter* inoculation recorded was 11.6 and 10.6 % under soybean-wheat and sorghum-wheat rotation, respectively, over uninoculated control. After harvest of wheat in soybean-wheat plots, there was 18.3 ppm additional mineral-N in comparison to soils under sorghum (non-fixing cereal control) - wheat rotation. The wheat yields obtained due to 120 kg N ha^{-1} without biofertilizers was close to the yield with 90 kg N ha^{-1} with biofertilizers indicating a saving of 30 kg N/ha.

3.3.5 Mixed biofertilizers

Mixed Biofertilizer formulations consisting of nitrogen fixing organisms and phosphate solubilizing bacteria (PSB) proved superior to individual inoculants. In wetland rice inoculation of 'Azophos' (*Azospirillum* and PSB) and PGPR (*Pseudomonas*) along with 75 % NPK gave maximum rice grain yield (6250 kg/ha), an increase of 5.8 % over 100 % NPK alone (5905 kg/ha) at Coimbatore. In on-farm trials on large field plots on rice at Amaravathi and Coimbatore, dual inoculation of *Azospirillum* and PSB at 100 % N, gave 5.35 tones/ha yield, whereas 100 % N alone yielded 4.77 tones/ha. The 75 % N + dual inoculation gave 5.03 tones/ha and, thus, saved 25 % nitrogen. In case of cotton, use of *Azotobacter* + PSB resulted in a significant increase in seed cotton yield (287 kg/ha) in the presence of chemical fertilizers (100 and 75% RDF) in a Vertisol at Parbhani. In black gram, there was highly significant improvement in nodulation, nodule mass and grain yield (122 kg/ha) over control. In sunflower, seed yield increase due to bacterization was 367 kg/ha. For growth of diazotrophs and PSB in mixed biofertilizer formulations, a new medium-yeast extract glucose molybdate agar (YEGMA) was developed which had glucose as carbon source instead of mannitol and 50 ppm sodium molybdate. In field trials on inoculation of mixed biofertiliser on pearlmillet, inoculation was found to save 25 % of the N dose (DARE/ICAR Annual Report 2003–2004).

3.3.6 Seaweed biofertilizer

Seaweeds are one of the most important marine resources of the world. Seaweed extracts have been marketed for several years as fertilizer additives and beneficial results from their use have been reported (Booth, 1965). The possibility of using seaweed in modern agriculture has been investigated by many workers (Thivy, 1961; Aitken and Sen 1965; Boney, 1965). Different forms of seaweed preparation such as LSF (Liquid Seaweed Fertilizers), SLF (Seaweed Liquid Fertilizers), LF (Liquid Fertilizers), and either whole or finally chopped powered algal manure have been used and all of them have been reported to produce beneficial effects on cereals, pulses and flowering plants. Seaweed manures have the advantage of being free from weeds and pathogenic fungi. Promising increased crop yield, nutrient uptake, resistance to frost and stress, improved seed germination of reduced incidence of fungal and inspect attack have been resulted by application of SLF. Seaweeds are known to contain appreciable quantities of plant growth regulators (Mooney and Van Staden 1985), Cytokinin (Smith and Van Staden 1984), IAA (Abe et al. 1972), gibberllins and gibberllin like substance (Bentley, 1960; Radley, 1961; Sekar et al. 1995). Hence marine algae, particularly seaweeds have a vital role to play in agriculture, especially in the third world countries where irrational use of chemical fertilizer and pesticides is a cause of concern. Extensive regional trials would need to be conducted with the product to determine the environmental limits on biological activity and monitor the survival and dispersal of the inocula (Davison 1988).

4. Dry Seeding

This option involves sowing rainy-season crops before the monsoon rains. Dry seeding overcomes the constraint of wet sowing and facilitates early germination. It results in early maturity of the crop, which eventually escapes terminal drought. Dry seeding was reported to increase yields of sorghum and cotton by 200-300 kg ha^{-1} (Joshi et al. 1998). Farmers obtained higher yields of sorghum (27 %) and cotton (38 %) by adopting dry seeding in Maharashtra (Bhole et al. 1998). It was further reported that cotton seed can remain in the soil for a longer time, and unlike seed of many other crops, is not affected by ants and other insects. Similarly, dry seeding of sorghum enabled farmers to double crop.

4.1 Use of improved seeds and moderate amounts of fertilizer

Improved varieties are an important component technology, and have gained popularity during the last two decades. The genetic potential of improved cultivars is utilized to increase production in Vertisols. Dry sowing involves less risk, higher profitability, is more responsive to fertilizers, matures early, and involves less shattering. Available farmyard manure (FYM) is generally not enough to meet crop requirement and nutrient mining. To improve soil fertility and overcome nutrient mining, Vertisol technology recommends the application of moderate amounts of chemical fertilizer that could be complemented by manure. Farmers who apply chemical fertilizer are convinced of the quick results and overcome the supply constraint involved with FYM.

4.2 Seed and fertilizer placement

Proper placement of seed and fertilizer can increase crop yields substantially. While seed is broadcast in traditional dry land agriculture, in Vertisol technology seed and fertilizer are placed together in order to minimize nutrient loss and make optional use of available moisture. Dibbling and using a seed drill are common practices adopted by farmers sowing seed and placing fertilizer, in order to maintain proper row arrangements in cotton and groundnut. Seed drills are commonly used to sow sorghum, wheat, pigeon pea, and chickpea. Joshi et al. (1998) observed a unique practice in Begumgunj village, where a mixture of seed and fertilizer was placed using either a drill or plough. Although unusual, this modified form of the technology went to prove that farmers accepted the placing of seed and fertilizer together. However, this practice may be avoided in sensitive crops like soybean, which may have adverse effect on placement of seed and fertilizer together.

4.3 Plant protection measures

Since rainfed agriculture is highly prone to diseases and pests, application of modest doses of insecticides/pesticides was advocated to protect plants. With the advent of High-Yielding Varieties (HYVs), the incidence of diseases and insect/pests has increased. Though several insect species attack during various stages of crop growth, those doing so at the reproductive stage are of major economic significance. Since the indiscriminate use of pesticides raises issues of sustainability and environmental pollution, improved plant protection measures are considered one of the essential components of Vertisol technology.

The components of Vertisols technology are basically a collection of innovations from which a farmer can choose, based on his need and benefit. Farmers may either adopt the complete package or part of it, or even a modified form of its recommendation (Byerlee and Polanco, 1986; Ryan and Subramanyam, 1975). Despite substantial investments in time, resources, and capital, there is insufficient research on the nature and extent of adoption of Vertisols technology, its benefits, and constraints.

4.4 Double cropping

Double cropping involves utilizing land for two growing seasons instead of one. This technology option helps in using the land for production for up to eight months in summer instead of four. In Vertisol areas where rainfall is greater than 750 mm year^{-1}, two crops are feasible without irrigation. Double cropping makes effective use of other fixed costs and production resources such as moisture, human labor, bullock time, and cultivation tools. During on-farm testing, several new cropping systems were recommended and verified for feasibility and adaptability.

Both sequential crops and intercrops that require two seasons to mature are considered options; however, farmers view moisture limitation as a major constraint to double cropping. There is, however, an opportunity to increase intensity through an intercrop (Table 15).

Table 15 : Effect of strip cropping on maize and wheat at 4% slope

Cropping system	Water loss(%)	Soil loss (kg/ha)	Maize yield (kg/ha)	Wheat yield (kg/ha)
Strip cropping 3:1	40.7	13.74	2,361	2,935
Strip cropping 2:1	37.3	11.29	2,373	2,761
Contour sowing	42.9	20.90	2,107	2,406
Cultivated fallows	51.8	54.08		

Source : Bhardwaj (1994)

5. Intercropping

Intercropping has been recognized as a potentially beneficial system of sustainable crop production in semi-arid tropics. Subsequent evidences affirm the utility of the concept in realizing substantial yield advantages over their sole cropping. The advantages may especially be important as these are achievable not by means of costly inputs, but by simple expedient of growing crops together (Willey, 1979a). In addition to making optimum use of available resources, the potentials of intercropping have been highlighted as late as 40s (Donald, 1946 and Aiyer 1949). Various aspects of intercropping systems were also in-depth looked into by Donald (1961 and 1963) and Francis and Stern (1987). Willey (1979a, b) made a critical analysis of the advantages accrued from the system. Salter et al. (1985) further focused the attention on the advantages of intercropping under mechanized farming. Chatterjee and Mandal (1992) also reviewed this subject and documented the trends in research on intercropping in early 90s.

The provision of species diversity (Holdridge, 1959 and Igbozurike, 1971) by mixed cropping is considered as major advantage over sole cropping. This diversification tends to promote yield stability because all the crops in a mixed cropping culture are not likely to be affected by weather vagaries or pests and diseases. Yield stability and protection against crop failures are the primary reasons prompting a farmer to resort to mixed cropping (Doutt and Nakata 1965; Dikinson, 1972; Flinn and Lagemann 1980; Jodha, 1976; Olukosi, 1976). An interfering row strips in mixed cropping can act as physical barrier to insects (Altieri et al. 1978; Gerard 1976; Suryatra and Harwood 1976; Satpathy et al. 1977) and decrease in the spread of diseases further confirming the yield stability of the system (Koli, 1975 and Shoyinika, 1976).

Billore et al. (2004) revealed that the planting of soybean and maize either in 4:2/ 2:2 row ratio (30 cm) or 100 + 50% seed mixture proved to be better over other treatments of seed mixture by giving highest total productivity, monetary advantage and land equivalent ratio with better companion indicating low competition interference and high energy output and energy use efficiency and energy productivity (Table 16). Of the six soybean varieties tested, NRC 37 (Ahilya 4), PK 1029 and PK 1024 were found to be most compatible with pigeon pea variety ICPL 871 19 in 4:2 row ratio as adjudged by higher yield levels, soybean equivalent yield, land equivalent ratio (LER), relative crowding coefficient (RCC), monetary returns and income equivalent ratio (IER) with low competition ratio. Aggressivity values indicated that pigeon pea dominated soybean varieties except JS 335, PK 416 and JS 90 41. This system also produced higher energy, energy use efficiency and energy productivity. PK 1024 and PK 416 with pigeon pea ICPL 871 19 was found the most energy intensive system. Soybean-pigeon pea intercropping system is most suited where no rabi cropping is feasible on account of limitations of irrigating water (Table 17).

The maximum soybean and pigeon pea yield was with FYM @ 5 t/ha + 75% of RDF and Zn @ 5 kg/ha +RDF (Table 18). Significant reduction in seed yield of soybean (18 to 36%) and pigeon pea (32 to 47%) was noticed when planted in intercropping systems at different fertility levels as compared to their sole plantings. The maximum soybean equivalent yield and land equivalent ratio was recorded with RDF + Zn during both the years. The application of Zn + RDF produced the maximum net returns and remained at par with Zn + 75% of RDF and RDF alone. While the highest B:C ratio was associated with Zn + 75% of RDF (Billore et al. 2009).

Table 16 : Effect of inter and mixed cropping of soybean + maize on productivity and competition functions

Treatment	Seed yield (kg/ha)		Soybean equivalent yield (kg/ha)	LER	CR	A	RCC	Monetary advantage (Rs/ha)	IER
	Soy-bean	Maize							
Sole soybean	1696	-	1696	1.00	-	-	-	15264	1.00
Sole maize	-	2685	1492	1.00	-	-	-	7460	1.00
Soybean + Maize (4:2)	1164	1433	1960	1.22	1.30	-0.09	2.48	3181	1.16
Soybean + Maize (2:2)	1300	1254	1997	1.24	1.64	0.12	2.85	3479	1.18
Soybean + Maize (100 + 10 %)	1438	636	1791	1.09	3.54	0.64	1.72	1331	1.05
Soybean + Maize (100 + 20 %)	1388	803	1834	1.12	2.73	0.49	1.94	1768	1.08
Soybean + Maize (100 + 30 %)	1261	1125	1885	1.16	1.76	0.18	2.08	2340	1.11
Soybean + Maize (100 + 40 %)	1228	1034	1802	1.10	1.85	0.22	1.62	1474	1.06
Soybean + Maize (100 + 50 %)	1219	1269	1924	1.19	1.53	0.06	2.28	2765	1.13
C D (P=0.05)	299	1228	252	0.14	0.87	0.27	0.47	7332	0.05

Table 17 : Yield, yield attributes and competition functions of soybean with pigeon pea intercropping

Treatment	Seed yield kg/ha		Soybean equivalent Yield (kg/ha)	LER	CR	Agre-ssivity	*RCC*	CI	Monetary advantage (Rs/ha)	IER	
	Soy-bean	Pigeon-pea								Soy-bean	Pigeon-pea
Sole JS 335	1653		1653	1.00						1.00	
Sole PK 1024	1198		1198	1.00						1.00	
Sole PK 1029	1199		1199	1.00						1.00	

Sole NRC 37	2070		2070	1.00						1.00	
PK 416	1315		1315	1.00						1.00	
Sole JS 90-41	1480		1480	1.00						1.00	
ICPL 87119		2422	3726	1.00							1.00
Intercropping											
JS 335 + ICPL- 87119	1366	898	2747	1.19	2.22	0.21	7.56	0.11	3706	1.66	0.73
PK 1024+ ICPL- 87119	927	1220	2804	1.27	1.54	-0.07	8.90	0.11	5037	2.34	0.75
PK 1029 + ICPL- 87119	1067	1250	2990	1.41	1.71	-0.03	18.77	0.05	7346	2.49	0.80
NRC-37 + ICPL- 87119	1837	1308	3338	1.43	1.65	-0.07	19.32	0.05	8609	1.64	0.90
PK 416 + ICPL- 87119	1046	1006	2797	1.20	1.93	0.06	8.36	0.12	3935	2.12	0.75
JS 90-41 + ICPL - 87119	1022	925	2445	1.07	1.81	0.10	5.23	0.19	1352	1.65	0.65
CD (P=0.05)	212	953	1390	0.26	0.37	0.17	9.36	0.08	4049	0.59	0.13

Table 18 : Effect of different levels of fertilizer on productivity, competition functions and monetary advantages from soybean + pigeon pea intercropping system

Treatment	Yield (kg/ha)			LER	CR	A	RCC	Monetary advantage (Rs/ha)	Income equivalent ratio	
	Soy-bean	Pigeon - pea	Soybean equivalent						Soybe an	Pigeon-pea
RDF	1860	1708	3994	1.40	1.29	0.03	5.99	13694	1.70	1.15
75% of RDF	1657	1699	3645	1.31	1.15	-0.07	3.73	10351	1.55	1.05
50% of RDF	1503	1471	3342	1.17	1.21	-0.03	1.98	5827	1.42	0.96
FYM @ 5 t/ha	1678	1549	3614	1.27	1.27	0.03	3.11	9220	1.54	1.04
FYM + RDF	1873	1646	3955	1.39	1.36	0.08	5.65	13316	1.68	1.14
FYM + 75% of RDF	1937	1632	3492	1.40	1.41	0.12	6.60	11973	1.49	1.00
FYM + 50% of RDF	1664	1506	3547	1.25	1.31	0.04	2.86	8513	1.51	1.02
FYM @ 5 t/ha + Zn @ 5 kg/ha	1642	1508	3528	1.24	1.30	0.03	2.74	8194	1.50	1.01
FYM + Zn + RDF	1848	1697	3970	1.40	1.30	0.03	5.73	13611	1.69	1.14
FYM + Zn + 75% of RDF	1920	1567	3875	1.38	1.46	0.15	5.70	12804	1.65	1.11
FYM + Zn + 50% of RDF	1642	1532	3556	1.25	1.27	0.02	2.83	9134	1.62	1.09
Zn + RDF	1839	1907	4223	1.46	1.15	-0.09	7.79	15966	1.80	1.21
Zn + 75% of RDF	1899	1775	4117	1.45	1.27	0.02	7.44	15332	1.75	1.18
Zn + 50% of RDF	1601	1565	3557	1.24	1.21	-0.03	2.73	8261	1.51	1.02
Sole soybean –RDF	2351	-	2351	1.00	-	-	-	-	1.00	-
Sole pigeon pea - RDF	-	2787	3484	1.00	-	-	-	-	-	1.00
CD (P=0.05)	232	237	517	0.16	0.98	0.07	2.19	3410	0.13	0.09

CR = Competition ratio, A= Aggressivity, RCC= Relative crowding coefficient

5.1 Supplemental irrigation

Supplemental irrigation is a method implied in low rainfall areas to assure enough water to the crops to survive, sustain growth and yield (Oweis et al. 1999; Perrier and Alinki 1987). While water harvesting is apt in areas that receive between 100 – 300 mm of rainfall annually, supplemental irrigation is suitable in areas with a slightly greater annual rainfall approximately 300–600 mm The goal of supplemental irrigation is to provide enough water during critical growth stages to produce optimal yield per unit of water, not to provide stress-free conditions throughout the growing season with the aim of producing maximum yield (Oweis et al. 1999).

This differs from conventional irrigation in that the amount of water applied in supplemental irrigation would not by itself be sufficient to ensure crop growth. Conversely, conventional irrigation supplies the entire water needs to the crop because rainwater may not provide sufficient water for plant growth for all or part of the season (Perrier and Alinki 1987). Conventional irrigation is used in regions where water is plentiful, while supplemental irrigation is often used in places where water is often scarce. Timing of water application is one of the most important factors to be determined when using supplemental irrigation. Supplemental water applications are especially important when water is scarce during critical growth periods. Oweis et al. (1999) present a rule of thumb that for crops, supplemental irrigation is needed when the soil water content drops to a rate of 50 % of available water in the root zone. Other crops such as potatoes and vegetables will produce better when soil water is kept within the top 35 % of available water.

Potential benefits that can be achieved through the use of supplemental irrigation include increased yields, stabilization of yields across years, and creating conditions that allow for the use of higher technology inputs such as high- yielding varieties, herbicides and fertilizers (Oweis et al. 1999). Research at the ICARDA has shown that water use efficiency can be greater under supplemental irrigation than under rainfed agriculture. The research finding has revealed that the application of a cubic meter of water at a time of water stress, combined with good management increased water use efficiency more than twice over that of rainfed production (Oweis et al. 1999).

6. Loss of Biodiversity due to Horizontal Expansion

Biodiversity losses can develop from the clearing of areas to be used for agriculture. When these areas are cleared, many plants native to the area may be lost, and disease and pest problems may also develop due to changes in the ecosystem. Soil erosion is also often a significant problem in areas of agricultural expansion. Many of the marginal areas to which agriculture expands in the developing world include hillsides and arid areas, which make soil erosion a particular concern. Three primary ways to enhance rainfed yields are examined, increasing effective rainfall use through improved water management, particularly water harvesting; increasing crop yields in rainfed areas through agricultural research; and reforming policies and increasing investments in rainfed areas.

7. Agricultural Research to improve Rainfed Crop Yield

It is a common perception that rainfed areas did not benefit much from the Green Revolution. On the contrary, breeding improvements have enabled modern varieties to reach in many rainfed regions. It is considered that over the past 10-15 years, most of the area expansion through the use of modern varieties has occurred in rainfed areas, beginning first with wetter areas and proceeding gradually to more marginal areas. Although adoption rates of modern varieties in rainfed areas are catching up with irrigated areas, the yield gains in rainfed areas remain lower. The high heterogeneity and erratic rainfall of rainfed environments make plant breeding a difficult task. For faster growth, it is warranted to target rainfed environment for development of varieties as the ones developed for irrigated environment may not pay off to that extent.

Both conventional and non-conventional breeding techniques are being used to increase rainfed yields. Three major breeding strategies encompass research to increase harvest index, to increase plant biomass, and to increase stress tolerance (particularly drought/thermo resistance). The first two methods increase yields by altering the plant architecture, while the third focuses on increasing the ability of plants to survive stressful environments. The first of these may have only limited potential for generating further yield growth due to physical limitations, but there is considerable potential from the latter two. If agricultural research investments can be sustained, the continued application of conventional breeding and the recent developments in non-conventional breeding offer considerable potential for improving cereal yield growth in rainfed environments. Use of biotechnological approach coupled with conventional breeding for the development of stress tolerant varieties for rainfed regions can help in increasing the yield levels of crops in rainfed regions.

Participatory plant breeding is likely to play a key role for successful yield increases through genetic improvement in rainfed environments. Farmer participation and his input in the very early stages of selection may help to fit the crop to a multitude of target environments and user preferences. Participatory plant breeding may be the only possible type of breeding for crops grown in remote regions, particularly for minor crops not attended by the formal breeding.

In order to assure effective breeding for high stress environments, the availability of diverse genes will be essential. It is therefore essential that the tools of biotechnology, such as marker-assisted selection, cell and tissue culture techniques and transgenic breeding are to be employed for crops in developing countries. Presently, by and large, the molecular biotechnology and the transgenic breeding is confined to the crops of commercial importance and by few life science companies at global level. There is a need to diversify these efforts by the public and private companies in the developing countries to facilitate farmers with varieties suitable for rainfed areas.

Breeding strategies for both irrigated and rainfed areas in the past have mainly emphasized on yield maximization. For example, the maize seed presently distributed comes from varieties selected without sufficient attention to agronomic and socioeconomic factors that limit production: periods of drought and/or frost during the crop's vegetative stage, soil exhaustion, grain type and color, and forage qualities (Hibon et al. 1992). More recent crop genetic improvements through both conventional and non-conventional breeding have placed more of an emphasis on altering specific characteristics of the crop. Conventional breeding uses whole plants to select desired characteristics. These techniques have generally focused on maximizing yields through increased plant productivity and resistance to stress. Non-conventional breeding uses cellular and molecular biology techniques (such as marker-assisted selection and cell and tissue culture techniques), which allow plants to be screened more quickly in the laboratory rather than the field. These techniques can also more rapidly and efficiently select for particular characteristics that may increase tolerance to diseases, pests, and adverse weather conditions, thus increasing yields. Non-conventional breeding also includes the transfer of genetic material from one species into another, creating transgenic or genetically modified organisms (GMOs) with beneficial characteristics that cannot be achieved through conventional breeding. Above described techniques are essentially required to increase the traits responsible for increasing the productivity.

7.1 Increasing the harvest index

One method to improve yield is to increase the harvest index (HI), but this technique may have only limited potential for generating further yield growth (Cassman 1999; Evans 1998). The harvest index is defined as "the ratio of grain to total crop biomass" (Cassman 1999).

7.2 Increasing total plant biomass

Another strategy to increase yield is to increase the total dry matter (biomass per unit of land and inputs such as water and nitrogen) by increasing plant productivity. The development of hybrid varieties and increased photosynthesis are the techniques that focus on increased biomass.

7.3 Hybridization

Hybrid varieties increase plant productivity and provide greater yields than conventional varieties, making use of hybrid vigor, the phenomenon in which the progeny of two distinctly different parents grow faster, yield more, and resist stress better than either parent. Hybrid varieties of all of the major cereals have been developed throughout the past several decades.

7.4. Increased photosynthesis

Increased photosynthesis or radiation use efficiency aims to improve carbon accumulation productivity, contributing to greater biomass production. One example of an attempt to increase photosynthesis is research to create C_4 rice plants. Tropical species such as maize, sorghum, and sugarcane are C_4 plants, which have evolved a more efficient photosynthesis mechanism than C_3 plants such as soybean, rice, barley, and wheat. C_3 plants may lose up to 50 % of their recently fixed carbon through photorespiration, while the carbon loss of C_4 plants is greatly reduced or completely inhibited; therefore, C_4 plants are more advantageous in their photosynthetic productivity. It should be noted, however, that C_4 species do not always have higher water use efficiency. Under non- ideal conditions such as drought, there are more tolerant C_3 species such as cowpea and cotton than the comparatively sensitive C_4 maize and sorghum cultivars (Ong et al. 1996).

7.5. Breeding for the target environment by increasing stress tolerance

Earlier research focused primarily on changing the plant architecture for enhancing yields. However, the ongoing research approaches involving genetic diversity in plant breeding aims at increasing yields, which can last over a wide range of environments. This can be observed in the CIMMYT/ICARDA wheat program, in which about 35 % of the bread wheat crossing program is devoted to abiotic stress tolerance (i.e. heat, drought, or cold) and about 44 percent towards pests and diseases. Varieties are selected under multi- location testing that represent wider range of biotic and abiotic stresses (ICARDA, 1997).

Genotype by environment (G x E) interaction is influential in limiting breeding program efficiency (Ceccarelli et al. 2000). Two strategies to address G x E interactions include: 1) avoidance by selecting cultivars that are adapted to the complete range of target environments, or 2) exploitation through the selection of several different cultivars, each of which are specially adapted to a subset of target environments. Selection for specific adaptation is especially significant in breeding crops for unfavorable conditions, because unfavorable environments tend to be more heterogeneous than favorable environments (Ceccarelli et al. 2000). Successful breeding with consideration of environmental adaptability includes Sooty-Rascon, a new variety of durum wheat bred by CIMMYT for drought-prone environments in WANA regions. This variety produced at least 3.4 tons per hectare regardless of drought severity (CIMMYT 2000).

7.6. Drought resistance

Breeding for drought resistance has a substantial positive impact on rainfed crop yield, as moisture stress is the most pronounced constraint throughout rainfed environments. The trade-off between yield and stress tolerance has been a difficult task for plant breeding. For example, a crop variety with a short growing season escapes the drought as may mature before it occurs, but during extended rainy years its yields are likely to be less than that of a long growing season variety. This complexity has slowed the development of drought-resistant crop varieties (OTA commissioned paper 1983). However, early maturing varieties may compensate for the yield loss by enabling double cropping. For instance, in the Bihar plateau in India, a switch to short duration varieties (85-105 days) from medium duration local varieties (120-130 days) in the medium and uplands helped the plants escape periodic drought during the rainy season and helped increase cropping intensity, as they fit well in sequential or mixed/intercropping patterns (Bagchi et al. 1995).

Introducing drought tolerance is probably one of the most difficult tasks for plant breeders. This difficulty originates from the on farm diversity and unpredictability drought conditions and also from the diversity of drought tolerance strategies developed by plants, which may be targeted and used as selection criteria. However, substantial progress has been made in recent years related to the physiology, genetics and molecular biology of drought tolerance in different plant species. Functional genomics increases understanding of drought tolerance control and defines strategies for crop improvement.

In several crop species, genetic maps have helped to identify chromosomal regions controlling some traits related to drought stress response. Cereal crops such as maize, sorghum, rice, wheat and barley have been studied to identify regions controlling characteristics such as phenology, root characteristics, plant architecture and growth, photosynthesis, chlorophyll amount or "stay green" character, and water-use efficiency (This and Teulat-Merah 2000).

The limiting factors in rainfed production are drought, temperature extremes, low nitrogen, low phosphorus, soil acidity, high aluminum saturation, and micronutrient deficiencies or excesses, making farm- level yields and aggregate production low and unstable (CIMMYT 1999b). While half the area sown to wheat in developing countries and up to 70 % of that grown in developed countries suffers from periodic drought (Trethowan and Pfeiffer 2000), wheat has been found to be relatively drought hardy, unlike maize, which may fail completely if the flowering stage is delayed beyond a critical threshold due to drought (Bolaños and Edmeades 1993). Breeding for drought tolerance in wheat, therefore, should focus more on improving overall radiation use efficiency under stress rather than reproductive stages of growth and partitioning (Reynolds et al. 2000).

7.7 Combining desirable traits

Combining desirable traits such as increased biomass production, stress tolerance, and quality traits such as cooking quality are becoming a common strategy of breeders. With attention placed on breeding the plant to fit the particular environments and farmers' preference, a high adoption rate can be expected. The major problem with the resource-poor farmers of developing countries is inability of using the farm saved seed for sowing in the next season and investment in purchasing fresh seed for sowing. This can be addressed by the breeders by developing a technique to produce seed through 'apomixis' so that the hybrid trait can be retained form one year to next. .

7.8 Modern farming methods

In addition to the water harvesting and conservation methods discussed above, the use of more modern farming techniques are to be utilized to help conserve soil/moisture and make more effective use of incipient rainfall. Conservation tillage measures such as conservation till and no till are effective to conserve soil water and decrease the rate of soil water evaporation. Precision agriculture has been suggested for use in developing countries. Along with research on integrated nutrient management, applied research to adapt conservation tillage technologies for use in unfavorable rainfed systems in developing countries could have a large positive impact on local food security and increased standards of living.

Conservation and no-tillage techniques may be especially helpful in areas where farmers do not have the capital or labor required for other techniques. The usefulness of these methods will also depend, however on the soil texture and structure. Fall and Faye (1999) discussed the use of three seedbed preparation techniques (tillage in dry soil conditions, scarification and different sizes of sweeps, and direct seeding with no tillage) in dry soil for better soil-water management. Tillage in dry soil conditions is not likely to be sustainable over time as it can promote erosion from wind and first rain events. Scarification of the soil surface, a method adopted in semi-arid region, allows for protection against erosion and runoff due to the addition of crop residues to the soil. Direct seeding in dry soil has been used in groundnut production. Advantages to this method include: reduced production costs, diminished soil erosion, decreased runoff, less soil compaction, and better timeliness in seeding (Fall and Faye 1999). Smallholders have also adopted dry seeding, planting on ridges, and minimum and no tillage, although some of the farmers do not seem to be aware of the soil and water conservation properties of these techniques as the methods were passed down from previous generations (Misika and Mwenya 1999). The commercial farming sector has experienced even greater success than the small–scale farmers, having practiced conservation tillage methods for over 15 years (Misika and Mwenya 1999). Increased maize yields from 1.8 to 4.8 tons per hectare were found when breaking up the plough-pan (FAO 2000).

No-tillage technology (often used with mulching) has been found to improve soil moisture conservation and thus reduce crop failure in dry years, particularly in arid or semi-arid areas. Additional soil improvements such as enhanced soil structure and increased organic matter content have also resulted. Ekboir *et al.*, (2001) found that farmers using no-till with mulch were able to reduce cash and labor investments and also experienced greater yields. However, despite the apparent benefits of conservation tillage, the use of mulch and related technologies for rainfed agriculture, there has been very little farm level adoption of these technologies. In many instances the technology is very location specific and requires high levels of investments for tailoring it to specific conditions and to disseminate it to particular groups of farmers.

Precision agriculture has had some success in developed countries. While traditional agricultural techniques have tended to apply the same management to an entire field, precision agriculture methods focus on information technology using site-specific soil, crop and other environmental data to determine specific inputs required for certain sections of a field. Many of these methods involve the use of technologies such as geographic information systems (GIS), satellites, and remote sensing.

Precision agriculture can directly increase crop yields, and also improve water availability through greater relative infiltration of rainfall. In developing countries, the smaller farm sizes could allow for management on a field basis. Precision agriculture may hold significant promise in the future for agriculture in developing countries, as nutrient levels can vary greatly from field to field. A recent study showed that variations in rice yields from 2400 to 6000 kg/hectare in 42 different fields were attributed to differences in soil nitrogen (Cassman 1999). More accurate analysis of soil nutrient levels could assist farmers in determining fertilizer levels specific to different areas in the field. However, a huge hurdle to overcome in implementing precision agriculture in developing countries is the availability of necessary data to determine these site-specific inputs, and the investment cost of obtaining and utilizing this data.

8. Prospects for the Future

If agricultural research investments can be sustained, the continued application of conventional breeding and the recent developments in non-conventional breeding offer considerable potential for improving yield growth in rainfed environments. Crop yield growth in farmers' fields will come both from incremental increases in the yield potential in rainfed and irrigated areas and from improved stress resistance in diverse environments, including improved drought tolerance (together with policy reform and investments to remove constraints to attaining yield potential, as discussed in the next section). The rate of growth in yields will be enhanced by extending research both downstream to farmers and upstream to the use of tools derived from biotechnology to assist conventional breeding, and, if concerns over risks can be solved, from the use of transgenic breeding. Participatory plant breeding plays a key role for successful yield increases through genetic improvement in rainfed environments (particularly in dry and remote areas).

Farmer participation in the very early stages of selection helps to fit the crop to a multitude of target environments and user preferences (Ceccarelli et al. 1996; Kornegay et al. 1996). Ceccarelli et al. (1996) demonstrate that participatory plant breeding may be the only possible type of breeding for crops grown in remote regions; a high level of diversity is required within the same farm, or for minor crops that are neglected by formal breeding. The study also suggests that the breeder was more efficient in selecting higher yielding entries in a high rainfall area, while the farmers were more efficient in selecting under high stress conditions (Ceccarelli et al. 2000).

Moving upstream, with the progress in mapping crop genomes, the similarities between crops will facilitate the integration of genetic information and contribute to the faster identification of desirable traits. In order to assure effective breeding for high stress environments, the availability of diverse genes is essential. It is therefore essential that the tools of biotechnology, such as marker-assisted selection and cell and tissue culture techniques, be employed for crops in developing countries, even if these countries stop short of true transgenic breeding.

To date, however, application of molecular biotechnology has been limited to a small number of traits of interest to commercial farmers, mainly developed by a few life science companies operating at a global level.

If successfully tapped, biotechnology will make an extremely important contribution to future crop yield growth, particularly in difficult rainfed environments. The debate over genetically modified organisms may significantly delay current crop improvement efforts and move the release dates of many improved varieties further into the future. When biotechnology moves beyond the application of molecular biology to assist conventional breeding to the creation of transgenic crops, it may introduce risks associated with the release of genetically modified material into the environment. These risks could include genes 'jumping' from genetically modified plants to other plants through cross-pollination, rapid creation of new pest biotypes through adaptation to genetically modified plants, and allergic reactions to the consumption of genetically modified foods. These risks are not well understood and they provoke a great deal of anxiety among some segments of the public. National institutions must have the capacity to evaluate these risks, to adapt and regulate breeding and crop management strategies to minimize these risks, and to implement and rigorously enforce appropriate regulatory systems.

References

Abe H, Vchiyams M, and Sato R. 1972. Isolation and identification of nature action in marine algae. *Agro. Biol. Chem.* 36:2259-2260.

Abrol I P. 1999. Sustaining rice-wheat system productivity in the Indo-Gangetic Plain: Water management related issues. *Agriculture and Water Management* 40: 31–35.

Abu-Awwad A M and Shatanawi M R. 1997. Water harvesting and infiltration in arid areas affected by the surface crust: examples from Jordan. *Journal of Arid Environments* 37(3): 443-452.

Aitken J E and Senn J L. 1965. Seaweed products as fertilizer and sal conditions. *Bot. Mar.* 8:144-148.

Aiyer A K Y N. 1949. Mixed cropping in India. *Indian Journal of Agricultural Sciences* 90(4): 439-543.

Altieri M A, Francis C A, Schoonhoven Van A and Doll J D. 1978. A review of insect prevalence in maize and bean poly-cultural system. *Field Crop Research* 1(1): 33-49.

Anonymous. 2008. http/www. fao.com.

Bagchi D K, Bnik P, and Sasmal, T.1995. Selection of appropriate technologies for upland rainfed rice growing on the Bihar Plateau, India. *In:* Proceedings of the International Rice Research Conference, February 13-17, 1995, Manila, Philippines.

Barghouti S M. 2001. Enhancing natural assets in less favourable areas: the case of the semi-arid tropics. *Sustainable Development International*, Edition 1.

Bentley J A. 1960. Plant hormones in marine phytoplankton, zooplankton and seawater. J. *Mar. Biol.Ass.* U.K., 39:433-444.

Beet W C. 1990. Raising and sustaining productivity of smallholder farming systems in the tropics. Alkmaar, Holland: Agbe Publishing.

Bhalla G S, Hazel P and Kerr J.1999. *Prospects for India's cereal supply and demand for 2020.* Food, Agric. and Env. Discussion Paper 29, IEPRI, Washington, DC, 24 p.

Bhale V M and Wanjari S S. 2009. Conservation Agriculture: A new paradigms to increase resource use efficiency. *Indian Journal of Agronomy* 54(2): 167-177.

Bhandari A L, Ladha J K, Pathak H, Padre A T, Dawe, D, and Gupta R K. 2002. Yield and soil nutrient changes in a long-term rice-wheat rotation in India. *Soil Science Society of America Journal* 66: 848–856.

Bhardwaj S P.1994. Bunding and strip cropping for erosion control in agricultural land of Doon valley. *Indian Journal of Soil Conservation* 22 (3):15-19.

Bhattacharyya T, Pal D K, Easter M, Williams S, Paustian K, Milne E, and Chandran P. 2007. Evaluating the century carbon model using long-term fertilizer trials in the Indo-Gangetic Plains, India. *Agricultural Ecosystems and Environment* 122: 73–83.

Bhaskar, K.S. 2002. Watershed concept and yield optimization under dry land farming. In A.M. Dhopte (ed.) *Agrotechnology for dry land farming*, Scientific Publishers, Jodhpur, India: 27–60.

Bhole B D, Alshi M R, Joshi P K, Bantilan M C S, and Chopde V K. 1998. Dry seeding in the Vidarbha region of Maharashtra. Pages 197-201 In: *Assessing joint research impacts: proceedings of an international Workshop on Joint Impact Assessment of NARS/ICRISAT Technologies for the Semi-Arid Tropics*, 2-4 Dec 1996, ICRISAT, Patancheru, India (Bantilan M C S, and Joshi P K, eds.). Patancheru 502 324, Andhra Pradesh, India: International Crops Research Institute for the Semi-Arid Tropics.

Billore S D, Joshi O P and Vyas A K. 2004. Biological and economical efficiency of inter and mixed cropping of soybean and maize under Malwa plateau conditions. In: *National Symposium on Resource Conservation and Agricultural Productivty*. Organised by ISA, ICAR and PAU held at PAU, Ludhiana on 22-25.11.2004. Pp.470-1.

Billore S D and Joshi O P. 2004. Screening of soybean varieties for their suitability for intercropping with pigeon pea. *Journal of Oilseeds Research*, 21(2): 354-6.

Billore S D, Vyas A K, and Joshi OP. 2009. Effect of integrated nutrient management in soybean and pigeon pea intercropping on productivity, energy budgeting and competition functions. *Journal of Food Legumes* 22(2):124-126.

Billore S D, Vyas A K, and Joshi O P. 2003. Prformance of co-inoculation of plant growth promoting rhizobacteria (PGPR) with Bradyrhizobium japonicum on soybean symbiotic characteristics and productivity. In: Abstracts and Short papers, 6th International workshop on PGPR held on 5-10 October 2003 at IISR, Calicut organized by IISR and ICAR. pp.113-115.

Billore S D, Vyas A K and Joshi O P. 2006. Symbiotic attributes and productivity of soybean as influenced by co-inoculation of plant growth promoting rhizobacteria with B*radyrhizobium japonicum*. *Journal of Oilseeds Research*. 23(2):327-328.

Boers Th M and Ben-Asher J. 1982. A review of rainwater harvesting. *Agricultural Water Management* 5: 145-158.

Bolaños J and Edmeades G O. 1993. Eight cycles of selection for drought tolerance in lowland tropical maize II. Response in reproductive behavior. *Field Crops Research* 31: 253-268.

Boney A D. 1965. *A biology of marine algae*. Hutchinson on Educational Ltd., London.

Bouwan B A M and Tuong T P. 2001. Field water management to save water and increase its productivity in irrigated lowland rice. *Agriculture and Water Management* 49: 11–30.

Botwright, TL., Condon, AG., Rebetzke, GJ. and Richards, RA. 2001. Improving grain yield by selection for greater early vigour in wheat. *In:* Proceedings of the 10th Australian Agronomy Conference, January 2001, Habart, Australia.

Booth E. 1965. The manorial value of seaweed. *Bot. Mar.* 8:138-143.

Bruins H J, Evenari M, and Nessler U. 1986. Rainwater-harvesting agriculture for food production in arid zones: The challenge of the African famine. *Applied Geography* 6(1): 13-32.

Byerlee D and Polanco E H de. 1986. Farmers' stepwise adoption of technological packages: evidence from the Mexican Altiplano. *American Journal of Agricultural Economics* 68:519-527.

Cassman K G. 1999. Ecological intensification of cereal production systems: yield potential, soil quality, and precision agriculture. *Proceedings of the National Academy of Sciences of the United States of America* 96(11): 5952-5959.

Ceccarelli S, Grando S and Booth R H. 1996. International breeding programmes and resource-poor farmers: Crop improvement in difficult environments. In *Participatory Plant Breeding*, ed. P Eyzaguire and M Iwanaga. In*:* Proceedings of a Work Shop in Participatory Plant Breeding, July 26-29, 1995, Wageningen, The Netherlands. Rome, Italy: International Plant Genetics Research Institute.

Ceccarelli S, Grabdo S, Tutwiler R, Baha J, Martini A M, Salahieh H, Goodchild A, and Michael M. 2000. A methodological study on participatory barley breeding: I. Selection phase. *Euphytica* 111: 91-104.

CIMMYT 1999. Medium-term plan of the international maize and wheat improvement center (CIMMYT) 2000-2002; Increasing wheat productivity and sustainability in stressed environments: abiotic Term_Plans/mtp2000_2002/htm/g5.htm> Updated March (accessed August 2001).

CIMMYT 2000. *CIMMYT in 1999-2000: Science and Sustenance*. Mexico, D.F.: CIMMYT.

Critchley W and Siegert K.1991. Water harvesting: A manual for the design and construction of water harvesting schemes for plant production. Rome, Italy: Food and Agricultural Organization of the United Nations.

Chatterjee B N and Mandal B K.1992. Present trends in research on intercropping. *Ind. Jl. of Agri. Sci.* 62(8): 507-18.

DARE/ ICAR Annual Report 2003-04.

Davison J. 1988. Plant beneficial bacteria. *Biotechnology*, 6 March, 1988, 282-286.

De Haen H. 1997. Environmental consequences of agricultural growth in developing countries. In: *Sustainability, growth, and poverty alleviation*, ed. S A Vosti and T Reardon. Baltimore, MD: Johns Hopkins University Press.

Desai N C, Ardeshna R B, and Intwala C G. 2000. Increasing pigeon pea productivity by providing land configuration in vertisols. *International Chickpea and Pigeonpea Newsletter* 7:66-67.

Donald C M. 1946. Competition between pasture species with reference to the hypothesis of harmful root interactions. *Journal of Council of Scientific and Industrial Research* 19: 32-7.

Donald C M. 1961. Competition for light in crops and pastures. *In*: Proceedings of Symposia for Society for Experimental Biology XV - Mechanisms in Biological Competition: pp 282-313.

Donald C M. 1963. Competition among crop and pasture plant. *Advances in Agronomy* 18: 1-118.

Doutt R L and Nakata J. 1965. Parasites for control of grape leaf hopper. *California Agriculture* 19(4): 3.

Dikinson J C. 1972. Alternatives to monoculture in the humid tropics of Latin America. *Professional Geography* 24: 217-22.

Ekboir J, Boa K, and Dankyi A A. 2001. *Impact of no-till technologies in Ghana*. CIMMYT Economics Paper, CIMMYT, D.F. forthcoming. Evans, L. T. 1998. *Feeding the ten billion: Plants and population growth*. Cambridge: Cambridge University Press.

El-Swaify S A, Pathak P, Rego T J, and Singh S. 1985. Soil management for optimized productivity under rainfed conditions in the semi-arid tropics. *Advances in Soil Science* 1:1-64.

Evans L T. 1998. *Feeding the ten billion: Plants and population growth*. Cambridge: Cambridge University Press.

Fall A and Faye A.1999. Minimum tillage for soil and water management with animal traction in the West-African region. In: *Conservation tillage with animal traction*, ed. P G Kambutho and T E Simalenga. Harare, Zimbabwe: A resource book of Animal Traction Network for Eastern and Southern Africa (ATNESA).

FAO. 1978. Report on the agro-ecological zones project:Results for southweat Asia. *World Soil Resources Report* 48/2. FAO, Rome.

FAO (Food and Agriculture Organization of the United Nations). 2000. *Crops and Drops*. Rome, Italy.

Flower D J. 1994. Vertisol technology in India: technology development, extension and impact assessment. Pages 58-66 *in* Evaluating ICRISAT research impact: summary proceedings of a Workshop on Research Evaluation and Impact Assessment, 13-15 Dec 1993, ICRISAT Center, India (Bantilan M C S, and Joshi P K, eds.). Patancheru 502 324, Andhra Pradesh, India: International Crops Research Institute for the Semi-Arid Tropics.

Flinn J C and Lagemann J. 1980. Evaluating technical innovations under low resource farmer conditions. *Experimental Agriculture* 16(1): 91-101.

Francis Ofori and Stern W R. 1987. Cereal-legume intercropping system. *Advances in Agronomy* 41: 41-85.

Gerard B M. 1976. Measuring plant density effects on insect pests in intercropped maize and cowpea. *In*: Intercropping in Semi-arid Areas. Report of a Symposium held at University of Dar esSalaam, Morogoro, Tanzania, 10-12th May, 1976, IRDC, Ottawa.

Gunnell Y and Krishnamurthy A. 2003. Past and present status of runoff harvesting systems in dry land peninsular India: A critical review. *Ambio*. 30: 320–324.

Gupta D K and Bhan S.1997. Effect of *In-situ* moisture conservation and fertilization on yield, quality and economics of maize mustard cropping system under rainfed condition. *Indian Journal of Soil Conservation* 25 (2):133-135.

Gupta G N. 1995. Rain-water management for tree planting in the Indian Desert. *Journal of Arid Environments* 31(2): 219-235.

Gupta R K, Tomar S S, and Tomar A S. 1979. Improved management practices for maize and sorghum grown on Vertisols of central India. *Journal of Agronomy and Crop Science* 148:478-483.

Gupta R K and Sharma R A. 1990a. Effect of land configurations on field water balance and productivity of rainfed crops grown on Vertisols under rainfed conditions. International Symposium on Natural Resource Management for a Sustainable Agriculture, February 6-10, 1990, New Delhi, Abst. Vol; 1 and 2, Pp. 28.

Gupta R K and Sharma R A. 1990b. Physical aspects of management of black clay soils on central and peninsular India. In*:* Proc. Of the International Agricultural Engineering Conference and Exhibition. Bangkok, Thailand, 3-6 December, 1990, Pp. 1163-1172.

Gupta R K, Sharma S G, Tembe G P, and Tomar S S. 1978. A new approach in farming system for problem soils of Madhya Pradesh. *JNKVV Research Journal* 12: 73-79.

Gupta R K and Sharma R A. 1994. Influence of different land configurations on *in-situ* conservation of rain water, soil and nutrients. *Crop Research* 8(2): 276-282.

Holdridge L R. 1959. Ecological indications of the need for a new approach to tropical land use. *Economic Botany* 13(4): 271-80.

Habitu N and Mahoo H. 1999. Rainwater harvesting technologies for agricultural production: A case for Dodoma, Tanzania. In: *Conservation tillage with animal traction*, ed. P.G. Kambutho, and T.E. Simalenga. Harare, Zimbabwe: A resource book of Animal Traction Network for Eastern and Southern Africa (ATNESA).

Hibon A, Triomphe B, López-Pereira M A, and Saad L.1992. Rainfed maize productions in Mexico: Trends, constraints, and technological and institutional Challenges for Researchers. CIMMYT Economics Working Paper 92-03. Mexico, D.F.: CIMMYT.

Hofwegen P van and Swendsen M. 2000. A vision of water for food and rural development. World Water Vision Report. Paris: United Nations Educational Scientific and Cultural Organization (UNESCO).

ICARDA (International Center for Agricultural Research in the Dry Areas). 1997. Caravan: Review of agriculture in the dry areas. Issue No. 6, Spring/Summer 1997. Aleppo, Syria.

Ines A V M, Honda K, Gupta A D, Droogers P, and Climente R S. 2006. Combining remote sensing-simulation modeling and genetic algorithm optimization to explore water management options in irrigated agriculture. *Agriculture and Water Management*. 83: 221–235.

Igbozurike M U. 1971. Ecological balance in tropical agriculture *Geographical Review* 61; 519-29.

Jodha N S. 1976. Resource base as a determinant of cropping patterns. *In*: Proceedings of a symposium on Cropping Systems Research and Development for the Asian Rice Farmers, 21-4, September, International Rice Research Institute, Los Banos, Philippines.

Joshi P K and Bantilan M C S. 1996. Evaluating adoption and returns to investment on crop and resource management research and technology transfer: a case of groundnut production technology. ECON 1 Progress Report. Patancheru 502 324, Andhra Pradesh, India: International Crops Research Institute for the Semi-Arid Tropics. (Limited distribution).

Joshi P K, Bantilan M C S, Ramakrishna A, Chopde V K, and Raju P S S.1998. Adoption of Vertisols technology in the semi-arid tropics. Pages 182-196 *in* Assessing Joint Research Impacts: *Proceedings of an international Workshop on Joint Impact Assessment of NARS/ICRISAT* Technologies for the Semi- Arid Tropics, 2-4 Dec 1996, ICRISAT, Patancheru, India (Bantilan M C S, and Joshi P K, eds.). Patancheru 502 324, Andhra Pradesh, India: International Crops Research Institute for the Semi-Arid Tropics.

Kampen J. 1982. An approach to improved productivity on deep vertisols. Information Bulletin No. 11. Patancheru, Andhra Pradesh 502 324, India: International Crops Research Institute for the Semi-Arid Tropics.

Kanwar J S, Kampen J, and Virmani S M. 1982. Management of Vertisols for maximized crop production- ICRISAT Experience. In: Vertisols and Rice Soils. Symp. Papers II, 12th National Congress Soil Science, New Delhi Pp.94-118.

Kaur R, Singh O, Srinivsan R, Das S N, and Mishna K. 2004. Comparison of a subjective and physical approach for identification of priority areas for soil and water management in a watershed—A case study of Nagwan watershed in Hazaribagh District of Jharkhand, India. *Environmetal Modeling & Assessment* 9: 115–127.

Kaushik S K and Gautam R C. 1994. Response of pearl millet (*Pennisetum glaucum*) to water harvesting, moisture conservation and plant population in light soils. *Indian Journal of Agricultural Sciences* 64(12): 858-60.

Kaushik S K and Lal K.1998. Effect of water-harvesting techniques on productivity and water-use efficiency of rainy season crops. *Indian Journal of Agronomy* 43(4): 747-750.

Koli S E. 1975. Pure cropping and mixed cropping of maize and groundnut in Ghana. *Ghana Journal of Agriculture Science* 8: 23-30.

Kornegay J, Beltran J A, and Ashby J. 1996. Farmer selections within segregating populations of common bean in Columbia: Crop improvement in difficult environments. In *Participatory plant breeding*, ed. P Eyzaguire and M Iwanaga. *Proceedings of a Work Shop in Participatory Plant Breeding*, 26-29 July 1995,

Kumar P. 1998. Food demand and supply projections for India. Agric. Econ. Policy Paper 98–01. IARI, New Delhi, India.

Lal R. 2006. Dryland farming in South Asia. In: *Dryland Agricutlure*, 2nd edition, Agronomy Monograph 33:527–576. Madison, WI: American Society of Agonomy.

Lal R. 2009. Soil Water Management in India. *Journal of Crop Improvement* 23: 55–70.

Li X L, Gong J D, and Wei X H. 2000. *In-situ* rainwater harvesting and gravel mulch combination for corn production in the dry semi-arid region of China. *Journal of Arid Environments* 46(4): 371-382.

Mamo T, Abebe M, Duffera M, Kidanu S Y, and Erkosa T. 1994. Response of chickpea to method of seed bed preparation and phosphorus application on Ethiopian vertisols. *International Chickpea and Pigeonpea Newsletter* 1:17-18.

McNeely J A and Scherr S J. 2001. Common ground, common future: How eco-agriculture can help feed the world and save wild biodiversity. Gland, Switzerland: International Union for the Conservation of Nature and Natural Resources.

Misika P and Mwenya E. 1999. Conservation tillage with animal traction for soil-water management and environmental sustainability in Namibia. In *Conservation tillage with animal traction*, ed. P. G. Kambutho, and T. E. Simalenga. Harare, Zimbabwe: Animal Traction Network for Eastern and Southern Africa (ATNESA).

Mooney P A and Van Staden J. 1985. Effect of seaweed concentrate on the growth of wheat under condition of water fern. *South African Journal of Science* 8:632-633.

Nabhan G P. 1984. Soil fertility renewal and water- harvesting in Sonoran desert agriculture, The Papago example. *Arid Lands Newsletter* 20: 21-38.

Ojasvi P, Goyal R K, and Gupta J P. 1999. The micro-catchment water harvesting technique for the plantation of jujube (*Zizyphus mauritiana*) in an agroforestry system under arid conditions. *Agricultural Water Management* 41(3): 139-147.

Olukosi J O. 1976. Decisions of farmers under risks and uncertainty - the case of Ipetu and Odo-ore farmers in Kwara State. *Somaru Agriculture Newsletter* 18(3): 108-22.

OTA (Office of Technology Assessment) commissioned paper. 1983. Technologies affecting water-use efficiency of plants and animals. In Water related technologies for sustainable agriculture in U.S. Arid/Semiarid Lands. Washington, D.C.: U.S. Congress, Office of Technology Assessment, OTA-F-2I2, October.

Ong, C. K., C. R. Black, P. M. Marshall, and J. E. Corlett. 1996. Principles of resources capture and utilisation of light and water. In *Tree-crop interactions: A physiological approach*, ed. C.K. Ong and P. Huxley. Wallingford, U.K.: CAB International.

Oweis T, Hachum A, and Kijne J. 1999. Water harvesting and supplementary irrigation for improved water use efficiency in dry areas. SWIM Paper 7. Colombo, Sri Lanka: International Water Management Institute.

Pacey A and Cullis A. 1986. Rainwater Harvesting: The collection of rainfall and runoff in rural areas. London: Intermediate Technology Publications.

Paroda R S, Woodhead T, and Singh R B. 1994. Sustainability of rice-wheat system in Asia and Pacific. Publication #1894-11. Bangkok, Thailand: FAO. Downloaded By: [CeRA]. 22 December 2010 *Soil Water Management* 69.

Patil R H and Jawaregowda. 2000. Pre-sowing agronomic practices for disease management. *The Hindu,* 2 Nov 2000.

Perrier E R and Alkini A B.1987. Supplemental irrigation in the Near East and North Africa. In*: Proceedings of a Workshop on Regional Consultation on Supplemental Irrigation*. ICARDA and FAO, Rabat, Morocco, December 7-9, 1987. Dordrecht, The Netherlands: Kluwer Academic Publishers.

Radley M. 1961. Gibberllin-like substances in plants. *Nature*, London 191: 684-685.

Ranade D H, Sharma R A, Gupta R K, and Patel A N. 1995. Effect of mechanical and vegetative barriers on conservation of runoff, soil and plant nutrients. *Crop Research* 9(2): 218-223.

Reddy J and Kumar D N. 2008. Evolving strategies of crop planning and operation of irrigation reservoir system using multi-objective differential evolution. *Irrigation Science* 26: 177–190.

Reij C, Mulder P, and Begemann, L.1988. Water harvesting for agriculture. *World Bank Technical Paper Number 91*. Washington, D.C.: The World Bank.

Reynolds M, Skovmand B, Trethowan R, and Pfeiffer W. 2000. Evaluating a conceptual model for drought tolerance. In: *Molecular approaches for the genetic improvement of cereals for stable production in Water-limited environments*, ed. J M Ribaut and D Poland. A Strategic Planning Workshop held at CIMMYT, El Batan, Mexico, June 21-25, 1999. Mexico, D.F.: CIMMYT.

Rees D J, Qureshi Z A, Mehmood S, and Raza S H. 1991. Catchment basin water harvesting as a means of improving the productivity of rain- fed land in upland Balochistan. *Journal of Agricultural Science* 116(1): 95-103.

Rockström J. 1993. Biomass production in dry tropical zones: How to increase water productivity. *Proceedings of an FAO informal workshop* January 31– February 2, 1993, Rome, Italy.

Rockström J and Falkenmark M. 2000. Semiarid crop production from a hydrological perspective: gap between potential and actual yields. *Critical Reviews in Plant Sciences* 19(4): 319-346.

Rosegrant M W. 1997. Water resources in the twenty-first century: Challenges and implications for action. Food, Agriculture, and the Environment Discussion Paper No. 20. Wasington, D.C.: International Food Policy Research Institute.

Ryan J G, Virmani S M, and Swindale L D. 1982. Potential technologies for deep black soils in relatively dependable regions of India. Pages 41-61 In: *Proceedings of the seminar on Innovative technologies for integrated rural development*, 15-17 Apr 1982, New Delhi, India. New Delhi, India: Indian Bank.

Ryan J G and Subramanyam K V. 1975. Package of practices approach in adoption of high-yielding varieties: an appraisal. *Economic and Political Weekly* 10(52): A 110.

Salter P J, Akehurst M Jayne and Morris G E L. 1985. An agronomic and economic study of intercropping Brussels sprouts and summer cabbage. *Experimental Agriculture* 21: 153-67.

Samra J S, Bijay-Singh, and Kumar K.2003. Managing crop residues in the rice-wheat system of the Indo-Gangetic plain. In: *Improving the productivity and sustainability of rice-wheat systems: Issues and impacts*, eds. J K Ladha, 173–195. ASA Spec. Publ. 65. Madison, WI: ASA, CSSA, and SSSA.

Satpathy J M, Das M S and Naik K. 1977. Effect of multiple andmixed cropping on the incidence of some important pests. *Journal of Entomological Research* 1(1): 78-85.

Sekar R, Thangaraju N, and Rangasamy R. 1995. Effect of seaweed liquid fertilizer from Ulva lactuca . on vigna Unguiculata L. (Walp). *Phykos.* 34:49-53.

Sengar S S. 1998. Effect of surface drainage and phosphorus levels on yield of pigeon pea. p. 155-156 In: *Abstracts: National Symposium on Management of Biotic and Abiotic Stresses in Pulse Crops*, 26-28 Jun 1998, Kanpur, India. Indian Society of Pulse Research and Development and Indian Institute of Pulses Research. Kanpur.

Scherr S J. 2000. A downward spiral? Research evidence on the relationship between poverty and natural resource degradation. *Food Policy* 25(4): 479-498.

Sharma R A and Gupta R K. 1984. Soil water evaporation from clay and sandy loam soils as influenced by straw and dust mulching. *Indian Journal of Soil Conservation* 12(1): 49-57.

Sharma R A and Gupta R K. 1990. Conservation of soil moisture through crop residues management and cultural practices for rainfed agriculture. In: *Proceedings of International Symposium on Water Erosion, Sedimentation and Resource Conservation*. October 9-13, 1990. Central Soil and Water Conservation Research and Training Institute, Dehradun (U P), pp. 233-243.

Sharma RA, Raghu JS and Thakur HS.2005. Rainfed farming technology for black clay soils. AICRP on Dryland, JNKVV, College of A griculture, Indore. PP. 1-128.

Sharma R A, Upadhyay, M S, and Tomar R S S. 1985a. Water use efficiency of some rainfed some rainfed crops on Vertisols as influenced by soil and straw mulch. *Journal of Indian Society of Soil Science* 33: 387-391.

Sharma R A., Verma G P, and Gupta R K. 1985b. Modification of evaporation from from a Vertisols by straw mulch. *Journal of Indian Society of Soil Science* 33: 201-210.

Sharma P B S and Rao V V. 1997. Evaluation of an irrigation water management scheme – A case study. *Agriculture and Water Management* 32: 181–185.

Shetty S V R, Krantz B A, and Obion S R. 1977. Weed research needs of the small farmers. Pages 47-60. In: *Proceedings of the Conference and workshop on Weed science*, 17-21 Jan 1977, Andhra Pradesh Agricultural University, Hyderabad, Andhra Pradesh. India: Indian Society of Weed Science.

Shoyinika S A. 1976. Attempted control of virus incidence in cowpeas by the use of barrier crops. In: *Report of a symposium on Intercropping in Semi-arid Areas held at Morogoro*, Tanzania, 10-12th May, IRDC, Ottawa.

Singh Y, Singh B, Ladha J K, Khind C S, Gupta R K, Meelu O, and Pasuquin E. 2004. Long-term effect of organic inputs on yield and soil fertility in the rice wheat rotation. *Soil Science Society of America Journal* 68: 845–853.

Singh Y, B Singh, and Timsina J. 2005. Crop residue management for nutrient cycling and improving soil productivity in rice-based cropping systems in the tropics. *Advances in . Agronomy* 85: 269–407.

Smith F B C and Van staden J. 1984. The effect of seaweed concentrate and fertilizer on growth and endogenous Cytokinin content of Phaseolus vulsaris. *South African journal of Botany* 3:375-379.

Suleman S, Wood M K B, Shah H, and Murray L.1995. Development of a rainwater harvesting system for increasing soil moisture in arid rangelands of Pakistan. *Journal of Arid Environments* 31(4): 471-481.

Suryatra E S and Harwood R R. 1976. Nutrient uptake of traditional combinations and insect and disease incidence in three cropping combinations. Ph D Thesis, IRRI, Los Banos, Phillipines.

Tabor J A. 1995. Improving crop yields in the Sahel by means of water-harvesting. *Journal of Arid Environments* 30(1): 83-106.

This D and Teulat-Merah B. 2000. Towards a comparative genomics of drought tolerance in cereals: Lessons from a QTL analysis in barley. In: *Molecular approaches for the genetic improvement of cereals for stable production in waterlimited environments*, ed. J M Ribaut and D Poland. A Strategic Planning Workshop held at CIMMYT, El Batan, Mexico, June 21-25, 1999. Mexico, D. F.:CIMMYT.

Thivy F. 1961. Seaweed manure for perfect sal and smiling fields. *Salt. Res. Indust.* 1:1-4

Timsina J and Connor D J. 2001. Productivity and management of rice-wheat systems: issues and challenges. *Field Crops Research* 69: 93–132.

Trethowan R and Pfeiffer W H. 2000. Challenges and future strategies in breeding wheat for adaptation to drought stressed environments: A CIMMYT wheat program perspective. In: *Molecular approaches for the genetic improvement of cereals for stable production in water-limited environments*, ed. J M Ribaut and D Poland. A Strategic Planning Workshop held at CIMMYT, El Batan, Mexico, June 21-25, 1999. Mexico, D. F.: CIMMYT.

Vincent L F. 2003. Towards a small holder hydrology for equitable and sustainable water management. *Natural ResourcesForum* 27: 108–116.

Verma G P. 1981. Erosion control on Vertisols under comparatively high rainfall conditions. In: *Proceedings of the South East Asian Regional Symposium on Problems of Soil Erosion and Sedimentation*, January 27-29, 1981. Bangkok, Thailand. Pp. 301-310.

Vital KPR., Singh HP, Rao KV, Sharma KL, Victor US, Chary GR, Sankar GRM, Samra JS and Singh Gurbachan.2003. Guidelines on Drought Coping Plans for Rainfed Production Systems. All India Co-ordinated Research Project for Dryland Agriculture, Central Research Institute for Dryland Agriculture, Indian Council of Agricultural Research, Hyderabad 500 059. Pp.39.

Vyas A K, Billore S D, Joshi O P, Pandya N, and Pachlania N.2005. Effect of lone and dual inoculation of AM and PSB with phosphorus application on symbiotic attributes and soybean yield. *International workshop on PROM. In PROM Review-2005, organized by MLSU, Udaipur, PROM Society, Udaipur and MPUAT, Udaipur held at MPUAT, Udaipur on 28. 12. 2005.pp.36-38*

Vyas A K, Billore S D and Joshi O P. 2004. Land configuration and nutrient management for sustainable soybean production. In: *National Symposium on Resource Conservation and Agricultural Productivity*, organized by ISA, ICAR and PAU held at PAU, Ludhiana on 22-25.11.2004. pp. 114-5.

Willey R W. 1979a. Intercropping: its importance and research needs. I Competition and yield advantages. *Field Crop Abstracts* 32: 1-10.

Willey R W. 1979b. Intercropping: its importance and research needs. II Agronomy and research approaches. *Field Crop Abstracts* 32:73-85.

□□□

Green Agriculture : Newer Technologies, 2012

New India Publishing Agency, New Delhi (India)
e-mail : info@nipabooks.com; website : www.nipabooks.com

Chapter-8

Eco-Safety and Agricultural Sustainability through Organic Agriculture

Ram A Jat, Suhas P Wani, Kanwar L Sahrawat and Piara Singh
International Crops Research Institue for the Semi-Arid Tropics,
Patancheru -502 324
E-mail : jat@cgiar.org

SUMMARY

Conventional agriculture has been contributing significantly to several modern day environmental problems like global climate change, soil degradation, water pollution, biodiversity loss etc. directly or indirectly. Organic agriculture has been identified as one of the eco-friendly agriculture techniques and is being promoted worldwide to meet the increasing demand for organic products and conserve the natural eco-systems. It helps mitigating climate change through reduced emission of green house gases (GHGs), greater C sequestration and less use of fossil fuel. Besides, it has been found helpful to reduce export of harmful agro-chemicals to water bodies, biodiversity conservation, restoring soil health etc. Organic agriculture can help to meet the objectives of sustainable and eco-friendly agriculture.

Introduction

Conventional agriculture is one of the foremost factors responsible for ecological degradation all over the world. Besides, degrading other resources like land, water, forest, biodiversity, it has been responsible to contribute to global climate change through emission of GHGs like CO_2, CH_4 and N_2O. This underlines the need to develop agriculture management practices that take care of resource conservation thus ensuring sustainability of agriculture. This is particularly important in the scenario of challenging task to feed the burgeoning population and almost no chance of enhancing the extra land under cultivation. Agriculture not only contributes to global climate change but also is victim of it. Therefore agriculture needs to contribute its share of climate change mitigation through adopting eco-friendly production techniques like organic farming. During the last two decades, there has been a significant sensitization of the global community towards environmental conservation and assurance of quality food. Ardent promoters of organic farming consider that it can meet both the demands. After almost a century of neglect, organic agriculture is now finding place in the mainstream of development and shows great promise commercially, socially and environmentally. While there is continuum of thought from earlier days to the present, the modern organic movement is radically different from its original form. It now has environmental sustainability and productivity at its core, in addition to the founders concerns for healthy soil, healthy food and healthy people.

Organic agriculture is a holistic production management system which promotes and enhances agro-ecosystem health, including biodiversity, biological cycles, and soil biological activity. It emphasizes the use of management practices in preference to the use of off-farm inputs, taking into account that regional conditions require locally adapted systems. This is accomplished by using, where possible, cultural, biological, and mechanical methods, as opposed to using synthetic materials to fulfill any specific function within the system. An organic production system is designed to a) enhance biological diversity within the whole system; b) increase soil biological activity; c) maintain long-term soil fertility; d) recycle wastes of plant and animal origin in order to return nutrients to the land, thus minimizing the use of non-renewable resources; e) rely on renewable resources in locally organized agricultural systems; f) promote the healthy use of soil, water and air as well as minimize all forms of pollution that may result from agricultural practices (Codex Alimentarius,1999).According to International Federation of Organic Agriculture Movement (IFOAM) organic agriculture is a production system that sustains the health of soils, ecosystems and people. It relies on ecological processes, biodiversity and cycles adapted to local conditions, rather than the use of inputs with adverse effects. Organic agriculture combines tradition, innovation and science to benefit the shared environment and promote fair relationships and a good quality of life for all involved. Organic agriculture dramatically reduces external inputs by controlling pests and diseases naturally, with both traditional and modern methods, increasing both agricultural yields and crop resistance to pest and diseases and exploiting organic nutrient sources. Keeping this hypothesis in view this chapter has been prepared to focus on the environmental benefits of organic agriculture to promote it as an important environment conservation strategy.

Ecological problems due to conventional agriculture

Direct impacts of conventional agriculture on the environment arise from farming activities which contribute to soil erosion, land salination, biodiversity loss due to habitat loss and harmful agrochemicals and release of greenhouse gases. The spread of green revolution in the Indo Gangetic Plains has been accompanied by over exploitation of land and water resources, and use of fertilizers and pesticides have increased many fold. Shifting cultivation has also been an important cause of land degradation. Leaching from extensive use of pesticides and fertilizers is an important source of contamination of water bodies. Intensive agriculture and irrigation contribute to land degradation particularly salination, alkalization and water logging. Therefore, conventional agriculture has been found unsustainable on long term basis due to various ecological consequences associated with it.

Emission of green house gases from agricultural activities

Modern industrial agriculture of the "Green Revolution" era contributes a great deal to climate change by contributing to emission of GHGs directly or indirectly. It is the main source of nitrous oxide and methane gases released into the atmosphere; besides contributing significant amount of CO_2 due to large scale use of fossil fuels and loss of soil carbon to the atmosphere especially through deforestation to make more land available for crops and plantations. Our planet's soils contain one thousand six hundred billion tonnes of carbon, more than twice as much as is contained in the atmosphere. Much of this will be released in the coming decades; unless there is a rapid switch to sustainable, largely organic, agricultural practices.

According to the fourth assessment report of the Intergovernmental Panel on Climate Change (IPCC), greenhouse gas emissions from the agricultural sector account for 10–12% or 5.1–6.1 Gt of the total anthropogenic annual emissions of CO_2-equivalents (IPCC, 2007 a). The total emissions were 6.1Gt CO2e, made up almost entirely of CH_4 (3.3 Gt) and N_2O (2.8Gt). However, this accounting includes only direct agricultural emissions; emissions during the production of agricultural inputs such as nitrogen fertilizers, synthetic pesticides and fossil fuels used for agricultural machinery and irrigation are not calculated. Agriculture production practices emit at least one-third of global anthropogenic GHG emissions, if forest land conversion to agriculture, food handling and processing activities are accounted (Scialabba and Mu¨ller-Lindenlauf 2010). Currently it is responsible for 25% of the world's carbon dioxide emissions, 60% of methane gas emissions and 80% of nitrous oxide; all powerful greenhouse gases (Bunyard, 1996). Nitrous oxide is generated through the action of denitrifying bacteria in the soil when land is converted to agriculture. When tropical rainforests are converted into a pasture, nitrous oxide emissions increase by three times. All in all, land conversion is leading to the release of around half a million tonnes a year of nitrogen in the form of nitrous oxide (Simon et al. 2007). Nitrogenous fertilisers are another major source of nitrous oxide emissions. N_2O is emitted during both the production of nitrogenous fertilizers (Figure.1) as well as after their application in the field. With continuous increase in fertilizer applications, especially in developing countries, nitrous oxide emissions from agriculture could double over the next 30 years (Bunyard, 1996). In the Netherlands, which has the world's most intensive farming, as much as 580 kilograms per hectare of nitrogen in the form of nitrates or ammonium salts are applied every year as fertiliser, and at least 10% of that nitrogen goes straight back into the atmosphere, either as ammonia or nitrous oxide (Moser et al.1991).

The growth of agriculture is also leading to increasing emissions of methane. In the last few decades, there has been a substantial increase in livestock numbers, cattle in particular which has resulted into conversion of tropical forests to pasture. Cattle emit large amounts of methane and the destruction of forests to raise cattle contributes to increased emissions of CO_2. Worldwide, the emissions of methane by livestock amount to some 70 million tonnes (Goldsmith, 2010). Even the fertilisation of grasslands with nitrogen fertilisers can both decrease methane uptake by soil bacteria and increase nitrous oxide production, thereby increasing atmospheric concentrations of both these gases. The expansion of rice paddies has also seriously increased methane emissions which accounts for 11% of the global agricultural GHGs emissions.

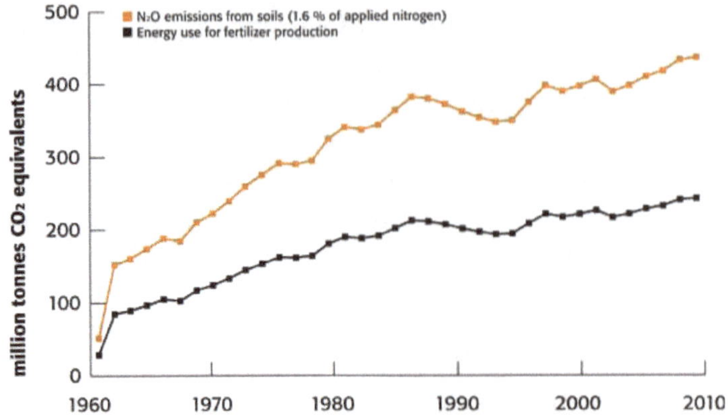

[**Figure 1 :** Energy used for production of synthetic fertilizers and emissions of nitrous oxide (N_2O) from soils after application of fertilizer (in million tons CO_2 equivalent), based on data from the International Fertilizer Industry Association IFA (Source: http://www.fertilizer.org/)]

In the United States, agriculture contributes 7.4 percent of the national greenhouse gas emissions (Anon 2007b). Livestock enteric fermentation and manure management account for 21 percent and 8 percent, respectively of the national methane emissions. Agricultural soil management, such as fertilizer application and other cropping practices, accounts for 78 percent of the nitrous oxide emitted in the USA (Anon 2007c). In the UK, agriculture is estimated to contribute directly 7.4 percent to the nation's greenhouse gas emissions, with fertilizer manufacture contributing a further 1 percent (Moser et al. 1991) and is comprised entirely of methane at 37.5 percent of national total and nitrous oxide at around 95 percent of the national total (Anon 2007a). Enteric fermentation is responsible for 86 percent of the methane contribution from agriculture, the rest from manure; while nitrous oxide emissions are dominated by synthetic fertilizer application (28 percent) and leaching of fertilizer nitrogen and applied animal manures to ground and surface water (27 percent). Figure 2 shows sectorwise contribution to GHG emission from agriculture in world. In India agriculture accounts for 28 % of total GHG emission and is second only to energy production and transformation (Table 1).

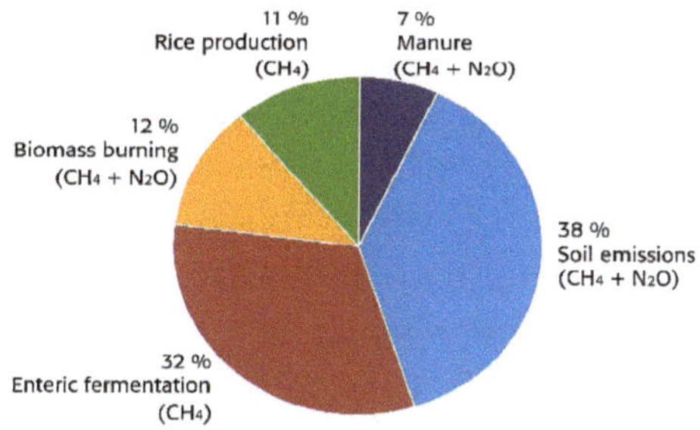

Figure 2 : Main sources of greenhouse gas emissions in the agricultural sector in 2005 *(Source : Smith et al. 2007).*

Table 1 : Contribution to GHG emissions by sectors in India

Sector	**Percent**
Energy production & transformation	35
Agriculture	28
Industry including	20
Industrial combustion	12
Other industrial processes	8
Transport	8
Residential sector	5
Land Use, Land use change and Forestry	2
Others (including waste)	1

(Source : NATCOM, MoEF, 2004)

Among the immediate options available organic agriculture must be considered seriously until suitable substitutions to fossil fuel based activities are developed. The energy used for the chemical synthesis of nitrogen fertilizers, which are totally excluded in organic systems, represent up to 0.4–0.6 Gt CO_2 emissions (FAOSTAT 2009; EFMA 2005). This is as much as 10% of direct global agricultural emissions and around 1% of total anthropogenic GHG emissions. Williams et al. calculated the total primary energy burden of conventional wheat production in the UK to be allocated by 56% to mineral fertilizers and by 11% to pesticides (FAOSTAT, 2009). Pimentel calculated similar results for corn in USA and allotted 30–40% energy burden for fertilization and 9–11% for plant protection for wheat and corn (Williams et al. 2006).

Pollution of water bodies

In recent years many reports have identified agricultural non-point source pollution as the leading cause of water quality deterioration in rivers and lakes. The global mean N use efficiency is estimated to be about 50% (Mosier, 2002). The efficiency of fertilizer use in India is only 30-35 percent and the part of balance 65 -70 percent may reach the water bodies. The remaining quantity of nitrogen is lost into the environment and part of this can possibly end up in water bodies. The current public health standards for safe drinking water require that maximum contaminant level (MCL) should not exceed nitrate concentrations of 10 ppm as nitrate-N or 45 ppm as nitrate (10 ppm nitrate-N is the same as 45 ppm nitrate). Nitrates in drinking water are associated with a number of health problems. Recurrent acute respiratory tract infections in some areas of Rajasthan have been attributed to high nitrate concentrations in drinking water (Gupta et al. 2000).Nitrate contamination of ground water depends upon climate, fertilizer or manure management, soil, crop, and farming systems. Eutrophication leading to suffocation of aquatic plants and animals due to rapid growth of algae, referred to as "algae blooms", is literally killing lakes, rivers and other bodies of water. Similarly heavy metals present in the fertilizers and sewage sludge leach into ground water.

Loss of biodiversity

Agriculture has been responsible for wildlife habitat degradation worldwide. There has been a 500% expansion in the extent of crop and pasturelands worldwide in the last 300 years. Habitat loss is now identified as the main threat to 85 - 90% of all species described by IUCN as 'threatened' or 'endangered' and is the most commonly recorded reason for species extinction during the last 20 years (Anon d). Growing high-yielding, uniform cultivars has been responsible for reducing the number of genetically viable species used in agriculture; 75% of agricultural crop diversity (agrobiodiversity) has been lost in the last 100 years (FAO 1998). Native animal breeds are also declining; it has been estimated that every week at least one breed of domestic animal becomes extinct, and over 25% of listed breeds are at risk (Anon D.). The contamination of cultivated and wild species by invasive exotic genes introduced through genetically modified organisms (GMOs) is causing pollution of the natural gene pool (Anon 2007d). Almost all pesticides are toxic in nature and pollute the environment leading to grave damage to ecology as well as animal health. Their indiscriminate use leaves toxic residues in food grains, fodder, vegetables, meat, milk, milk products etc. besides in soil and water. High doses of pesticide residues in water bodies lead to mass killing of aquatic life. The application of weedicides may lead to killing of non-target plants which reduces the biodiversity. The contamination of cultivated and wild species by invasive exotic genes introduced through genetically modified organisms (GMOs) is causing pollution of the natural gene pool. Similarly chemical fertilizers are also source of heavy metals which cause deleterious effects on animal health (Table 2) and with regular application of fertilizers the concentration of heavy metals increase to harmful levels. Deb and Joshi (Deb and Joshi, 1994) have also reported about presence of heavy metals in the chemical fertilizers and sludge.

Table 2 : Concentration of selected heavy metals in commonly used fertilizers (ppm on dry weight basis)

S. No.	Source	Arsenic	Cadmium	Lead	Nickel
1.	Urea	<0.04	<0.2	<0.4	<0.2
2.	DAP	9.9-16.2	4.6-35.5	2.1-3.7	7.4-222
3.	MOP	0.4	<0.2	<0.4-10	<0.2
4.	TSP	10.3	15.0	11	17

(Source : Chhonkar *2003)*

Besides, pesticide use in conventional agriculture is responsible for harmful effects on human health. According to WHO, 14000 people die every year in the third world countries due to pesticide poisoning. Pesticides are becoming less effective against target pests due to their indiscriminate use. Five hundred species of insect-pests have already developed genetic resistance to pesticides, as have 150 plant disease pathogens, 133 kinds of weeds and 70 species of fungus (Bunyard 1996). The reaction today is to apply evermore powerful and more expensive pesticides, which in the US, cost 8 billion dollars a year, not counting the cost of spreading them on the land. Similarly, repeated use of herbicides over a period of time results in shift of the weed flora. The weeds of minor importance, often, become major weeds. Repeated use of weedicides leads to development of resistance in weed at alarming proportions.

Soil health degradation

Conventional agriculture alongwith deforestation is the primary cause of devastating soil degradation all over the globe. On a global scale, soils hold more than twice as much carbon (an estimated 1.74 trillion U.S. tons) as does terrestrial vegetation (672 billion U.S. tons) but this pool is dwindling fast due to conventional agricultural practices leading to degradation of soil health. The carbon loss from soil contributes to soil erosion by degrading soil structure, increase vulnerability to drought by greatly reducing the level of water-holding carbon in the soil; and the loss of soil's native nutrient value. Surprising analysis of the USA's oldest continuous cropping test plots in Illinois showed that contrary to long-held beliefs, nitrogen fertilization does not build up soil organic matter (Timothy 2008). Conventional industrial agriculture is getting increasingly unproductive due to soil degradation and the stark evidence of this can be found in the Indo-Gangetic plains South Asia. Long term continuous use of high doses of chemical fertilizers badly affects the physical, chemical and biological properties of the soil. A study carried out at the University of Agricultural Sciences, Bangalore confirmed the deterioration of soil health because of the reduction in water holding capacity, soil pH, organic carbon content and the availability of the trace elements such as zinc in case of ragi crop even with the application of normal doss of fertilizer in the long run (Hegde et al. 1995).

Thus, with foregoing it can be concluded that conventional agriculture is one of the major environmental polluters and therefore to promote the eco-safety as well as agricultural sustainability farmers need to adopt eco-friendly production technologies. Organic agricultural practices are based on a maximum harmonious relationship with nature aiming at the non-destruction of the environment. Naturally, organic agriculture is looked upon as one of the means to remedy these problems associated with conventional agriculture.

Environmental benefits of organic agriculture

Organic agriculture help mitigate and adapt climate change, conserve water quality, promotes biodiversity, soil health etc. Through its holistic nature, organic farming integrates wild biodiversity, agro-biodiversity and soil conservation, and takes low-intensity farming one step further by eliminating the use of chemical fertilizers, pesticides and genetically modified organisms (GMOs), which is not only an improvement for human health (food quality) and agrobiodiversity, but also for the associated off farm biotic communities. Organic agriculture restores the environmental balance and reduces or avoids deleterious effects on the environment.

Climate change mitigation and adaptation through organic agriculture

Organic agriculture could be one of the most powerful strategies in the fight against global warming but so far agriculture is being considered as an undervalued and underestimated climate change mitigation and adaptation tool. Even though climate and soil type affect sequestration capacities, the multiple research efforts verify that regenerative agriculture, if practiced on the planet's 3.5 billion tillable acres, could sequester up to 40 percent of current CO_2 emissions (Timothy, 2008). As stated in the 2002 report of the United Nations Food and Agriculture Organization (FAO), organic agriculture enables ecosystems to better adjust to the effects of climate change and has major potential for reducing agricultural greenhouse gas emissions (Erisman et al.2008). The global potential of nitrogen supply through organic waste recycling and biological nitrogen fixation; which are basic ingredients of organic systems is far bigger than the current production of synthetic nitrogen and, as such can contribute significantly to reduce the GHGs emissions (Williams et al.2006).

Organic agriculture help mitigate climate change in the following ways

Reduces emission of greenhouse gases, especially nitrous oxide, as no chemical nitrogen fertilizers are used and nutrient losses are minimized. Sequesters carbon in soil and plant biomass by promoting more and diversified vegetation on organic farms, encouraging agro-forestry and forbidding the clearance of primary ecosystems. Minimizes energy consumption by 30-70% per unit of land by eliminating the energy required to manufacture synthetic fertilizers, pesticides, growth hormones etc. and by using internal farm inputs, thus reducing fuel used for transportation

Organic agriculture helps farmers adapt to climate change because it

Prevents nutrient and water loss through high organic matter content and soil covers, thus making soils more resilient to floods, droughts and land degradation processes. Preserves seed and crop diversity, which increases crop resistance to pests and diseases. Maintenance of diversity also helps farmers evolve new cropping systems to adapt to climatic changes. Minimizes risk due to stable agro-ecosystems and yields, and lower production costs.

Impact of organic agriculture on greenhouse gas emissions

Organic agriculture is an important climate change mitigation strategy as it reduces emissions of green house gases. N_2O emissions are less as release of N from organic sources is slow enough to be absorbed by growing plants and therefore free N is not available for denitrifying activities. It is usually assumed that 1–2 percent of the nitrogen applied to farming systems is emitted as N_2O irrespective of the form of the nitrogen input. The default value currently used by the Intergovernmental Panel on Climate Change (IPCC) is 1.25 percent, but new research finds considerably lower values, such as for semi-arid areas (Barton et al. 2008). From organically managed fields CO_2 emission is less due to reduced soil erosion because of better soil structure and plant cover. However, the effect of erosion on CO_2 emissions is still controversial (IPCC 2007b; Lal 2004; Van Oost et al. 2004; Renwick et al. 2004). As chemical inputs like fertilizers and pesticides are prohibited on the organic farms, CO_2 emission during their manufacturing using fossil fuels is also precluded. Similarly, methane emission from organically managed livestock farms is less due to less livestock density. Assuming half of all nitrous oxide emissions come from N fertilizers, phasing them out would save 11.56 Mt of CO_2e which is equivalent to another 1.5 percent of the USA's green house gases emissions (Anon 2007c). The total green house gases savings from phasing out N fertilizers amount to 2.5 percent of UK's national emissions (Anon2007c). The UK is not a prolific user of N fertilizers compared to other countries, so globally, it seems reasonable to estimate that phasing out N fertilizers could save at least 5 percent of the world's green house gases emissions. Similar views about reduced N_2O emissions under organic agriculture have also been expressed by FAO (Scialabba and Hattam 2002). This is due to lower chemical N inputs, less N release from organic manures due to lower livestock densities under organic agriculture; higher C:N ratios of applied organic manures giving less readily available mineral

N in the soil as a source of denitrification; and efficient uptake of mobile N, P, K in soils by using cover crops in organic agriculture. Greenhouse gas emissions were calculated to be 48-66 percent lower per hectare in organic farming systems in Europe (Stolze et al.1999), and were attributed to no input of chemical N fertilizers, less use of high energy consuming feedstuffs, and elimination of pesticides as characteristic of organic agriculture. Dabbert (2006) reported that on a per-hectare scale, the CO_2 emissions are 40 to 60% lower in organic farming systems than in conventional ones, whereas on a per-unit output scale CO_2 emissions tend to be higher in organic farming systems. Similar results are expected by experts for N_2O and CH_4 emissions, although to date, no research results exist. Diversifying crop rotations with green manure improves soil structure and diminishes N_2O emissions. Soils managed organically are more aerated and have significantly lower mobile nitrogen concentrations, which further reduces N_2O emissions. Mathieu, et al. (2006) pointed out that higher soil carbon levels may lead to N_2 emission rather than N_2O which is peculiar in organic agriculture (Mathieu et al. 2006). Calculations of NH_3 emissions in organic and conventional farming systems conclude that organic farming bears a lower NH_3 emission potential than conventional farming systems. Nevertheless, housing systems and manure treatment in organic farming should be improved to reduce NH_3 emissions further.

Impact of organic agriculture on energy use

The FAO report found that, "Organic agriculture performs better than conventional agriculture on a per hectare scale, both with respect to direct energy consumption (fuel and oil) and indirect consumption (synthetic fertilizers and pesticides)", with high efficiency of energy use. Since 1999, the Rodale Institute's long-term trials in the United States have reported that energy use in the conventional system was 200 percent higher than in either of two organic systems - one with animal manure and green manure, the other with green manure only - with very little differences in yields (Timothy, 2008).Research in Finland showed that while organic farming used more machine hours than conventional farming, total energy consumption was still lowest in organic systems (Lötjönen, 2003); that was because in conventional systems, more than half of total energy consumed in rye production was spent on the manufacture of pesticides. Similarly, the energy used per kilogram of milk produced was lower in the organic than in the conventional dairy farm, and it also took 35 percent less energy to grow a hectare of organic spring barley than conventional spring barley (Lim Li Ching,2010). Figure 3 illustrates that food energy output per fossil energy input is higher in organically grown maize and soybean than conventionally grown maize and soybean.

Lim Li Ching (2010) reported that on average in the US, about 2 units of fossil fuel energy is invested to harvest a unit of energy in crop. Counting all energy inputs in fossil fuel equivalents in an organic corn system, the output over input ratio was 5.79 (i.e., one can get 5.79 units of corn energy for every unit of energy spent), compared to 3.99 in the conventional system. There was also a total energy input reduction of 31 percent, or 64 gallons fossil fuel saving per hectare. The total mitigation potential of organic sustainable food systems is 29.5 percent of global greenhouse gases emissions and 16.5 percent of energy use, the largest components coming from carbon sequestration and reduced transport from localising food systems. On the most conservative estimates based on these examples, localising food systems alone could save at least 10 percent of CO_2 emissions and 10 percent of energy use globally (Anon 2007c). It takes 35.3 MJ of energy on average to produce each kg of N in fertilizers (Biermann et al. 1999).

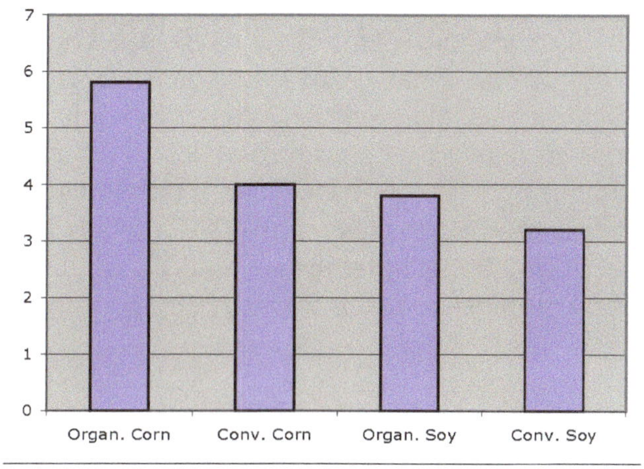

Figure 3 : The food energy output (kcal) per fossil energy input (kcal) for organic corn, conventional corn, organic soybean, and conventional soybean (Source: Pimentel 2006)

Impact of organic agriculture on carbon sequestration : Soils are an important sink for atmospheric CO_2, but this sink has been increasingly depleted by conventional agricultural land use especially due to deforestation. The Stern Review on the Economics of Climate Change commissioned by the UK Treasury published in 2007 (Anon, 2007d) highlights the fact that 18 percent of the global greenhouse gas emissions (2000 estimate) comes from deforestation, and that putting a stop to deforestation is by far the most cost-effective way to mitigate climate change, for as little as $1/ t CO_2 (Saunders, 2007).

Many studies have reported that among the agricultural production systems organic system have greater C sequestration potential. The Sustainable Agriculture Farming Systems (SAFS) Project at University of California Davis in the United States (Clark 1998, 1999) found that organic carbon content of the soil increased in both organic and low-input systems compared with conventional systems, with larger pools of stored nutrients. This was also true in the Rodale Institute trials, where soil carbon levels had increased in the two organic systems after 15 years, but not in the conventional system (Timothy 2008). Based on another study Pimentel (2006) reported that after 22 years organic farming systems averaged 30 percent higher in soil organic matter than the conventional systems (Figure 4).

It is estimated that up to 4 tonnes CO_2 could be sequestered per hectare of organically managed soils each year (Garrett, 2007). On this basis, a fully organic

UK could save 68 Mt of CO_2 or 10.35 percent of its green house gases emissions each year. Similarly, if the United States were to convert all its 65 million hectares of crop lands to organic, it would save 260 Mt CO_2 a year. Globally, with 1.53 billion hectares of crop land fully organic, an estimated 6.13 Gt of CO_2 could be sequestered each year, equivalent to more than 11 percent of the global emissions, or the entire share due to agriculture. Table 3 gives comparative account of carbon loss and gains in different farming systems in long-term field experiments. In the organic plots, carbon was sequestered into the soil at the rate of 875 lbs/ac/year in a crop rotation utilizing raw manure, and at a rate of about 500 lbs/ac/year in a rotation using legume cover crops (Timothy, 2008). He also reported that results from the Compost Utilization Trial (CUT) at Rodale Institute- a 10-year study comparing the use of composts, manures and synthetic chemical fertilizers, revealed that the use of composted manure with crop rotations in organic systems can result in carbon sequestration of up to 2,000 lbs/ac/year. By contrast, fields under standard tillage relying on chemical fertilizers lost almost 300 pounds of carbon per acre per year. Storing or sequestering up to 2,000 lbs/ac/year of carbon means that more than 7,000 pounds of carbon dioxide are taken from the air and trapped in field soil.

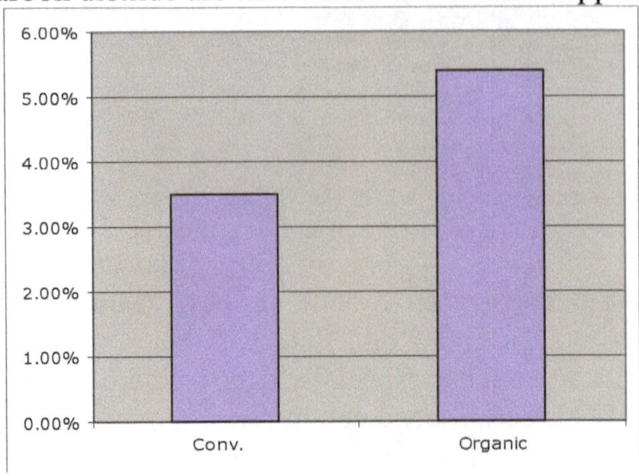

Figure 4 : Soil organic matter in the conventional (a)and organic farming (b)systems after the 22 -year experiments at the Rodale Institute (Source: Pimentel, 2006)

Table 3 : Comparison of soil carbon gains and losses in different farming systems in long-term field experiments

Field trial	**Components compared**	**Carbon gains (+) or losses (-) kg C ha^{-1} yr^{-1}**	**Relative yields of the respective crop rotations**
Frick2 Reduced Tillage Trial, Research Institute FiBL, (Switzerland) Running since 2002	Organic, with ploughing	0	100 %
	Organic,with reduced tillage	+ 879	112 %
Rodale FST, Rodale Institute, Kurtztown,	Organic, with farm yard manure	+ 1 218	97 %

Pennsylvania (USA,) (Hepperly et al.2006; Pimentel et al.2005) Running since 1981	Organic, with legume based green manure.	+ 857	92 %
	Conventional	+ 217	100 %
Scheyern[3] Experimental Farm, University of Munich, Germany, running since 1990 (Rühling 2005)	Organic	+ 180	57 %
	Conventional	- 120	100 %

Besides, climate change mitigation organic agriculture reduces the vulnerability of the farmers to climate change and variability. First, organic agriculture comprises highly diverse farming systems and thus increases the diversity of income sources and the flexibility to cope with adverse effects of climate change and variability, such as changed rainfall patterns. This leads to higher economic and ecological stability through optimized ecological balance and risk-spreading. Second, organic agriculture is a low-risk farming strategy with reduced input costs and, therefore, lower risks with partial or total crop failure due to extreme weather events or changed conditions in the wake of climate change and variability (Eyhorn 2007; Scialabba and Hattam 2002). Higher farm incomes are thus possible due to lower input costs and higher sale prices which eventually enhance coping capacity of the farmers and lowers risk of farmers getting indebted. Risk management, risk-reduction strategies, and economic diversification to build resilience are also prominent aspects of adaptation, as mentioned in the Bali Action Plan (UNFCC 2007). Crops and crop varieties used in organic agriculture are usually well adapted to the local environment. Notwithstanding this potential, more research is needed on how organic agriculture systems perform under increased disease and pest pressures, which are important effects of climate change on agriculture (IPCC 2007c), and on how local crop varieties adapt to climate change and variability. Organic agriculture also seems to perform better than conventional agriculture under water stress conditions (Hepperly et al. 2006; Badgley et al. 2007). Organic agriculture increases soil organic matter content. In consequence, soils under organic agriculture capture and store more water than soils under conventional cultivation (Niggli et al. 2008). Organic agriculture accordingly addresses key consequences of climate change, namely increased occurrence of extreme weather events, increased water stress and drought, and problems related to soil quality (IPCC 2007).

Reduction in water pollution through organic agriculture : It has been observed that organic farming contributes significantly to improve water quality of aquifers and other surface water bodies through reduced non-point source pollution from agriculture. Dabbert (2006) found upto 57% lower leaching rates on per hectare basis from organic fields, however, the rates per ton of produced output were similar or slightly higher in organic fields. As chemical fertilizers and pesticides are not used in organic systems, nutrient and chemical pollution in waterways is significantly reduced. Not only does this translate into long-term cleaner waterways, but it will also save in environmental cleanup costs. Many experiments have found reduced leaching of nitrates from organic soils into ground and surface waters, which are a major source of nitrous oxide. Stoppes et al. (2002) made a comparison between organic and conventional farming and found that nitrate losses from arable crops averaged 47 and 58 kg N ha–1 for organic and conventional farming systems respectively, and concluded that under similar cropping, losses from organic systems are similar or slightly smaller than those from conventional farms following best practice. It has been found that in organic farming only 64 % N, 4 % pesticides, 87% fossil fuel are used compared to integrated farming (Table 4). Kramer et al found that cumulative nitrate leaching was highest from the conventionally fertilized treatment after spring fertilization, followed by the integrated and then the two organic treatments, where leaching rates were not significantly higher than the control (Table 6). However, it has been pointed out that ploughing legume at the wrong time, unfavourable crop rotations, and composting farmyard manure on unpaved surfaces increase the possibility of nitrate leaching in organic farming. However, awareness of the problem and alternative measures have been developed and introduced in practice (CABI). In the emerging scenario of organic farming in India, nitrogen leaching losses have to be assessed and appropriate measures have to be taken, on a case to case basis, to reduce harmful effects on water quality.

Table 4 : Input and output of organic and integrated farming systems of the DOK trial, DOK long-term field trial in Therwil Switzerland (Data for the years 1977 to 2005)

Parameter Nutrient input	**Unit**	**Organic farming**	**Integrated farming with FYM**	**Organic in % of Integrated farming**
	Kg N total ha^{-1} yr^{-1}	101	157	64
	Kg P ha^{-1} yr^{-1}	25	40	62
	Kg K ha^{-1} yr^{-1}	162	254	64
Pesticide applied	Kg ha^{-1} yr^{-1}	1.5	42	4
Fuel use	L ha^{-1} yr^{-1}	808	924	87
Total yield output for 28 years	%	83	100	83
Soil microbial biomass output	tons ha^{-1}	40	24	167

(Source: Mäder 2006)

Table 5 : Comparative NO_3 leaching and N_2O emission from different management systems.

Treatments	NO_3 leaching (fall), µg of NO_3 -Nat 100 cm	N_2O (fall) g/ha N_2O-N	NO_3 leaching (spring), µg of NO_3 -Nat 100 cm	N_2O (spring) g/ha N_2O-N	Annual NO_3 leaching µg of NO_3 -Nat 100 cm
Organic					
Compost	9.66 ^{a,b}	88.57 ^{b,c}	180.13 ^{a}	330.83 ^{b}	241.26 ^{a}
Alfalfa	9.38 ^{a,b}	55.65 ^{b}	234.11^{a}	316.10 ^{b}	309.84 ^{a}
Control	3.73 ^{a}	16.83 ^{a}	68.06 ^{a}	282.28 ^{a,b}	108.47 ^{a}
Integrated					
$CaNO_3$+compost	14.08 ^{b}	124.57 ^{c}	608.26 ^{b}	327.25 ^{b}	772.83 ^{b}
Control	4.43 ^{a}	19.24 ^{a}	97.50 ^{a}	269.03 ^{a,b}	154.85 ^{a}
Conventional					
$CaCNO_3$	13.08 ^{b}	125.87 ^{c}	1092.24 ^{b}	325.98 ^{b}	1352.52^{c}
Control	3.41 ^{a}	30.24 ^{a}	73.38 ^{a}	175.70 ^{a}	130.96^{a}

(Source : Kramer *et al. 2006)*

Note : Significant differences at the 0.05 level (least significant difference) are indicated by different letter in the rows.

Biodiversity promotion through organic agriculture : Organic farming performs better than conventional farming in respect to floral and faunal diversity due to prohibition of synthetic pesticides and N-fertilizers, with secondary beneficial effects on wildlife conservation and landscape. With respect to habitat and landscape diversity, research deficits were identified. Nevertheless, in productive areas, organic farming is currently the least detrimental farming system with respect to wildlife conservation and landscape (Dabbert, 2006). Diverse crop rotations in organic farming provide more habitats for wildlife due to the resulting diversity of housing, breeding and nutritional supply. However, direct measures for wildlife and biotope conservation depend on the individual activities of the farmers. Bengtsson et al. (2005) concluded that "organic farming usually increases species richness, having on average 30% higher species richness than conventional farming systems". However, they also pointed to the high variability among studies. Of the studies examined, 16% showed organic farming to have a negative effect on species richness. They concluded that "the attitude of individual farmers, rather than which farming systems is used, is probably the most important factor determining biodiversity at the farm level". Similarly, Hole et al. (2005) summarized their synopsis on biodiversity stating "The majority of the 76 studies reviewed clearly demonstrates that species abundance and/or richness, across a wide range of taxa, tend to be higher on organic farms than on locally representative conventional farms."

Restoration of soil health through organic agriculture

Organic farming tends to conserve soil fertility better than conventional farming systems. This is mainly due to higher organic matter content and higher biological activity (Dabbert 2006). Therefore, organic farming seems to control erosion more effectively. A more continuous soil cover due to close crop rotations also supports this. As Pimentel (2006) stated that "high level of soil organic matter in organic systems is directly related to the high energy efficiencies observed in organic farming systems; organic matter improves water infiltration and thus reduces soil erosion from surface runoff, and it also diversifies soil-food webs and helps cycle more nitrogen from biological sources within the soil." Organic matter is restored through the addition of manures, compost, mulches and cover crops. Auerswald et al. (2003) performed a large scale modeling exercise to compare the effects of organic and conventional farming on erosion. They found that "on average organic agriculture will cause about 24% less erosion than conventional agriculture." They also pointed to a large variation in extent of erosion for both systems, showing that within both systems erosion could be reduced considerably. This finding is completely in line with the description of Stolze et al. (2000) on the topic of soil erosion. There is the high correlation between increased soil carbon levels and very high amounts of mycorrhizal fungi. These fungi work to conserve organic matter by aggregating organic matter with clay and minerals. The biological support system of mycorrhizal fungi is more prevalent and diverse in organically managed systems than in soils that depend on synthetic fertilizers and pesticides (Timothy 2008). In soil aggregates, carbon is more resistant to degradation than in free form and thus more likely to be conserved. Mycorrhizal fungi produce a potent glue-like substance called glomalin that stimulates increased aggregation of soil particles. This results in an increased ability of soil to retain carbon. Table 6 gives comparative account of soil properties under all bio farms where no chemicals inputs were used and conventionally managed farms. Kramer et al. (2006) recorded higher total N, organic matter, microbial biomass C, microbial biomass N, and enzymatic activities under organic treatment compared to integrated and conventional treatments (Table 8).

Table 6 : Mean value of aggregated soil data from 16 pairs of farms each with organic (bio) and conventional (con) farming.

S.No.	Soil property	All 'bio'farm	All 'conven-tional'farm
1.	Bulk density (Mg m^{-3})	1.07	1.15*
2.	Penetration resistance (0 to 20 Mpa)	2.84	3.18*
3.	Carbon (%)	4.84*	4.27
4.	Respiration ($\mu l O_2 h^{-1}$)	73.7*	55.4
5.	Mineralization N (mg kg^{-1})	140.0*	105.9
6.	Ratio of mineralization N to C (mg kg^{-1})	2.99*	2.59
7.	CEC (cmol kg^{-1})**	21.5*	19.6

* = significantly different at $p<0.01$, **= Cation exchange capacity in centimeters of cation charge (=) per kilogram of soil. *Source :* adapted from (Reganold 1995).

Table 7 : Soil biological and chemical properties under different management systems.

Treatment	**Total soil N, ppm**	**Organic matter, %**	**Microbial biomass C**	**Microbial biomass N**	**L- asparaginase**	**b-glucosidase**
Organic	1955	3.40	512.7	61.0	91.9	192.9
Integrated	1755	3.10	420.8	39.7	63.7	134.4
conventional	1242	2.23	357.7	34.0	59.8	131.4

(Source : Kramer, *2006)*

Thus we can see organic agriculture works in greater harmony with nature and assures quality food on sustainable basis with much less adverse impacts on natural ecosystem health. Thus promoting organic agriculture means promoting eco-safety, sustainability and better health of human beings and animals.

Conclusion

In the process of producing more food for the burgeoning human population through conventional agriculture, we have created several ecological problems like natural ecosystem degradation, loss of biodiversity, declining soil health, release of harmful agrochemicals in the environment, besides releasing emission of harmful greenhouse gases. As a regenerative production technique organic agriculture can avoid such deleterious effects to varying degrees provided when adopted to larger areas. But efforts should be made so that productivity does not suffer and food security can be ensured. Governments on their parts need to should acknowledge organic agriculture as an effective strategy for environment conservation and agricultural sustainability particularly more so due to the imminent threat of climate change, dwindling biodiversity, water pollution, soil health deterioration etc. Organic agriculture should be included in Kyoto Protocol carbon credit mechanisms so that farmers can get economic incentives and motivated to increasingly follow the organic farming systems. At the same time more research activities need to be undertaken to develop and disseminate organic production techniques.

References

Anon (2007a).Observatory monitoring framework indicator data sheet, Environmental impact: Climate change Indicator DD1: Methane emissions. http:// statistics. defra.gov.uk/esg/ace/dd1_data.htm

Anon (2007b). Executive Summary Inventory of U.S. Greenhouse Gas Emissions and sinks: 1990-2005. http:// www. epa.gov/climatechange/emissions / downloads06 / 07ES.pdf

Anon (2007c) Mitigating Climate Change through Organic Agriculture and Localized Food Systems. Institute of science in society, science society sustainability. http://www.i-sis.org.uk/mitigating ClimateChange.php

Anon (2007d).Organic Agriculture and Biodiversity.http://www.iforgan agriculturem.org / growing_organic/1_arguments_for_organicriculture /environmenta l_benefits / pdfs/ Biodiversity_Leaflet.pdf

Simon Dietz, Chris Hope, Nicholas, Stern, and Dimitri, Zenghelis (2007). A Robust Case for Strong Action to Reduce the Risks of Climate Change. *World Economics*. 8: (1)121-168

Auerswald K., Kainz M., Fiener P. (2003) Soil erosion potential of organic versus conventional farming evaluated by USLE modeling of cropping statistics for agricultural districts in Bavaria. *Soil Use Man*. 19:305-311.

Badgley, C., Moghtader J., Quintero E., Zakem E., Chappell M., Avilez-Vazquez K., Samulon A. and Perfecto. I. (2007) Organic agriculture and the global food supply. *Renewable Agriculture and Food Systems,* 22(2): 86–108.

Barton L, Kiese R, Gatter D, Butterbach-Bahl, K., Buck, R., Hinz, C. and Murphy, D. (2008) Nitrous Oxide Emissions from a Cropped Soil in a Semi-arid Climate. *Global Change Biology* 14: 177–92.

Bengtsson, J., Ahnström, J., and Weibull, A.-C. (2005) The effects of organic agriculture on biodiversity and abundance: A meta-analysis. J. Appl. Ecol. 42:261-269.

Berner, A., Hildermann, I., Fliebach, A., Pfiffner, L., Niggli, U., Mäder, P. (2008) Crop yield and soil fertility response to reduced tillage under organic management. *Soil and Tillage Research* 101 : 89-96.

Biermann S, Rathke, G.W., Hülsbergen, K.J. and Diepenbrock, W. (1999) Energy Recovery by Crops in Dependence on the Input of Mineral Fertilizer, Final Report, European Fertilizer Manufacturers Association, http://www.efma.org/publications/ Energy%20Study/endbericht.pdf

Bunyard, P. (1996) Industrial Agriculture - Driving Climate Change. *The Ecologist* 26 (6): 290-98.

Chhonkar, P.K. (2003). Organic Farming: Science and Belief. *J Soil Sci. Soc. India, Vol.* 51(4) : 365-377.

Clark, M.S., Horwarth, W.R., Shennan, C. and Scow. K.M. (1998) Changes in soil chemical properties resulting from organic and low-input farming practices, *Agronomy Journal*, 90: 662-671.

Clark, S., Klonsky, K., Livingston, P. and Temple, S. (1999) Crop-yield and economic comparisons of organic, low-input, and conventional farming systems in California's Sacramento Valley. *American Journal of Alternative Agriculture* 14: 109-121

Deb, D.L. and Joshi, H.C. (1994) Environment pollution and agriculture, in natural resource management for sustainable agriculture and environment, Angkor publications (P) Ltd., New Delhi, pp. 227-293.

Edward Goldsmith. (2010) How to feed people under a regime of climate change. http://www.edwardgoldsmith.org/page97.html

El-Hage Scialabba, N., and Hattam, C. (2002) Online document. "Organic Agriculture, Environment, and Food Security." Environment and Natural Resources Service, Sustainable Development Department, Food and Agriculture Organization of the United Nations (FAO). http: // www. fao.org/ docrep / 005 / y4137e /y4137e00.htm.

Erisman, J.W., Sutton, M.A., Galloway, J., Klimont, Z., Winiwarter, W. (2008) How a century of ammonia synthesis changed the world. *Nature Geoscience.1*: 636-639.

EFMA (European Fertilizer Manufactures Association). (2005) Understanding Nitrogen and Its Use in Agriculture. EFMA, Brussels, Belgium.

Eyhorn, F. (2007) Organic farming for sustainable livelihoods in developing countries: the case of cotton in india. Ph.D diss. Department of Philosophy and Science, University of Bonn. http://www.zb.unibe.ch/downlorganic agricultured /eldiss /06eyhorn_f.pdf.

FAO. (1998) Crop Genetic Resource, in Special: Biodiversity for Food and Agriculture, FAO, Rome, at: http://www.fao.org/sd/EPdirect/EPre0039.htm (accessed 29/6/05)

FAOSTAT (2009) FAO Statistical Database Domain on Fertilizers: ResourceSTAT-Fertilizers. Food and Agriculture Organisation of the United Nations (FAO) Rome, Italy. Available at http://faostat.fao.org/site/575/default. aspx#anchor

Garrett, T. (2007) Overall UK consumption related GHGs, April 2007, Food Climate change Research Network, http: //www. fcrn.org. uk/ frcnresearch /publications.

Gupta, S. K., Gupta, R. C., Seth, A. K., Bassin, J. R., Gupta, D. K. and Sharma, S. (2000) Recurrent acute respiratory infections in areas with high nitrate concentrations in drinking water. *Environ. Health Prospect*, 108: 363–366.

Hegde, B.R., Krishnegowda, K.T. and Parvathappa, H.C. (1995) organic residue management in red soils under dryland conditions, in Shivashankar, K. (ed.), Alternatives to fertilizers in sustainable agriculture, University of Agricultural sciences, Bangalore.

Hepperly, P., Douds Jr., D. and Seidel R. (2006) The Rodale Farming Systems Trial, 1981 to 2005: Long-term analysis of organic and conventional maize and soybean cropping systems. in long-term field experiments in organic farming, ed. J. Raupp, C. Pekrun, M. Oltmanns, and U. Köpke, 15–32. Bonn, Germany: International Society of Organic Agriculture Research (ISOFAR).

Hole, D. G., Perkins, A.J., Wilson, J. D., Alexander, I. H., Grice, P. V., Evans, A.D. (2005) Does organic farming benefit biodiversity? *Biol. Conserv.* 122:113-130.

IPCC. (2007a) Synthesis report. In Metz, O.R.D., Bosch, P.R., Dave, R. and Meyer, L.A. (eds). Fourth Assessment Report: Climate Change 2007. Cambridge University Press, Cambridge, UK.

IPCC. (2007b) Summary for Policy Makers. In: IPCC Fourth Assessment Report, Working Group III Report: Mitigation of Climate Change. See specifically on agriculture, chapter 8. http://www.ipcc.ch/pdf/assessment-report/ar4/wg3/ar4-wg3-spm.pdf and http://www.ipcc.ch/ipccreports/ar4-wg3.htm.

IPCC. (2007c) Summary for Policy Makers." In IPCC Fourth Assessment Report, "Working Group II Report: Impacts, Adaptation, and Vulnerability. See specifically on adaptation, chapter 17; on inter-relationships between adaptation and mitigation, chapter 18; on vulnerability, chapter 19. http://www.ipcc.ch/pdf/ assessment-report/ar4/wg2/ar4- wg2-spm.pdf and http://www.ipcc.ch/ ipccreports/ ar4-wg2.htm.

Kramer, S. B., Reganold, J. P., Glover, J. D., Bohannan, B. J. M. and Mooney, H. A. (2006) Reduced nitrate leaching and enhanced denitrifier activity and efficiency in organically fertilized soils. *PNAS.* 103 (12):4522–4527.

Lal, R. (2004) Soil Carbon Sequestration Impacts on Global Climate Change and Food Security, *Science* 304: 1623–27. (See also Lal, R., M. Griffin, J. Apt, L. Lave, and G.Morgan. 2004. "Managing Soil Carbon," *Science 304: 393*; and "Response to Comments on 'Managing Soil Carbon, *Science* 305: 1567.

Lim Li Ching. (2010) Global Conference on Agriculture, Food Security and Climate Change. http: // www. iatp. Org /climate /files /document / TWN % 20BP % 201_hague % 20ecol%20ag% 20dev%20smart.pdf

Lotjonen T.(2003). Machine work and energy consumption in organic farming. *Ecology and Farming*. 32: 7-8, IFOAM.

Mäder, P., Fließbach, A., Dubois, D., Gunst, L., Jossi, W., Widmer, F., Oberson, A., Frossard, E., Oehl, F., Wiemken, A., Gattinger, A., Niggli, U. (2006) The DOK experiment (Switzerland). In: *Long-term field experiments in organic farming*. Raupp, J., Pekrun, C., Oltmanns, M., Köpke, U. (eds.). pp 41-58. Koester, Bonn.

Mathieu, O., Lévêque, J., Hénault, C., Milloux, M.-J., Bizouard, F., Andreux, F. (2006) Emissions and spatial variability of N2O, N2 and nitrous oxide mole fraction at the field scale, revealed with 15N isotopic techniques. *Soil Biology & Biochemistry* 38: 941-951.

Moser, A., Schime, D., Valentine, D., Bronson, K., and Parton, W. (1991) Methane and nitrous oxide fluxes in native fertilized and cultivated grassland, *Nature* : 350: 330 - 332

Mosier, A. R. (2002) Environmental challenges associated with needed increases in global nitrogen fixation. *Nutr. Cycl. Agroecosyst.*, 63 : 101–106.

Nadia El-Hage Scialabba and Maria Mu¨ller-Lindenlauf. (2010) Organic agriculture and climate change. *Renewable Agriculture and Food Systems:* 25(2): 158–169

NATCOM, MoEF (2004) India's Initial National Communication (NATCOM) to the UNFCCC was implemented and executed by the Ministry of Environment and Forests, Government of India, funded by the Global Environment Facility under its enabling activities program through the United Nations Development Program, New Delhi. The data has been drawn from a variety of official GoI sources as well as industry association and stakeholder reports.

Niggli, U., Fliessbach, A., Hepperly, P. and Scialabba. N. (2008). Low greenhouse gas agriculture: mitigation and adaptation potential of sustainable farming systems. FAO.

Pimentel, D. (2006) Impacts of Organic Farming on the Efficiency of Energy Use in Agriculture. The Organic Center, Cornell University, Ithaca, NY.

Pimentel, D., Hepperly, P., Hanson, J., Douds, D., Seidel, R. (2005) Environmental, energetic, and economic comparisons of organic and conventional farming systems. *Bioscience* 55: 573-582.

Reganold, J.P. (1995). Soil quality and profitability of biodynamic and conventional farming systems: a review. *American Journal of Alternative Agriculture*. 10(1): 36-45

Renwick, W., Smith, S., Sleezer, R. and Buddemeier, R. (2004) Comment on "Managing Soil Carbon (II)," *Science* 305:5690 (Page 1567c).

Rühling, I., Ruser, R., Kölbl, A., Priesack, E., Gutser, R. (2005) Kohlenstoff und Stickstoff in Agrarökosystemen. In: Landwirtschaft und Umwelt - ein Spannungsfeld. Osinski, E., Meyer-Aurich, A., Huber, B., Rühling, v., Gerl, G., Schröder, P. (eds.). pp 99-154. Ökom Verlag, München.

Saunders, P,T. (2007) The economics of climate change. Science in Society, 33, 20-23.

Scialabba, NE-H and Hattam, C.(2002) Organic agriculture, environment and food security, FAO, Rome.

Smith, P., Martino, D., Cai, Z., Gwary, D., Janzen, H., Kumar, P., McCarl, B., Ogle, S., O'Mara, F., Rice, C., Scholes, B. and Sirotenko, O. (2007) Agriculture. In Climate Change:Mitigation. Contribution of Working Group III to the Fourth Assessment Report of the Intergovernmental Panel on Climate Change [B. Metz, O.R. Davidson, P.R. Bosch, R. Dave, L.A. Meyer (eds)], Cambridge University Press, Cambridge, United Kingdom and New York, NY, USA.. Available athttp// www. Mnp .nl/ ipcc /pages_media /FAR4docs/ final_pdfs_ar4 /Chapter08.pdf

Stolze, M, Piorr, A., Häring, A. and Dabbert, S. (1999). Environmental and resource use impacts of organic farming in Europe, Commission of the European Communities, Agriculture and Fisheries (FAIR) specific RTD programme, Fair3-CT96-1794, "Effects of the CAP-reform and possible further development on organic farming in the EU".

Stolze, M., Piorr, A., Häring, A., and Dabbert, S. (2000). The environmental impacts of organic farming in Europe, Organic Farming in Europe: *Economics and Policy*, Volume 6. Universität Hohenheim, Stuttgart-Hohenheim.

Stopes, C., Lord, E. I., Phillips, L. and Woodward, L.; (2002). *Soil Use Manage.* 18 (suppl.); 256–263.

Timothy J. L. (2008) Regenerative 21st Century Farming: A Solution to Global Warming. The Rodale institute. Accessed at http :// www. rodaleinstitute. org/ files / Rodale_Research_Paper.pdf

Van Oost, K., Govers, G., Quine, T. and Heckrath. G. (2004). Comment on 'Managing Soil Carbon. *Science.* 305 : 5690 P. 1567 (www.sciencemag.org. content1305/5690/ 1567.2.full.html)

Williams, A.G., Audsley, E. and Sandars, D.L. (2006) Determining the environmental burdens and resource use in the production of agricultural and horticultural commodities. Main Report. Defra Research Project IS0205. Bedford: Cranfield University and Defra./ www.silsoe.cranfield.ac.uk, and www.defra.gov.uk.

□□□

Green Agriculture : Newer Technologies, 2012

New India Publishing Agency, New Delhi (India)
e-mail : info@nipabooks.com; website : www.nipabooks.com

Chapter-9

Seed Quality Improvement through Seed Priming for Better Crop Stand Establishment : A Practice towards Sustainable Farming

K. Bhanuprakash, Roopa A Reddy and G.Sarika
Seed Science and Technology, IIHR, Hessaraghatta, Bangalore-560089
E-mail : dr.bp100@gmail.com

SUMMARY

Seed quality comprises the sum of all properties or characteristics, which determine the potential level of the seed or seed lot performance and crop establishment. Once the seed is sown many factors (biotic, abiotic etc.,) influence its survival and potential. Seed priming is one of the short–term and most programmatic approaches followed to improve the quality of seed so as to exhibit maximum possible potential even under adverse conditions. Seed priming is a method that well fit in to the one or more principles of sustainability, as it is practiced as a low-cost, low-risk interventions that increase and stabilize yields to have a large and direct impact on the livelihoods of farmers. Various priming methods viz., hydro priming, halo priming, matrix priming, bio priming are in practice for advancement in seed germination, better crop growth and yield. The adoption of seed priming as a successful technology is widely practiced across the Globe for various benefits at farm level. On-farm seed programme is one such successful programmes adopted under UK aid across 11 countries towards sustainable agriculture campaign.

1. Introduction

The production of high-quality seed is the basis for sustainable and profitable agriculture and horticulture. Seed quality comprises the sum of all properties or characteristics, which determine the potential level of the seed or seed lot performance and crop establishment. Seed is the carrier of new technologies like increased yield potential, resistance to pests and diseases. Once the seed is sown many factors (biotic, abiotic etc.,) play crucial role in its germination, development and growth. Following a set of crop specific package of practices (ex: application of pesticides, fungicides, weedicides, manures and fertilizers *etc.*) is a common practice to obtain good yields. However, repeated and excess use of these leads to soil health deterioration and environment pollution. Hence, methodologies that minimize deterioration of soil, plant and environment health are to be adopted as a sustainable practice without comprising the expected yields. Sustainability rests on the principle that we must meet the needs of the present without compromising the ability of future generation to meet their own needs. While conventional agriculture is driven almost solely by productivity and profit, sustainable agriculture integrates biological, chemical, physical, ecological, economical and social sciences in a comprehensive way to develop new farming practices that are safe and don't degrade our environment. This definition can be supplemented by some fundamental principles of sustainable agriculture i) that farm productivity is enhanced over the long term ii) that adverse impacts on the natural resource base and associated ecosystems are ameliorated, minimized or avoided iii) that residues resulting from the use of chemicals in agriculture are minimized iv) that net social benefit (in both monetary and non-monetary terms) from agriculture is maximized and that farming systems are sufficiently flexible to manage risks associated with the vagaries of climate and markets. Seed priming is one of the short–term and most programmatic approaches well fit in to the one or more principles outlined above. The technique of seed priming has gained popularity since ages due to its beneficial effects like enhancement in seed germination, advancement of germination, pest and disease resistance, better crop stand establishment and stable yields even under adverse conditions.

2. Importance

There is ample evidence that poor crop stand establishment is a widespread constraint of crop production in developing countries, particularly in the marginal environments farmed by poor people. Patchy plant stands are common, and yields are often reduced simply because there are not enough plants in the field. In addition, plants that do eventually emerge often grow slowly, and are highly susceptible to stresses such as drought, pests and diseases. Farmers can choose to re-sow, although this entails severe yield penalties and increased labour and financial costs, and there is evidence from our country that borrowing to pay for replacement seed can initiate or add to a spiral of indebtedness. *Clearly, anything that can be done to increase the proportion of seeds that emerge, and the rate at which they do so, will have a large impact on farmers' livelihoods.* Low, unstable yields are a major contributor to the fragile lives of poor farmers in marginal areas. Low-cost, low-risk interventions that increase and stabilize yields will have a large impact on the livelihoods of such farmers. At this juncture, seed priming technology plays a crucial and vital role in providing right seedling for better crop establishment and stable yields even under adverse conditions. The present scenario of erratic rainfall, monsoon failures and alarming change in the global climate that has become a terror to crop growth and threat to livelihood, can be tackled to major extent by following simpler techniques like seed priming and seedling dipping to obtain expected yields

3. Definition and Concept

Seed priming is a pre-sowing treatment, that involves exposure of seeds to a low external water potential that limits hydration,(controlled hydration of seed) to a level that permits pre-germinative metabolic activity to proceed, but prevents actual emergence of the radical. The low vigour seeds can be made vigorous by priming with various chemicals. This will ensure better field emergence & disease resistance under various adverse conditions. The purpose of priming is to reduce the germination time, and improve stand and percentage germination under adverse environmental conditions. Primed seeds are used immediately, but may be dried and stored for short time for later use. Basic objective of seed priming is to ensure rapid seed germination and faster growth and to achieve successful and uniform stand establishment in the field. A seed priming treatment can be pre – sowing or pre – storage (or mid – storage) treatment. For higher productivity/production, it is always desirable to have a very high proportion of germinable seeds in the samples at the time of sowing. Pre sowing treatments enhance crop tolerance to adverse conditions (Ashraf et al. 2008).

Many priming methods are in practice for seed quality enhancement (Bhanuprakash 2010).

4. Methods

Broadly the seed invigoration treatments followed to enhance seed quality can be classified in to:-

1. Pre sowing treatment: This is practiced for improved field performance by enhancing germination potential of the seed and early germination with high seedling vigour. Ex: a) Seed hardening/hydro-priming; b) Seed priming; c) Seed fortification and Seed pelleting
2. Pre-storage treatments: This method is in practice for better storability and better field performances of stored seeds. Ex: a) Seed halogenations b) Dry permeation
3. Mid-storage treatments: This is practiced to treat the stored seeds during storage for better storage performance (vigour and viability). Ex: Hydration and dehydration treatment b) Use of antioxidants and botanical agents

   However, for convenience and better understanding, the priming methods are classified in to:-

1. Hydro priming 2. Halo priming 3. Osmo priming 4. Matrix priming 5. Thermo priming 6. Bio priming 7. Drum priming 8. Priming using growth regulators

4.1 Hydro priming : This technique implies soaking the seeds in water for about specific duration. This terminology is currently used both in the sense of steeping (imbibition in water for a short period), and in the sense of 'continuous or staged addition of a limited amount of water'. Hydro priming methods have practical advantages of minimal waste material produced when compared to osmo and matrix priming.

4.2 Hydro priming -"steeping" : This is one of the simplest methods and it is being practiced over many centuries. On-farm steeping was advocated in many parts of the world as a pragmatic, low cost/low risk method for improved crop establishment. Steeping can also remove residual amounts of water soluble germination inhibitors from seed coats. It can also be used to infiltrate crop protection chemicals for the control of deep-seated seed borne disease, etc. This type of seed treatment usually involves immersion or percolation (up to 30° C for several hrs.), followed by draining and drying back to near original moisture content. Short 'hot-water steeps' (thermotherapy), typically $<50^{O}$ C for 10 to 30 min, are used to disinfect or eradicate certain seed borne fungal, bacterial, or viral pathogens. However, extreme care, precision needed to avoid loss of seed quality.

4.3 Halo and Osmo priming : In halo-priming, seeds will be soaked in various solution of inorganic salts such as KCl, KNO_3, $CaCl_2$, Ca $(NO_3)_2$, KH_2PO_4 *etc.* This method is practiced for higher germination and plant emergence in salt-affected soils. Incase of osmo priming, substances like polyethylene glycol (PEG), sugars, glycerol, sorbitol, mannitol etc., are used as osmotic solutes to develop lower water potential. As this process, unlike hydro priming, regulates water movement in much controlled fashion for longer period ,this method is preferred in those crops where soaking in treatment solutions, even for shorter period, leads to germination (ex: onion, beans etc).

4.4 Matrix priming : Solid matrix priming is done using solid carriers with low matrix potentials eg: vermiculite, peat moss, sand, celite etc., for slow imbibition process. In this case, seeds slowly imbibe and reach an equilibrium hydration level. After priming, the moist matrix material is removed by sieving or screening, or may be partially incorporated into a coating. This process mimics the natural uptake of water by the seed from soil. Seeds generally mixed into carrier at matric potentials from -0.4 to -1.5 MPa at 15-20° C for 1-14 days.

4.5 Thermo priming : It is a kind of pre soaking seed treatment with high and low temperature to improve germination and emergence under different environmental (low and high temperatures) conditions. This process enables seeds to germinate at temperatures, lower or higher, than those at which they would have been able to germinate untreated.

4.6 Bio priming : Treating the seed with some of microbial agents like rhizobium, azospiriullum *Pseudomonas aureofaciens, Bacillus, Trichoderma, Gliocladium* etc., is practiced in this method, for improving seed viability or vigour . Beneficial microbes are included in the priming process, either as a technique for colonizing seeds and/or to control pathogen proliferation, during priming. Compatibility with existing crop protection seed treatments and other biologicals need to be looked in to while practicing this method. Costs of registration, other factors currently limit commercial use of bio priming. Bio –priming as seed treatment that integrates the biological and physiological aspects of disease control was recently used as alternative method for controlling many seed and soil borne pathogens.

4.7 Drum priming : Seed are hydrated in a tumbling drum using precise volume of water. The amount of water is limited so that, it is less than the amount needed for natural imbibition and seed germination to occur. In this method, the seeds are evenly and slowly hydrated to a predetermined moisture content (typically <25-30% fresh weight basis) by misting, condensation, or dribbling. Drum priming enhances seed performance without the loss of additional materials associated with the conventional osmotic priming technique.

4.8 Priming using growth regulators : In this method, seeds are primed using solutions containing minute quantities of plant growth regulators like Giberellic Acid, Indole Acetic Acid, Benzyl Adenine, Methyl Jasmontite, 1-amino cyclo-propane 1-carboxylic acid etc. This method is usually followed to address seed dormancy problems or to enhance seed germination under adverse soil conditions or to re-activate of impaired metabolism of aged and deteriorated seeds. Soaking papaya seeds in GA_3 250 ppm enhanced seedling emergence even at low temperature conditions(Bhanuprakash et al., 2010)

5. Seed Priming as a ITK : Today it is widely accepted among agricultural scientists throughout the world that the re-assessment of indigenous technical knowledge is an indispensable part of the introduction of new agricultural technology. It is recognized that the knowledge of farmers must be taken into account before any new technology is developed and disseminated since the farmers have a wealth of knowledge pertaining to their environment and have developed specific skills designed to make the best use of that knowledge. e.g., Use of neem leaves, dry chilli powder, red earth, wood ash, edible and non-edible oil (castor oil). Many government and non government organizations are popularizing these technologies after validation for benefit of agricultural community. Some indigenous seed treatments observed in cereals, pulses and vegetables are listed below.

Table 1 : Some indigenous treatments observed in various crops

S.No.	Crop	Method	Attribute
A. Cereals			
1.	Paddy	Paddy seeds filled in gunny bag are kept immersed in a water trough for 12 hours and later in diluted biogas slurry for another 12 hours before sowing.	Increases the resistance of seedlings to pest and diseases.
2.	Wheat	seeds are immersed in milk before sowing	To control wheat rust
3.	Castor	seeds are soaked in whey before sowing	To give resistance Pests
4.	Maize	Seeds are soaked in cow urine for 12 hours before sowing	For better germination
5.	Sorghum	Soaking in cow's urine	To induce drought tolerance
		Treating with salt solution	To ensure better germination
B. Pulses			
1.	Soybean	Seeds are treated with leaf powder of 'Usil' (*Albizia amara*) @ 150 g leaf powder/kg of soybean.	To increase seed germination in saline/alkaline lands
2.	Bean	Soaking in milk for a day before planting	Results in healthy plants with good yield
3.	Any pulse crop	soaking at the rate of 40 kg of seeds in 5 litres of butter milk for 12-24 hours before sowing	For better germination
C. Vegetables			
1.	Chilli	Seeds are immersed in biogas slurry for half an hour/overnight in fermented buttermilk	For better germination
2.	Bottle Gourd	Soaking in water for 24 hours before sowing.	For better germination
3.	Cucumber and bean seeds	seeds are dipped in kerosene and then sown	To avoid seed loss by ants.
4.	Watermelon	Soaked in Kumkum (vermilion) water for a day before sowing.	Germinate faster and grow into healthy plants.
5.	Bitter gourd seeds	soaked in milk for a day prior to sowing	germinate faster and develop well

Having noticed the benefit of seed priming, particularly under adverse conditions, the DFID (Development For International Development) crop protection organization has conducted Global level "on farm seed priming" trails (13 crops in eight countries), and recommended seed priming as a safe and beneficial method as the response to priming in different test locations was generally positive. Soaking / priming solanaceous seeds in fermented butter milk, before sowing enhanced germination speed, vigour and other attributes (Bhanuprakash et al. 2003)

6. On Farm Seed priming : Farmers can prime their own seed if they know the safe limits. These safe limits are calculated for each variety so that germination will not continue once seeds are removed from the water. Primed seed will only germinate if it takes up additional moisture from the soil after sowing. It is important to note this distinction between priming and pre-germination–sowing pre-germinated seed under dry land conditions can be disastrous. In most cases seed can be primed overnight and is simply surface-dried and sown the same day. Occasionally, sowing may be unavoidably delayed – by heavy rain for example. If primed seed is surface-dried and kept dry it can be stored for several days, then sown as usual and this still perform better than non-primed seed. Farmers can prime their own seeds if they know the maximum length of time for which their seeds can be soaked before seed or seedling damage occurs. After the seeds have been soaked for the appropriate length of time, the water is drained off and the seeds are surface-dried by placing them on a cloth or plastic sheet on the ground for 15 to 30 minutes or, for small amounts of seeds, rolled gently in a dry cloth so that they do not stick together (Harris, 2006).

7. Advantages of Seed Priming:

1. Increases germination rate
2. Early and uniform emergence
3. Germination under broader environment (Drought and high salt)
4. Improves performance of low vigour seeds
5. Improves vigour of immature seeds
6. Breaks seed dormancy
7. Permits germination in suboptimal temperature
8. Reverse seed deterioration effect
9. Increases enzyme activity, protein content and ATP level
10. Synchronization in flowering among hybrid and parental lines

Table 2 : Different seed priming techniques and their effectiveness in improving growth of various crops under adverse conditions

S. No.	**Method**	**Priming agent**	**Crop**	**Attributes improved**	**References**
A	**Osmo priming**				
1.		PEG	Tomato and Asparagus	Germination under saline condition	Pill et al., 1991
2.		Mannitol	Cucumber	Germination under saline condition	Passam and Kakouriotis (1994)
B	**Hydro priming**	Water	Wheat	Germination rate under saline condition	Roy and Srivastava (1999)
1.		Water	Triticale sps.	Drought tolerance	Mehmet and Digdem 2008

2.		Water	Maize	Drought tolerance	Janmoha mmadi et al. 2008
C	**Thermo priming**	Chilling treatment	Brassica sps	Germination under saline condition	Sharma and Kumar (1999)
1.		PEG at 20^{0} C	Carrot	Low and high temperature stress	Marcio et al, 2009
D	**Chemical priming**	gibberellins	Okra and Perlmillet	Growth and yield under saline condition	Vijayaraghavan (1999)
1.		Solution of inorganic salts	Broadbean	Growth under saline condition	Sallam (1999)
2.		Salicyclic acid	Muskmelon	Growth under drought stress	Ahment et al, 2007
3.		NaCl	musk melon	Growth under saline stress	Yeoung et al, 1996
4.		Salts	Amaranthus	Drought tolerance	Moosavi et al, 2009
5.		Brassinosteriods	Tomato	Drought tolerance	Behnamnia et al, 2009
6.		NaCl	Sugarcane	Drought tolerance	Vikasyadav et al, 2009
7.		GA_3	Rapeseed	Drought tolerance	Li et al, 2010
8.		Polyamines	Rice	Drought tolerance	Muhammad et al, 2009
9.		Salts	Onion	Drought tolerance	Arvin and Kazemipoor 2003
10.		Paclobutrazol	Tomato	Drought tolerance	Souza Machado et al, 1999
11.		Chitosan	Maize	Low temperature tolerance	Ya-jing et al, 2009
12.		Putrescine	Tobacco	Chilling tolerance	Shengchunxu et al, 2010
E	**Bio priming**				
1.		*Pseudomonas fluorescens-strain-* Pf_1 *Bacillus subtilis*-strain EPB5	Green gram	Enhance the water stress resistance	Saravan kumar et al, 2010
3.		inoculation with Rhizo-bium and *Pseudomonas*	Maize	Salt tolerance	Bano and Fatima, 2009
4.		Pseudomonas sps. Strain GAP –P45	Sunflower	Increased the survival, plant biomass and root adhering soil/root tissue ratio of sun-flower seedlings sub-jected to drought stress	Sandhya et al, *2009*

Contd. ...

5.		*Pseudomonas* sp. Strain-AKM –P6	Sorghum	Enhanced the tolerance of seedlings to elevated temperatures	Ali et al. *2009*
6.		*Pseudomonas fluorescens* isolates	Pearl millet	Improved growth of the plants and resistance against downy mildew	Niranjan Raj et al. 2004
7.		*Trichoderma sps*	Soybean	Control of damping of soyabean caused by *C. truncatum* of soyabean	Begum et al. 2010
8.		*pseudomonas fluorescens*	Sunflower	Control of *Alternaria* blight	Rao et al. *2009*
9.		*pseudomonas fluorescens*	Chickpea	Control *fusarium* wilt in tomato	Vidhyasekaran and Mutamilan, 1995

Conclusion

The adoption of seed priming as a successful technology is widely practiced across the Globe for various benefits at farm level. DFID (Development For International Development) a program under UK aid has multi billion projects across 11 countries on on-farm seed priming under sustainable agriculture campaign. These seed technological approaches enhance physiological quality, vigour and synchrony to establish a crop in the field under diverse environmental conditions.

References

Ahment, K., Murat, U and Ali Riza, D. (2007). Treatment with acetyl salicyclic acid protects muskmelon seedlings against drought stress. *Acta physiologiae plantarum*. 29: 503-508.

Ali, S.K.Z., Sandhya,V., Minakshi Grover., Kishore .N., Venkateswara Rao,L and Venkateswarlu, B. (2009). *Pseudomonas sp.* Strain AKM –P6 enhances tolerance of sorghum seedlings to elevated temperatures. *Biology and fertility of soils* . 46: 45-55.

Arrvin, M.J and Kazemi Poor, N. (2003). Response of onion cultivation to drought and salinity stress at germination stage and seed priming by chemicals to improve germination. *Iranian journal horticultural sciences and technology* 4: 95-104.

Ashraf, H.R., Athar, P.J.C., Harris,D and Kwon, T.R. (2008). Some prospective strategies for improving crop salt tolerance *Advances in agronomy* 97: 45-92.

Bano, A and Fatima, M.(2009). Salt tolerance in *Zea mays* (L.) following inoculation with *Rhizobium and Pseudomonas. Biology and fertility of soils* . 45: 405-413.

Begum,M.M., Sariah,M., Puteh,A.B., Zainal Abidin, M.A., Rahman,M.A. and Siddiqui, Y.(2010). Field performance of bio primed seeds to suppress *Colletotrichum truncatum* causing damping off and seedling strand of soybean. *Biological control.* 53:18-23.

Behnamnia, M., Kalantari, M and Rezanejad, F. (2009). Exogenous application of Brassinosteriods alleviates drought induced oxidative stress in *Lycopersicon esculentum. General and applied plant physio*logy. 35 : 22-24.

Bhanuprakash, K., Ganeshan, S and Yogeesha,H.S.(2003). Effect of soaking of solanaceous vegetable seeds in buttermilk on germination. *In*: *Validation of indigenous Technical Knowledge in Agriculture Document.* 3:207-213. NATP mission mode project on collection, documentation and validation of indigenous Technical Knowledge.

Bhanuprakash,K., Yogeesha,H.S and Arun,M.N.(2010). Physiological and biochemical changes in relation to seed quality in ageing bell pepper seeds. *Indian journal of agricultural sciences.* 80 (9): 10-13.

Bhanuprakash,K. (2010). Seed quality enhancement through seed priming. *in* : souvenir on National conference on production of quality seed and planting material -health management in horticultural crops. Pp 199-204.

Harris,D. (2006). Development and testing of onfarm seed priming. *Advances in agronomy*. 90:129-178.

Janmohammadi, M., Moradi Dezfuli, P and Sharifzadeh, F. (2008). Seed invigoration techniques to improve germination and early growth of inbred line of maize under salinity and drought stress. *General and applied plant physiol*ogy 34: 215-226.

Li, Z.,Lu, G.Y.,Zhang,Y.K.,Zou, C.S.,Cheng,Yand Zheng,P.Y.(2010) Improving drought tolerance of germinating seeds by exogenous application of GA_3 in rape seed. *Seed science and technology*. 38: 432-440.

Mao,W,, Lumsden,R.D., Lewis, J.A., Hebbar, P.K. (1998). Seed treatment using pre-infiltration and biocontrol agents to reduce damping –off of corn caused by species of *Pythium and Fusarium. Plant disease*. 82: 294-299.

Marcia, D.P., Denise, C., Luiz, A and Eusdarardo, F.A. (2009). Primed carrot seeds performance under water and temperature stress. *Scientia agricola*. 662:174-179.

Marulanda, A., Jose-Mguel Barea, Rosario, A. (2009). Stimulation of plant growth and drought tolerance by native microorganisms (AM Fungi and Bacteria) from dry environments : Mechanisms Related to Bacterial Effectiveness . *Journal of plant growth regulation.* 28:115-124.

Mehmet, Y and Digdem, K. (2008). Alleviation of osmotic stress of water and salt in germination and seedling growth of triticale with seed priming treatments. *African journal of Biotechnology.* 7(13):2156-62.

Mohamedy , R.S.R., Abdalla, M.AA., Badiaa, R.I.(2006). Soil amendment and seed bio priming treatments as alternative fungicides for controlling root rot diseases on cowpea plants in Nobaria Province. *Research journal of agriculture and biological sciences. 2*(6): 391-398.

Moosavi, A., Tavakkol Afshari, R., Sharif-Zadeh, F and Aynehband, A. (2009). Seed priming to increase salt and drought stress tolerance during germination in cultivated sps of amaranthus. *Seed science and technology*. 37: 781-785.

Muhammad, F., Abdul, W and Dong, J.L. (2009). Exogenous applied polyamines increases drought tolerance of rice by improving leaf water status, photosynthesis and membrane properties. *Acta physiological plantanam* . 31: 937-945.

Niranjan raj,S., Shetty N.P., Shetty, H.S. (2004). Seed bio- priming with *Pseudomonas fluorescens* isolates enhances growth of pearl millet plants and induces resistance against downy mildew. *International journal of pest management.* 50(1): 41-48.

Passam, H.C and Kakouriotis, D. (1994). The effect of osmo conditioning on germination and emergence and early plant growth of cucumber under saline conditions. *Scientia horticulturae*. 57: 233-240.

Pill, W.G., Freett, J.J and Morneau, D.C. (1991) .Germination and seedling emergence of primed tomato and asparagus seeds under adverse conditions. *Horticultural science* 26: 160-1162.

Rao, M.S.L., Kulkarni S., Lingaraaju ,S. and Nadaf, H.L.(2009). Bio priming of seeds: A potential tool in the integrated management of *alternaria* blight of sunflower. *Helia.* 32:107-114.

Rashmi,S., Abdul Khalid., Singh, U.S. and Sharma, A.K. (2010). Evaluation of arbuscular mycorrhizal fungus, fluorescent *Pseudomonas* and *Trichoderma harzianum* formulation against *Fusarium oxysporum* f.sps lycopersici for the management of tomato wilt. *Biological control.* 53:24-31.

Roy, N.K and Srivastava, A.K. (1999). Effect of presoaking seed treatment on germination and amylase activity of wheat under salt stress condition. *Rachis*.18: 46-51.

Sallam, H.A. (1999). Effect of some seed soaking treatments on growth and chemical components on faba bean plants under saline conditions. *Annals of agricultural sci*ences 44: 159-171.

Sandhya, V., Ali, S.K.Z., Minakshi Grover, Gopal Reddy and Venkateswaralu,B. (2009). Alleviation of drought stress effects in Sunflower seedlings by the exopolysaccharides producing *Pseudomonas putida* strain GAP – P45. *Biology and fertility of soils* . 46: 17-26.

Saravanakumar, D., Kavino, M., Raguchander,T., Subbiah, P and Samiyappan, R. (2010). Plant growth promoting bacteria enhance water stress resistance in green gram plants. *Acta physiologiae plantarum*. DOI10.1007/s11738-010-0539-1.

Selvakumar, G., Piyush Joshi., Sehar Nazim., Pankaj, K., Mishra., Jaideep, K., Bisht and Hari,S.G.(2009). Phosphate solubilization and growth promotion by *Pseudomonas fragi* CS11RH1 (MTCC 8984). A psychrotolerant bacterium isolated from high altitude himalyan rhizosphere. *Boilogia*. 64:239-245.

Sharma, P.C and Kumar, P. (1999). Alleviation of salinity stress during germination in *Brassica juncea* by pre sowing chilling treatments to seeds. *Biologia plantarum*. 42: 451-455.

Shongchessxu., Jin Hu., Yongping Li., Wenguang Ma., Yune zheng and Shujii zhu. (2010). Chilling tolerance in *Nicotiana tabacum* induced by seed priming with putrescience . *Plant growth regulation*. Online publication DOI : 10.1007/s 10725-010-9528-Z.

Souza –Machado V., Pitblado, R., Ali, A and May, P.(1999). Paclobutrazol in tomato for improved tolerance to early transplanting and early harvest maturity. *Acta horticulture (*ISHS).487: 139-144.

Vidhyasekharan, P and Muthamilan, M. (1995). Development of formulations of *pseudomonas fluorescens* for control of chickpea wilt. *Plant diseases*.79:782-786.

Vijayaraghavan, H.(1999). Effect of seed treatment with PGR on bhendi grown under sodic soil condition. *Madras agricultural journal*. 86: 247-249.

Vikas Yadav P., Sujata , B and Surasanna. (2009). Halopriming imparts to tolerance to salt and PEG induced drought stress in Sugarcane. *Agriculture, ecosystem, environment*. 134: 24-28.

Ya-Jing Gnan., Jin Hu., Xian-Ju Wang and Chen-xia Shao. (2009). Seed priming with chitosan improves maize germination and seedling growth in relation to physiological enhances under low temperature stress. *Journal of zhejiang university sciences*. 10(6):427-433.

Yeaoung, Y.R., Wilson, J.R., Murray, G.A. (1996). Germination performance and loss of LEA proteins during muskmelon seed priming. *Seed science and technology* 24: 429-439.

□□□

Green Agriculture : Newer Technologies, 2012
© Kambaska Kumar Behera (ed.), 231-257
New India Publishing Agency, New Delhi (India)
e-mail : info@nipabooks.com; website : www.nipabooks.com

Chapter-10

Drainage for Sustainable Agriculture

R.M. Singh[1], D.K. Singh[2*]
1 Department of Farm Engineering, Institute of Agricultural Sciences, Banaras Hindu University, Varanasi- 221005, U.P.,
2 Indian Institute of Vegetable Research, Shahanshahpur, PO: Jakhini, Varanasi- 221305, UP, India
*E-mail : dharmendradksingh@rediffmail.com

SUMMARY

The natural drainage systems are severely affected by the development processes and thus increased in waterlogged and salt affected areas. In major and medium irrigation projects due to inadequate designs coupled with poor management practices has raised groundwater table and in turn sizeable command areas are being affected both by water logging and soil salinization, Water logging and salt problem have been experienced in irrigation projects all over the country. Both adversely affect the growth and the yield of the crops. An area with water table within 2 m from the land surface is called as water logged area. It is potential to water logging if water table is between 2-3 m from the land surface. It has been estimated that around 16.71 million hectare land is affected by salt and waterlogging in India. The drainage is the remedy to these problems for sustainable agriculture.

Drainage is defined as the natural or artificial removal of surplus ground- and surface water and dissolved salt from the land in order to enhance agriculture production. In the case of natural drainage the excess waters flows from the fields to lakes, swamps, streams and rivers. However, in an artificial system surplus ground or surface water is removed by means of sub surface or surface conduits. Improved drainage create a healthier environment for plant growth, It conserve soil and water; and provide drier field conditions for ease in farm operations for the crop production. Agricultural drainage is must to realize the full benefit of irrigation.

Certain drainage criteria must be used to determine drainage need. Drainage is broadly divided into two types i.e., Surface and subsurface drainage. The excess surface or sub surface water of an agricultural field can be removed by applying the appropriate drainage method. Drainage system may consists of field drains system and main drainage system. Water of field is drained through field drainage and sent to main drainage for moving towards outlet. Field drainage system may be divided into surface drainage by gravity flow and subsurface drainage by gravity or pumped flow. The main drainage may be divided into deep collectors consisted of pipe or ditches for subsurface drainage and shallow collectors consisted of channels or ditches for surface drainage. The collector drains flows to disposal drains and to outlets of the drainage system to some stream or depressions.

Drainage coefficient is defined as the amount of water that runs off from a given area and is to be removed in 24 hours. While designing surface drainage system, a low value of the drainage coefficient will lead to partial improvement in drainage though the cost of design may be relatively low, whereas a high value would increase the cost substantially without any additional gain in the removal of surface congestion. Estimation of 24 hr rainfall depth that might occur with a probability level generally of 20% or a return period of 5 years should be considered for agricultural drainage.

Field drains for a surface drainage system have a different shape from field drains for subsurface drainage. Those for surface drainage have to allow farm equipment to cross them and should be easy to maintain with manual labour or ordinary mowers. Surface runoff reaches the field drains by flow through row furrows or by sheet flow. In the transition zone between drain and field, flow velocities should not induce erosion. Field drains are thus shallow and have flat side slopes. Simple field drains are V-shaped. Their dimensions are determined by the construction equipment, maintenance needs, and their cross ability by farm equipment. Side slopes should not be steeper. Nevertheless, long field drains under conditions of high rainfall intensities, especially where field runoff from both sides accumulates in the drain, may require a transport capacity greater than that of a simple V-shaped channel. Without increasing the drain depth too much, its capacity can be enlarged by constructing a flat bottom, thereby creating a shallow trapezoidal shape.

All field drains should be graded towards the collector drain with grades between 0.1 and 0.3%. Open collector drains collect water from field drains and transport it to the main drainage system. In contrast to the field drain, the cross-section of collector drains should be designed to meet the required discharge capacity. Besides the discharge capacity, the design should take into consideration that, in some cases, surface runoff from adjacent fields also flows directly into the field drains, which then require a gentler side slope. When designing the system, maintenance requirements must be considered. Attention must also be given to the transition between the field drains and the collector drains, because differences in depth might cause erosion at those places. For low discharges, pipes are a suitable means of protecting the transition. For higher discharges, open drop structures are recommended. A free board of 25% of designed depth is kept. Permissible values for average velocity of flow to avoid scouring may be adopted

Subsurface drainage improvement is designed to control the water table level through a series of drainage pipes that are installed below the soil surface. The subsurface drainage network generally outlets to an open ditch or stream. Subsurface drainage requires some minor maintenance of the outlets and outlet ditches. For the same amount of treated area, subsurface drainage improvements generally are more expensive to construct than surface drainage improvements

Subsurface drainage may be achieved by tubewell drainage, open drains or subsurface drains (pipe drains or mole drains). Tubewell drainage and mole drainage are applied only in very specific conditions. Subsurface (groundwater) drainage for water table and soil salinity in agricultural land can be done by horizontal and vertical drainage systems. Horizontal drainage systems use open ditches (trenches) or buried pipe drains. Parallel, herringbone, targeted and double main system layout could be adopted for subsurface drainage system. The spacing od drains could be evaluated using Hooghoudt or Ernst or Child method. Drainage system requires several materials to be used such as tiles, pipes, and envelope materials for its better functioning. Envelope materials such as gravel envelop and filter including geotextile filters can be used as per requirement. Various coefficients developed may be used for proper selection of the materials to realize effective drainage for sustainable agriculture.

Introduction

The natural geo-physiographical and agro-ecological situations of India are one of the major factors in causing surface water logging and development of salt affected areas in. The natural drainage systems are severely affected by the development processes and thus increased in waterlogged and salt affected areas. The other major factor is the development of man-made major and medium irrigation systems, where huge quantity of water is being transported into new geo- hydrological arid and semi-arid regions. The lack of working experiences in these regions caused inadequate designs coupled with poor management practices has raised groundwater table and in turn sizeable command areas are being affected both by water logging and soil salinization. National Commission on Agriculture, Govt. of India (NCA 1976) defined an area as waterlogged when the water table causes saturation of crop root zone soil, resulting to restriction to air circulation, decline in oxygen and increase in carbon dioxide levels.

The Working Group on Problem Identification in Irrigated area, constituted by the Ministry of Water Resources, Govt. of India (MOWR 1991) adopted the following norms for identification of waterlogged areas:

(i) Waterlogged area : Areas with water table within 2 m from the land surface
(ii) Potential area for waterlogging : Areas with water table between 2-3 m from the land surface
(iii) Safe area : Areas with water table below 3 m from the land surface

The physical effects of waterlogging are lack of aeration in the crop root zone, difficulty in soil workability and deterioration of soil structure. Its chemical effect is soil salinisation. Both adversely affect the growth and the yield of the crops. The extent of drop damage depends upon the magnitude, duration and frequency of the waterlogged condition and the degree of soil salinity. Salt problem is a major cause of decreasing agricultural production in many of the irrigation project areas. Salinity may be a major problem in many non-irrigated areas where cropping is based on limited rainfall. The various agencies evaluated the status of water-logging and soil salinization problems in these areas. However, the officially accepted one is the estimates of Working Group, (1991) (Table 1).

Table 1 : Water logged and Salt affected areas in million hectares

Source	**Irrigated Command Area**				**Country as a whole**			
	Water logged	**Salt affected**		**Total**	**Water logged**	**Salt affected**		**Total**
		Saline	**Alkali**			**Saline**	**Alkali**	
Working Group of MoWR (1991)	2.46	3.06	0.24	5.76	-	-	-	-
MoA, GoI	-	-	-	-	8.53	5.50	3.58	17.61

Water logging and salt problem have been experienced in irrigation projects all over the country. The examples are Chambal Command areas in Rajasthan and M.P., Indira Gandhi Canal Project in western Rajasthan, Kosi and Gandak Project Commands in Bihar, the Tungabhadra Project area in Karnataka, the Nagarjunasagar Project area in Andhra Pradesh and the Kakrapar Project area in Gujarat. Construction of drainage canals, field drains and avoiding wastage of canal supplies have been adopted as remedial measures. However, lack of maintenance, operational constraints of large irrigation projects, construction of highways, railway embankments and other obstructions, without providing for adequate drainage facility are still the major factors for water logging (Singh et al., 2011). In the Chambal Command area soils became water logged with a few years of introduction of irrigation. In many coastal areas excessive groundwater exploitation has caused seawater intrusion, worsening the salinity problem. Extent of waterlogged and salt affected areas for some states in India has been presented in Table 2.

There are extensive low lying areas in the rice growing coastal belts of eastern and south eastern regions of India where poor drainage seriously affects crop production in the monsoon season. The agricultural drainage is the remedy to these problems for sustainable agriculture. Reclamation of water logged/saline affected land by scientific and cost-effective methods should form a part of command area development programme. The drainage system should form an integral part of any irrigation project right from the planning stage. Some examples are the Sardar Sarovar Project in Gujrat the Narmada Canal Project in Rajasthan, Madhya Pradesh, the Indira Sagar Project in Madhya Pradesh, the Subarnarekha

Barrage project in West Bengal, the Arjun Sahayak Pariyojana in UP, the Bodwad Parisar Sinchan Yojana in Maharashtra and many others (Singh et al. 2011). Sustainable agriculture could be achieved if components of sustainable development, environment, society and economy remain in balance (Ott 2003 and Adams, 2006). Different types if agricultural drainage systems to manage water logging and soil salinity have been discussed in this chapter.

Table 2 : Geographical, waterlogged and salt affected areas of some states in India

State	Geographical area, million hectares	Waterlogged area, million hectares	Salt affected area, million hectares
Andhra Pradesh	27.44	0.339	0.813
Bihar	17.40	0.363	0.400
Gujrat	19.60	0.484	0.455
Haryana	4.22	0.275	0.455
Karnataka	19.20	0.036	0.404
Kerala	3.89	0.012	0.026
Madhya Pradesh	44.20	0.057	0.242
Maharashtra	30.75	0.111	0.534
Orissa	15.54	0.196	0.400
Punjab	5.04	0.199	0.520
Rajasthan	28.79	0.348	1.122
Tamilnadu	12.96	0.128	0.340
Uttar Pradesh & Uttaranchal	29.40	1.980	1.295
Total	258.43	4.528	7.006

(*Source :* Ghosh 1991 and Tyagi 1999.)

1. Drainage

Drainage is defined as the natural or artificial removal of surplus ground- and surface water and dissolved salt from the land in order to enhance agriculture production. In the case of natural drainage the excess waters flows from the fields to lakes, swamps, streams and rivers. However, in an artificial system surplus ground or surface water is removed by means of sub surface or surface conduits (*Source: FAO Glossary of Land and Water Terms).* Improved drainage create a healthier environment for plant growth, It conserve soil and water; and provide drier field conditions for ease in farm operations for the crop production. Adequate drainage is required to improve soil health and soil water plant interaction for enhanced water productivity to ensure sustainability (Singh et al, 2009). Agricultural drainage is must to realize the full benefit of irrigation.

1.1. Drainage criteria

Certain drainage criteria must be used to determine drainage need. A groundwater balance of the drainage area is the most accurate tool to calculate the volume of the water to be drained. Besides agricultural drainage criteria, technical drainage criteria (relating to minimization of the cost of installation and operation of system while maintaining the agricultural criteria), environmental criteria (relating to the minimization of the environmental damage), and economic drainage criteria (relating to the maximization of the net benefits i.e. the difference between benefits and costs, and damages) should also be taken into account.

1.2. Types of drainage system

Drainage is broadly divided into two types i.e., Surface and subsurface drainage. The excess surface or sub surface water of an agricultural field can be improved by applying the appropriate drainage method. Drainage system may consists of field drains system (internal) and main drainage system (external). Water of field is drained through field drainage and sent to main drainage for moving towards outlet. Field drainage system may be divided into surface drainage by gravity flow and subsurface drainage by gravity or pumped flow both divided into regular system and checked system (Figure 1).

2. Surface Drainage System

Surface drainage uses the potential energy due to land elevation to provide a hydraulic gradient for the movement of water. Surface drainage improvements are designed to minimize crop damage resulting from water ponding on the soil surface following a rainfall event, and to control runoff without causing erosion. Surface drainage can affect the water table by reducing the volume of water entering the soil profile. This type of improvement includes: land leveling and smoothing; the construction of surface water inlets to subsurface drains; and the construction of shallow ditches and grass waterways, which empty into open ditches and streams. These have disadvantage of requiring annual maintenance; and extensive and expensive earthmoving activities, and land grading might expose less fertile and less productive subsoils. Also, open ditches may interfere with moving farm equipment across a field.

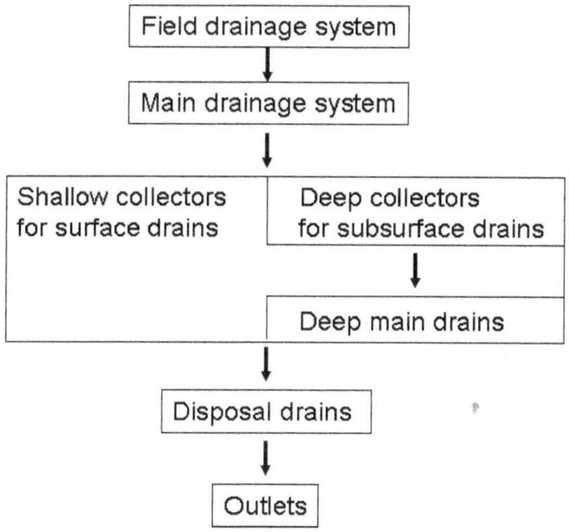

Figure 1 : Agricultural drainage system. The main drainage may be divided into deep collectors consisted of pipe or ditches for subsurface drainage and shallow collectors consisted of channels or ditches for surface drainage. The collector drains flows to disposal drains and to outlets of the drainage system to some stream or depressions.

2.1 Causes of surface drainage

Low-lying flat areas, heavy soils with low permeability and lands in humid tropical or sub-tropical regions with high intensity storms, are subject to surface inundation and therefore, require surface drainage. Surface drainage problem in many cases might not be of local origin as in the case of low lands where problem is caused due to the flow of surface runoff from uplands. Overflow from rivers or natural channels sometime contribute to the drainage problem of an area. The following reasons could be ascribed to the problem:

(i) Flat land surfaces causing hindrance in the natural runoff from the upper catchment area. The problem is severe in heavy textured soils and in humid climate.

(ii) Inadequate capacity of the drainage channels particularly during critical periods could cause surface stagnation. It is one of the main reasons of surface stagnation in many parts of the Indo-Gangetic plains. During intense storms the main drainage channels are full to the brim thereby reducing the capacity of the lateral and collector channels causing inundation upstream.

(iii) Inadequate outlet conditions partly due to developmental works, which obstruct the flow and partly due to choking of the outlet of the natural drainage system.

(iv) Non-availability of outlet due to backwater flow particularly in coastal regions.

(v) Waterlogging in agricultural lands is called as critical, if water table fluctuates between 0-2 m below ground surface. It is treated as semi-critical, if water table fluctuates between 2-3 m below round surface.

2.2. Factors affecting drainage

Generally, climate, soil, depths of water table and crop affect the drainage requirement. Climatic conditions in a particular region decide the degree and frequency of surface water stagnation. Thus, besides the seasonal and annual rainfall, frequency analysis procedures are used to determine various return period rainfall events, number of storms and the dry periods to plan an effective drainage and reuse strategies.

Kind of the soils and topography determines the degree of surface congestion. For the same event, degree of stagnation would be more in a heavy than a light textured soil. A flat terrain would be more prone to congestion than a rolling topography. Water absorption into the soil profile would depend upon the depth to water table. An area with relatively high water table would be subjected to relatively greater depth of water stagnation and for periods larger than the area with deep water table. Drainage is influenced to a great extent on the kind of crops grown and their tolerance to surface stagnation. The crop tolerance to water stagnation varies from 1-7 days depending upon which group of crops is grown.

2.3. Open-surface drainage

Open-surface drainage is defined as the diversion or orderly removal of drainage water by means of improved natural or constructed channels, supplemented when necessary by the shaping and grading of land surfaces of such channels.

2.4. Parts of surface drainage system

The surface drainage system consists of three parts, (1) collection system, (2) conveyance or disposal system, and (3) outlet (Figure 2).

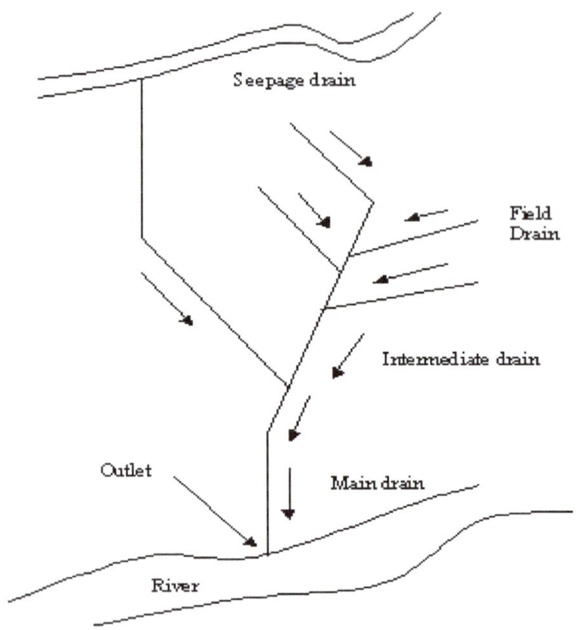

Figure 2 : A typical drainage system. Water to be drained from the individual field is collected through collection system (consists of field drains) and moves through disposal system (consists of intermediate and main drains) to outlet.

2.5. Design criteria for field drainage

The very purpose of a good surface drainage system is to prevent the harmful effects of waterlogging on crops. Selection of an appropriate drainage coefficient is the key to design a successful surface drainage system. Drainage coefficient is defined as the amount of water that runs off from a given area and is to be removed in 24 hours. While designing surface drainage system, a low value of the drainage coefficient will lead to partial improvement in drainage though the cost of design may be relatively low, whereas a high value would increase the cost substantially without any additional gain in the removal of surface congestion. The waterlogging tolerance of a crop should be considered while estimating the drainage coefficient for a surface drainage project. Although there are several methods for estimation of drainage coefficients, a simple method involves following steps for its estimation:

1. Estimation of 24 hr rainfall depth that might occur with a probability level generally of 20% or a return period of 5 years should be considered for agricultural drainage
2. Evaluate the basic infiltration rate of the soil and determine the expected potential evapotranspiration.
3. Estimate the crop tolerance to surface congestion at sensitive growth stages in days. Multiply the infiltration rate and expected potential evapotrans-piration with crop tolerance period.
4. Subtract the value calculated in step (3) from estimated probable rainfall value. The resulting value when divided by crop tolerance period (days) will give the drainage coefficient in depth of water/day.

These steps could be described mathematically in the form of an equation as:

$$q = \{R - n(E + I)\} / n \qquad \ldots\ldots (1)$$

Here q is the drainage rate in mm/day, R is the rainfall in mm, E and I are potential evapotranspiration and infiltration rate in mm/day and n is number of days. When n is greater than one, it may be useful to increase the duration of rainfall in step 1 also from 1-day maximum to n-day maximum. The drainage coefficient thus calculated will be closer to the actual field values. For rice crop, the drainage coefficient would be less as at the end of drainage period some depth of water is allowed to stand in the fields. This depth is subtracted before dividing by number of days, n.

2.6. Design of surface drains

Field drains for a surface drainage system have a different shape from field drains for subsurface drainage. Those for surface drainage have to allow farm equipment to cross them and should be easy to maintain with manual labour or ordinary mowers. Surface runoff reaches the field drains by flow through row furrows or by sheet flow. In the transition zone between drain and field, flow velocities should not induce erosion. Field drains are thus shallow and have flat side slopes. Simple field drains are V-shaped. Their dimensions are determined by the construction equipment, maintenance needs, and their cross ability by farm equipment. Side slopes should not be steeper. Nevertheless, long field drains under conditions of high rainfall intensities, especially where field runoff from both sides accumulates in the drain, may require a transport capacity greater than that of a simple V-shaped channel. Without increasing the drain depth too much, its capacity can be enlarged by constructing a flat bottom, thereby creating a shallow trapezoidal shape.

All field drains should be graded towards the collector drain with grades between 0.1 and 0.3%. Open collector drains collect water from field drains and transport it to the main drainage system. In contrast to the field drain, the cross-section of collector drains should be designed to meet the required discharge capacity. Besides the discharge capacity, the design should take into consideration that, in some cases, surface runoff from adjacent fields also flows directly into the field drains, which then require a gentler side slope. When designing the system, maintenance requirements must be considered. Attention must also be given to the transition between the field drains and the collector drains, because differences in depth might cause erosion at those places. For low discharges, pipes are a suitable means of protecting the transition. For higher discharges, open drop structures are recommended. Permissible values for average velocity of flow to avoid scouring may be adopted from Table 3

Table 3 : Maximum allowable velocities in channels for different soil textures.

S No.	Soil texture	Max. Allowable velocity (m/sec)
1.	Very light silty sand	0.30
2.	Light loose sand	0.50
3.	Coarse sand	0.75
4.	Sandy and sandy loam	0.75
5.	Silty loam	0.90
6.	Firm clay loam	1.00
7.	Stiff clay or stiff gravelly soil	1.50
8.	Coarse gravel	1.50
9.	Shale, hardpan, soft rock etc.	1.80
10.	Hard cemented conglomerates	2.50

Slightly higher velocities are allowed if water contains colloidal silt. If the land slopes are steeper to create scouring velocity, the same has to be reduced by a gentle slope of the drain through provision of suitable drops/falls in the channels.

2.6.1. Side slopes

The Side slopes of the drains in general are recommended as given below:-

Firm soil1.0:1 (horizontal: vertical)

Loam soil1.5:1

Sandy soil2.5:1

However, it is desirable to design the side slope of a channel from consideration of angle of repose of the soil.

2.6.2. Channel grade

The channel should be as uniform as possible. The grade should be as steep as possible provided the maximum allowable flow velocities are not exceeded. Design grades should be from 1- 0.3 % and should never be less than 0.05%.

2.6.3. Channel depth and width

The depth of channel should be sufficient to carry the design discharge. In general the depths of main and sub main should be kept between 2-3 m. The bed width depends on peak design discharge at different points.

2.6.4. Channel bottom width

The bottom width can be computed for a given discharge after the channel grade, grad depth and side slopes are selected. The bottom width for most efficient cross section (hydraulic radius is one half the depth) The minimum bottom width should be 1.2 m except in small lateral. The cross section is designed to meet the requirement of the capacity, velocity, and side slope, bottom width. The parabola has the smallest wetted perimeter and it is well suited for concrete channels. A trapezoidal section with minimum recommended bottom width of 0.60 m is recommended for earthen channels. The following factors may be considered for designing: A deeper ditch gives a higher velocity than a shallow one and also may provide future opportunity for pipe drainage it will remain effective for a longer period sediment bars may cause less obstruction, it requires less waterways than a shallow one. It may uncover unstable layers of soil. Shallow drain may be more practical to maintain by pasturing or by mowing flat side slopes, the depth should be related to a good outlet condition ns, Design velocity should be selected so as to maintain the ditch cross s section with time. In channels that flow intermittently, some scouring may be desirable at high flows to counteract sediment deposition that occurs at low flows.

2.6.5. Berms and spoil banks

Berms are required to provide for work areas and facilitate spoil bank spreading, prevent excavated materials falling back into the ditch. The berm width for drains should not be less than the depth of cutting. The minimum berm widths are given in the following Table 4. The berm width should be increased in unstable soil where it is feared that the drain will enlarge in the section. The spoil is spread until the height is reduced to an economical figure usually not more than 1m.

Table 4 : Minimum berm widths

Depth of drain, m	Minimum berm width, m
0.6-1.2	1.2
1.2- 1.8	1.8
1.8 – 2.4	3.0
> 2.4	4.5

2.6.6. Free board

The free board is additional depth above the design water level used to provide a safety factor for the design storm. A free board of 25% of designed depth is kept.

2.7 Location, spacing and alignment

Drain ditches should be located in a way to provide the most effective drainage and to cause the least interference with irrigation system and farm operations. These serve as outlet for surface runoff frond rainfall, as outlet for excess irrigation water, or as disposal ditches for pipe drains. They may be located parallel to canal embankments to collect seepage water. For controlling water table they may be installed parallel and at regular intervals with the same depth and spacing as pipe drains. Ditches should be normally close to the low point depression. Crossing with irrigation watercourses should be avoided. Grade control and crossing structures should be minimized.

2.8. Patterns

Two main types of surface drainage patterns are random and parallel. Each includes lateral ditches that permit water to flow from drainage system to a suitable outlet. The chosen pattern depends upon the soil type and topography of the land.

2.8.1 Random

The random ditch pattern is practiced to slowly permeable soils having depression areas that are too large to be eliminated by land smoothing or grading (Figure 3). Soil from the ditches can be used to fill minor low spots in the field. Field ditches should extend through most of the depressions for complete drainage, and they should follow the natural slope of the land.

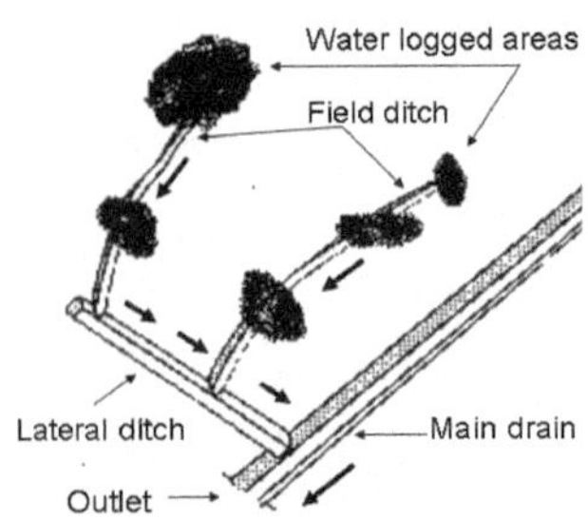

Figure 3 : Random drain system. Field ditches connect the major low spots and remove excess surface water from them. They are generally shallow enough to permit frequent crossing by farm machinery

2.8.2 Parallel

The parallel ditch pattern is suitable for flatter, poorly drained soils that have numerous shallow depressions (Figure 4). Dead furrows are neither desirable nor necessary. Although the ditches must be parallel, they need not be equi-distant. The spacing between them depends upon the permissible length of row drainage for the soil type and upon the amount of earth and the distance it must be moved to provide complete row drainage. The maximum length of the grade draining to a ditch should be 200 m. The success of a parallel pattern depends largely upon proper spacing of the parallel ditches and the smoothing or grading between them. During the grading operation, fill all depressions and remove all barriers. Excavated material from ditches can also be used as fill for establishing grades.

2.9. Shaping the surface

2.9.1. Grading

Land grading (also termed precision land forming) is the reshaping of surface of land with tractors and scrapers to planned grades. Its purpose is to provide excellent surface drainage although the amount of grading will depend upon the soil and costs. To do a good job of land grading, you need a detailed engineering survey and construction layout. To assure adequate surface drainage, eliminate all reverse surface grades that form depressions. The recommended surface grades range from 0.1 to 0.5 percent and may be uniform or variable. The cross slopes normally should not exceed 0.5 percent. Minimum grade limits should include a construction tolerance that will permit the elimination of all depressions either in original construction or in post-construction touch up.

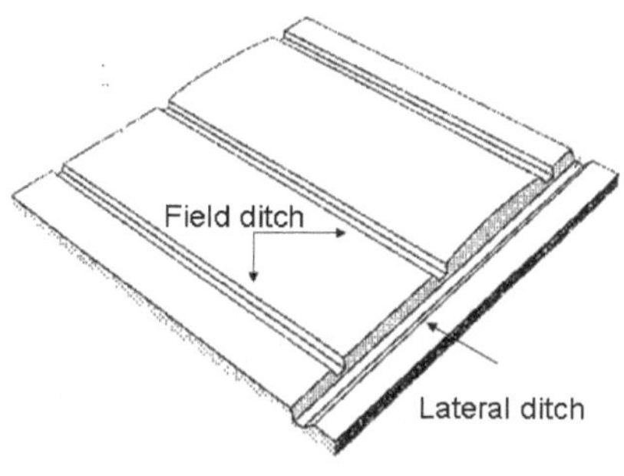

Figure 4 : Parallel drain system. In fields that can be cultivated up and down slope, parallel field ditches are installed across the slope to break the field into shorter units of length and make it less susceptible to erosion. The field should be farmed in the direction of the greatest slope.

Reverse grades can be eliminated with relative ease in a field that has minimum grades of 0.2 percent. Unusual precision in construction is required to eliminate reverse surface grades in fields that have 0.1 percent and flatter grades. Land grading is hampered by trash and vegetation. This material should be destroyed or removed before construction and kept under control while the work is being done. The fields should be chiseled before construction if there are hard pans. The field surface should be firm when it is surveyed so that rod readings taken at stakes will reflect true elevation. Do not grade fields when they are wet because working wet soil impairs physical condition of soil.

2.9.2. Smoothing

Land smoothing removes irregularities on the land surface and should be done after land grading and may be useful in other situations. Special equipment such as a land plane or land leveler should be used (Figure 5).

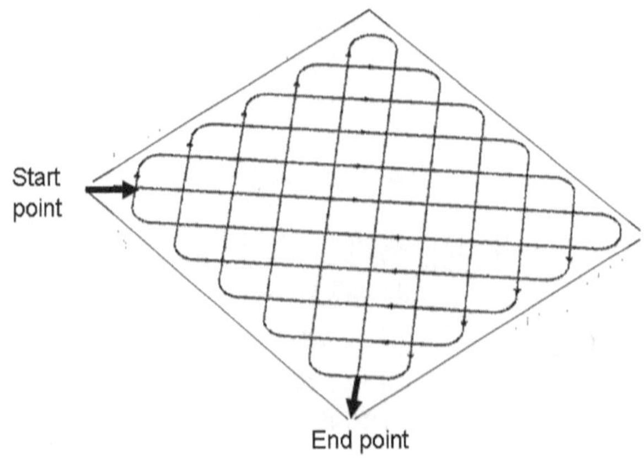

Figure 5 : Smoothing operation. The purpose of land smoothing is to improve surface drainage. The smoothing operation may ordinarily be directed in the field without detailed surveys or plans, although grid surveys may be needed for some critical parts of the field. A smoothing operation consists of a minimum of three passes with a land leveler.

Make the first two passes on opposite diagonals as noted in Figure 5 and the last pass in the direction of cultivation. Either before or after the final land smoothing operation chisels fields to loosen the cut surfaces and to blend the fill material with the underlying soil. The finished surface should be free from minor depressions so that runoff will flow unobstructed to field or lateral ditches.

2.10. Good surface drainage practice

Following are the points to be remembered for good drainage:

i) Gravity is the primary vehicle for carrying runoff away. There must be a continuous minimum fall in the ground level to assure drainage, and a minimum slope of 1% to 5% for grass swales. 1% minimum slope for smooth interior pipe is a general guideline for pipe conveying runoff water to a discharge point

ii) Large amounts of water should not cross a sidewalk to reach the street storm drain. Use drains or install piping to cross walks or other pedestrian walkways to prevent hazards.. Consult an engineer or architect for minimum slope in critical applications.

iii) Break up one large drain to several smaller drains to:

a) Prevent erosion on steep landscapes by intercepting water before it accumulates too much volume and velocity.

b) Provide a safety factor. If a drain inlet clogs, other surface drains may pick up water

c) Improve aesthetics. Several smaller drains will be less obvious than one large drain.

d) Spacing smaller drain inlets will give surface runoff a better chance of reaching the drain. Water will have farther to travel to reach one large drain inlet.

iv) Erosion is a big problem in drainage - slopes must be carefully calculated to ensure continuous flow, yet not steep enough to erode.

v) Slow moving water will create a bog, while water moving too fast will cause erosion, form gullies and weaken foundations. Design a drainage system that will eliminate both extremes.

vi) Design paved areas so they are graded almost level - avoid wildly sloping paved areas or dramatic changes in slope.

vii) Runoff water must never be directed purposefully from one property onto another property. It is acceptable for water that flows naturally from one property to the other to continue, but you must never increase this flow artificially through grading and piping.

viii) Check local code requirements and their applications

ix) When designing a system, work from the discharge point towards the highest elevations.

x) Design a secondary drain route to allow for overflow conditions during severe rainfall or in case the primary drain system fails.

xi) Many systems require a grate or "clean out" fitting every 50 to 100 feet or at alignment changes of 45 degrees or greater to clean out the pipeline. Clean outs are normally constructed at grade.

xii) Keep it simple. Over-design in storm water systems is expensive.

3. Subsurface Drainage

The objective of subsurface drainage is to drain excess water and salt from the plant root zone of the soil profile by artificially lowering the level of water table (Figure 6). Subsurface drainage improvement is designed to control the water table level through a series of drainage pipes (or tubing) that are installed below the soil surface (Figure 7). The subsurface drainage network generally outlets to an open ditch or stream. Subsurface drainage improvement requires some minor maintenance of the outlets and outlet ditches. For the same amount of treated acreage, subsurface drainage improvements generally are more expensive to construct than surface drainage improvements. The main objective of drainage is to remove excess water quickly and safely to reduce the potential for crop damage.

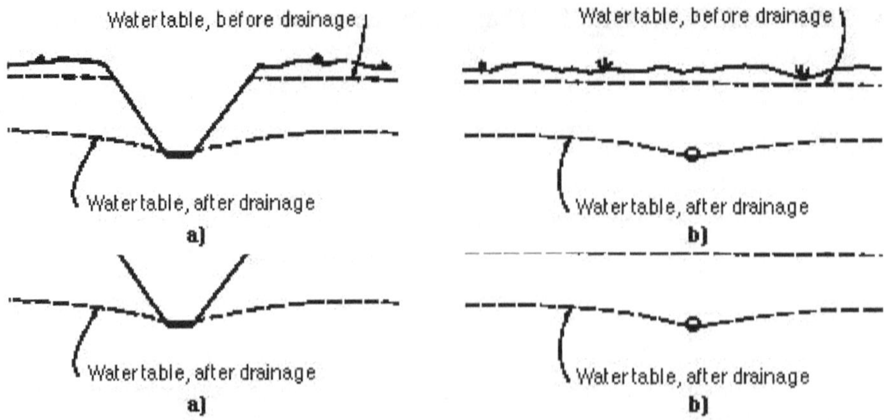

Figure 6 : Level of water table before and lowered level after drainage improvement: a) surface drainage ditch; b)subsurface drainage pipe. (USDA-ERS, 1987)

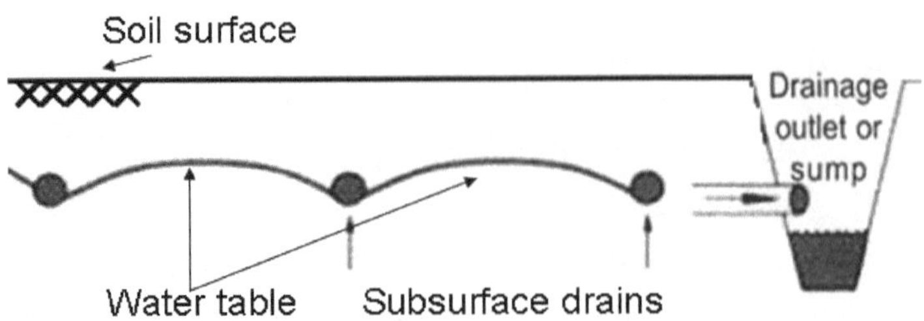

Figure 7 : A controlled level of water table profiles under subsurface drainage system. Draining excess water from the soil profile where plant roots grow helps aerates the soil and reduces the potential for damage to the roots of growing crops. It produce soil conditions more favorable for farming operations.

3.1. Types of subsurface drainage systems

Subsurface drainage aims at controlling the water table and a control that may be achieved by tubewell drainage, open drains or subsurface drains (pipe drains or mole drains). Tubewell drainage and mole drainage are applied only in very specific conditions. Subsurface (groundwater) drainage for water table and soil salinity in agricultural land can be done by horizontal and vertical drainage systems. Horizontal drainage systems use open ditches (trenches) or buried pipe drains.

3.1.1 Open drains

Open drains have the advantage that they can receive overland flow directly, but the disadvantages often outweigh the advantages. The main disadvantages are the loss of land, interference with the irrigation system, the splitting-up of the land into small parcels, which hampers mechanized farming operations, and a maintenance burden.

3.1.2. Tile drainage

Tile drainage is an agriculture practice that removes excess water from subsurface soil. Whereas irrigation is the practice of adding additional water when the soil is naturally too dry, drainage brings soil moisture levels down for optimal crop growth. While surface water can be drained via pumping and/or open ditches, tile drainage is often the best recourse for subsurface water. Too much subsurface water can be counterproductive to agriculture by preventing root development, and inhibiting the growth of crops. Too much water also can limit access to the land, particularly by farm machinery.

3.1.3. Mole drainage

Heavy soils of low hydraulic conductivity (less than 0.01 m/day) often require very closely spaced drainage systems for satisfactory water control. With conventional pipe drains, the cost of such systems is usually uneconomic and hence alternative techniques are required. Surface drainage is one possibility; the other is mole drainage (Figure 8).

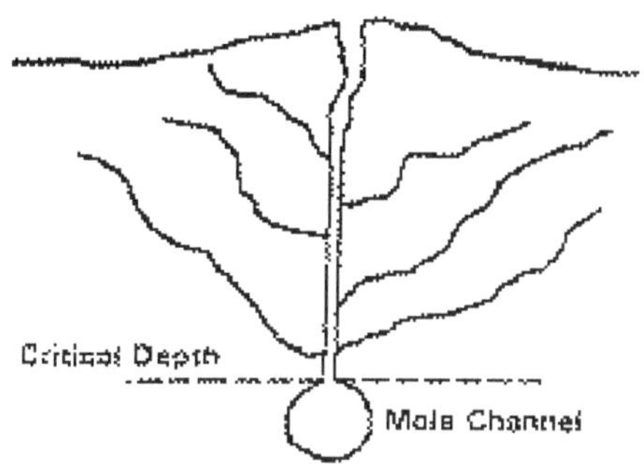

Figure 8 : Mole drains are unlined circular soil channels which function like pipe drains.

Advantage of mole drainage is their low cost and hence they can be installed economically at very close spacing. Their disadvantage is their restricted life but, providing benefit/cost ratios are favourable, a short life may be acceptable. The success of a mole drainage system is dependent upon satisfactory water entry into the mole channel and the mole channel stability. Mole drains are formed with a mole plough (Figure 9), which comprises a cylindrical foot attached to a narrow leg, followed by a slightly larger diameter cylindrical expander. The mole plough is attached to the drawbar of a tractor and the mole channel is installed at depths between 0.4 and 0.7 m. Common lengths of run vary from 20 to 100 m.

Figure 9 : A mole plough. The foot and expander form the drainage channel and the leg generates a slot with associated soil fissures which extends from the surface down into the channel.

3.1.4. Vertical drainage

Vertical drainage systems use pumped wells, either open dug wells or tube wells. Agricultural land is drained by pumped wells to improve the soils by controlling water table levels and soil salinity. Tubewell drainage refers to the technique of controlling the water table and salinity in agricultural areas by pumping, from a series of wells, an amount of groundwater equal to the drainage requirement. The success of tubewell drainage depends on many factors, including the hydrological conditions of the area, the physical properties of the aquifer to be pumped and those of the overlying fine-textured layers.

3.2. Subsurface drainage coefficient

Drainage coefficient is the volume of water per unit area to be removed in 24 h. The drainage coefficient is important in subsurface drainage design and a dependable drainage coefficient is difficult to obtain. It depends on land use, rainfall, runoff, infiltration and evapotranspiration. Incorporation of all these factors in a single physical parameter is difficult to measure effectively.

In arid regions, drainage coefficients are calculated on the basis of irrigation management and leaching requirements for salinity control. Since excess salts in the root zone are critical to the reclamation process than improvement in the aeration. In humid areas, the drainage coefficient is taken to be the depth of rainfall removed in 24 h. Drainage coefficients can be calculated from a soil water balance method. Selection and use of suitable drainage coefficients have always been problems in the design of subsurface drainage system. Use of a low value will reduce the effectiveness of the drainage system whereas use of a high value will raise its cost.

Drainage coefficient is the most important parameter that decides the lateral drain spacing, size of the laterals and collectors and capacity of the pump to dispose off the drainage effluent. The drain spacing is less in cases where drainage coefficient is more as compared to a case where drainage coefficient is less. As such, the cost of the system depends largely upon this parameter. Therefore, the need to select an appropriate value for this parameter has always been emphasized. The drainage coefficients for some of the sites in India have been observed to be in the range of 1-5 mm. In the case of subsurface drainage design based on non-steady state conditions, the drainage criterion is based on time to lower the water table from a predecided original to the final level. Usually the original water table is considered at the soil surface while the final level is taken as 30 cm below the soil surface. As per the recommendation the capacity of the drainage system should be sufficient to lower the water table by 30 cm in 2 days time. In general, design rates area likely to be in the following DC ranges given in Table 6.

Table 6 : Range of drainage coefficient suitable for various conditions

DC, mm/d	Suitable conditions
< 1.5	Soils low infiltration rate
1.5 – 3.0	Moist soils, with higher rate for more permeable soils and where cropping intensity is high
3-4.5	Extreme conditions of climate, crop and salinity managements, an under poor irrigation practices
> 4.5	Special conditions, e.g. rice irrigation on light textured soils

3.3. Drainage system layout

Although there may be many possible layout for a given field specific drainage objectives should be evaluated to find the best layout (Figure 10). System layout and drainage needs should be based anticipating future needs where possible. Additions to a system will be much easier to make if the established mains are already large enough and located appropriately.

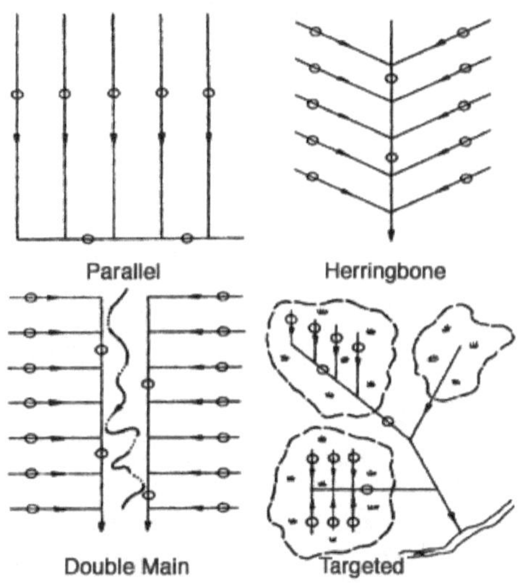

Figure 10 : Various drainage system layout. The objectives of layout include removing water from an isolated area, improving drainage in an entire field, intercepting a hillside seep, and so on.

3.4. Spacing between drains

When there is a linear flow of ground water, spacing between drains is given by the relationship,

$$S^2 = 4\,K\,\frac{\left(H^2 - h^2 + 2dH - 2dh\right)}{Q} \quad (2)$$

where S = spacing between drains,
K = hydraulic conductivity,
Q = rate of discharge,
H =Maximum height of water table above bottom of drains and
h= height of water in the drain.
d = depth of impervious layer below the drain.

3.5. Measurement of in situ hydraulic conductivity

Measurement of hydraulic conductivity on disturbed soil samples is not a very reliable approach. Similarly, determining these values on undisturbed samples drawn with a core sampler is also not a flawless method. Considerable errors are caused due to presence of root holes. Methods have, therefore, been developed to measure hydraulic conductivity in situ. Three simple methods in this respect are those of hooghoudt, Ernst and Childs. These are described below.

3.5.1. Hooghoudt's method

The methods is suitable for homogeneous soil. In this method a hole is made in the soil below a water table with the help of post hole digger. The water level in the bore is allowed to come in equilibrium with the water table in the soil and the water is pumped out from the bore. The rate of rise of water in the bore is recorded and the hydraulic conductivity is calculated with the following formula

$$K = \frac{2.3aS}{(2d + a)\Delta t}\log_{10}\frac{y_0}{y_1} \quad \textit{when auger hole is drilled upto the impermeable layer} \ldots\ldots (3)$$

Where,

a = hole radius
Δt = time in seconds,
y_0 = initial water level in cm,
y_1 = final water level in cm,
K = hydraulic conductivity in meter/second,
d = distance from bottom of the hole to water table and
S = constant of proportionality which is a distance from bottom of the hole to the impermeable layer and is equal to ad/0.19.

3.5.2 Ernst's method for layered soils

In this method two auger holes B_1 and B_2 of different depths are made in the soil. The bottom of the first hole is kept 10 cm above the second layer. The second hole should be deep enough in the lower layer. The entity (d-h) should be greater than 15.

The conductivity in the first hole, K_1 is determined as usual. The mean conductivity K is determined for two layers by recording rate of rise in the second hole. Hydraulic conductivity in the 2nd layer, K_2 is then calculated by the following relationship.

$$K_1d_1 + K_2(d_2 - d_1) = Kd_2 \quad .... (4)$$

Where, K_1 and K_2 = hydraulic conductivity of 1st and 2nd layer, respectively.

K = mean hydraulic conductivity of two layers

d_1 and d_2 = distance between water table and bottom of the 1st and 2nd bore, respectively.

The results are satisfactory if K_2 is greater than K_1

3.5.3 Child method with two auger holes

Two auger holes, w_1 and w_2 of equal diameter and depth, are made below the water table. Water from w_1 is pumped out and poured into w_2 to create a difference in hydraulic head ΔH. Hydraulic conductivity K is then calculated as follows-

$$K = \frac{Q}{\pi L \Delta H} Cosh^{-1} \frac{b}{2a} \quad ... (5)$$

Where Q = rate of pumping,

ΔH = hydraulic head difference between the two holes,

L = depth of the hole below the water table,

a = radius of the hole,

b = distance between the two holes,

4. Drainage Materials

The subsurface field drainage systems consist of horizontal or slightly sloping channels made in the soil; they can be open ditches, buried pipe drains, or mole drains; also a series of wells. Drainage system requires several materials to be used such as tiles, pipes, and envelope materials for its better functioning. Envelope materials such as gravel envelop and filter including geotextile filters can be used as per requirement. Various coefficients developed may be used for proper selection of the materials.

4.1. Tiles materials

Concrete, burnt clay and PVC are the materials used for tiles. Concrete tiles are easy to manufacture and can be transported without breaks. Concrete tiles can also be locally made using simple moulds made of mild sheets. Concrete 1: 2:4 proportions should be used and tiles should be properly cured. Clay tiles should be well burnt and without any defects. Good clay tiles have distinct ring when tapped with metal objects. These have advantage of being cheaper than concrete tiles and also can be used in acid and alkali soils. However, they are difficult to transport without breakage and will not be able to with stand as much load as concrete tiles can take.

Plastics pipes are increasingly being popular for tile drainage. The common material is the polyvinyl chloride and polyethylene (PVC). PVC pipes are somewhat more resistant to outside pressure. Plastic pipes come in both smooth and corrugated types. The smooth pipes are rigid and come in standard length. Corrugated pipes are flexible and come in coils. They can be conveniently be used with tile lying machines. They have greater hydraulic resistance than smooth pipes and hence larger size pipe is required to drain the same amount of water. Circular hole in the corrugation allows the water enter into tile drains. Around 20 holes per 30 cm length are provided. The Use of envelope material increases the effective diameter and hence fewer perforations per unit length will be required to get the desired flow. Pipes require less plastic material per unit length and have greater resistance to outside pressure.

4.2. High density polyethylene

Polyethylene is an extremely versatile piping material and has some properties that make ideal for use in underground drainage systems. Relatively lightweight, polyethylene allows for easier and less costly transportation and installation costs. It is not easily susceptible to cracking during pipe handling and installation activities. A corrugated HDPE pipe is resistant to abrasion, corrosion, chemical scouring and is structurally strong with the ability to support large loads. This characteristic offers advantages for use as underground drainage systems.

Corrugated perforated HDPE pipes are generally used for subsurface drainage. Corrugated both inside and outside, perforated pipe is ideal for subsurface drainage used in everything from draining garden to major highways. Lightweight and durable, can handle chemically abrasive environments, is easy to install and not easily susceptible to cracking during pipe handling and installation. Flexibility and perforation are its most important features. Due to its flexibility, perforated pipe is better suited for trenches or where the ground is uneven, and easy to install. These have slots around the entire circumference or half-perforated, i.e., have holes on only one side of the pipe, non -perforated, and perforated covered in filter sock. Perforations allow subsurface water to be collected and transported to desired locations for discharge. The flexibility and lightweight of perforated pipe becomes an advantage for use in harsh conditions.

4.3. Drain envelopes and filters

Drain envelopes and filters are two different techniques used to solve different problems. Drain envelopes are permeable materials, such as gravel place around the drains for the purposes of improving flow conditions. Filters for drains are permeable materials, such as geotextiles, placed around the drains for the purpose of preventing fine-grained materials in the surrounding soil from being carried into the drain by groundwater. Drain envelope is a material placed around a drain pipe to provide either hydraulic function, which facilitates flow into the drain, *or* barrier function, which prevents certain sized soil particles from entering the drain. It is termed as sock also. Drain envelopes are not filters. Filters become clogged over time, but, drain envelopes do not.

Many types of envelope material such as thick gravel and organic fiber to thin geotextiles are in use. The useful life of a synthetic drain envelope is long, if not left in the sun for a long time and exposed to much ultraviolet radiation. Fine-textured soils with a clay content of 25 to 30 percent are generally considered stable, so they don't need drain envelopes. A geotextile sock is recommended for coarse-textured soils free of silt and clay. These soils are considered unstable even if undisturbed, so that particles may wash into pipes. The need for an envelope in intermediate soils (clay contents less than 25 to 30 percent) is best left to a professional contractor or soil and water engineer because soil movement is more difficult to predict.

In general gravel, coarse and very coarse sand, silty clay loam, sandy clay and peat soil do not require any filter or envelop materials. Gravelly fine sand, medium sand, fine sand, clay sand, sandy loam, loam and silt require filter materials in low to high degree of urgency. However, silty clay and clay require envelope materials.

4.4. Envelope material

In soils where there is a chance of heavy movement of soil particles into the tile drains and consequently clogging them, filter materials are placed around the tiles. The filter or envelope materials prevent the entry of the relatively coarser particles as well. They act as bedding material and thus allow the tiles to take greater loads. They also increase the effective diameter of the tiles and allow quantities of water to flow into the tiles. A variety of materials have been used as envelope materials in tile. These consist of gravel coarse sand organic materials like straw, coir matting etc. However, gravel has been the most widely used material because of its efficiency and also of permanent nature.

4.4.1. Gravel envelope

The function of a drain envelope is to improve permeability in the surrounding of the drain. For this reason, the envelope material should have a hydraulic conductivity 7 times higher than the base material. Since envelopes are not designed for their filtration capacity, they do not need to be well graded. The gravel envelope should prevent movement of soil particles into the tile drains. The thickness should be the same as the sand and gravel filter (i.e. 100 mm around the pipe). All the envelope material should be smaller than 38 mm, $D90 < 19$ mm and the $D10 > 0.250$ mm.

The gravel envelope should be designed taking into consideration the nature of the soil around the tile drains. The gravel should be clean; free from organic matter and bentonite clays. The United States Bureau of Reclamation recommends the following criteria for the design of gravel envelopes. For uniform soils, the ratio of D_{50} of envelope material and $D_{50}$of soil should range from 5 to 10. However, the ratio ranged from 12 to 58 for graded soils.

4.4.2. Filters

Filters can be either geotextile or well graded gravel and sand. Filters are necessary only where there is something in soil that needs to be filtered out, namely fine sand. If sand is not present, filters are not necessary. Not only do filters add to the cost of drainage but they also constitute an additional barrier to inflow of water and can, therefore, reduce the effectiveness of the drain. Few soils present the danger of sand particles clogging the drainage system. Most soils contain sufficient amounts of clay or organic matter to form relatively stable aggregates of individual soil particles. Filters in these soils are of no benefit and may have reduced drain performance. Instead of filters, a porous envelope is appropriate to ensure good flow conditions at all times at the interface between drain and soil. Drain rock, pea gravel and similar materials meet these requirements and are used extensively in some drainage applications. The first step in design of a proper filter system, either geotextile or sand and gravel, is to perform a particle size analysis of soil at drain depth in field. Usually soils with more than 30% clay content do not require a filter.

4.4.2.1. Geo textile filters

Suitable filters may be used to restrict entry of fine particles of silt and sand from entering the drains in soils with poor cohesion. A properly designed filter stabilizes the soil around the drain and allows free entry of water. There are two basic types of geotextile filters, knitted and non-woven. Knitted geotextiles are usually made of polyester or polypropylene filaments that are knitted or woven together. The most common type has a thickness of 1 mm, a weight of 150 g/m^2 and an Apparent Opening Size (AOS) of 300 microns. For applications requiring more filtration capacity, a sock knitted with velour or pile on one side, that is thicker (< 2 mm), heavier (250 g/m^2) and has an AOS of about 100 microns is needed.

Non-woven geotextiles are made from several layers of randomly distributed fibers that are rolled pressed and usually interconnected by needle punching. It has been found that fabrics about 2 mm thick are very good for silty soils. The general guide for designing a geotextile filter is that the ratio of O_{95} Fabric material and D_{85} Soil should be =< 2.5. Where O_{95} is the apparent opening size (AOS) of the geotextile filter D_{85} is the size of which 85% of the particles are finer.

4.4.2.2. Gravel and sand filters

In arid areas sand and gravel filters are used to some extent instead of geotextile filters. Drains usually run deeper and the sand and gravel filters also act as an envelope to improve bedding and permeability characteristics. Filter materials should be well graded. If more than one gradation is used, the layers should be from coarsest to finest material, starting at the pipe. A minimum thickness of 100 mm is recommended for each layer of the filter.

Limits for the filter material are such that ratio of D_{50} Filter material and D_{50} Base material ranges between 12 to 58 and that of D_{15} Filter material and D_{15} Base material from 12 to 40. D_{50} (the size of which 50% of particles are passing through the screen) of the base material times 12 and 58 will yield the lower limit and upper limit for D_{50} filter. Provided the filter has no more than 5% finer than 0.074 mm and is relatively well graded.

The chosen filter material should be checked for stability by adopting the criterion that the ratio of D_{15} Filter material and D_{85} Base material should be < 5. The D_{85} size of the filter material with respect to the opening of the drainpipe should be verified using ratio of D_{85} Filter material and Maximum drainpipe opening as > 2.

A well graded filters material is required for good performance. A filter material is considered well graded when all particle sizes from the largest to the smallest are present in a balanced way. As given below, the coefficient of uniformity can be used to verify how well graded the material is.

$$C_u = D_{60} \text{ Filter material} / D_{10} \text{ Filter material} \qquad (6)$$

where C_u = Coefficient of uniformity

$$C_c = (D_{30})^2 / (D_{10})(D_{60}) \qquad (7)$$

where C_c = coefficient of curvature.

In general, for well graded filter materials maximum size of aggregates should be 38 mm. However, D_{90} upto 19 mm and D_{10} upto 0.25 mm. $C_u > 4$ for sand and >6 for gravel may be recommended. However, C_c might range in between 1 and 3 for well graded filter materials.

5. Conclusions

An area with water table within 2 m from the land surface is called as water logged area. It is potential to water logging if water table is between 2-3 m from the land surface. It has been estimated that around 16.71 million hectare land is affected by salt and waterlogging in India. The drainage is the remedy to these problems for sustainable agriculture. Drainage is defined as the natural or artificial removal of surplus ground- and surface water and dissolved salt from the land in order to enhance agriculture production. In the case of natural drainage the excess waters flows from the fields to lakes, swamps, streams and rivers. However, in an artificial system surplus ground or surface water is removed by means of sub surface or surface conduits.

Improved drainage create a healthier environment for plant growth, It conserve soil and water; and provide drier field conditions for ease in farm operations for the crop production. Agricultural drainage is must to realize the full benefit of irrigation. Certain drainage criteria must be used to determine drainage need. Drainage is broadly divided into two types i.e., Surface and subsurface drainage. The excess surface or sub surface water of an agricultural field can be removed by applying the appropriate drainage method.

The drainage system may consists of field drains system and main drainage system. Water of field is drained through field drainage and sent to main drainage for moving towards outlet. Field drainage system may be divided into surface drainage by gravity flow and subsurface drainage by gravity or pumped flow. The main drainage may be divided into deep collectors consisted of pipe or ditches for subsurface drainage and shallow collectors consisted of channels or ditches for surface drainage. The collector drains flows to disposal drains and to outlets of the drainage system to some stream or depressions.

Drainage coefficient is defined as the amount of water that runs off from a given area and is to be removed in 24 hours. While designing surface drainage system, a low value of the drainage coefficient will lead to partial improvement in drainage though the cost of design may be relatively low, whereas a high value would increase the cost substantially without any additional gain in the removal of surface congestion. Estimation of 24 hr rainfall depth that might occur with a probability level generally of 20% or a return period of 5 years should be considered for agricultural drainage.

Field drains for a surface drainage system have a different shape from field drains for subsurface drainage. Those for surface drainage have to allow farm equipment to cross them and should be easy to maintain with manual labour or ordinary mowers. Field drains are shallow and have flat side slopes. Simple field drains are V-shaped. A free board of 25% of designed depth is kept. Permissible values for average velocity of flow to avoid scouring may be adopted

Subsurface drainage improvement is designed to control the water table level through a series of drainage pipes that are installed below the soil surface. The subsurface drainage network generally outlets to an open ditch or stream. Subsurface drainage requires some minor maintenance of the outlets and outlet ditches. For the same amount of treated area, subsurface drainage improvements generally are more expensive to construct than surface drainage improvements

Subsurface drainage may be achieved by tubewell drainage, open drains or subsurface drains (pipe drains or mole drains). Tubewell drainage and mole drainage are applied only in very specific conditions. Subsurface (groundwater) drainage for water table and soil salinity in agricultural land can be done by horizontal and vertical drainage systems. Horizontal drainage systems use open ditches (trenches) or buried pipe drains. Parallel, herringbone, targeted and double main system layout could be adopted for subsurface drainage system. The spacing od drains could be evaluated using Hooghoudt or Ernst or Child method. Drainage system requires several materials to be used such as tiles, pipes, and envelope materials for its better functioning. Envelope materials such as gravel envelop and filter including geotextile filters can be used as per requirement. Various coefficients developed may be used for proper selection of the materials to realize effective drainage for sustainable agriculture.

References

Adams, W.M. (2006). The Future of Sustainability: Re-thinking Environment and Development in the Twenty-first Century. Report of the IUCN Renowned Thinkers Meeting, 29–31 January 2006

FAO Glossary of Land and Water Terms. (2001). FAO Glossary of Land and Water Terms http://www.fao.org/landandwater/ glossary/lwglos.jsp? keyword1= &subject= %25&term_e= Drainage+&search= Display. (FAO/IPTRID, drainage and sustain-ability, issues paper No. 3 October 2001).

Gosh, S.P. (1991). Agro-Climatic Zone specific Research-Indian perspective under NATP. ICAR, Krishi Bhawan, New Delhi.:539.

MOWR (1991) Report of the Working Group on Problem Identification in Irrigated Areas with Suggested Remedial Measures. Ministry of Water Resources, Govt. of India. New Delhi.

NCA (1976). Report of the National Commission on agriculture. Ministry of Agriculture. (Department of Agriculture), Govt. of India, New Delhi.

National Water Policy. (2002). Ministry of Water Resources, Government of India, New Delhi

Ott, K. (2003). The case for strong sustainability. In: Greifswald's Environmental Ethics. Ott, K. and P. Thapa (eds.). Greifswald: Steinbecker Verlag Ulrich Rose.

Singh, D.K., Chandola, V.K. and Singh, R.M., (2009). Sustainable management of irrigation water, An approach to address impact of climate change in agriculture. Flair Book Publication, New Delhi.

Singh, R, Singh, R.M., Kishore, R., and rao, K.V.R. (2011). Agricultural drainage technologies-status and Indian experience. Proceedings of National Seminar on Sustainable Management of Water Resources. Banaras Hindu University, Varanasi, 14-15 January, 2011: 7-28.

Tyagi, N.K. (1999). Management of salt affected Soil. In : 50 Years of Natural Resource Management Research.(Ed. G.B. Singh and B.R. Sharma.). Division of Natural Resource Management. ICAR, Krishi Bhawan, New Delhi.365.

United Nations. (1987). Report of the World Commission on Environment and Development. General Assembly Resolution 42/187, 11 December 1987.

□□□

Green Agriculture : Newer Technologies, 2012

New India Publishing Agency, New Delhi (India)
e-mail : info@nipabooks.com; website : www.nipabooks.com

Chapter-11

Soil and Sustainable Agriculture

Sucharita Mohapatra
Directorate of Water Management
Chandrasekharpur, Bhubaneswar, Orissa, 751023
E-mail: sucharita.mohapatra@yahoo.com

SUMMARY

To meet the food requirement of spirally growing population we need to produce more crops per unit area of land. Soil is the raw material for land and is the outer covering of earth surface. It is a very important resource and gift of nature for the substrate where plant can grow. Agriculture is the backbone of our country, and agriculture mainly dependent on soil quality. The success management of soil quality depends on the mechanistic approach, how soil responds to agricultural use and practices over times. As soil degradation grows day by day there is a big challenge against agriculture to sustain. Now days the main cause for soil degradation is non-planed use of land resources. This is the time when we think about this big challenge and face it for the future generation which we need to manage the soil quality for self sustainability. Because adaptation of appropriate soil management scientific methodologies and land use planning could replenish the degraded soil properties for sustainable agriculture which is need of the hour.

Introduction

Soil is the outer cover of earth surface. It is a very important substrate where plant grows. Healthy soils are vital to a sustainable environment. They store carbon, produce food and timber, filter water, support wild life and the urban and rural landscapes. The soil also preserves the records of ecological and cultural pasts. It serves as a major link between climate and biogeochemical systems. However, there are increasing signs that the condition of soils has been neglected and that soil loss and damage may not be recoverable. Thus the soil must not be neglected in any development endeavour either at local, regional or global level. Soil is used in agriculture, where it serves as the primary nutrient base for plants; however, as demonstrated by hydroponics .It is not essential to plant growth if the soil-contained nutrients could be dissolved in a solution. The types of soil used in agriculture (among other things, such as the purported level of moisture in the soil) vary with respect to the species of plants that are cultivated.

Soil material is a critical component in the mining and construction industries, serves as a foundation for most construction projects. Massive volumes of soil can be involved in surface mining, road building and dam construction. Earth sheltering is the architectural practice of using soil for external thermal mass against building walls. The important soil resources are critical to the environment, as well as provide minerals and water to plants. It absorbs rainwater and releases it later, thus preventing floods and drought also cleans the water as it percolates. Soil is the habitat for many organisms: the major part of known and unknown biodiversity is in the soil, in the form of invertebrates (earthworms, woodlice, millipedes, centipedes, snails, slugs, mites, springtails, enchytraeids, nematodes, protists), bacteria, archaea, fungi and algae; and most organisms living above ground have part of them (plants) or spend part of their life cycle (insects) belowground. Above-ground and below-ground biodiversities are tightly interconnected, making soil protection of paramount importance for any restoration or conservation plan.

The biological component of soil is an extremely important carbon sink since about 57% of the biotic content is carbon. Even on desert crusts, cyano bacteria lichens and mosses capture and sequester a significant amount of carbon by photosynthesis. Poor farming and grazing methods have degraded soils and released much of this sequestered carbon to the atmosphere. Restoring the world's soils could offset some of the huge increase in greenhouse gases causing global warming while improving crop yields and reducing water needs.

Waste management often has a soil component. Septic drain fields treat septic tank effluent using aerobic soil processes. Landfills use soil for daily cover. Land application of wastewater relies on soil biology to aerobically treat BOD. Agriculture is the back bone of our country we need better management of soil. We need more production to fulfil the requirement of the vastly increasing population. As more advanced creation of god we human beings have to keep the natural habitat at a ecofriendly condition. It should be done in priority basis otherwise it will create a drastic situation for our future generation.

Sustainable agriculture

India covers an area of 329 million ha, extending from the snow covered Himalayan heights of the North to the tropical rain forests of the South. It stands apart in Asia as the seventh largest country in the world. In the North it has the Great Himalayas as a wall that spreads Southwards and at the tropic of cancer, tapers off into the Indian Ocean between the Bay of Bengal on the East and the Arabian Sea on the

West. Being an agriculture dominant country, the agricultural land use assumes a great significance for optimizing land utilization for future development. From the ancient time the basis of our food is agriculture. We badly need the improvement of agriculture. Instead of improvement, we exploited it in the name of modernisations. Agriculture has changed dramatically, especially since the end of World War II. Food and fibre productivity soared due to new technologies, mechanization, increased chemical use, specialization and government policies that favoured maximizing production. These changes allowed fewer farmers with reduced labour demands to produce the majority of the food and fibre in the U.S.

Although these changes have had many positive effects and reduced many risks in farming, there have also been significant costs. Prominent among these are topsoil depletion, groundwater contamination, the decline of family farms, continued neglect of the living and working conditions for farm labourers, increasing costs of production, and the disintegration of economic and social conditions in rural communities. As chemical fertilizers are used in more quantity, the quality of soil depleted. The fertilizers leached through lower soil level and cause contamination of groundwater. A growing movement has emerged during the past two decades to question the role of the agricultural establishment in promoting practices that contribute to these social problems. Today this movement for sustainable agriculture is garnering increasing support and acceptance within mainstream agriculture. Not only does sustainable agriculture address many environmental and social concerns, but it offers innovative and economically viable opportunities for growers, labourers, consumers, policymakers and many others in the entire food system.

Sustainable agriculture refers to the development of agriculture using new technologies without hampering the needs of the future generation. But the way we want to improve agriculture leads to the depletion of soil quality. Which create a very big challenge in front of us?. If it continues there will be disastrous situation for the future generation. So the time comes when we have to face this challenge which is the need of the hour. To save agriculture we have to protect the soil quality. For which we have to know what soil quality is?

Soil quality : Soil quality is the capacity of a specific kind of soil to function, within natural or managed ecosystem boundaries, to sustain plant and animal productivity, maintain or enhance water and air quality, and support human health and habitation. Soil quality reflects how well a soil performs the functions of maintaining biodiversity and productivity, partitioning water and solute flow, filtering and buffering, nutrient cycling, and providing support for plants and other structures. Soil management has a major impact on soil quality.

Soil quality index : Soil quality index is a useful tool for assessing the overall soil condition and response to management or resilience towards natural and anthropogenic forces.

Soil properties that can be changed in a short time by land use dynamic are considered as soil quality indicators. There is a considerable amount of International literature on the derivation and use of Soil Quality Indicators (SQIs). Selecting indicators for soil quality monitoring is a potentially onerous task. Soil quality indicators should be based on the soil function; e.g:- environmental interaction, production of food etc.

Soil quality degradation : Soil quality degradation is a major consequence of ecosystem destruction and soil disturbances. It refers to the decline in soil productivity through adverse changes in nutrient states, organic matter structural stability and concentration of electrolytes and toxic chemicals. Now a days soil degradation is influenced by a number of factors namely soil characters, land uses, socio- economic and political controls. Out of which land use is one of the most important issues. The causes of soil degradation (World) are described in the following figure(Fig.1).

Soil degradation is a major threat to our food and environmental security. In the past due to lower attention towards the land care many thriving civilizations vanished. Soil degradation can either be as a result of natural hazards or due to unsuitable land use and inappropriate land management practices. Natural hazards include land topography and climatic factors such as steep slopes, frequent floods and tornadoes, blowing of high velocity wind, rains of high intensity, strong leaching in humid regions and drought conditions in dry regions. Deforestation of fragile land, over cutting of vegetation, shifting cultivation, overgrazing, unbalanced fertilizer use and non-adoption of soil conservation management practices, over-pumping of ground water (in excess of capacity for recharge) are some of the factors which comes under human intervention resulting in soil degradation. So emphasis needs to be placed on sustainable rather than exploitative land uses. Following deforestation, soil organic carbon (SOC) and total nitrogen (TN) decreased as a power function of years cultivated. Human activities also cause soil degradation. This can be seen in the figure given below.

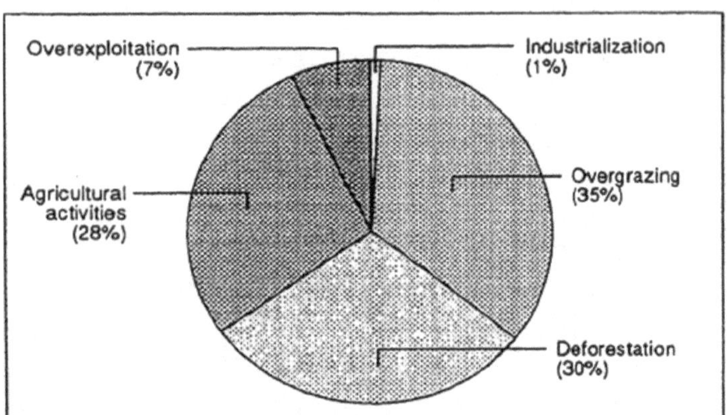

Figure 1 : *Source :* World Resources 1992-93 (World Resources Institute)

The rates of soil organic carbon (SOC) and total nitrogen (TN) degradation apparently decrease with time in cultivation. The initial increase of soil erosion is almost certainly due to the removal of the canopy and surface litter that protects the soil surface from the energy of raindrop impact and surface detachment. As evidenced from the initial 3 order of magnitude increase, it is obvious that surface cover is probably the greatest single management effect on soil erosion and this effect is documented well in the literature (Nearing et al. 1994).

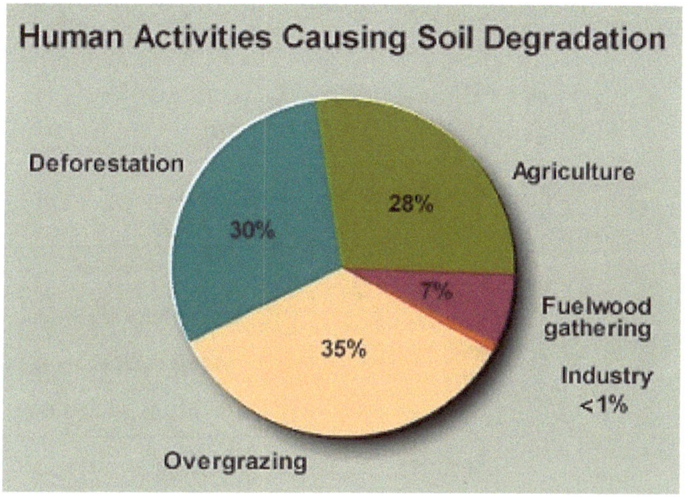

Figure 2 : Human activities causing soil degradation.

Native structure is destroyed by the plough and the stabilizing effects of root fibres become insignificant as the roots are shredded by the tillage microbial decomposed following deforestation. Organic matter, particularly Particulate Organic Matter (POM) derived from root decomposition, has a strong influence on the stability of soil aggregates (Gale et al. 2000). It is probable that in the first few years following deforestation, larger, more slowly decomposed organic matter remained in the soil and, coupled with the relatively high crop residue production resulting from the mining of soil nutrients, provided sufficient organic matter input to slow the rate of mean weight diameter (MWD) degradation. As the larger pieces of organic matter were decomposed and the productivity declined, apparently about the twelfth year following deforestation, the rate of MWD degradation increased markedly. Rasmussen (1998) reported that all deforestation may induce some nitrogen leaching during a short period before revegetation by herbs or trees. However, during the rotation period of most forests, this leaching is far less than that from intensively managed agricultural fields. The exception to this would be forests in decline due to stresses related to age, climate, or insect attack. The TN content reflects the basic fertility of the soil and some may be tied up in organic matter and unavailable for leaching or uptake by plants. Available N, including NO_3–N and NH_4–N, is available for plant uptake and the NO_3–N is readily leached.

As a result of the rapid decomposition of forest floor litter and tree roots following deforestation, much available N is in the soil for plant uptake or leaching losses.

Many practices of management can be threatening to soil sustainability. These practices are: over cultivation, decreased or increased water abstraction, under fertilization or over fertilization, careless use of biocides, failure to maintain soil organic matter levels and clearing natural vegetation. During the use of soils for such purposes, soils can suffer various types of degradation that can primarily reduce their ability to produce food resources. The figure below illustrates the main types of soil degradation:

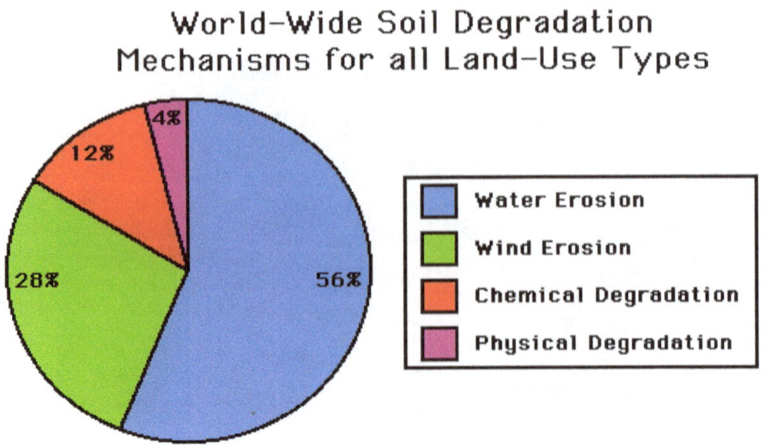

[*Source of Data* : Oldeman, L.R., R.T.A. Hakkeling, and W.G. Sombroek. 1990. World Map of the Status of Human-Induced Soil Degradation. An Explanatory Note, rev. 2nd edition. International Soil Reference and Information Centre, Wageningen, the Netherlands]

Soil Pollution

Soil is the foundation of the worlds' ecosystems. As industry increases, though, pollution increases too. Soil pollution is particularly dangerous for the environment and our health because soil, either in the mountains or in the plains, contains the largest part of the water we drink and produces all the food we need. The following figure shows how industrial waste pollutes soil along with ground water.

(1) Pollution from pesticides

Today, agriculture has become an industry, named intensive farming that produces more on quantity than on quality to maximize profits. So, a huge of pesticides is used to fight parasite insects, moulds and herbs that can destroy part of all our crops. The problem is that the residues of these pesticides are toxic for human beings when present in the vegetable products we consume and when they remain and accumulate in the soil. Here, pesticides can be absorbed by the following crops or be carried by rains to the nearest rivers and to ground-waters. Ground-waters pollution is particularly dangerous for the water we drink, coming from wells and natural sources of the areas where pesticides are used, given that pesticides, after reaching the deep layers of the soil and the ground-waters, are protected from the oxidation by the air and are more persistent.

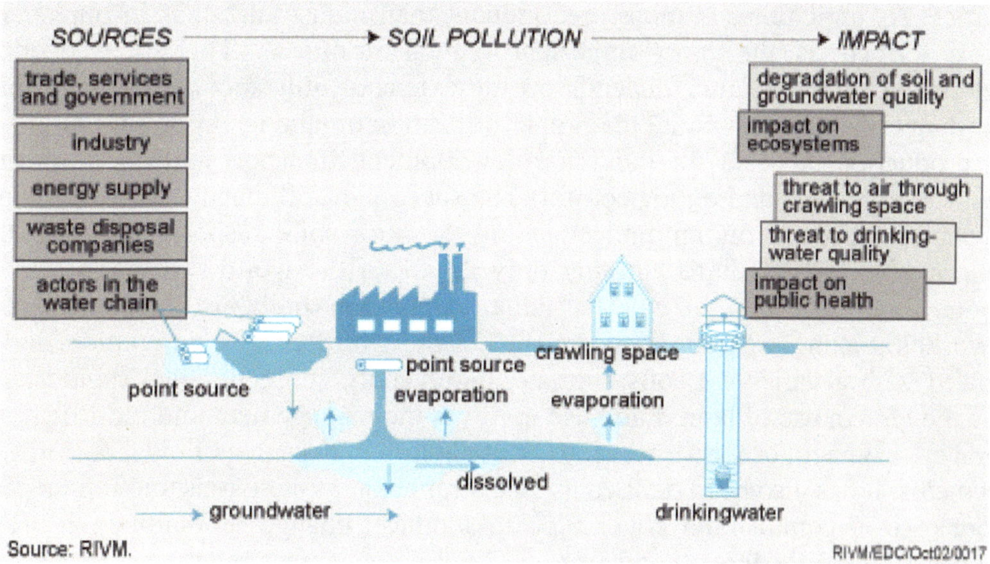

Figure 3 : Soil pollution

To give an example, in Italy one of the most developed farming Countries of Europe, the herbicides pollution in the soil has reached worrying levels: from a monitoring campaign about pesticides made by the APAT public authority in the period 2003-2005, there were 119 different pesticides detected as pollutants; 112 of them were found in superficial waters and 48 in ground-waters. The pesticides residues were found, only in 2005, in 485 monitoring sites (47% of the total sites) and the levels of pesticides were above the limits for drinking water in 27.9% of sites, about underground waters, 630 monitoring sites were contaminated and were 24.8% of the total, with 7.7% of cases above the limits for drinking water. In addition, intensive farming tends to deplete the soil of its mineral content and against this; it uses artificial fertilizers containing phosphorus and nitrogen. Also this is pollution because the soil is exploited too much and changes its features.

To remedy against a situation like this, the only solution is a definitive and massive conversion of all cultures to organic farming. This is not a fashion, but a sustainable cultivation system that doesn't use synthetic pesticides and fertilizers, practices natural cultivation systems that respect the equilibrium of soils and the natural environment and uses a wider variety of cultures to preserve bio-diversity and the typical regional products. This farming system is not intensive; so, it doesn't deplete and pollute the soil. As you can see, pollutants can impact soil everywhere. And the effects are potentially disastrous: small chemical changes in soil can render an area inhospitable to plants, virtually eliminating the food chain's foundation.

Also, chemicals that runoff through soil into rivers and streams can contaminate drinking water. For farmers, pollutants can reduce crop yields and lead to heavy erosion.

Land use : Land resource takes the fundamental position in the bases used for human activities. As agriculture is more predominant than industrialization in our country, land is termed as the most important natural resources. The socio economic development of the country depends on the extent of utilization of land resources. According to land use potential the proper utilization of land is important not only for food production but also for industrial development, transport and communication network, social need and public comfort.The success in soil management to maintain soil quality depends on an understanding of how soils respond to agricultural practices over time. India occupying only 2.4 % of the world's geographical area supports about 16.2% of the world's human population. India also has only 0.5% of the world's grazing area but supports 18% of the world's cattle population. India is endowed with a variety of soils, climate, biodiversity and ecological regions. Land use is the human use of land. Land use involves the management and modification of natural environment or wilderness into built environment such as fields, pastures, and settlements. It has also been defined as "the arrangements, activities and inputs people undertake in a certain land cover type to produce, change or maintain it" (FAO, 1997a; FAO/UNEP 1999). Land use and land management practices have a major impact on natural resources including water, soil, nutrients, plants and animals. Land use information can be used to develop solutions for natural resource management issues such as salinity and water quality. For instance, water bodies in a region that has been deforested or having erosion will have different water quality than those in areas that are forested. Most of the important soil quality indicators were significantly influenced by different land use systems, particularly at the surface horizon. The bulk density, structure, OC, soil pH, CEC, total N, different forms of P, exchangeable bases, and available micronutrients were affected due to intensive cultivation and use of acid forming inorganic fertilizers for the past three decades. Except for the available P, nearly all of the chemical properties of the soil under the abandoned land were very poor as compared to the other land use systems. The virgin land was superior in most of the soil quality indicators. In general, the continuous intensive cultivation and use of acid forming inorganic fertilizers on acid soils for crop production without appropriate soil management has degraded most of the important soil quality indicators. Therefore, reducing intensive cultivation, and integrated use of inorganic and organic fertilizers could replenish the degraded soil quality parameters for sustainable agricultural production and productivity.

The improper use of land leads to a considerable wastage and progressive deterioration of the production and productivity, it should be avoided. The present land use pattern seems to be in a stage of static harmony and adjustment with the other main characteristics of agro- climatic features and the economy of the region. Keeping the natural endowments and recent advanced technology in view, the country may need certain modifications and changes to the existing land use pattern. The land use category is changing day by day. The following figure shows the land use categories of 1988-89.

Forestry

Although India is the seventh largest country in the world, only 1.8 % of the world's forests are found here. Based on recent Indian Remote Sensing Satellite data, forests in India cover approximately 21 % of the area. However, a more realistic estimate shows that forests cover only 11 % of the land base. Out of India's population of one billion, 360 million live in or around forest areas, exerting tremendous pressure on limited forest resources. This is in addition to the need to fulfil the requirements of urban population and wood-based industries (FAO, 2000).

Area under broad land use / land cover categories in India (1988 -89)

Category	Area (00 ha.)	%
Builtup Land	139137	4.24
Agricultural Land	1652443	50.26
Forest	476226	14.49
Wastelands	755153	22.98
Waterbodies	106108	3.22
Others	74899	2.28
Unsurveyed	83294	2.53
TOTAL	32872630	100.00

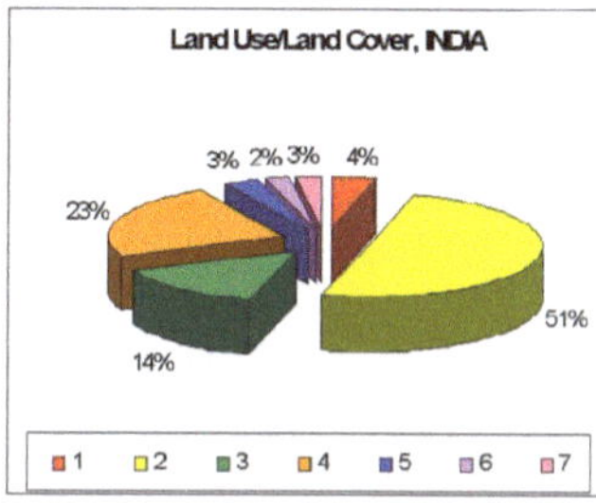

Figure 4 : Landuse categories in India (1988-89)

Many native forests are changed to agricultural fields also to grass fields, rice fields, industrial sectors and households for which land use is the ultimate cause. The soil organic carbon content remains highest in the grass land than coniferous forest. The conversion of native pasture and forest soils into cropland during the 18 years period increases bulk density by 16%, decreases soil organic matter and total nitrogen by 50%. Total porosity also decreases in cultivated lands.

Agriculture

Out of 328.7 million ha of geographical area of India, about 141 million ha is Net Cultivated area and of this, about 57 million ha (40 %) are irrigated and the remaining 85 million ha (60%) are rainfed. This area is generally subject to wind and water erosion and is in different stages of degradation. Therefore it needs improvement in terms of its productivity per unit of land and per unit of water for optimum production. According to the 2005 report of National Bureau of Soil

Survey and Land Use Planning, an area of 146.82 million ha is suffering from various kinds of land degradation in India.

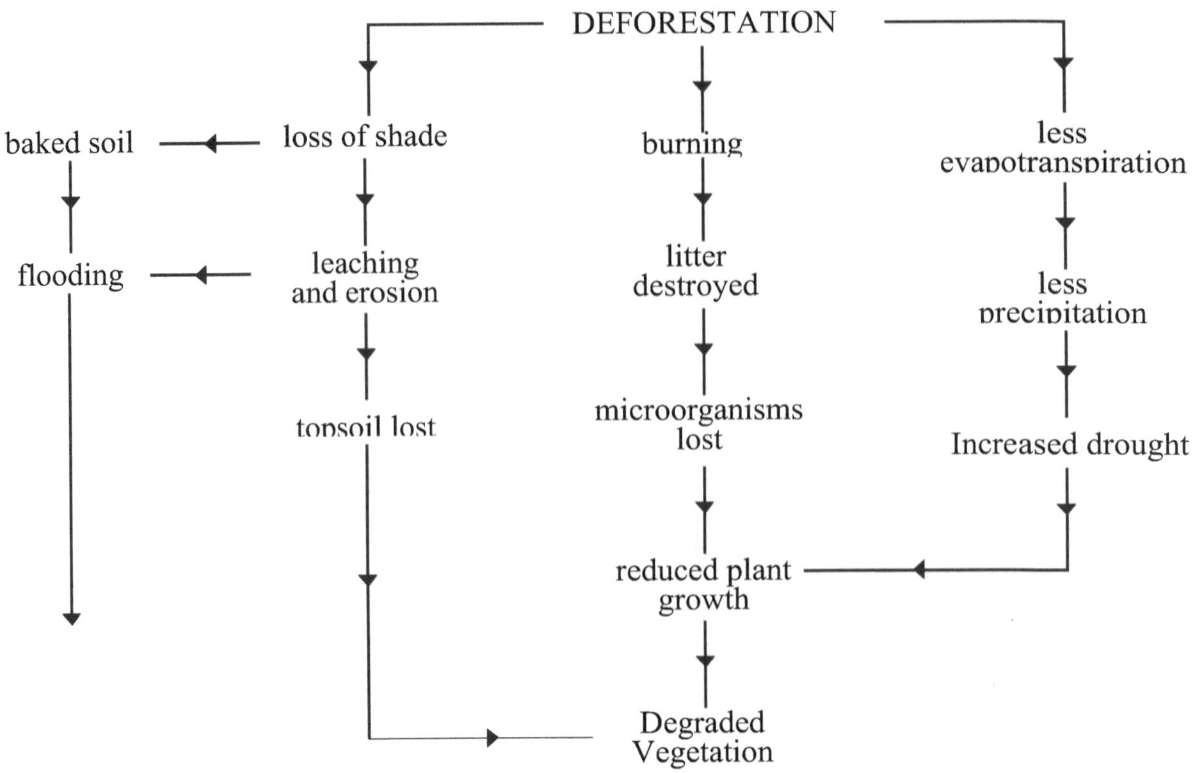

Figure 5 : A graphic presentation of results of deforestation

The overwhelming effort in increasing agricultural production in India is attributed to several changes facing intensive tilling of land, water and fertilizer application as well as risk in environmental pollution and degradation soil and water resources. Too much disturbance in soil through tillage operations is actually not required to obtain good crop yields. In agriculture a major portion of energy (25-30%) is utilized for either field preparation or crop establishment where conventional tillage is mostly followed.

Impact of conventional tillage on soil physical properties is always not positive. It varies from soil to soil and between the crop systems also. Soil physical parameters mostly showed significant improvement under conventional tillage with residue incorporation. It also facilitated better root development in maize and mustard and resulted in more leaf area, biomass and subsequently higher yields of the crops. Residue incorporation under conventional tillage also reduces soil compaction significantly.

Cultivation over the years caused a net decrease in soil organic carbon stock but with balanced fertilization it increases.

Organic matter, bulk density decreases in cultivated land as compared to the forest soil.

Rice field crop

The recycling of its residues has the great potential to return a considerable amount of plant nutrients to the soil in the rice based crop production systems. Particularly the rice-wheat cropping system is the most intensive production system in the country. The yield stagnation consequent upon the declining soil organic carbon is a major threat to this system. Therefore it is a great challenge to the agriculturists to manage rice residues effectively and efficiently for enhancing sequestration of carbon and maintaining the sustainability of production

Rice residue management is also important as machines are being increasingly used for harvesting of grains, and this mechanical harvesting leaves huge amount of residues in the field. There are several options for management of rice residues: burning, incorporation, surface retention etc. Every management options have its advantages as well as disadvantages. Now it is the location, soil and situation, which will govern the practice to be selected. Of course, intensive research is required to solve this problem of managing rice residues. Sometimes surface retention may be the best option in many situations. For sowing/ planting of subsequent crops having rice residues, both stubbles and loose straw in the field needs to be managed, for that intensive investigation in different rice growing areas is required. No tillage technology may be adopted which is fitted to the location and soil condition. Research has shown that diversification of the rice-wheat cropping system is also essential. If rice residues are managed properly, then it can warrant the improvements in soil physical, chemical and biological properties and sustain productivity of rice-wheat cropping system.

Non Rice field crop

The integrated nutrient management practices like use of organic resource materials and inorganic fertilizers are known to modify the fertility status of soil as well as soil quality. Use of chemical fertilizers regarded as a source of nutrient for modern agricultural technology. A factor in agricultural production is fertilizer that changed the world from food scarcity to food sufficiency region. The fertilizer production is largely dependent on non renewable energy resources. Till now very little attention was paid to maintain soil productivity, while reducing dependence on fertilizers.

Agriculture includes not only rice crop but also non rice crops like vegetables, mustards, fruits etc. In non rice crop fields 'N' decline is found due to use of organic manures and green manures to supply fertilizer. The practise of continuous cropping raises a question of sustainability of the system highly dependent on chemical fertilizer input and the crops have started giving signals of decline in yield. However integrated approach of plant nutrient management improves soil fertility and maintains soil health without affecting the yield of crops.

In the non rice crop fields combined application of zinc and organic matter along with recommended levels of increase the status of major and micro nutrients along with enhancement of organic carbon and other physical properties of soils.

Industrialization

Agriculture is the main stake of India. Agriculture is the backbone of our country. For political leaders, this is the main platform– as majority of voters are from agriculture background. But what we are doing? Every day agricultural land is being converted into industrial land. Industrialization is the indicator of development of the country. In Indian context, if we talk about current scenario, the leaders are preparing plans without any plans. 80% of Indian population is directly or indirectly related to agriculture and cottage industries. The plan will be done as per the situation. Indian context must be taken into consideration before making plans. Large industrial houses which have set up their plants in various states –must have the basic idea about our infrastructure and the requirements of people. They are not thinking about the problems of common people in India. Our population is dependent on agriculture, animal husbandry and food industries. Industrialization is a factor for declining of agricultural sector. Now this is a difficult time. Agriculture growth rate is between 2 to 7%. Due to industrialization, our agriculture land is shrinking. Indian population has many challenges. There is a big challenge of providing food and shelter to the people. Due to fast industrialization, agriculture land is being converted into industrial land. Some restrictions must be put on this conversion. This process is going on at a fast pace. With this pace, we will be left with less agricultural land. Raw material is also available from agriculture sector. Agricultural lands should not be used for industrialization. Government should give waste land or barren land to the industries. There is land which has very high fertility; industries are also being set up there. For proper development, the government must promote the backward areas and those areas where agriculture is not possible. Our economy is totally based on agriculture. Our social system is also based on agriculture. 80% population is dependent on agriculture and related sector. Our growth achieve equitable development, we should not use agriculture land for industrial development. Our agriculture growth rate is falling, exports are also increasing. Due to all these we will have shortage of food. We have a growing population and we need massive production of food items. Now we are importer of most of the food items. We have to change our policies and formulate appropriate policies. We have to improve our situation In India, agriculture is based on natural rain and natural weather. Industrialization also uses natural water. They use land which has easy access to water. Thus there is a clash between the interests of industries and agriculture. Industries should be set up far away from the city. Industries cause pollution and other problems. Therefore there is a need for analysis of the situation. Food production is a major concern for any country. Due to increasing industries, we should look at agriculture set up and develop it further. We do have to set up plants, malls and all those that are required. But we have to also consider the interest of agriculture. Industries should be scattered throughout the country. However, industries are developing only in a few areas. Industries are being developed in areas which are also very fertile.

Aquaculture

Aquaculture means culture of fish, which is done in ponds especially designed for it. Shrimp culture is an important part of aquaculture. Selection of suitable site for aquaculture plays an important role in the shrimp farming, since it is helpful in proper planning, designing, construction, operation and maintenance of ponds. Traditionally aquaculture ponds are earthen ponds, constructed in areas where the soil has sufficient clay content, to prevent leakage of water (Singh, 1993). The setbacks witnessed in coastal aquaculture disease outbreak all along the Indian cost are mainly due to the unplanned growth of aquaculture farms. Proper designing of aqua farm is very important for better management and to overcome or to minimize the adverse impact on environment. Most of the existing farms are not constructed as per technical designs or standards and there are many deficiencies in design, construction, and management of farms (Omprakash 1994). As a consequence, majority of the farms constructed are not able to with stand the hazardous encountered in coastal region (Mukharjee and Ghosh 1987). The lack of attention to soil properties results in aquaculture ponds being not used to either full potential. Soil and water quality parameters of shrimp ponds are important factors affecting production. Soil condition is more crucial than water condition in brackish water productivity (Hickling 1971). According to Banerjea (1967) the fish production is more in decreased organic carbon content.

For fish growth organic carbon content has to be reduced and accordingly soil becomes less fertile. So the waste lands have to be used for aquaculture. It is a means of lively hood for many peoples. So it should not be banned. But we have to use the alternatives.

Domestic households

As population is increasing day by day the requirements are also increasing. To fulfil the requirements clever man invents new ways. Some people live on agriculture, some are working in administrative sectors, and there is a class of people who lives on business. So large unit of land is used market complexes, shopping malls, business sectors etc. Industries and factories take a great place in that. As population is increasing, the requirement of shelter is also increasing. So agricultural fields, forests are converted into building, concert roads. Where we need more cultivation to fulfil requirement of food, in another way we also need more place to make our shelter. This is a big challenge in front of us, at the present time. We also create situations that cause soil, water, air and environmental pollution. By clearing the forest we destroy the native place of wild animals to make our native secure. At the present time we human beings are behave like selfish. We have to think about the current situation; otherwise the native of our future off springs will be in danger only because of our selfishness. The business sectors are also the requirements of the time. So these can't be avoiding. The uncultivated waste lands should be used for the domestic purpose. Human want to change the environment around as per his comfort. But it gradually comes to the way of destruction. But safety is in our hands. We can save our surrounding. The hazardous weast of the industries that cause soil pollution can be controlled by appropriate management steps.

How land use related to soil degradation

The development of mankind over the past decades has gone through a number of historical stages. The development process includes exploitation of natural resources for the purpose of converting it into usable form. For example human activities such as land tillage, forest clearing, and irrigation practices are done in order to increase food production aimed to feed the population ever increasing. In many cases, in the name of development unplanned activities are done which results serious land degradation. Land degradation means a reduction or loss in arid and dry sub humid areas of biological or economic productivity of rainfed cropland, irrigated cropland forest, pasture etc., resulting from land uses. These processes include those arising from human activities and habitation patterns such as soil erosion caused by wind and water, deterioration of soil physical, chemical and biological properties of soils as well as loss of natural vegetation (Dejene et.al.1997; Majule et al.1997; Boesen et.al.1999).

Soil degradation mainly refers to loss of production. Soil degradation is a most serious environmental problem in many countries. One of the most important driving forces of soil degradation is human activities. Human activities often influence the natural processes in soil. Deforestation is a perceptible aspect of human activities in environment. This change has many interlink effects that can appear through the reduction of chemical and physical qualities of the soil resources (Doran et al.1998; Cavelier et al.1999; Chew, 2001; Liu et al. 2002; Johnson and Lewis 2007; Seeger and Ries 2008).

Geological, climatic, biological and human factors all these are responsible for soil degradation. These lead to the degradation of both physical, chemical and biological potential of soil which endanger biodiversity, land use and survival of human communities also (FAO 1993). According to Lal (1997), decline in soil quality with a continuing reduction of productivity are the results of soil degradation. Besides this overgrazing, deforestation, inappropriate agricultural practices and other human activities are also responsible for loss of soil productivity. Soil quality degradation to some extent related to cropping and cultivation type. Variability of soil properties especially soil chemical properties arise due to different land use practice. Mainly properties like organic matter, soil texture, cation exchange capacity and exchangeable acidity appear to be the dominant indices of soil degradation in this environment. Land use and soil quality are correlated. If land is used in unplanned way soil quality degradation will be continuing.

Soil quality parameters

Soil the base of earth surface. It is as important as air and water. It is the duty of mankind to protect this substrate. Now a days land use gives threatening to the sustenance of soil quality. Protection of soil quality means to protect soil quality indicators in its original situation. For which we have to understand what are the sensitive indicators how to know it and the way to protect it. Soil quality indicators are the properties of soil that can be change in short time by land use dynamics. So the indicators are very much sensitive towards land use change. So before applying any change first we have to understand how the soil responds to the change and then go through a proper planning. Because unplanned land use causes serious conditions and that can't be recoverable.
Basically soil quality indicators are of three types.

1. Chemical parameters
2. Physical parameters
3. Biological parameters

Chemical parameters

Some important and sensitive chemical parameters are given below.

pH

pH is generally known as –ve logarithm of hydrogen ion concentration. It is a measure of the concentration of hydrogen ions in a solution (Sorensen,1909). In pH 'p' stands for potenz which means power or concentration, and H for hydrogen ion. pH value is a numerical number between 0 to 14, that indicates whether the solution is acidic ($pH<7$), neutral ($pH=7$) or alkaline ($pH>7$). Distilled water is the only neutral solution. pH is an important parameter to determine the soil quality because all other parameters like organic carbon, hydraulic conductivity, soil texture change with change in pH. To measure pH of a solution there are two methods, addition of pH indicator and using pH meter. Universal pH indicator is a prepared solution that changes colour of the solution it is added to according to the pH of it. pH meter is an electronic device in which an electric current is created due to the hydrogen cations completing the circuit. Now a days pocket pH meters are also available. To measure pH we have to standardize the pH meter with pH buffer solutions. pH is the primary and very important parameter of soil. It helps in soil management by indicating that whether it is acidic or alkaline because management practice is different for acidic and alkaline soils.

Electrical conductivity

Electrical conductivity (EC) is the measurement of soil salinity. Salinity refers to the amount of soluble salt in soil. Plants are detrimentally affected, both physically and chemically, by excess salts in soils. By agricultural standards, soils with an EC greater than 4 dS/m are considered saline. In actuality, salt-sensitive plants may be affected by conductivities less than 4 dS/m and salt tolerant species may not be impacted by concentrations of up to twice this maximum agricultural tolerance limit. Thus, the reclamation scientist must exercise care in interpretation of salinity standards. Salinity should be defined in terms of the pre disturbance land use potential, the proposed post disturbance land use, and the plant species to be seeded on the site (Munshower 1994). Soil Electrical conductivity affects soil pH and accordingly it is as important as soil pH. To measure soil electrical conductivity electronic EC meters are available. Pocket EC meters are available which makes its measurement much easier.

Organic carbon

The organic matter in soil derives from plants and animals. In a forest, for example, leaf litter and woody material falls to the forest floor. This is sometimes referred to as organic material. When it decays to the point in which it is no longer recognizable it is called soil organic matter. When the organic matter has broken down into a stable humic substances that resist further decomposition it is called humus. Thus soil organic matter comprises all of the organic matter in the soil exclusive of the material that has not decayed. Soil organic carbon is the biggest part of soil organic matter. It is considered as most important indicator of soil quality and productivity. Change in this indicator affects soil's structure, water storage capacity and nutrient supply. It helps in maintaining soil quality and fertility. It affects the soil colour, physical and biological properties. There are laboratory methods to determine soil organic carbon percentage, i.e. Walkly and Black method (Walkly and Black1934).

Total nitrogen

Total nitrogen is as important as organic carbon. It is a part of soil organic matter. Nitrogen is an essential elemant for plant growth. It is important that nitrogen quantity should be maintained, because as too much availability of nitrogen is harmful for soil and plant as too low quantity. In soil nitrogen is found both in organic and inorganic forms. But 90% 'N' is found associated with organic matter. Nitrogen is in compounds identifiable as part of the original organic material such as proteins, amino acids, or amino sugars, or in very complex unidentified substances in advanced stages of decomposition. These uncharacterized substances resist further microbial degradation and account for the very slow availability of soil N. Conversion in soil 'N' content affect soil moisture condition, soil acidity, temperature and soil microbial activities. Nitrogen can be permanently removed from soil by erosion, leaching, de-nitrification, or volatilization. As nitrogen is a very important index of soil should be in its measurable condition. It is necessary for maintenance of soil condition. Many laboratory methods like kjeldhal method are available now to test and maintain soil 'N' content.

Except the above properties exchangable P, available K, available Na, SO_4 are some parameters of soil that should be taken into consideration to protect soil and maintain its quality.

Physical parameters

The physical parameters also important as chemical parameters. Some important ones are given below.

Bulk density

Bulk density (BD) refers to the mass per unit volume of soil (sampled as a clod or core), dried to constant weight at 105°C. The bulk density of soil depends greatly on the mineral make up of soil and the degree of compaction. Bulk density is not an intrinsic property of a material; it can change depending on how the material is handled. So it is very sensitive towards land use change. Change in bulk density can affect gravity, porosity, organic and inorganic particles of soil. Bulk density is the indicator of soil porosity and structure that govern the movement of O2 and water in soil. It indicates the srength of soil as it a measurement of degree of compaction of soil. An important characteristic of bulk density is plant growth. Use of tractors directly affects the bulk density causing extreme compaction in soil. As it is so important property the care full management of land is nessesary for optimum plant growth and healty soil. To measure bulk density core sampling method is done in laboratories.

Moisture retention capacity

The space exist between the soil particles are called pores. These are the passage for water, air with in the soil particles. The amount of water absorb by the pore is the moisture retention or water retention capacity. This character is strongly related to particle size of soil. Because the water molecules hold tightly to the fine particles of a clay soil than the coarser particles of sandy soil.so the water retention capacity is more in clay soil than in sandy soil (Leeper and Uren 1993). Water retention capacity of soil is highly influenced by soil type, organic matter content and soil structure (Charman and Murphy 1977). The maximum amount of water that a given soil can retain is called field capacity, whereas a soil so dry that plants cannot liberate the remaining moisture from the soil particles is said to be at wilting point (Leeper & Uren 1993). Available water is that which the plants can utilise from the soil within the range of field capacity and wilting point. Soil water retention is essential to life. It provides an ongoing supply of water to plants between periods of replenishment (infiltration) so as to allow their continued growth and survival. As it is so important it need management of soil. Soil water retention capacity can de estimated by the instrument known as Pressure Plate Apparatus.

Soil texture

Soil texture is an important soil characteristic that drives crop production and field management. The textural class of a soil is determined by the percentage of sand, silt, and clay. Soils can be classified as one of four major textural classes: (1) sands; (2) silts; (3) loams; and (4) clays. Soil texture refers to the relative percentage of sand, silt and clay in a soil. Natural soils are comprised of soil particles of varying sizes. Texture is an important soil characteristic because it will partly determine water intake rates, water storage in the soil, the ease of tillage operation, aeration status etc. and combindly influence soil fertility. As the soil is a mixture of various sizes of soil separates, it is therefore, necessary to establish limits of variation for the soil separates with a view to group them into different textural classes. Texture is a basic property of soil and it should not be altered or changed. Soil texture can be determined in laboratory by using two methods, i.e., International pipette method and hydrometer method. This property of soil can affect, organic matter, hydraulic conductivity of soils.

Hydraulic conductivity

Saturated hydraulic conductivity is a quantitative measure of a saturated soil's ability to transmit water when subjected to a hydraulic gradient. It can be thought of as the ease with which pores of a saturated soil permit water movement. Saturated hydraulic conductivity is affected by both soil and fluid properties. It depends on the soil pore geometry as well as the fluid viscosity and density. The hydraulic conductivity for a given soil becomes lower when the fluid is more viscous than water. Practically it is the flow of water in soil through pore space. It can be measured in laboratory using the sophisticated instrument called Permeameter. The unit used for its determination is cm/h. It is affected by salinity of soil and water. It is symbolically represented as 'K'. It is high for sandy soil and low for clay soil. So it is also affected by soil texture.

Other physical characteristics like porosity, structure, water holding capacity are also very much important. These properties should be maintained for their originality. Because the quality, productivity of soil highly dependent on these.

Biological parameters

Soil biological parameters mean the availability of soil micro organisms and their nature of function towards soil quality. There is a number of micro organisms live in soil. They help in organic matter decaying. Soil structure is greatly affected by the animals and microbes in the soil. For example, the chemical and physical nature of the soil is changed as it passes through the intestines of worms. Soil animals and microbes can directly impact the availability of certain nutrients. There are more organisms in a teaspoon of topsoil than there are people on Earth. Soil organisms are an intimate part of the organic fraction of soil, and contribute significantly to soil fertility and soil structure. Plant residues have little value in the form we return them to the soil. The soil organisms, whether large (macro) or small (micro), feed on this residue and break it down in a continuous process. Virtually all topsoil has passed through the gut of soil animals. Although we might think of burrowing animals such as groundhogs, moles, and shrews as having a large impact on soil because they are relatively visible, they are far less important to soil processes than the much more numerous, tiny animals and microbes. The effect of soil animals on soil structure is considerable. Topsoil is basically composed of animal feces of varying ages. Soil animals ingest organic matter and mineral components of soil, and mix them together before depositing the combined material as fecal pellets or casts. Reduce tillage and add organic matter: this will increase soil organism populations and improve soil structure.Highly specialized microbes, mostly bacteria, are involved in the transformation of nitrogen through the N cycle. Nitrogen is essential for plant growth and microbial activity. The rate of the decomposition is governed by the relative availability of a few key nutrients: carbon (C) and nitrogen (N).

Need of soil management

Declining yields and environmental problems associated with many agricultural systems around the world. have resulted in an on-going global call for adoption of sustainable wa in such a way that its productivity is maintained or preferably, ys of agricultural production. Good soil management has always required that the soil be used enhanced. This requires that the chemical and physical condition of the soil does not become less suitable for plant growth than when cultivation commences. Cultivation normally means that the soil will, in fact, deteriorate due both to nutrient removal when harvesting crops, and to physical damage to the soil structure. What is essential is that the deterioration is reversible, by chemical additions to the soil, mechanical manipulation, or natural processes of fertility restoration under pasture or trees This implies that the soil must be resilient, i.e. after being subjected to the stresses involved in crop production, it must have the ability to return to its former condition, or an improved condition (Greenland and Szabolcs 1994). The land must produce on a secure basis, the natural resources must be protected, and the management system must be economically viable and socially acceptable. However it must also be recognized that land cannot be managed sustainable unless the soil, which is a component of the land, is properly managed. This requires maintaining and improving soil productivity, avoiding and rectifying soil degradation, and avoiding environmental damage. Soil is the basic resource for us. It gives us both food and shelter. It is our duty to protect it. The need of the time is food with settlement. For food we have to increase the food production. For settlement industrialization and other market complexes are also needed. So we have to manage both in a simultaneous way. Continuous use of land for crop production without appropriate soil management has degraded most of the important soil physicochemical properties. Therefore adoption of apprpriate soil management and land use planning could replenish the degraded soil properties for sustainable agricultural productivity.

Conclusion

Soil is a vital component for the sustainance of lively hood. It the natural resource that we can and have to hand over to our future generation. So it is our responsibility to protect it and keep in good condition. Good condition of soil means maintain its properties to be productive and fertile. Now a days population is increasing in a vast number. So the requirement of the time is also increasing with population. So the lands are moving from their original condition and converted into another condition. Agricultural lands are used for industrialization, forest lands are used for rice fields crops, grass lands. Native grass lands are converted to agricultural fields. By this activity we are lossing the originality of the soil. The success of soil management to maintain soil quality depends on an understanding of how soils respond to agricultural use and practices over times. If we use fertile lands for industrialization it causes soil pollution and degradation of soil quality and that loss cannot be recoverable. Industrialization is as important as agriculture. But maintainance of soil quality is also very important. We have to use land by through planning. Doing aquaculture may cause change in soil structure and physical properties. So we have to think about this factor and take appropriate steps that should be in favour of both agricultural manner and industrialization and that will be manage soil to maintain its quality.

References

Banerjea, S.M. 1967. Water quality and soil condition of fish ponds in some states of India in relation to fish production. *Indian J. Fish.*, 14: 113- 144.

Boesen, N.J., Kikula, I.S., Maganga, F.P.1999. Sustainable Agriculture in Semi Arid Tanzania, ISBN, 99-60-311-8.

Cavelier, J., Aide, T. M., Dupuy, J. M., Eusse, A. M.and Santos, C.1999. Long-term effects of deforestation on soil properties and vegetation in a tropicallowland forest in Colombia. *Ecotropicos.* 12, 57-68.

Chew, S.C. 2001. World Ecological Degradation: Accumulation, Urbanization, and Deforestation, 3000 B.C.-A.D. 2000. Rowman & Littlefield. 216 pp.

Charman, P.E.V.and Murphy, B.W. 1998.Soils, their properties and management, 5th Edn, Oxford University Press, Melbourne.

Dejene, A., Shishira, E.K., Yanda,P.Z. and Jhnsen, F.P. 1997. Land Degradation in Tanzania Perception from the Village, United States of America, ISSN: 0253-7994.

Doran, J. W., Leibig, M. And Santana, D. P. 1998. Soil health and global sustainability. In: proceedings of 16th World Congress of Soil Science, Montpellier, France, pp. 20–26.

FAO. 1993. Sustainable development of dry lands and combating desertification. Land and water Div. FAO, Rome, Italy.

FAO. 1994. Soil Management for Sustainable Agriculture and Environmental Protection in the Tropics. Introduction ànd legestion conservatoire de leau, de la biomasse et de la fertilité des sols (GCES).

FAO . 2000.Global Forest Assessment. *FAO Forestry paper* 140.

FAO.2010. Land and Water Division retrieved 14 September 2010.

Gale, W.J., Cambardella, C.A. and Biley, T.B.2000 . Root-derived carbon and the forma-tion and stabilization of aggregates. *Soil Science Society of American Journal* 64: 201–207.

Greenland, D.J. and Szabolcs, I. (eds) .1994. Soil Resilience and Sustainable Land Use. CAB International, Wallingford, UK.

Hickling, C.F. 1971. The fish culture. Faber and Faber London.

Johnson, D. L. and Lewis, L. A. 2007. Land Degradation: Creation and Destruction. Lanham: Rowman & Littlefield, 306 pp.

Kaihura, F.B.S., M.Stocking and E. Kahembe.Soil Management and Agrodiversity: A Case Study from Arumeru, Arusha, Tanzania. *Managing biodiversity in agricultural systems*, Montreal, 8-12 November, 2001.

Krishna G. Mandal*, Arun K. Misra, Kuntal M. Hati, Kali K. Bandyopadhyay, Prabir K. Ghosh and Manoranjan Mohanty.2004. Division of Soil Physics, Indian Institute of Soil Science (ICAR), Nabibagh, Berasia Road, Bhopal-462 038, Madhya Pradesh, India. *Food, Agriculture & Environment* 2 (1): 224-231.

Lal, R.1997. Research and Development priorities. In: Methods for assessment of soil degradation. Advances in soil science (Edited by Lal, R, & Blum W, H, &, Valentine, C, &, Stewart, B, A). CRC Press.

Leeper, G.W. and Uren, N.C. 1993. Soil science, an introduction, 5thEdn, Melbourne University Press, Melbourne.

Liu, S.L., Fu, B.J., Lü,Y.H. and Chen, L.D 2002.. Effects of reforestation and deforestation on soil properties in humid mountainous areas: a case study in Wolong Nature Reserve, Sichuan province, China. *Soil use and management.* 18: 376 – 380.

Mukharjee, A.B. and Ghosh, A. 1987. Engineering aspects of designing prawn farms in tidal regions of sunderbans. *J. Ind. Soc. Coastal Agri. Res* 5(1): 257- 265.

Munshower, F.F. 1994. Practical Handbook of Disturbed Land Revegetation, Lewis Publishers, Boca Raton, Florida.

Nearing, M.A., Lane, L.J., Lopes, V.L.1994. Modeling soil erosion. In: Lal, R. (Ed.), Soil Erosion Research Methods (2nd Ed.). Soil and Water Conservation Society, Ankeny, IA, pp. 127–156.

Omprakash, K. 1994. Recent advances in aquaculture engineering on shrimp farm design, construction, management and maintenance. *Fishing Chimes*, 32- 34.

Rasmussen, L.1998. Effects of afforestation and deforestation on the deposition, cycling and leaching of elements. *Agriculture, Ecosystems and Environment 67*: 153–159.

Seeger, M.and Ries, J.B.2008. Soil degradation and soil surface process intensities on abandoned fields in Mediterranean mountain environments. Land degradation & development. 19: 488-501.

Shaoshan A.N., Fenli Zheng, A., Feng, Zhang., Scott, Van Pelt., Ute, Hamer.and Franz, Makeschin.2008.Soil quality degradation processes along a deforestation chronosequence in the Ziwuling area, China Elsevier. *Catena* 75: 248-256.

Singh, T. 1993. The use of pond liners in aquaculture. *Info fish International*, 5: 49- 51.

Sorenson, S.P.L.1909. The Measurement andMeaning of Hydrogen Ion Concentra-tion.*Biochemische Zeitschrift*, 21:131-200. (hppt :// dbhs .wvusd .k12 . ca. us /webdocs.Chem-History/Sorenson-article.html).

Walkly, A., Black, C.A., 1934. An examination of degtjareff methods for determining soil organic matter and proposed modifications of the chromic acid titration method. *Soil Sci*. 37:29-38.

□□□

Green Agriculture : Newer Technologies, 2012
© Kambaska Kumar Behera (ed.), pp. 281-302
New India Publishing Agency, New Delhi (India)
e-mail : info@nipabooks.com; website : www.nipabooks.com

Chapter-12

Nutrient Management in Vegetable Crops for Sustainable Production

S.N.S. Chaurasia, R. N. Prasad, R.B. Yadava and D. K. Singh
Indian Institute of Vegetable Research, Varanasi
E-mail : chaurasiaiivr@yahoo.com

SUMMARY

Vegetables play a pivotal role in food and nutritional security of the ever increasing population of our country. The vegetarian society largely depends on vegetables for the want of their nutritional requirements. Vegetables are exhaustive crop with an exceedingly high turnover of plant nutrients, Thus , nutrient element required for optimum growth and development of vegetables should supplied to soil and plants in right time and in a balanced and integrated mode. Neither the chemical fertilizers are alone nor the organic sources exclusively able to sustain the soil fertility/productivity. The national average productivity of different vegetable crops of our country at present is much below the potential productivity. It may be increased up to 2-3 times more by adoption of appropriate nutrient management practices. The majority of our Indian soils are deficient in about 7-8 macro and micronutrients which leads lower productivity. Integrated nutrient management (IPNM) is one of the recent methods of supplying nutrients to the plants by organic as well as inorganic means together to fulfill the nutrient requirements. Organic manures are the key characteristics including the long term fertility of soil by maintaining organic matter levels, fastening soil biological intervention, nitrogen fixation as well as effective recycling of organic materials including crop residues and live stock wastes and weed. Micro - organisms play an important role in various chemical transformations in soil and thus, influence the availability of major nutrient like N, P, K. and S to the plant. The IPNM is a system approach to nutrient management by tapping all the possible sources of organic, inorganic and biofertilizers in judicious and synchronous way to maintain soil fertility and crop productivity. In the present chapter the same has been discussed in detail.

Introduction

In India more than 100 kinds of vegetables are being grown from sea level to snow line. Among them about 60 are being grown on commercial scale and produces large amount of vegetables to meet daily need of the country. Since independence, India has emerged as a leading vegetable producing country and ranks second in the world. At present (2010-11) India produces 141.35 million tons of vegetables with annual growth rate of 5.8 percent. The total vegetable production in the country in 1993 was 63.8 million tons to 129.07 million tons in 2008-09 and 133.54 million tons in 2009-10. Vegetables play a pivotal role in food and nutritional security of the ever increasing population of our country where vegetarian society largely depends on vegetables for want of their nutritional requirements. However, the national average productivity of different vegetable crops is much below the potential productivity (Table-1). One of the major reasons for the low realization of the production potential of most of the vegetable crops in the country is the non adoption of appropriate nutrient management practices. According to the Liebig first law the yield of the field crop reduces in exact proportion to the reduction in the quantity of minerals applied.

Table 1 : Average vegetable productivity scenario

Crop	**India (Average)**	**World (Average)**	**Potential productivity**	**Maximum productivity**
Tomato	17.08	26.69	60-80	70.45 (USA)
Eggplants	16.08	17.48	40-50	34.7(Japan)
Chilli	9.18	14.4	30-40	44.5 (Spain)
Okra	9.59	6.47	15-20	17.78 (Jordan)
Peas	9.14	8.35	18-20	20 (Lithuania)
Melons	20.48	20.95	30-40	45.83 (Cyprus)
Cucurbits	9.72	12.97	25-30	41.33 (Israel)
Cucumber	6.67	16.98	40-50	67.67 (Korea)
Watermelon	12.75	27.13	30-40	40.96 (Spain)
Cabbage	21.43	21.1	30-40	42.59 (Japan)
Cauliflower	17.14	18.36	35-40	45.25 (New Zealand)

For better utilization of resources and to produce good crops with less expenditure, integrated nutrient management is the best approach. All the possible sources of nutrients are applied based on economic consideration and supplemented with organic, inorganic and biofertilizers. The contribution from crop residues is generally ignored. To realize higher yields with good quality of the crops, 17 essential elements are required. The micronutrients like Zinc, Iron, Manganese, Copper, Molybdenum and Boron could be made available to the plants either by soil application or by foliar sprays. They play a major role in plant metabolic activities like chlorophyll syntheses, photosynthesis of enzymes and hormones, cell division and other physiological functions.

Iron Acts

Iron acts as a catalyst in formation of chlorophyll. It is required for electron transport in photosynthesis. It also acts as an activator for several enzymes and is necessary for almost all metabolic functions of plant directly or indirectly. Whereas, manganese is an essential micronutrient, which is responsible for synthesis of chlorophyll and many other important enzyme. Zinc is involved in enzymes in plant metabolism. Deficiency of zinc reduces the auxin levels in the plant. It is involved in protein syntheses. Boron plays an important role in cell wall biosynthesis and structure and plasma membrane integrity. Copper is very essential for chlorophyll formation and many other plant enzymes. The molybdenum affects the activity of several enzymes in the plant system; it promotes nitrogen metabolism and synthesis of proteins in plants. The farmers of our country are applying nutrients in vegetables (only NPK through chemical fertilizers) in unbalanced manner. This leads poor yield of the vegetables, lower down the quality and also deteriorates soil. Nitrogen from natural and commercial sources is vital to plants and animals. It is the main nutrient required for growth in the plants and for building protein in the animals. It biologically fixes the atmospheric nitrogen symbiotically on the roots of the leguminous plants and leads enormous contribution to productivity. In poor soils where alternate sources of fertilizer are either unavailable or unaffordable, biological nitrogen fixation is vital to crop production. Approximately140-170 million tones of nitrogen valued US$90,000 million are fixed every year by microorganisms every year. The living microbes, fungi and invertebrates found in the soil are responsible for decomposing carbon and nitrogen and making them available for plant growth while at the same time contributing the rate of production and consumption of carbon dioxide, methane and nitrogen. Although the P requirement is not very high for any particular crop but plants require more P for better growth in onion, potato , tomato and cauliflower. K required in high quantities by root crops like carrot, turnip and garden beet, tuber crops and bulb crops.

1. Integrated nutrient management

The basic concept underlying the principles of integrated nutrient management is the maintenance, and possible improvement, of soil fertility for sustaining crop productivity on a long-term basis. Sustained productivity may be achieved through the combined use of various sources of nutrients, and by managing these scientifically for optimum growth, yield and quality of different crops, in a way adapted to local agro-ecological conditions. In vegetable production in Asian countries, farmers have been using organic manures for centuries, together in recent decades with chemical fertilizers, to meet the nutrient demands of crops. Integrated nutrient use has assumed great significance in recent years in vegetable production, for two reasons. Firstly, the need for continued increases in per hectare yields of vegetables requires that applications of nutrients increase. Not enough chemical fertilizer is available to meet crop nutrient requirements. Secondly, the results of a large number of experiments on manures and fertilizers conducted in several countries reveal that neither chemical fertilizers alone, nor organic sources used exclusively, can sustain the productivity of soils under highly intensive cropping systems (Singh and Yadav 1992). Subbiah *et al.* (1985) obtained higher yields of tomato and eggplant with combined use of FYM and fertilizers. For eggplant, applications of 100 kg N/ha, half in urea (50%) and half in poultry manure (50%), resulted in higher yields (45.8 mt/ha) than the same level of nitrogen applied in urea alone (37.8 mt/ ha). The integrated use of urea and poultry manure also resulted in a higher nutrient uptake (Jose *et al.* 1988). Jablonska (1990) reported that the combined use of rye straw and nitrogen resulted in higher yields of tomato, eggplant and pepper than either N fertilizer or FYM used alone. Hosmani (1993) also reported higher yields of chili with integrated use of chemical and organic fertilizers than with the use of either of these separately.

Integrated Nutrient Management (INM) in vaegetables refers to maintenance of soil fertility and plant nutrient supply to an optimum level for sustaining the desired crop productivity through optimization of the benefits from all possible sources of plant nutrients in an integrated manner. Another important aspect of INM is the enhancing of the fertilizer use efficiency (FUE) by proper placement of fertilizer in close proximity to the rhizosphere of the highest root activity. Integrated Nutrient Management has become one of the common practices among progressive horticulture producers today

Integrated nutrient management is one of the recent methods of supplying nutrients to the plants by organic as well as inorganic means together to fulfill the nutrient requirements. At the same time the main aim of integrated nutrient management is to minimize the use of chemical fertilizers without sacrificing the yield. Application of organic manures improves the soil productivity, soil structure and moisture holding capacity of the soil. Organic manures are the key characteristics including the long term fertility of soil by maintaining organic matter levels, fastening soil biological intervention, nitrogen fixation as well as effective recycling of organic materials including crop residues and live stock wastes and weed. Micro - organisms play an important role in various chemical transformations in soil and thus, influence the availability of major nutrient like N, P, K. and S to the plant. A few microorganisms such as nitrogen fixing bacteria and phosphate solubilizers can be utilized to partially augment the supply of major nutrients Rhizobium, Azotobactor, Azospirllum, Blue Green Algae and Phosphate Solublizing Bacteria (PSB) can be used as biofertilizers increases the crop production.

The main objective of Integrated Nutrient Management (INM) is to ensure adequate availability of quality fertilizers to farmers through periodical demand assessment and timely supply, promoting integrated nutrient management, which is soil test-based judicious and balanced use of chemical fertilizers in conjunction with organic manures and bio-fertilizers, promotion of organic farming and ensuring quality control of fertilizers through implementation of Fertilizer

2. Concept of integrated nutrient management

Integrated nutrient management means the supply of nutrients to the plants from various sources (organic and inorganic). According to Ange (1991) the Integrated Nutrient Management is concept which aims to adjust or maintain the soil fertility and plant nutrient supply to an optimum level for sustaining the desired crop productivity through optimization of benefit from all possible sources of plants nutrients in an integrated manner. The main objective is to efficiently utilize all the sources of plants nutrients to adequately take care of nutrient needs of plants for optimum biological productivity and economic gain to the farmers. It must adequately ensure that the processes, the soil and the medium on which plant are grown. It must be relevant to soil, crops, system and environment and relevant to the local situations, social, geographical, economic situations and farm/ farmer specific.

3. Approaches to integrated nutrient supply

Vegetables are very exhaustive to the nutrients. To supply the nutrient to the plants neither organic nor inorganic alone able to sustain the soil fertility and crop productivity. The Integrated nutrient management has proved superior to the use of its component separately. A system approach to supply nutrients to the plants from all possible sources (organic, inorganic and biofertilizers) is a judicious and synchronous way to maintain soil fertility and crop productivity is the essence of Integrated nutrient management.

4. Possible sources of integrated nutrient management

The main sources of nutrient supply in vegetables are:

- **Chemical Fertilizers** : Urea, SSP, MOP, etc.
- **Organic sources** : FYM, vermicompost, NADEP compost, Biodyanamic compost, poultry manure, pig manure, biogas slurry, rural urban compost, press mud, sugarcane filter cake comopost, sea weed, spent mushroom, carpet waste, leaf compost and other compost from plant and animal origin.
- **Biofertilizers** : Products containing living cells of different types of micro-organisms.
- **Gareen manuring**: Different types of legume crop like Sesbania, Sunhemp *etc.*

5. Nutrient uptake

The amount of nutrients taken up by these crops depends on the number of fruit and the amount of dry matter produced. This in turn is influenced by a number of genetic and environmental variables. In tomato yielding 38 mt/ha fruit removes 104 kg N, 9.5 kg P and 116 kg K from the soil. Varieties which take a long time to mature require more nutrients than short-duration ones, mainly because of their higher production of dry matter and fruit. Hegde and Srinivas (1990) studied NPK uptake in tomato in soils with different levels of soil potential and applied N. They observed that nutrient uptake declined with increasing soil moisture stress, and increased with higher levels of N application. They reported that a crop yielding 60.8 mt/ha of fruit removed 147.8 kg N, 19.8 kg P and 156.2 kg K. Large variations between tomato varieties in N uptake were reported by Chakraborty *et al.* (1990). They reported that varieties "Pusa Early Dwarf" absorbed 20.8, 87.6, 421.9, and 672.2 and mg N/plant at 6, 33, 47 and 58 days after transplanting (DAT). In contrast, Pusa Ruby absorbed 14.2, 69.5 308, and 893.2 mg N/ plant. They further observed significant differences in N use efficiency between varieties, in terms of dry matter produced per unit of N absorbed. The highest concentrations of NPK in tomato were found in the fruit, and the lowest in the roots (Maestrey et al. 1987). Eggplant is a long duration crop, with high yields which remove large quantities of plant nutrients. An eggplant crop yielding about 60 mt/ha of fruit removes 190 kg N, 10.9 kg P and 128 kg K. Nutrient uptake in eggplant partly depends on the source of nutrients (Jose et al. 1988). Integrated use of both organic and inorganic sources results in higher uptake and increased fruit production. Pepper needs to absorb more nutrients than tomato or brinjal to produce a unit of dry matter or fruit yield. Concentrations of NPK are highest in the leaf, followed by that in the fruit and the stem (Hegde 1989). However Ca and Mg contents are highest in the leaf, followed by those in the stem and fruit.

Nutrient uptake and dry matter production (fruit yield) are closely related (Hegde 1988). A crop yielding 18.02 mt/ha of fruit removed 55.5 kg N, 13.2 kg P, 73.1 kg K, 22.3 kg Ca and 20.9 kg Mg . According to the data on nutrient uptake from different studies, to produce one ton of fresh fruit, plants need to absorb 2.5 - 3 kg N, 0.2 - 0.3 kg P and 3 - 3.5 kg K in the case of tomato; 3 - 3.5 kg N, 0.2 - 0.3 kg P and 2.5 - 3 kg K in the case of eggplant; and 3 - 3.5 kg N, 0.7 - 1 kg P and 5 - 6 kg.

5.1 Growth and nutrient uptake

The dry matter accumulationin vegetables during the initial 30 days after transplanting (DAT) is low, less than 5% of the total dry matter produced by the end of the growth. Later, there is an almost linear increase in dry matter production up to 90 DAT. It then slows, and during the final stages of the growth there is decline in dry matter, due to leaf fall. The rate of dry matter accumulation in the stem and fruit continues to increase until the crop reaches full maturity. The proportion of dry matter distributed in fruits ranged from 51% in crops without N fertilization, to 39% in crops which had recieved 240 kg N/ ha (Hegde and Srinivas 1989a). Dry matter production and nutrient uptake are very closely related. During the four months after transplanting, about 5%, of total nutrient uptake will be achieved by 30, DAT, 12-15% by 45 DAT, 35-40% by 60 DAT, 60- 65% by 75 DAT, 85-90% by 90 DAT, and 95% by 105 DAT (Hegde and Srinivas, unpublished data). Thus, about 50% of the total nutrient uptake takes place between 60 and 90 DAT, a period coinciding with peak fruit development. In the case of pepper, dry matter production continues to the end of the life cycle (Hegde 1987a). Growth in terms of dry matter production is very slow until 30 DAT. It then picks up between 45 and 105 DAT, later slowing down, mainly due to a reduction in leaf dry matter from leaf fall. In this crop also, nutrient uptake and dry matter production are closely related. Around 5, 35 - 40, 75 - 80 and 90% of total nutrient uptake was achieved by 30, 60, 90 and 105 DAT. Thus, about 40% of nutrient uptake takes place during a period of 30 days, between 60 and 90 DAT (Hegde, unpublished data). In tomato, the period when plants have the greatest requirement for K, N, Ca and P is just before the fruit begin to ripen. In tomato, 28 - 45% of the total nutrient uptake was during the night.

5.2 Partitioning of nutrient uptake

Generally, the proportion of total nutrients found in the fruit declines with an increase in the level of nutrients applied. Hegde and Srinivas (1990) partitioned total nutrient uptake by tomato into different plant parts, and found that 45.8 - 59.2% N, 56.5 - 63.6% of P and 62 - 69.6% of K was partitioned into the fruit. The stem contained the lowest proportion of N and P, while the leaf contained the lowest proportion of K. A similar trend was observed except that the stems contained a higher proportion of P and K than the leaves. Concentrations of P, K, and to a lesser degree N, tend to decrease with age in the vegetative tissues of tomato. Maestrey et al. (1987) observed the highest and lowest concentrations of N, P and K in the fruit and roots, respectively. Shakhazizyan (1989) reported that N concentration declined with age, while P increased initially and then declined as the fruit ripened. A small proportion of the N, and an even smaller proportion of the P and K, which has accumulated in the leaves and stem tends to be translocated into the fruit.

6. Nutrient management

The quantity of nutrients which the farmer needs to apply depends on the yield potential of the cultivar, the level of available plant nutrients already in the soil, and growth conditions. Since vegetative and reproductive stages overlap in this group of crops, they need a continuous and steady supply of nutrients throughout their life span. It is necessary to adopt appropriate nutrient management practices which help to supply nutrients in quantities adequate to just meet crop demand and minimize losses, thereby increasing the nutrient use efficiency. Such practices will be environmentally friendly, and lead to sustainability in vegetable production.

6.1 Application of nutrients in splits

Application of N in four splits at 30-day intervals has been recommended to achieve maximum yields and profits in chilli production. Subhani et al. (1990) obtained the highest yield of chili when both N and K were applied

6.2 Application of nutrients through fertigation

An application of 50% of total NPK in trickle irrigation resulted in a higher nutrient uptake in tomato than when the same NPK was applied before. Goyal *et al.* (1985) reported that fertigated pepper, tomato and eggplant receiving 9, 18 or 30 g urea/ plant all had higher yields than plants receiving a side application of urea (15 + 15 g/plant at planting and first harvest). Fertigation with acidifying N fertilizers such as urea may sometimes have an adverse effect on growth and productivity. Mulching, especially when an overhead irrigation system was used, considerably increased the total recovery of applied nitrogen in tomato (Sweeny et al. 1987).

6.3 Application of nutrients through Slow-release fertilizers

Slow-release fertilizers hold great promise for the production of vegetables such as eggplant and tomato low-release fertilizer (Plantacote) and conventional fertilizers at 100:80:90:30 kg/ha N P K Mg. They have found that slow-release fertilizers produced 92 mt/ha of tomato, compared to only 42mt/ha when ordinary commercial fertilizers were used.

7. Nutrient Management Studies Conducted at IIVR, Varanasi

7.1 Use of major nutrients (NPK)

Studies conducted on the response of vegetable crops to fertilizer use revealed that the application of 180 kg N/ha and 120 kg P_2O_5/ha gave maximum yield (503 q/ha) and C:B ratio (1:2.07) in tomato hybrid ARTH-3 and thus, has been recommended for Varanasi conditions. In capsicum hybrid Bharat, application of 180 kg N /ha and 120 kg P_2O_5/ha resulted into maximum yield (328 q/ha) and C:B ratio (1:5.15). For French bean variety Arka Komal, fertilizer dose of 120 kg N and 60 kg P_2O_5/ha has been recommended for maximum yield (96.5 q/ha) and CB ratio (1:1.38).

7.2 Use of liquid fertilizers

Experiments conducted on the efficacy of liquid fertilizers in vegetable crops showed that five foliar sprayings of water-soluble fertilizer NPK (19:09:19) @

0.5% at 10 days interval after 30 DAT over and above the recommended dose of NPK (150:80:100 kg/ha) resulted into maximum yield of tomato hybrid cv. Tolstoi (745.12 q/ha) and C:B ratio (1:4.12) at IIVR, Varanasi. In capsicum cv. Indra highest yield (82.10 q/ha) and C:B ratio (1:4.08) were obtained with 5 foliar sprays of water soluble liquid fertilizer having a combination of NPK – 19:09:19 at 10 days interval after 40 DAP. Similar results have been obtained in cauliflower cv. Snowball-16 also with five foliar sprays of water soluble fertilizers (NPK-19:19:19) at 10 days interval after 40 DAP (IIVR, Ann. Rep. 2000-2010).

7.3 Response to micronutrients

In view of the growing deficiency of micronutrients on account of imbalanced and inadequate use, negligible application of organic manures and excessive mining of these elements from the soil reserves, use of micronutrients has become inevitable for sustaining higher crop yields as well as the quality of the produce. Tolstoi revealed that maximum yield (827.85 q/ha) and C:B ratio (1:4.81) were recorded with 3 foliar application of micronutrients (mixture of B, Zn, Cu, Fe, Mn each @ 100 ppm and Mo @ 50 ppm) at 10 days interval 30 DAT over and above the recommended dose of NPK (150:80:100 kg/ha) Chaurasia *et al.(*2006). In brinjal, foliar spray of Zn, Mo, and B @ 50 ppm at three critical crop growth stages i.e. at active growth stage, at flowering and at fruiting stage increased the yield by 12-39% over control. In cauliflower cv. Snow Ball-16, foliar spray of Zn @ 50 + Mo @ 25 ppm thrice at 10 days interval from 30 days after transplanting significantly increased the yield and quality over control. In case of chilli, spray of 100 ppm of commercial formulation of micronutrients has been found suitable for higher yield of better quality seeds. In French bean cv. Swarn Priya, three foliar sprays of Zn, Cu and B @ 50, 5 and 5 ppm respectively, at 30 days after sowing at 10 days intervals gave significantly highest pod and seed yield over control.

Fig. *Lycopenseion esculanteem*

7.4 Integrated nutrient management

The imbalanced and indiscriminate use of chemical fertilizers coupled with negligible application of organic manures have resulted into multinutrient deficiencies, deteriorated soil physical, chemical and biological health and environmental pollution. Under such situations, adoption of integrated nutrient management practices is being emphasized to boost the agricultural productivity and environmental sustainability.

The studies conducted at IIVR, Varanasi revealed that application of recommended doses of NPK (120:60:60) + FYM @ 10 t/ha + Sulphur @ of 25 kg / ha + Azotobacter + mixture of all micro nutrients (Zn, B, Mo, Fe, Cu and Mn) resulted in the maximum yield (413.83 q/ha) as well as highest C: B ratio (1:3.65) in tomato cultivar H-86. In case of tomato hybrid Avinash-2, use of 20 t FYM + ½ recommended dose of NPK (150:80:100 kg/ha) was found more effective in maximizing the yield (773. 0 q /ha) and C: B ratio (1:4.00) Chaurasia *et al.(*2001)..

Application of 120 kg N + 60 kg P_2O_5 +60 kg K_2O + 5 t/ha press mud and root dipping treatment with Azotobacter as well as foliar spray of S @ 20 ppm at 30, 45 and 75 days after transplanting significantly increased the yield of tomato hybrid Avinash-2 over control Chaurasia et al. (2002). In another study, application of Pressmud @ 5 t/ha + rest NPK through chemical fertilizers gave maximum yield and C:B ratio in brinjal cv. IVBL-9 and okra cv. VRO-6. In case of chilli cv. LCA-235, application of 75% recommended dose of N (150 kg/ha) + Azospirillum as seed treatment, seedling dip and soil incorporation gave maximum yield of green (117.52 q/ha) along with maximum C:B ratio (1:1.77) followed by application of 50% recommended dose of N + Azospirillum. Application of poultry manure@ 5t/ha + 50% of the recommended dose of NPK resulted in the maximum yield of capsicum var. Indra (142.3 q/ha)with average fruit weight (130.5 g), fruit length (9.3 cm) and fruit width (7.3 cm) Chaurasia et al. (2002).

Studies conducted on bottle gourd variety Kashi Ganga (IVRBTG-2) involving three different organic sources alone and in combination with half of the recommended dose of NPK, *Azospirillum* and micronutrient mixture revealed that the average fruit weight (1095.72 g), number of fruits /plot (76.65) and yield (335.95 q/ha) were maximum with Vermicompost @ 2.5 t/ha + ½ NPK + micronutrient mixture.

In cucumber, application of half NPK (60:30:30 kg/ha) + FYM @ 10 t/ha + Biofertilizer has been found effective for higher yield and cost: benefit ratio. In garden pea cv. Azad P-3, application of FYM @ 10 t/ha + half of the recommended dose of NPK resulted into maximum yield (80.7 q/ha) under Varanasi conditions.

For carrot cv. Early Nantes, application of half of the recommended dose of NPK (80:60:60 kg/ha) + vermicompost @ 2t /ha + biofertilizer resulted into maximum fruit length (23.6 cm) and yield (327.7 q/ha). In broccoli hybrid Fiesta, integrated use of poultry manure @ 5 t/ha + half of the recommended dose of NPK gave maximum head yield (366.00 q/ha) over 100% NPK alone (IIVR Ann Rep. 2005-06).

An experiment comprising four cauliflower hybrids and four nutrient sources i.e., FYM, sewage sludge, press mud and inorganic fertilizers alone and in combination with half of the recommended dose of NPK (120:60:60 kg/ha) revealed that the maximum yield (389.3 q/ha) was recorded under press mud applied @ 20 t/ha. Among the cauliflower hybrids, the maximum yield was noted under Amazing (367.8 q/ha) Chaurasia, *et al*. (2008), Chaurasia, et al. (2009), and Rakesh Singh et al. (2007).

In cowpea variety Kashi Kanchan (IVRCP-4) the maximum yield (150.41 q/ha) was recorded with the application of FYM @ 10t/ha + NPK (30:30:30 kg/ha) + PSB.

7.5 Quality of vegetable crops in relation to nutrient management

The quality of fruits in terms of acidity and vitamin C improved in tomato cv. H-86, DVRT-1, DVRT-2 and Sel-7 under organic farming. Similarly, soil application of Azotobacter @ 15 kg/ha along with NPK @ 150:60:80 kg/ha influenced the pericarp thickness and shelf life in tomato hybrids. In broccoli, application of FYM + digested sludge (each @ 10t/ha) and seedling inoculation with VAM or PSB significantly improved the carotenoid content. Vitamin C content significantly improved in broccoli by application of FYM or digested sludge @ 20 t/ha.

In Chinese cabbage cv. Solan Band Sarson, vitamin C content increased by 36.5% with the application of digested sludge @ 20t/ha over control (recommended NPK). In case of FYM application @ 20 t/ha, this increase was to the tune of 43.8%. Application of organic manures and biofertilizers independently or in combination improved the total carbohydrate (5.6%), vitamin C (22.5%) and total carotenoids (11.6%) in pea over the use of recommended NPK fertilizers. In lettuce cv. Great Lake, a significant improvement in vitamin C content (9.25%) was noticed with the combined application of FYM and digested sludge @ 10 t/ha each.

8. Nutrient management studies conducted in other parts of the country

The salient findings of the nutrient management studies conducted at different centers of All India Coordinated Research Project on Vegetable Crops are summarized below (AICRP Ann. Rep. 2000-2010):

8.1 Tomato

8.1.1 Use of major nutrients (NPK)

Application of nitrogen @240kg/ha and P_2O_5 120 kg/ha has been recommended for tomato hybrid ARTH-3 under Kanpur conditions. For variety Pusa Ruby, Sioux and KS-2, NPK @ 150:60:60 kg/ha has been recommended to get maximum return and highest cost/benefit ratio. For Hisar conditions, application of nitrogen @180kg/ha and P_2O_5 @60kg/ha is recommended for obtaining highest yield (372q/ha) and C:B ratio (1:4.9) in the same hybrid. Application of N @ 180 kg/ha and P_20_5 @ 120 kg/ha resulted in maximum C:B ratio (1:2.99) in determinate hybrid ARTH-3 whereas use of NPK@ 150:60:60/ha has been found suitable for varieties HS-101 and Pusa Ruby in Sabour and for Pusa Ruby, Pusa Early Dwarf and Arka Vikas varieties of tomato under Bhubaneshwar conditions. The varieties Pusa Ruby (indeterminate), Punjab Chhuhara (semi-determinate) and Pusa Early Dwarf (determinate) under Jorhat conditions responded to NPK use up to 75:60:60 kg/ha.

8.1.2 Use of liquid fertilizers

At Durgapura, the maximum yield of tomato cv. Pusa hybrid-2 (486.9 q/ha) and C:B ratio (1:1.5) were recorded with 5 foliar sprays of water-soluble fertilizer NPK (15:15:30 @ 0.5%) at 10 days interval over and above to the recommended dose of NPK (120:80:60 kg/ha). Similar response was observed in tomato cv. Marutham at Hyderabad also. At Kalyanpur, Kanpur 5 foliar applications of water soluble fertilizer NPK (17: 10:27) at 10 days interval resulted in maximum yield (241.5 q/ha) and C:B ratio (1: 2.97) in tomato cv. Type –1. At Jabalpur, the maximum yield of tomato (269.28 q/ha) along with the C:B ratio (1:2.08) were obtained with 5 sprays of Multi K 13:0:45 applied at 10 days interval after 40 days transplanting. Under Sabour condtions, the highest yield (494. 11q/ha) and C:B ratio 1: 2.13 of tomato was recorded with the five foliar sprays of NPK (15:15:15) @ 5g / litre in addition to recommended dose of NPK. At Coimbatore, the highest mean fruit yield (712.50 q/ha) and the maximum C: B ratio (1: 4.78) were recorded in hybrid tomato with five foliar sprays of NPK (19:19:19).

8.1.3 Application of micronutrients

Three foliar sprays of Borax @ 100 ppm over and above to the recommended dose of NPK at 10 days interval 30 DAT at Hisar gave maximum yield (372.8 q/ha) and C:B ratio (1:4.13) in cv. Hisar Arun. At Pantnagar, maximum yield (314 q/ha) and C:B ratio (1:2.15) along with high TSS (5.7%) and shelf life (7.6 days) were recorded in Pusa Hybrid –1 tomato with three foliar sprays of micronutrient mixture (B, Zn, Cu, Fe, Mn, each @ 100 ppm and Mo @ 50 ppm) at 10 days interval starting from 40 days after transplanting. 1under Coimbatore conditions, foliar application of ZnSo4 at 100 ppm concentration thrice starting from 40 days after transplanting resulted in the maximum yield (645.6q/ha) along with C:B ratio (1:4.86) in case of tomato hybrid – 1. Under Durgapura conditions, three foliar applications of Ferrous sulphate @ 100 ppm at 40, 50 and 60 DAT gave the maximum mean yield of 454.11 q/ha along with highest C:B ratio 1:2.44 of tomato cv. Pusa Hybrid-2.

8.1.4 Integrated nutrient management

At Hisar, application of recommended dose of NPK + FYM @ 10 t/ha + S @25 kg/ha + mixture of all micronutrients (Zn, B, Mo, Fe, Cu, Mn) + *Azotobacter* gave maximum yield (432.3 q/ha) and C: B ratio (1:2.66. The same treatment recorded the highest yield (336.20 q/ha) and C: B ratio of 1:2.63 under Kalyanpur conditions also.

At Faizabad, the highest mean yield (368.35 q/ha) and C.B. ratio (1:3.02) was recorded in tomato var. Narendra Tomato-6 with the application of NPK @ 120:60:60 kg / ha + FYM 10 t/ha + Sulphur 25 kg/ha + *Azotobactor* + MM (mixuture of all micronutrients, Zn, B, Mo,Fe, Cu, & Mn). Whereas in case of

Narendra Tomato-2, green manuring + recommended dose of NPK (60:30:30 kg/ha) was found more suitable under the same conditions.

The highest yield (346.87 q/ha) and C:B ratio (1:2.35) was obtained with the application of FYM @ 20 t/ha + full recommended dose of NPK (150-60-60 kg/ha) in tomato cv.S-7 at Sabour. At Jaipur, use of FYM @ 40 t/ha + full recommended dose of NPK (180:120:80 kg/ha) in tomato hybrid ARTH-3 gave maximum yield (329.69 q/ha) and C:B ratio (1:1.49). Under Pantnagar conditions, application of PSB + recommended dose of NPK (150:90:60 kg/ha) gave the highest average yield of tomato i.e., 653.4 q/ha along with the maximum C:B ratio 1:3.19. At Kalyanpur, application of Azospirillum with 75% N and 100% PK gave the highest average yield 267.44 q/ha of tomato var. TYPE-1 with the maximum C:B ratio 1:2.20 . Root dipping of tomato seedlings with Azotobacter along with 75 per cent N + 100 per cent PK recorded highest fruit yield (635.38 q/ha) and C:B ratio (1:3:26) under Srinagar conditions. Under Coimbatore conditions, seedling root dip with the Azospirillum in addition to 75% N + 100% PK recommended resulted in highest yield 427.2 q/ha along with the maximum C:B ratio of 1:4:11 in tomato. At Jabalpur, maximum yield (310.68 q/ha) and C:B ratio (1:2.58)of cv. Jawahar Tomato-99 was obtained with the application of 20 t/ha FYM and full dose of N:P:K (180:120:80 kg/ha)through fertilizers.

8.2 Brinjal

8.2.1 Use of NPK fertilizers

In variety Pusa Purple Long, application of 150 kg/ha of nitrogen and 100 kg/ha of phosphorous (P_2O_5) were found optimum at IIHR, Babgalore. Fertilizer dose of 100 kg N + 60 Kg P_2O_5/ha has been recommended for varieties H-4 and Pusa Purple Long under Hisar conditions. For Pantnagar region, application of 50 kg N /ha as basal dose has been recommended for the variety Pusa Purple Long, while for PBR 129-5, 100 kg N /ha as basal dose has been recommended. Application of 120 kg N /ha in variety BB-1 has been recommended for Bhubneshwar conditions.

8.2.2 Use of liquid fertilizers

At Hisar, 3 foliar sprays of water soluble fertilizer NPK (19:19:19 @ 0.5 % at 10 days interval after 30 DAT over and above to the recommended dose of NPK) in brinjal cv. Hisar Shyamal gave maximum yield (259.8 q/ha) and C:B ratio (1:2.89).

The highest yield of brinjal cv. Pusa hyrid-6 (329.0 q/ha) along with C:B ratio of 1:3.20 were recorded with 5 foliar sprays of water-soluble fertilizer NPK (17-10-27 @ 0.5 %) at 10 days interval after 30 DAT over and above to the recommended dose of NPK-150:90:90 kg/ha under Pantnagar conditions. At Hyderabad, maximum yield of brinjal (365.6 q/ha) with C:B ratio (1:3.8) were obtained with 5 sprays of water soluble fertilizer having the combination of NPK 15 : 15 :30. At Coimbatore foliar application of water soluble fertilizer NPK (19

:19: 19) 5 times at 10 days interval starting from 40 days after transplanting resulted in maximum yield (556 q/ha) and C:B ratio (1:5.05) in Brinjal Hybrid –1 (COBH-1).

Under Faizabad conditions, in addition to the recommended dose of NPK (150:80:80) five foliar applications of water soluble NPK fertilizer (19:10:27) after 30 days of transplanting at 10 days interval resulted in maximum yield (395 q/ha) and highest C;B ratio (1:3.53) of brinjal hybrid Suchitra. At Vellanikkara, 5 foliar applications of Multi-K at 10 days interval beginning 30 DAP resulted maximum yield (496 q/ha) and C:B ratio (1:1.6) in brinjal hybrid Neelima.

8.2.3 Application of micronutrients

Spray of 100ppm ferrous sulphate on brinjal under Kanpur conditions is recommended for higher yield of better quality seeds.

Foliar application of a mixture of micronutrients (i.e. 100ppm each of Zinc sulphate, manganese sulphate, copper sulphate, ferrous sulphate, boric acid and 50ppm ammonium molybdate) has been found optimum for higher seed yield of brinjal in Maharashtra.

8.2.4 Integrated nutrient management

At Jorhat application of FYM @ 10 t/ha + rest NPK through fertilizers resulted in higher yield of brinjal (244.5 q/ha) as well as okra (131.5 q/ha) along with maximum C:B ratio (1:3.35) for the whole cropping system. Application of neem cake @ 5 q/ha and rest of the recommended NPK through chemicals fertilizer gave the highest mean yield of 415.86 q/ha along with the maximum C:B ratio of 2.34 in brinjal at Kalyanpur. At Bhubaneswar, application of FYM @ 10 t/ha and recommended doses of NPK (125 : 50 : 75 kg/ha) gave the highest fruit yield of brinjal (229. 36 q/ha) along with the maximum C:B ratio of 1: 4.05. At Faizabad, the maximum mean yield of brinjal i.e., 269.55 q/ha and C:B ratio 1:1.57 was recorded with the application of neem cake @ 5 q/ha + recommended dose of NPK through chemical fertilizers.

8.3 Chilli & Capsicum

8.3.1 Use of NPK fertilizers

Application of NPK @ 90:60:40 kg/ha has been recommended for getting economic red ripe yield in variety Pant C-1 under agro-climatic conditions of Faizabad. Application of 120 kg N, 30 kg P_2O_5 and 10 kg K_2O/ha has been recommended for chilli Jawahar – 219 under Jabalpur conditions.

Application of 120 kg N and 60 kg P_2O_5/ha has been recommended for variety Pusa Jwala under Kanpur conditions. For Tarai region of U.P , application of NPK @ 120:60:60 kg/ha resulted in the highest yield (82.64 q/ha) from variety

Pant C-1. For Durgapura conditions, application of NPK @ 90:60:60 kg/ha has been recommended for getting the maximum yield from variety Pusa Jwala. Application of 240 kg N, 60 kg P_2O_5 and 60 kg K_2O/ha under Hyderabad condition gave maximum dry chilli yield (16.42 q/ha) of cv G-4 and cost benefit ratio (1:1.92). For Coimbatore conditions, application of 180 kg/ha nitrogen and 120 kg/ha P_2O_5 is recommended for the highest yield (63.7 q/ha) and C:B ratio (1:3.98) in the same hybrid. Application of Nitrogen @ 100kg N/ha is recommended for obtaining maximum seed yield in Paprika under Solan conditions with a cost: benefit ratio of 1: 3.08. Application of 200 kgN/ha has been recommended to obtain highest seed yield of 366 kg/ha of Paprika (KTPL-19) under Srinagar conditions.

8.3.2 Use of liquid fertilizers

At Pantnagar, the highest yield of capsicum hybrid Bharat (167.45 q/ha) along with C:B ratio (1:1.52) were recorded with three foliar sprays of water soluble fertilizers (15:15:30 @0.5%) at 10 days interval starting from 30 days after transplanting. At Srinagar maximum yield (229 q/ha) of capsicum cv. Nishat-1 with C:B ratio of 1:2.31 was obtained with three sprays of water soluble fertilizers having the combinations of NPK 17:10:27.

8.3.3 Application of micronutrients

At Kalyanpur three foliar sprays of mixture of micronutrients (B, Zn, Cu, Fe, Mn each @ 100 ppm and Mo @ 50 ppm) at 10 days interval starting from 40 days after transplanting. resulted in maximum yield (83.89 q/ha) and C:B ratio (1:3.43) in capsicum cv. California Wonder. Foliar application of mixture of all micronutrients (i.e. 100ppm Zinc sulphate, Manganese sulphate, Copper sulphate, Ferrous sulphate, Borax & 50ppm Ammonium molybdate) at ten days interval starting from forty days after transplanting is recommended in Capsicum var. Nishat-1 under Srinagar conditions. Foliar application of a mixture of micronutrients (i.e. 100ppm each of Zinc sulphate, manganese sulphate, copper sulphate, ferrous sulphate, boric acid and 50ppm ammonium molybdate) is recommended for higher seed yield of Chilli in Karnataka. For enhanced seed production of bell pepper cv. California Wonder, spray of Borax @ 0.5% at vegetative, flowering and fruit setting stage has been recommended for Solan conditions.

8.3.4 Integrated nutrient management

Application of *Azospirillum* @ 2 kg/ha as basal application in combination with 75% recommended dose of N_2 (i.e. 56 kg/ha), has been recommended for Tamil Nadu conditions. At Kalyanpur, the maximum yield (88.38 q/ha) of red ripe chilli cv. Azad Mirch-1 along with C:B ratio (1:2.91) was recorded with the application of 100% recommended dose of NPK (120:60:80 kg/ha) + Azospirillum.

8.4 Garden Pea

8.4.1 Use of NPK fertilizers

A fertilizer dose of 25 kg N + 60 kg P_20_5 + 40 kg K_20 /ha was found optimum for seed production of pea cv. Arkel in Himachal Pradesh. For Jabalpur conditions, application of 45:90:60 kg NPK/ha is recommended for higher seed yield.

8.4.2 Integrated nutrient management

At IIHR, Bangalore, the garden pea variety Arka Kartik recorded the highest yield with the application of 10 t/ha FYM plus half dose of N:P:K (20:30:25 kg / ha). At Faizabad, application of neem cake @ 2.5 q/ha plus half of the recommended dose of NPK (15:30:20 kg / ha) recorded the highest mean yield of pod in garden pea cv. Azad Pea-3. At Bhubaneswar, the integrated application of poultry manure @ 2.5 t/ha + half of the recommended dose of NPK (25:37.5:25 kg/ha) has been recommended.

8.5 Okra

8.5.1 Use of NPK fertilizers

A fertilizer dose of 150 kg each of N, P_2O_5 and K_2O /ha during rainy season for variety Pusa Sawani has been recommended for Jabalpur conditions.

8.5.2 Use of liquid fertilizers

In okra cv. Aruna, maximum yield (161.4 q/ha) and C:B ratio (1:1.3) were obtained with 3 foliar sprays of water soluble fertilizer having the NPK formulation of 19:19:19 at Vellanikkara. At Hyderabad, maximum yield (78.3 q/ha) and C:B ratio (1:1.68) were obtained in okra cv. Arka Anamika with 5 foliar sprays of water soluble fertilizers having NPK formulation – 15:15:30. At Coimbatore, maximum yield (202 q/ha) and C:B ratio (1:3.82) were obtained with 5 foliar sprays of water soluble fertilizers having NPK formulation – 19:19:19 in okra hybrid Mahyco No.10

8.5.3 Use of micronutrients

Foliar application of a mixture of micronutrients (i.e. 100ppm each of Zinc sulphate, manganese sulphate, copper sulphate, ferrous sulphate, boric acid and 50ppm ammonium molybdate) has been recommended for higher seed yield of Okra in Maharashtra.

8.6 Cauliflower

8.6.1 Use of NPK fertilizers

Application of 150 kg N and 60 kg P_2O_5/ha has been recommended for mid-season variety Pant Shubhra at Pantnagar. For Faizabad conditions, recommendation of 100 kg N, 60 kg P_2O_5 and 60 kg K_2O/ha has been made for the same variety. Whereas, 120 kg N, 60 kg P_2O_5 and 60 kg K_2O/ha has been recommended for this variety under Kalyanpur conditions.

Application of N @ 120 kg/ha, 60 kg each of P_2O_5 and K_2O has been recommended in variety Improved Japanese at Jaipur. For Srinagar conditions, application of 100 kgN/ha has been recommended for cauliflower variety Snowball-16.

8.6.2 Use of liquid fertilizers

At Bhubaneswar, the maximum curd yield (178.03 q/ha) and C:B ratio 1:2.11 of cauliflower was obtained with three foliar sprays of NPK (19:19:19) in addition to recommended dose of NPK.

8.6.3 Use of micronutrients

At Faizabad, the maximum C:B ratio (1:3.15) and yield of cauliflower cv. Pusa Snowball K-1 was obtained with the foliar application of boron @ 100 ppm + Molybdenum @ 50 ppm. At Bhubaneswar, the higher yield of cauliflower (166.6 q/ha) along with maximum C:B ratio 1:4.09 was obtained with the foliar sprays of boron @ 100 ppm in addition to recommended dose of NPK.

At Kalyanpur, the higher curd yield of cauliflower (317.09 q/ha) along with the maximum C: B ratio 1:3.17 were recorded with the soil application of borax @ 10 kg/ha + Ammonium Molybdate @ 2.0 kg/ha.

8.6.4 Integrated nutrient management

The maximum yield of cauliflower cv. Pusa Snowball K-1 (304.5 q/ha) and C:B ratio (1:3.88) were recorded with the application of PSB @ 500 g/ha as seedlings root dip along with recommended dose of NPK through fertilizers at Pantnagar. At Hyderabad, the maximum yield (226 q/ha) along with C:B ratio (1:2.96) were obtained in late Cauliflower with the application of VAM @ 15kg/ha. + recommend dose of NPK. At Hisar, the highest yield (347.8q/ha) and C:B ratio (1:5.53) were obtained with the application of PSB +75% P and recommended dose of nitrogen and potassium in late cauliflower. At Faizabad, the maximum yield (257.75 q/ha) and C;B ratio (1:3.19) were obtained in cauliflower cv. Snowball-16 with the application of *Azospirillum* plus recommended dose of NPK. At Kalyanpur, the application of *Azospirillum* with 75% of N and full dose P and K gave the maximum yield (293.95 q/ha) and C:B ratio (1:1.60) of cauliflower cv. Snowball-16.

8.7 Cabbage

8.7.1 Use of NPK fertilizers

Application of 180 kg N + 50 kg P_2O_5 + 50 kg K_2O/ha has been recommended for variety Pride of India under agro-climatic conditions of Jabalpur. Use of 180 kg N/ha + 60 kg/ha each of P_2O_5 and K_2O has been recommended for variety Pride of India under Kanpur conditions. Application of 120 kg N/ha is recommended for cabbage variety Pride of India under Kymore Plateau and Satpura hill conditions of M.P.

8.7.2 Use of liquid fertilizers

Highest seed yield of 11q/ ha in Cabbage cv Golden Acre was obtained at Srinagar with five weekly foliar applications of poly feed (19:19:19) @ 5 g/l starting from 30 days of transplanting. Foliar application of water soluble fertilizer 19:19:19 at 0.5% concentration is recommended for higher seed yield of Cabbage in Himachal Pradesh.

8.7.3 Use of micronutrients

At Durgapura, 3 foliar sprays of commercial formulation of multiplex @ 100 ppm over and above the recommended dose of NPK at 10 days interval after 30 DAT gave maximum yield (321.7 q/ha) and C:B ratio (1: 0.62) in cv. Bajrang. At Jabalpur, the maximum yield (360.40 q/ha) and C:B ratio (1: 2.75) of hybrid cabbage Krishna was recorded with 3 foliar sprays of $ZnSO_4$ at 100 ppm concentration. At Bhubaneswar, the maximum yield (431.6q/ha) of hybrid cabbage along with C:B ratio (1:1.83) was obtained with 3 sprays of 100 ppm $ZnSO_4$ at 10 days interval starting from 40 days after transplanting. The maximum head yield of 557 q/ha and C:B ratio (1:5.93) were obtained with three foliar sprays of micronutrient mixture (B, Zn, Cu, Fe, Mn, each @ 100 ppm and Mo @ 50 ppm) at 10 days interval starting from 40 days after transplanting in cabbage cv. Golden Acre at Srinagar. The highest yield (400.7 q/ha) along with the maximum C:B ratio (1:2.92) was obtained in cabbage hybrid cv. Meenakshi with the application of Ferrous sulfate @ 100 ppm in addition to recommended dose of NPK (180:60:60 kg/ha) under Hyderabad conditions.

8.7.4 Integrated nutrient management

For getting optimum yield (526 q/ha) and C:B ratio (1:2.09), seed treatment with *Azospirillum* (500 g/ha) + soil application (5 kg/ha) + seedling dipping (1.0 kg/ha) +

application of 60 kg N/ha, has been recommended for variety Pride of India under Solan conditions. Maximum yield (274 q/ha) from the variety Pride of India was obtained with the application of *Azospirillum* (seedling dipping @ 1.0 kg/10 lit. water + soil application @ 5 kg/ha) supplemented with 75 % of the recommended dose of N (180 kg/ha). Hence, these treatments have been recommended for Kaymore plateau and Satpura hills of M. P. Application of Azotobacter along with 75% of the recommended dose of nitrogen is recommended for cabbage variety Pride of India under Srinagar conditions. Application of Azotobacter as seed / seedling treatment soil application of 75% recommended dose of N (140 kg/ha) in cabbage variety Pride of India has been recommended for Tarai conditions of Pantnagar. At Hyderabad, application of Azotobacter + 75% recommended dose of nitrogen can save 25% of nitrogen per hectare.

8.8 Carrot

8.8.1 Integrated nutrient management

At Hyderabad, application of vermicompost @ 2 t/ha + biofertilizer + half of the recommended dose of NPK (25 : 20 : 25 kg /ha) has been recommended in carrot cv. improved Kuroda for Hyderabad conditions. At Durgapura, the maximum yield of carrot i.e., 557.09 q/ha and C:B ratio 1:2.59 was obtained under half recommended dose of NPK (30:20:60 kg/ha) + Vermi compost @ 2 tonnes /ha + biofertilizer. For Faizabad conditions, application of half NPK (40:30:30 kg/ha) + Legume green manure @ 2.5 t/ha + Biofertilizers is recommended in carrot cv. Nantes. Whereas for Bangalore conditions, the application of half NPK (40:30:25 kg/ha) + FYM 10 t/ha + biofertilizer has been recommended in the same variety. At Srinagar, application of FYM @ 10t/ha + Vermicompost @ 2t /ha was found suitable for organically produced of carrot var. Chamman.

8.9 Muskmelon

Application of NPK @ 100:60:60 kg/ha has been recommended for Durgapura Madhu variety under agro-climatic conditions of Durgapura. Similarly, application of NPK @ 100:60:60 kg/ha has been recommended for river bed areas of Faizabad.

8.10 Watermelon

The fertilizer dose of 100 kg N + 60 kg P_2O_5 + 60 kg K_2O/ha has been recommended for getting maximum net return from variety Sugar Baby under river bed conditions of Faizabad. At Sabour, maximum yield (176q/ha) and C : B ratio (1:3.85) were obtained from the variety Sugar Baby with the application of 100 kg N and 60 kg each of P_2O_5 and K_2O.

8.11 Other Cucurbits

8.11.1 Use of NPK fertilizers

In bitter gourd, N:P:K @ 60:60:30 kg/ha has been recommended under Coimbatore conditions. Application of N:P:K @ 90:60:60 kg/ha has been recommended under Faizabad conditions for local variety Jaunpuri Karela. Application of 60 kg /ha each of N, P_2O_5 and K_2O_5 has been recommended in bitter gourd variety Priya for Saurashtra region of Gujarat. For Rahuri conditions, application of 90 kg nitrogen, 60 kg phosphorus and 60 kg potassium/ha in bitter gourd variety Hirkani has been recommended.

8.11.2 Use of micronutrients

At Kalyanpur, the maximum mean yield (161.25 q/ha) of bitter gourd cv. Summer Green and highest C: B ratio (1:2.13) was recorded with the foliar application of mixture of micronutrients (Zn, B, Mo, Cu, Fe, Mn).

8.11.3 Integrated nutrient management

At Kalyanpur, the maximum mean yield (223.18 q/ha) along with the highest C: B ratio of 1:3.68 was recorded in cucumber with the application of half of the recommended dose of NPK + FYM @ 10t/ha + biofertilizer. At Hyderabad, the maximum fruit yield (111 q/ha) and C:B ratio (1:2.10) were obtained in cucumber cv. Poinsette with the application of FYM @ 10 t/ha + biofertilizers in addition to recommended dose of NPK (100:50:50 kg/ha).

9. Conclusion

Vegetables are exhaustive crop with an exceedingly high turnover of plant nutrients, Thus, nutrient element required for optimum growth and development of vegetables should supplied to soil and plants in right time and in a balanced and integrated mode. Neither the chemical fertilizers are alone nor the organic sources exclusively able to sustain the soil fertility/productivity. A system approach to nutrient management by tapping all the possible sources of organic, inorganic and biofertilizers in judicious and synchronous way to maintain soil fertility and crop productivity of the essence of integrated nutrient management.

10. Future thrusts

The research findings summarized above indicate that except some studies on quality aspects in few vegetable crops, most of the work has been confined to the yield response of vegetable crops to nutrient management practices. In view of this, there is a need to focus our efforts towards the following points.

- Nutrient dynamics studies in the soil – plant system for better understanding of transformations and flows of the nutrients.
- Monitoring of soil health (physical, chemical as well as biological) in relation to different vegetable production/ nutrient management systems including organic farming.
- Evaluation of carbon sequestration potential under different vegetable based cropping systems.
- Monitoring of emerging deficiencies of secondary and micronutrients in soils under different vegetable based production systems.
- Studies on heavy metals contamination in soil – plant system in peri-urban areas where sewage-sludge / sewage water / industrial effluents are used for vegetable production.
- Evaluation and utilization of different types of industrial wastes/ by products as supplemental source of nutrients in vegetable production.

References

AICRP (Vegetables) Annual Rep. 2000-2010.

Ange, A.L. (1997) Integrated plant Nutrient Management: An overview. In : *Proc. National workshop* on IPNS 1991, IFFCO, New Delhi PP:16-33.

Chakraborty, A.K., P.K. Maiti, and N.C. Chattopadhyay. 1990. Varietal difference, in growth, uptake of nitrogen and ield of tomato under low level of fertilizer application. *Indian Journal of Horticulture* 47: 89-92.

Chaurasia, S.N.S., Nirmal De., Singh, K.P. and Kalloo, G.(2002) *Azotobactor* improves shelf life of tomato. *Indian J. Agric. Sci.*71 (12):765-7.

Chaurasia, S.N.S., Nirmal De and Singh, K.P. (2002) Response of hybrid chilli to applied nitrogen in inceptisol. *Indian J. Hort.* 59 (4): 423-426.

Chaurasia, S.N.S., De, Nirmal and Singh K.P. (2001) Influence of graded dose of nitrogen on indeterminate tomato hybrids.*Veg.Sci*, 28 (2): 157-159.

Chaurasia, S.N.S., Singh, K. P. and Mathura Rai (2006) Response of tomato (*Solanum lycopersicum* Mill) to foliar application of micronutrients.*Veg. Sci.* 33 (1):96-97.

Rakesh Singh, Chaurasia, S.N.S., and Singh, S. N.(2007) Response of nutrient sources and spacing on growth and yield of broccoli (*Brassica oleracea* var. italica Plenck) *Veg. Sci.* 33(2):198-200.

Chaurasia, S.N.S., Singh, A.K., Singh, K.P., Rai A.K., Singh, C.P.N., and Mathura Rai (2008) Effect of Integrated Nutrient Management on Yield and quality of cauliflower *(Brassica oleracea* L. var. botrytis) variety Pusa Snow Ball K-1.*Veg. Sci.* 35(1):41-44.

Chaurasia, S.N.S., Rakesh Singh and Mathura Rai (2009) Effect of Integrated Nutrient Management and Spacing on Yield, quality and Economics of Broccoli *Brassica oleracea* var. Italica Plenck, *Veg. Sci.* 36(1):51-54.

Chaurasia, S.N.S., Singh, K. P. and Mathura Rai (2005) Effect of Foliar Application of Water Soluble Fertilizers on Growth, Yield and Quality of Tomato (*Lycopersicon esculentum L.). Sri Lankan J. Agric. Sci.* 42:66-70.

Goyal, M.R., L.E. Rivera, and C.L. Santiago. 1985. Nitrogen fertigation in drip irrigated peppers, tomatoes and eggplant. In: Drip/Trickle Irrigation in Action, Vol. 1. St. Joseph, Michigan, USA, ASAE, pp. 388-392.

Hegde, D.M. 1988. Irrigation and nitrogen requirement of bell pepper (*Capsicum annuum* L.). *Indian Journal of Agricultural Sciences* 58: 668-672.

Hegde, D.M. 1989. Effect of soil moisture and nitrogen on plant water relation, mineral composition and productivity of bell pepper (*Capsicum annuum* L.). *Indian Journal of Agronomy* 34: 30-34.

Hegde, D.M., and K. Srinivas. 1989a. Growth and yield analysis of tomato in relation to soil matric potential and nitrogen fertilization. *Indian Journal of Agronomy* 34: 417-425.

Hegde, D.M. 1987a. Growth analysis of bell pepper (Capsicum annuum L.) in relation to soil moisture and nitrogen fertilization. *Scientia Horticulture* 33: 179-187.

Hegde, D.M., and K. Srinivas. 1990. Effect of irrigation and nitrogen fertilization on yield, nutrient uptake, and water use of tomato. *Gartenbauwissenschaft* 55: 173- 177.

Hosmani, M.M. 1993. Chili Crop (Capsicum annuum L.). 2nd Edition. Mrs. S.M. Hosmani, Dharwad, Karnataka.

Jablonska, C.R. 1990. Straw as an organic fertilizer in cultivation of vegetables. Part II. Effect of fertilization with straw on the growth of vegetable plants. *Horticultural Abstracts* 63: 244.

IIVR Annual Rep. 2000-2010.

Jose, D., K.G. Shanmugavelu, and S. Thamburaj. 1988. Studies on the efficiency of organic vs. inorganic form of nitrogen in brinjal. *Indian Journal of Horticulture* 45: 100-103.

Lampkin N.(1990) Organic Farming. Farming Press, Ipswich, 701pp.

Maestrey, A., H. Cardoza, A.J. Tremols, and R. Gomez. 1987. Nutrient uptake by spring tomatoes. I. Changes in N, P, and K concentrations during the growth cycle. *Horticultural Abstracts* 59: 3054.

Singh, G.B., and D.V. Yadav. 1992. Integrated plant nutrition system in sugarcane. *Fertilizer News* 37: 15-22.

Subbiah, K., S. Sundararajan, S. Muthuswami, and R. Perumal. 1985. Responses of tomato and brinjal to varying levels of FYM and macronutrients under different fertility status of soil. *South Indian Horticulturelist* 33: 198-205.

Shakhazizyan, R.S. 1989. Physiological activity of tomato leaves in relation to ripening phases. *Biological Zhurnal of Armenii.* 42: 687-690.

Subhani, P.M., C. Ravishankar, and N. Narayan. 1990. Effect of graded levels and time of application of N and K2O on flowering, fruiting and yield of irrigated chili. *Indian Cocoa, Arecanut and Spices Journal* 14: 70-73.

Sweeny, D.M., D.A. Graetz, S.J. Locascio, and K.L. Campbell. 1987. Tomato yield and nitrogen recovery as influenced by irrigation method, nitrogen source and mulch. *Horticultural Science* 22: 27-29. Terebayashi, S., K. Takii, and T. Namiki.

□□□

Green Agriculture : Newer Technologies, 2012
© *Kambaska Kumar Behera, pp. 303-351*
New India Publishing Agency, New Delhi (India)
e-mail : info@nipabooks.com; website : www.nipabooks.com

Chapter-13

Global Scenario of Plant Tissue Culture and its Application in Agriculture

Kambaska Kumar Behera
Dept. of Bio-Science and Biotechnology,
Banasthali Unviersity, Rajasthan-304022.
E-mail : kambaska@yahoo.co.in

SUMMARY

Plant tissue culture mainly comprises a set of *in vitro* techniques, methods and strategies that are part of the group of technologies called plant biotechnology. Tissue culture has been exploited not only to create genetic variability from which crop plants can be improved but also to improve the state of health of the planted material and to increase the number of desirable germ plasms available to the plant breeder through a shorter duration in cost effective manner. Tissue-culture protocols are available for most crop species, although continued optimization is still required for many crops, especially cereals and woody plants. Tissue culture techniques, in combination with molecular techniques, have been successfully used to incorporate specific traits through gene transfer. *In vitro* techniques for the culture of protoplasts, anthers, microspores, ovules and embryos have been used to create new genetic variation in the breeding lines, often via haploid production. Cell culture has also produced somaclonal and gametoclonal variants with crop-improvement potential. The culture of single cells and meristems can be effectively used to eradicate pathogens from planting material and thereby dramatically improve the yield of established cultivars. Large-scale micropropagation laboratories are providing millions of plants for the commercial ornamental market and the agricultural, clonally-propagated crop market. With selected laboratory material typically taking one or two decades to reach the commercial market through plant breeding, this technology can be expected to have an ever increasing impact on crop improvement as we approach for the need of the hour.

Introduction

Tissue-culture techniques are part of a large group of strategies and technologies, ranging through molecular genetics, recombinant DNA studies, genome characterization, gene-transfer techniques, aseptic growth of cells,tissues, organs, and in vitro regeneration of plants, that are considered to be plant biotechnologies. The use of the term biotechnology has become widespread recently but, in its most restricted sense, it refers to the molecular techniques used to modify the genetic composition of a host plant, i.e. genetic engineering. In its broadest sense, biotechnology can be described as the use of living organisms or biological processes to produce substances or processes useful to mankind and, in this sense, it is far *from* new. The products of plant breeding and the fermentation industries (e.g.cheese, wine and beer), for example, have been exploited for many centuries (Zhong et al. 1995). What is new and what has changed in the last two decades is the available technology (Davis and Reznikoff,1992). We no longer have to rely on pollination and cross-fertilization as the only ways to genetically modify plants. That the newer molecular and cellular technologies have yet to make a broad based significant impact on crop production is not surprising since a plant-breeding process of 10 to 20 years duration is still required to refine a selected plant to the stage of cultivar release (Plucknett and Smith 1986; Kuckuck et al. 1991; Brown and Thorpe, 1995).Plant tissue culture, the growth of plant cells outside an intact plant, is a technique essential in many areas of the plant sciences. It relies on maintaining plant cells in aseptic conditions on a suitable nutrient medium. The culture can be sustained as a mass of undifferentiated cells for an extended period of time or regenerated into whole plants. The unique characteristics of the plant cell to regenerate itself as a complete plantlet which is known as totipotency. The growing demand for energy for mass propagation of trees led to developing the technique which is known as plant tissue culture. It allows whole plants to be produced from minute amounts of plant parts like the roots, leaves or stems or even just a single plant cell under *in vitro* laboratory conditions.

The father of plant tissue culture is French botanist George Morel who discovered the technique in 1965 while he was attempting to obtain a virus-free orchid plant. Thereafter, the commercial use of the technology started in the 1970s in advanced countries. In earlier stages, the concept was restricted to laboratory and academic interest and at best, it was earlier used to develop ornamental plants and flowering plants for export. But in most developing countries, the shortage of biomass and the ever-increasing energy requirements created the need to explore possibilities of mass propagation of trees by tissue culture. Tissue culture or mass cloning methods of elite tree species is done for increasing land productivity. They are being modified or adapted for large-scale modification and increasing the yield and productivity. The concept has great relevance for many developing countries such as India where agriculture still remains predominant profession and requires adoption of new technologies to increase production.

Generally the species are selected for tissue culture based on the following considerations (i) Species of plants that have regeneration problems, especially because of poor seed quality (as in banana, Irish potato and bamboo). In these cases, seeds collected from superior trees are used for initiating cultures and increasing production. (ii) Species where plants of any one particular sex is of commercial importance, for example female plants of papaya and male plants of *Asparagus*. In tissue culture cells, tissues, and organs of a plant are separated. These separated cells are grown especially in containers with a nutrient media under controlled conditions of temperature and light. The cultured plant requires a source of energy from sugar, salts, a few vitamins, amino acids, etc. that are provided in the nutrient media. From these cultured parts, an somatic embryo may develop, which then grows into a whole new plant. Even the somatic embryo may be encapsulated in sodium algenate bids which is known as synthetic seed and stored in cryopreservation for short or long term.

Tissue culture plants have poor photosynthesis efficiency and lack the proper mechanism to control water loss. They need to be hardened gradually by moving them along a humidity gradient in the greenhouse. Once these plants are in the research fields, they are evaluated under field conditions. A large number of tissue culture plants that have grown into trees are remarkably uniform and show an increase in biomass production over the conventionally raised plants. Tissue culture is used for rapid vegetative multiplication of plant material which is also known as micropropagation as well as for production of disease free and pest resistant plant. Because plant cell culture is not affected by changes in environmental conditions, improved production may be available in any place or season. Therefore, studies on the production of useful metabolite by plant cell culture have been carried out on an increasing scale since the end of the 1950's. Their results stimulated more recent studies on the industrial application of this technology in many countries. However, there are still a few barriers that must be overcome before commercialization. To overcome barriers hindering industrial application of plant cell cultures, however, it is required to conduct more fundamental research, including elucidation of biosynthetic pathways of many useful secondary metabolites in plants and mechanisms for their biosynthesis, collaboration with a number of researchers in other scientific fields is also very helpful. In this review, the background of research on plant cell culture, various approaches to improve the productivity of secondary metabolites and different techniques are discussed.

From centuries, mankind is totally dependent on plants as a source of carbohydrates, proteins and fats for their food & shelter. In addition, plants are valuable source of a wide range of secondary metabolites, which are used as pharmaceuticals, agrochemicals, flavors, fragrances, colors, biopesticides and food additives. Over 80% of the approximately 30,000 known natural products are of plant origin (Phillipson 1990, Balandrin and Klocke 1988). In 1985, 3500 new chemical structures have been identified and 2600 derived from the higher plants. Worldwide, 121 clinically useful prescription drugs are derived from plants (Payne *et al.* 1991). Even today, 75% of the world's population relies on plants for traditional medicine. In the US, where chemical synthesis dominates the pharmaceutical industry, 25% of the pharmaceuticals are based on plant-derived chemicals (Glaser 1999, Farnsworth 1985).

Plants will continue to provide novel products as well as chemical models for new drugs in the coming centuries (Cox and Balick, 1994). The advent of chemical analysis and the characterization of molecular structures have helped in precisely identifying these plants and correlating them with their activity under controlled experimentation. Despite advancements in synthetic chemistry, there is need of biological sources for a number of secondary metabolites including pharmaceuticals (Pezzuto,1995). Elaborative pathways from basic primary metabolites, which are synthesized immediately as a result of photosynthetic activity, produce secondary metabolites. Other techniques like hairy root culture, biotransformation, immobilization and elicitations are used for the increased production of secondary metabolites. Through plant tissue culture the totipotent characteristics of plant can be used for the *in vitro* regeneration of plant.

Tissue culture is an experimental technique through which mass of cell is produced from the explants tissue. The callus produced can be utilized directly to regenerate plantlets or to extract or to manipulate some primary and secondary metabolites. Callus culture and suspension culture are the basic technique used to produce the desired metabolites of plants (Vaniserce et al. 2004). The plant and tissue cultures have been enabled to increase the knowledge in many areas including differentiation, cell division, cell nutrition and cell preservation. But now cells are cultivated *in-vitro* in bulk or as clone from single cells to grow whole plants from isolated meristem, then induce callus and develop complete plantlets by organogenesis or by embryogenesis. The research needs are based on the elements of scientific progress and development of new techniques, which either enables more critical experiment to be undertaken, or rendering easy accessibility to complicated problems through experimental studies (Mantell and Smith 1983, Evans et al. 1983).

Need for a Biotechnological approach

Biotechnology offers an opportunity to exploit the cell, tissue, organ or entire organism by growing them *in vitro* and to genetically manipulate them to get desired compounds. Since the world population is increasing rapidly, there is extreme pressure on the available cultivable land to produce food and fulfill the needs. Therefore, for other uses such as production of pharmaceuticals and chemicals from plants, the available land should be used effectively. The development of micropropagation methods for a number of medicinal plant species has been already reported and needs to be adopted (Nayak 1989).

The selective, deliberate transfer of beneficial genes from one organism to another to create new improved crops. Examples of genetically engineered crops include cotton, maize, sweet potato, soybeans, tomato etc. Genetically modified plants and seeds are created by the process of genetic engineering, which allows scientists to move genetic material between organisms with the aim of changing their characteristics. All organisms including plants are composed of cells that contain the DNA molecule. Molecules of DNA form units of genetic information, known as genes. Each organism has a genetic blueprint made up of DNA that determines the functions of its cells and the characteristics that make it unique.

Prior to genetic engineering, the exchange of DNA material was possible only between individual organisms of the same species. With the advent of genetic engineering in 1972, scientists have been able to identify specific genes associated with desirable traits in one organism and transfer those genes across species boundaries into another organism. For example, a gene from bacteria, virus, or animal may be transferred into plants to produce genetically modified plants having changed characteristics. Thus, this method allows mixing of the genetic material among species that cannot otherwise breed naturally. After decades of research, plant specialists have been able to apply their knowledge of genetic engineering to improve various crops such as corn, potato, Bt brinjel and cotton.

Rapid strides in this field of biotechnology has allowed and opened new vistas of opportunities to scientists and companies to explore the possibilities of use of the technology in farming. Today even in developing countries, more and more land is being planted with genetically modified varieties of an ever-expanding number of crops. Research efforts are being made to genetically modify most plants with a high economic value such as cereals, fruits, vegetables, and floriculture and horticulture species.

Application of genetic engineering in plants has allowed the following benefits to mankind:

- Developing plants that are resistant to diseases and pests.
- Increasing the shelf life of fruits and vegetables.
- Producing plants that possess healthy fats and oils and that have increased nutritive value thus improving the lifestyle.
- Producing soy beans with a higher expression of the anti-cancer proteins naturally found in soy beans.
- Increasing the yield per hectare and leading to increased productivity.

History in plant tissue culture

Haberlandt (1902) first envisioned the concept of plant tissue culture when he cultivated cells of various species on a medium containing glucose and peptone. Although he did not observe cell division, he described an approach whereby cellular physiology could be followed without the complexity of the whole level intervening. Haberlandt's pioneering research provided the groundwork for the first instances of callus induction and sustainable cell growth nearly forty years after his original experiments. Three independent and nearly simultaneous reports described callus initiation from carrots (Gautheret et al. 1939) and tobacco (White,1939) *in-vitro* cultures.

Callus induction from other dicotyledonous species soon followed (Gautheret, 1983) and callus from the first monocot was reported by LaRue (1947) in *Zea mays* endosperm cultures. One of the reasons for their success in initiating and maintaining these cultures was the use of the plant growth regulators. The discovery of indole acetic acid (IAA) by Went (1926) provided a means to initiate callus from nontumorous tissues. Furthermore, the positive influence of coconut milk on cell proliferation was attributed to new class of growth regulator, kinetin and other cytokinins (Miller et al. 1955). The combined use of IAA and kinetin by Skoog and Miller (1957) illustrated the synergism association with these compounds regarding the development of synchronized shoot-root plant regeneration. It should be pointed out the phrase 'plant tissue culture' is a really a misnomer in that the only tissue of plant origin that can be cultured is wound callus. In other instance of *in-vitro* culture where differentiation has occurred into shoots or roots, mixed tissues are involved in the final organs produced. However, Murasigue and Skoog (1962) develop the MS mediua for plant tissue culture *in vitro* which is now the most widely used media in plant tissue culture.

In a historical perspective, Cocking (1983) described problems and frustrations associated for the initial attempt to isolate and culture protoplasts. However, there have been several advances in this technology since the early 1970s. This includes the first reports of regeneration from protoplasts of *Nicotiana tobaccum*. The work on *Solanum tuberosum* was also a milestone since it represented the first case of protoplast-derived plant regeneration from a major agronomic crop. Advances are occurring in several monocotyledonous species such as *Dactylis glomerata*, *Oryza sativa*, *Saccharum species, Triticum aestivum* and *Zea mays*, where plant regeneration is now possible. A survey of historical milestone in plant tissue culture is shown in Table 1.

Table 1 : Historical milestones in plant cell and tissue culture techniques.

Year	Authors	Results	Species
1892	Klercker	First attempts to isolate protoplasts mechanically.	-
1902	Haberlandt	First cultivation experiments with isolated plant cells; cell growth, but no cell division obtained.	*Tradescantia*
1904	Hanning	Establishment of embryo culture for the first time	*Cochleria raphanus*
1909	Kuster	First observation of fusing cells	-
1922	Kotte, Robins	*In vitro* cultivation of root tips, no permanent cultures obtained	*Zea, Pisum*
1924 1925	Dieterich Laibach	Embryo rescue- "artificial premature birth"	*Linum*
1934	Gautheret Nobecourt	First permanent callus culture using B-vitamins and auxins	*Daucus, Nicotiana glauca* × *N. longsdorffii*
1942	Gautheret	Observation of secondary metabolites in plant callus culture	-

Contd. ...

Year	Authors	Results	Species
1946	Ball	Micropropagation: first development of stem tips and sub adjjacent regions	*Tropaeolum*
1952	Morel and Martin	Stem tips and adjacent regins: plants free of viruses.	*Dahlia*
1954	Morel et al.	First suspension cultures of single cells or cell aggregates. Nurse culture	*Tagetes, Nicotiana, Daucus, Picea, Phaseolus*
1955	Mothes and Kala	First reports of secondary metabolite production in liquid media	-
1956	Routien and Nickell	US patent No.2747334 for the production of substances from plant tissue culture.	*Phaseolus*
1958	Wickson and Thimann Reinert, Steward et al.	Establishment of axillary's branching Somatic embryogenesis in tissue cultures	*Daucus*
1959	Tukecke and Nickell	First report of large-scale (1341) culture of plant cells: carboy system	*Ginkgo,Lolium, Rosa,llex*
1960	Bergmann	Cell clones obtained from single cultured cells plated in an agar medium	*Nicotiana, Phaseolus*
1960	Jones et al.	Hanging drop culture in conditioned medium	*Nicotiana*
1960	Cocking	Method for obtaining large number of protoplasts from plant tissue	*Lycopersicon*
1965	Morel	Clonal multiplication of horticultural plant (orchid) through tissue droplet	*Cymbidium*
1965	Vasil and Hildebrandt	Regeneration of a plant from one single cell cultivated in a hanging droplet	*Nicotiana*
1966	Kohlenbach	First cell division and culture of differentiated mesophyll cells.	*Macleaya*
1967	Kaul and stabe	Reports of the yields of certain of differentiated mesophyll cells.	*Ammi*
1967	Bourgin and Nisoh Guha and Maheshwari	*In vitro* production of haploid plant from immature pollen with cultured anthers.	*Nicotiana Datura*
1970	Carlson	Isolation of auxotrophic mutants from cultured cells	*Nicotiana*
1977	Noguchi et al.	Cultivation of tobacco cells in 200001reactors	*Nicotiana*

Contd. ...

Year	Authors	Results	Species
1978	Zenk	Manifold increase in product yields by selection over parent plant documented for a variety of plant metabolites	-
1979	Brodelius et al.	Alginate beads used to immobilize plant cells for biotransformation and secondary metabolite production	-
1981	Shuler	Use of hollow reactor for secondary metabolite production	-
1983	Mitsui petrochemi.	First industrial production of secondary plant products by suspension cultures	*Lithospermum*
1983	Barton Brill	Insertion of foreign genes attached to a plasmid	-
1983	Chilton	Production of transformed tobacco plants following single cell transformation	Tobacco
1985	Kohntoco	Somatic hybrids in tobacco mediated by electro fusion	Tobacco
1986	Sundberg glemelius	Somatic hybrids in Brassicaceae	Plant species of Brassicaceae.

Plant tissue culture in agricultural biotechnology

Green biotechnology is biotechnology applied to agricultural processes. An example is the designing of transgenic plants to grow under specific environmental conditions or in the presence (or absence) of certain agricultural chemicals. One hope is that green biotechnology might produce more environmentally friendly solutions than traditional industrial agriculture. An example of this is the engineering of a plant to express a pesticide, thereby eliminating the need for external application of pesticides. An example of this would be Bt corn, Bt brinjal. Green revolution in India is a successful example of green biotechnology, resulting in increased yield and productivity and self sufficiency in food production. Introduction of various biofertilsers and biopesticides are other examples of green biotechnology.

Plant breeding and biotechnology

Plant breeding can be conveniently separated into twoactivities (Kleese & Duvick 1980): manipulating genetic variability and plant evaluation. Historically, selection of plants was made by simply harvesting the seeds from those plants that performed best in the field. Controlled pollination of plants led to the realization that specific crosses couldresult in a new generation that performed better in the field than either of the parents or the progeny of subsequent generations, i.e. the expression of heterosis through hybridvigour was observed. Because one of the two major activities in plant breeding is manipulating genetic variability, a key prerequisite to successful plant breeding is the availability of genetic diversity (Kuckuck et al. 1991; Villalobos &Engelmann 1995). It is in this area, creating genetic diversity and manipulating genetic variability, that biotechnology (including tissue-culture techniques) is having itsmost significant impact. In spite of the general lack of integration of most plant-biotechnology and plant-breeding programmes, field trials of transgenic plants have recently become much more common. There are therefore reasons to believe that we are on the verge of the revolution, in terms of the types and genetic make up of our crops, that has been predicted for more than a decade (Bodde,1982).More than 50 different plant species have already been genetically modified, either by vector-dependent (e.g. Agrobacterium) or vector-independent (e.g. biolistic, micro-injectionand liposome) methods (Sasson, 1993; Anon, 1994). In almost all cases, some type of tissue-culture technology has-been used to recover the modified cells or tissues. In fact, tissue-culture techniques have played a major role in the development of plant genetic engineering. Tissue culture will continue to playa key role in thegenetic-engineering process for the foreseeable future, especially in efficient gene transfer and transgenic plant recovery (Hinchee et al. 1994).

Biotechnology in its broadest sense means the application of all natural sciences and engineering in the direct or indirect use of living organism or parts of organisms in their natural or modified form, in an innovative manner in the production of goods and services and/or to improve existing industrial processes. The market application of the modern biotechnology techniques is typically in the general areas of human health care, agriculture and food production, industrial bio processing and other public good and environment settings. Thus, biotechnology refers to a set of technologies that involve understanding, mapping, manipulation or change of the genetic characteristics of a living organism.Following is the application of modern biotechnology techniques in various fields ranging from agricultural application to industrial processes.

1. Plant biotechnology- Significance in food production

Green biotechnology which is more commonly known as Plant Biotechnology is a rapidly expanding field within Modern biotechnology. It basically involves the introduction of foreign genes into economically important plant species, resulting in crop improvement and the production of novel products in plants. Use of environment friendly and cost effective alternatives to industrial chemicals such as bio fuels, bio fertilizers and bio pesticides are not only resulting in enhanced crop output, improvement in health and safety standards, these new products are also leading to less environment pollution and use of green technology. The ever increasing demand of agricultural produce has given new impetus to research in the field and has resulted in great benefits for farmers and users alike.

The significance of the application of modern biotechnology in the area of food production and its resultant impact in terms of human health and development can not be undermined. As the world is faced with ever increasing population and more and more food shortage and regional imbalances, new technologies and techniques are being developed to enhance production and increase the shelf life of perishable items. It is in this direction that new research initiatives in the field of green biotechnology are being made to enhance productivity and nutrition value of food items. Foods produced through modern biotechnology can be categorized as follows:

1. Foods consisting of or containing living/viable organisms, e.g. maize.
2. Foods derived from or containing ingredients derived from Genetically Modified Organisms (GMOs), e.g. flour, food protein products, or oil from GM soybeans, wheat etc.
3. Foods containing single ingredients or additives produced by GM microorganisms (GMMs), e.g. colours, vitamins and essential amino acids.
4. Foods containing ingredients processed by enzymes produced through GMMs, e.g. high-fructose corn syrup produced from starch, using the enzyme glucose isomerase (product of a GMM).

The first genetically modified food item - GM food (delayed-ripening tomato) was introduced on the US market in the mid-1990s. Since then, GM strains of maize, soybean, rape and cotton have been adopted by a number of countries and marketed internationally. In addition, GM varieties of papaya, potato, rice, squash and sugar beet have been trialled or released. It is estimated that GM crops cover almost 4% of total global arable land.

The development of GM organisms has revolutionized the scenario of world food production. It has also offered the potential for increased agricultural productivity or improved nutritional value that can contribute directly to enhancing human health and development. From a health perspective, there may also be indirect benefits, such as reduced agricultural chemical use and enhanced farm income, and improved crop sustainability and food security, particularly in developing countries. While the introduction of GM crops has undoubtedly changed the agricultural scenario and led to significant impact on human development, it has also raised various social, cultural and ethical issues and reluctance on the part of various countries and governments to accept the GM food even in times of grave needs such as famine and drought. While some countries have established pre-market regulatory standards for risk assessment of each and every food item before being launched in the market for application, there may be a case of a consistent and uniform international regulatory structure so as to ensure that such food items conform to a set of standards which are fair and equitable. This will also help quell a number of misgivings and doubts in the minds of countries yet to benefit from these food items. For example, recent non-approval of Bt brinjal in India while pesticide residue load in the cultivated egg plant creates more health problems. Today plant biotechnology encompasses the following main areas of research and application.

1. Plant tissue culture

A technique that allows whole plants to be produced from minute amounts of plant parts like the roots, leaves or stems or even just a single plant cell under laboratory conditions. An advantage of tissue culture is rapid production of clean planting materials. Examples of tissue culture products in India include banana, sugarcane orchids, *Antherium,* teak, bamboo while in Kenya include banana, cassava, Irish potato, pyrethrum and citrus. The number of nutrient media is uses from time to time for growth and development of different explants. Their names and composition are given in Table.2.

Figs. 1a-1f. Different stages of tissue culture of *Musa accumunata. In vitro* shoot multiplication in banana (a), shoot elongation (b) rooting and completed plant regeneration (c), initial hardening in green house (d), secondary hardening plantlets in net house (e) flowering and fruiting in tissue culture banana (f)

Table 2 : Media and their compositions for Plant tissue and Cell cultures (mg/L)

Components	Murashige & Skoog - 1962 (MS)	White -1963	Gamborg -1968 (B_5)	Nitsch - 1951	Heller -1953	Schenk - Hildebrandt - 1972	Nitsch - Nitsch - 1967	Kohlenbach - Schmidt -1975	Knop -1865	Uchimiya & Murashige's - 1974
Inorganics										
$(NH_4)_2SO_4$	-	-	134	-	-	-	-	-	-	-
$MgSO_4\times 7H_2O$	370	720	500	250	250	400	125	185	250	370
Na_2SO_4	-	200	-	-	-	-	-	-	-	-
KCl	-	65	-	1,500	750	-	-	-	-	-
$CaCl_2\times 2H_2O$	440	-	150	25	75	200	-	166	-	440
$NaNO_3$	-	-	-	-	600	-	-	-	-	-
KNO_3	1,900	80	3,000	2,000	-	2,500	125	950	250	1900
$Ca(NO_3)_2\times 4H_2O$	-	300	-	-	-	-	500	-	1,000	-
NH_4NO_3	1,650	-	-	-	-	-	-	720	-	1650
$NaH_2PO_4\times H_2O$	-	16.5	150	250	125	-	-	-	-	-
$NH_4H_2PO_4$	-	-	-	-	-	300	-	-	-	-
KH_2PO_4	170	-	-	-	_	-	125	68	250	170
$FeSO_4\times 7H_2$)	27.8	-	27.8	-	-	15	27.85	27.85	-	27.8
Na_2EDTA	37.3	-	37.3	-	-	20	37.25	37.25	-	37.25
$MnSO_4\times 4H_2O$	22.3	7	10 (1 H_2O)	3	0.1	10	25	25	-	22.3
$ZnSO_4\times 7H_2O$	8.6	3	2	0.5	1	0.1	10	10	-	8.6
$CuSO_4\times 5H_2O$	0.025	-	0.025	0.025	0.03	0.2	0.025	0.025	-	0.02
H_2SO_4	-	-	-	0.5	-	-	-	-	-	-
$Fe_2(SO_4)_3$	-	2.5	-	-	-	-	-	-	-	-
$NiCl_2\times 6H_2O$	-	-	-	-	0.03	-	-	-	-	-
$CoCl_2\times 6H_2O$	0.025	-	0.025	-	-	0.1	0.025	-	-	0.025
$AlCl_3$	-	-	-	-	0.03	-	-	-	-	-
$FeCl_3\times 6H_2O$	-	-	-	-	1	-	-	-	-	-
$FeC_6O_5H_7\times 5H_2O$	-	-	-	10	-	-	-	-	-	-
KI	0.83	0.75	0.75	0.5	0.01	1	-	-	-	0.83
H_3BO_3	6.2	1.5	3	0.5	1	5	10	10	-	6.2
$Na_2M_0O_4\times 2H_2O$	0.25	-	0.25	0.25	-	0.1	0.25	0.25	-	0.25
Organics										
Sucrose	30,000	20,000	20,000	50000 or	20,000	30,000	20,000~ 30,000	10,000	-	30,000
Glucose	-	-	-	36,000	-	-	-	-	-	-
Myo-Inositol	100	-	100	-	-	1,000	100	100	-	-

Contd. ...

Nicotinic Acid	0.5	0.5	1	-	-	0.5	5	5	-	4.5
Pyridoxine HC1	0.5	0.1	1	-	-	0.5	0.5	0.5	-	9.5
Thiamine HC1	0.1-1	0.1	10	1	1	5	0.5	0.5	-	9.9
Ca-Pantothenate	-	1	-	-	-	-	-	-	-	-
Biotin	-	-	-	-	-	-	0.05	0.05	-	-
Glycine	2	3	-	-	-	-	2	2	-	2
Cysteine HC1	-	1	-	10	-	-	-	-	-	-
Folic Acid	-	-	-	-	-	-	0.5	0.5	-	-
Glutamine	-	-	-	-	-	-	-	14.7	-	-
Casein acid hydrolysate	-	-	-	-	-	-	-	-	-	2000
Agar agar	8000	8000	8000	8000	8000	8000	8000	8000	8000	8000

i. Plant tissue culture by micropropagation

Micropropagation is a combination of the arts and sciences of plant multiplication *in-vitro* and plant acclimatization. It is the true-to-type propagation of a genotype that comprises many steps-stock plant care, explants selection and sterilization, media manipulation to obtain proliferation, rooting, acclimation, and growing on of liners and is usually associated with commercial production. The purpose of micropropagation is to produce carbon copies of original unique plants or more simply put to grow clones in quantity. The micropropagation process is the culmination of all the inventions, theories, and discoveries man has made regarding the anatomy and physiology of the living world. When the long histories of other approaches to deliberate plant propagation, it is shortest but it is the most promising technique, today about 150 plants are commercially micropropagated. Out of them few are reported by us like in banana (Das and Das 1991, 1992, Palai and Das 2002), *Mussaenda* (Das et al. 1993), *Hardwickia* (Das *et al.* 1995), *Cymbopogon* (Das 1999, Das et al. 1997, 1998), *Typhonium* (Das and Das 1998), *Albizia* (Ghosh et al. 2010) Different steps of banana micropropagations *in vitro* is elaborated in Fig. 1. Many of these have reached the limits of their improvement by traditional methods. The emphasis on sustainable agriculture, increasing world population and the loss of prime land to housing and industry make this method of propagation indispensable. A few successfully micropropagted forest tree were included in Table 3.

Table 3 : Micropropagation of forest tree species

Plant species	Explant used	Medium for shoot prliferation	Medium for rooting	Reference
Acacia auriculiformis	AB	B5 + coconut milk (5-10%) + BA (0.25)	B5 + IAA (0.02) or NAA (0.02)	Mittal et al. 1989
Acacia catechu	NE	MS + BA (4) + NAA (0.5) or BA (1) + Kn (1) + Adenine sulphate (25) + ascorbic acid (20) + glutamine (150)	¼ MS + IAA (3)	Kaur et al. 1998
Acacia mearnsii	ST	¾ MS + BA (2) or + IBA (0.01)	½ MS + NAA (0.6)	Huang et al. 1994
Aegle marmelos	ST	MS + BA (0.05-2.25) + IAA (1)	½ MS + IAA (0.5) or + IBA (0.5)	Ajithkumar & Seeni 1998
Anacardium occidentale	CN	MS + sucrose + maltose + BA (5)	MS + IAA (0.5) + IBA (1)	D' Silva & D'Souza 1992
Annona squamosa × *Annona cherimola*	NE	MS + BA (0.5) + Kn (0.5) + CH (100)	½ MS + IBA (50) or + AC (0.25%)	Nair et al 1984
Annona cherimola	NE	MS + BA (0.15) + Z (0.5)	MS + AC (1%) or ½MS + IBA (100) + Citric acid (200)	Encina et al 1994
Artocarpus heterophyllus	ST, NE	MS + BA (1)	½MS + citric acid (200) or MS + sucrose (2%) + IBA (0.2)	Rahaman & Blake 1988
Artocarpus heterophyllus	NE	MS + BA (1) + Kn (0.5)	½MS + NAA (1) + IBA (1-2) or NAA (2)	Roy et al. 1990
Artocarpus heterophyllus	ASB	MS + BA (1-2) or Kn (1-2)	½MS +IBA (2) or NAA (2)	Amin & Jaiswar 1993
Bauhinia variegata	NE	MS + BA (0.5-7)	MS + IBA (1)	Mathur & Mukunthakumar 1992
Caesalpinia pulcherima	NE	MS + BAS/Kn (1) + NAA (0.5)	MS + BA / Kn (1) + IAA (1)	Rahman et al. 1993
Cleistanthus collinus	NE	MS + BA (0.2) + citric acid (10-50) + PVP (50-100)	MS + IAA (5) for 2 min	Quaraishi & Mishra 1998
Caesalpinia pulcherima	NE	MS + BA/Kn (1) + NAA (0.5)	MS + BA/Kn (1) + IAA (1)	Rahman et al. 1993
Cleistanthus collinus	NE	MS + BA (0.2) + Citric acid (10-50) + PVP (50-100)	MS + IAA (5) for 2 min	Quaraishi & Mishra 1998
Commiphora wightii	NE	MS + BA (4) + Kn (4) + Glutamine 100 + Thiamine (10) + AC (0.3%) or BA (0.3) + Kn (0.3)	IBA /IAA (2) + NAA (1) + AC (0.5%) + Sucrose (2%)	Brave and Meheta 1993
Cornus florida	NE	WPM + BA (0.5-1)	WPM + IBA (1)	Kaveriappa et al. 1997
Dalbergia latifolia	H	MS + BA (1) + Calcium pantothenate (1) + Biotin (1) + folic acid (1) or +Ba (0.5) + NAA (0.5) +calcium pantothenate (1) + Biotin (1) + Folic acid (1)	NAA (1) + IAA (1) + IBA (1)	Rai & Chandra 1989
Dalbergia latifolia	NE	MS + BA (1) + NAA (0.5) to MS (¾ major elements) + Ba 91) + NAA (0.5)	½ MS + IBA (2) +Sucrose (2%)	Raghvaswamy et al. 1992

Dalbergia latifolia	CN	MS + BA (2)	½ MS + IAA (1)+IBA (1)+IPA (1)	Pradhan *et al.* 1998a,c
Dalbergia sissoo	NE	MS (inorganic salts + B5 (amino acids) + Sucrose (2%) + agar 0.9% + ascorbic acid (5) + BA (1) / Kn (1)	MS (inorganic salts) + B5 (amino acids) + Sucrose (2%) + agar 0.9% + ascorbic acid (5) + IAA (0.5) + Kn (1)	Datta et al. 1983
Dalbergia sissoo	C	MS + BA (1)	½ MS + IBA (2) to ½ MS + IAA (1) or IBA (1)	Jaiswal et al. 1995
Dalbergia sissoo	CN	MS + BA (2)	½ MS + IAA (1) + IBA (1)+IPA (1)	Pradhan *et al.*1998b, Pattnaik *et al.* 2000
Dalbergia sissoo	NE	MS + BA (1) + NOA (1)	MS + IBA (2) to ½ MS + IBA (2) + AC (1%) to ½ MS	Gulati & Jaiswal 1996
Dipterocarpus alatus	CN	WPM (-NH_4NO_3) + 2 iP (2)	-	Linington 1991
Dipterocarpus intricatus	CN	WPM (-NH_4NO_3) + BA (0.15-3) or BA (1) + NAA (0.1)	WAP + IBA (200) just a dip	Linington 1991
Eucalyptus tereticornis	AB	MS + BA (1) +NAA (0.1)	Knop's medium + IBA (1)	Das and Mitra 1990
Feronia limonia	NE	¾ MS + Sucrose (3%) + Kn (2) + BA (0.5)+NAA (0.01) + citric acid (10) + Ascorbic acid (100)+ PVP (50)	¼ MS + sucrose (1%) + IBA (1.5%)	Purohit & Tak 1992
Ficus carica	ST	MS + BA (0.1) + NAA (0.18) + GA3 (0.03)	MS + IBA/NAA (0.5)	Muriithi et al. 1982
Ficus carica	AB	MS + BA (2) + NAA (0.2)	MS + IBA (2) + AC (0.2%)	Kumar et al. 1998
Fraxinus excelsior	CN	Driver & Kuniyuki Medium + BA (0.5)	WPM + IBA (0.5)	Hammatt & Ridout 1992
Gmelina arborea	NE	MS + BA (0.25)	MS + IBA (50) for 5 min	Kanan & Jasraj 1996
Lagerstroemia flos-reginae	NE	Ms + BA (7.5 - 20)	MS + IBA (1)	Paily & D'Souza 1986
Leucaena leucocephala K67	LB	MS + BA (3) + NAA (0.05)	½ MS + IBA (3) + Kn (0.05)	Goyal & Felker 1985
Malus domestica cv. Crab apple	ST	MS + BA (1-2)	MS + NAA (0.1-2)	Singha 1982
Malus domestica cv. Jonathan Delicious	NE	MS + BA (2)	½ MS + NAA (0.2) or IBA (150) to ½ MS	Sriskandarajah et al. 1982
M. baccata x M. pumila (almery crab apple)	ST	MS + BA (2) + 0.3% agar	-	Singha 1984
Maytenus emerginata	AB, NE	MS + IAA (0.1) + BA (2.5) or + BA (1)	½ MS + IBA (25)	Rathore et al. 1992
Mitragyna parviflora	ST	Ms + BA (1)	MS + NAA (1) + IAA (1) + IBA (1)	Roy et al. 1988
Morus alba	ST	MS + Ba (0.1-1)	MS + NAA (0.1-1)	Oka & Ohyama 1986
Morus australis	NE	MS + BA (1)+GA3 (0.3)	MS + IAA (1) + IBA (1) + IPA (1)	Pattnaik et al. 1996

Contd. ...

Morus cathayana, Morus ihou	NE	MS + BA (1) + GA3 (0.4)	MS + IAA (1) + IBA (1) + IPA (1)	Pattnaik & Chand 1997
Morus indica	AB	MS + BA (1)	MS besal	Mhatre et al. 1985
Morus indica	ASB	MS + BA (1)	½ MS + IBA (1)	Chand et al. 1995
Morus laevigata	NE	MS + BA (1)	MS + IBA (0.1) + NAA (0.1)	Hossain et al. 1992
Morus nigra	SA	Knop's macro + MS micro + BA (0.5)	MS + IBA (0.2) + NAA (0.2)	Ivanicka 1987
Morus nigra	ST, NE	MS + BA (1)	MS + IBA (0.25)	Yadav et al. 1990
Olea auropaea cv. Kalamon	NE	OM + Cysstine (0.5) + Ca-pantothenate (100 + GA3 (0.5) + IAA (0.1) + ZR (10/BA (7.5)/2iP (10)	Modified OM + NAA (1-3)	Rama & Pontikis 1990
Olea auropaea cv. Frantoio, Moraiolo Dolce Agogia	NE	OM + Z (4)	½ MS + ½ BN + ½ Knop's (macro) + Heller's (micro) + NAA (1)	Rugini 1984
Parkinsonia aculeata	NE	MS + BA (0.5 - 7)	MS + IBA (1)	Mathur & Mukuntha Kumar 1992
Prosopis cineraria	LB	MS + Kn (0.05) + IAA (3) to ½MS	WM + Kn (0.5) +IBA (3)	Goyal & Arya 1984
Prosopis cineraria		MS + BA (2.5 - 5) + IAA (0.1) + Ascorbic acd (100) + citric acid (50) + PVP (100) to MS + NAA (0.1) + BA (1) +Ascorbic acd (100) + citric acid (50) + PVP (100)	½MS + IBA (100) for 4 h	Sekhawat et al. 1993
Psidum guajava cv. Banaras local	NE	BA (1) + IAA/IBA/ GA3 (0.1)	½MS + IBA (0.2) + NAA (0.2) + AC (1g/l)	Amin & Jaiswal 1987
Psidum gujava cv. Chittidar	ST, NE	MS + BA (0.01-1)	½ MS + IBA (0.2) + NAA (0.2) + AC (1g/l)	Amin & Jaiswal 1988
Psidum guajava	ST, NE	MS + BA (0.01-1)	MS basal	Loh & Rao 1989
Psidum gujava	ST	OM + BA (2)	OM + NA (0.5-0.1) + BA (0.5-1)	Papadatou et al. 1990
Prunus amygdalus cv. Ferragnes	ST	TK + BA (0.5) + IBA (0.01)	BN + IBA/NAA (1)	Rugini & Verma 1982
Prunus armeniaca	ST	WPM + 2iP (2)	½ MS + NAA (0.5)	Snir 1984
Prunus avium (Sweet cherry)	ST	TK medium then Ms + BA (1) + IBA (1)	½ MS + NAA (0.5)	Snir 1982
Prunus persica (Peach)	SB	MS + ascorbic acid (50) + Vitamins (20 ml) + BA (2) + NAA (0.1)	MS + Ascorbic acid (50) + NAA (10.1)	Miller et al. 1982
Prunus persica	ST	MS + BA (2) + IBA (0.01)	½ MS + NAA (0.5-1) or ½MS + IAA (0.5-1) +PG	Hammerschlag et al. 1987
Prunus communis cv. William Bon, Chretien, Packhan's, Triumph, Beurre, Bosc	NE	Z(2) + 2iP (2) +BA (1.3/1.8/2)	MS + NAA/IBA (2)	Shen & Mullins 1984

Contd. ...

Prunus communis cv. William Bon, Chretien, Packhan's, Triumph, Beurre, Bosc	ST	MS + BA (2)		Singha 1984
Pyrus pyrfolia	ST	MS + BA (1.5) + NAA (0.02)	MS +NAA (2) +PG (162)	Bhojwani et al. 1984
Quercus robur	ST	BTM/WMP + BA (0.2) or TDZ (0.001)	BTM/WMP + NAA (0.2-0.5) + IBA (0.2-0.5)	Chalupa 1968
Quercus robur	ST, NE	BTM + BA (0.2) or WPM + BA (0.2)	BTM or WPM with low sugar percentage	Chalupa 1984
Sterculia urens	CN	MS + BA (2)	½ MS + IBA (500) for 5 min	Purohit & Dave 1996
Sassafras randaiense	ST	LS + BA (5) + Kn (5) + NAA (0.05) + sucrose 5% + coconut water 30% + malt extract (100) + glutamine & Arginine (50)	LS + IBA (5-10)	Wang & Hu 1984
Sapium sebiferum	NE	MS + Kn (0.1) + BA (0.2)	W + Na_2MoO_4 (0.25) + $CoCl_2$ (0.25) to IAA (1) + IBA (1) +IPA (1) +AC (0.25%)	Mridula et al. 1983
Tectona grandis	ST	Ms + Kn (1) + BA (1)	White's liquid medium 72 h to White's liquid medium 72 h to White's liquid medium + IBA (2)	Devi et al. 1994
Tecomella undulata	NE	MS + BA (2) + IAA (0.05)	½ MS + IBA (2.5) for 48h to ½MS	Rathore et al. 1991
Wrightia tinctoria	CN	MS + BA (5) + NAA (0.01) to MS + BA (1) + NAA (0.01)	¼MS + IBA (500) for 5 min	Purohit & Kukda 1994
Wrightia tomentosa	NE	MS + BA (1-2)	¼MS +IBA (100) for 15 min	Purohit et al. 1994
Zizyphus mauritiana	AB	MS + BA (1) +NAA (0.05)	½ MS + IBA (0.5) + Kn (0.05)	Goyal & Arya 1985
[AB= axillary bud, AC=activated charcoal, ASB=apical shoot bud, CN=Cotyledonary node, LB=lateral bud,				
NE=nodal explant, SA=Shoot apex, SB=shot bud, ST=Shhot tips, figures in parentheses = concentrations (mg.l)				
C=cotyledon, H=hypocoty]				

1.1 Plant cell cultures

Plant tissue culture provides an alternative way for the production of phytochemicals of therapeutic importance. Callus culture and suspension culture are the basic techniques used to produce the desired metabolites. Now a day techniques like hairy root culture, immobilization and elicitation are used to enhance the production of secondary metabolites. The plant tissue culture have some salient feature which includes, the clone generated through tissue culture which are identical in size, development stage and rate of metabolic activities and are capable of performing the transformative activity which involves biotransformation to produce primary and secondary metabolites in culture medium. Through plant tissue culture, the totipotent characteristic of plant cell lend them amenable to in-vitro regeneration into whole plant via embryogenesis, organogenesis or through micropropagation. Growing research in tissue culture technology provides the opportunities of biosynthesis of a variety of natural products using tissue culture. Different techniques of plant tissue culture are given below:-

1.2. Organ cultures

Since production of secondary metabolites is generally higher in differentiated tissues, there are attempts to cultivate shoot cultures and root cultures for the production of medicinally important compounds, these organ cultures are relatively more stable. There are a number of medicinal plants whose shoot cultures have been studied for metabolites. Similarly, root cultures are valuable sources of medicinal compounds, root systems of higher plants generally exhibit slower growth and are difficult to harvest.

Wide Hybridization

A critical requirement for crop improvement is the introduction of new genetic material into the cultivated lines of interest, whether via single genes, through genetic engineering, or multiple genes, through conventional hybridization or tissue-culture techniques. During fertilization in angiosperms, pollen grains must reach the stigma of the hostplant, germinate and produce a pollen tube. The pollentube must penetrate the stigma and style and reach the ovule. The discharge of sperm within the female gametophyte triggers syngamy and the two sperm nuclei must then fuse with their respective partners. The egg nucleus and fusion nucleus then form a developing embryo and the nutritional endosperm, respectively (Tilton & Russel 1984; Zenkteler 1990). This process can be blocked at any number of stages, resulting in a functional barrier to hybridization and the blockage of gene transfer between the two plants. Pre-zygotic barriers to hybridization (those occurring prior to fertilization), such as the failure of pollen to germinate or poor pollen-tube growth, may be overcome using in vitro fertilization (IVF; Yeung et al. 1981). Post-zygotic barriers (occurring after fertilization), such as lack of endosperm development, may be overcome by embryo, ovule or pod culture. Where fertilization cannot be induced by in vitro treatments, protoplast fusion has been successful in producing the desired hybrids (see below).

In vitro Fertilization

IVF has been used to facilitate both inter-specific and inter-generic crosses, to overcome physiological-based self incompatibility and to produce hybrids. A wide range of plant species has been recovered through IVF via pollination of pistils and self- and cross-pollination of ovules (Yeung et al.1981; Zenkteler 1990; Raghavan 1994). This range includes agricultural crops, such as tobacco, clover, com, rice, cole, canola, poppy and cotton. The use of delayed pollination, distant hybridization, pollination with abortive or irradiated pollen, and physical and chemical treatment of the host ovary have been used to induce haploidy (Maheshwari& Rangaswamy 1965; Zenkteler 1984).

Embryo Culture

The most common reason for post-zygotic failure of wide hybridization is embryo abortion due to poor endosperm development. Embryo culture has been successful in overcoming this major barrier as well as solving the problems of low seed set, seed dormancy, slow seed germination, inducing embryo growth in the absence of a symbiotic partner, and the production of monoploids of barley (Raghavan, 1994; Yeung et al. 1981; Collins & Grosser 1984; Zenkteler 1990). The breeding cycle of *Iris* was shortened from 2 to 3 years to a few months by employing embryo rescue technology (Randolph 1945). A similar approach has worked with orchids and roses and is being applied to banana and *Colocasia* (Yeung et al. 1981). Interspecific and intergeneric hybrids of a number of agriculturally important crops have been successfully produced, including cotton, barley, tomato, rice, jute, *Hordeum* X *Secale, Triticum* x *Secale, Tripsacum*x *lea* and some Brassicas (Collins &Grosser 1984; Palmer & Keller 1994; Zapata-Arias et al.1995).

Protoplast Fusion

Protoplast fusion has often been suggested as a means of developing unique hybrid plants which cannot be produced by conventional sexual hybridization. Protoplasts can be produced from many plants, including most crop species (Gamborg et al. 1981; Lal & Lal, 1990; Feher & Dudits, 1994). However, while any two plant protoplasts can be fused by chemical or physical means, production of unique somatic hybrid plants is limited by the ability to regenerate the fused product and sterility in the inter-specific hybrids (Pandeya et al. 1986; Schieder & Kohn, 1986; Evans &Bravo, 1988) rather than the production of protoplasts. Perhaps the best example of the use of protoplasts to improve crop production is that of *Nicofiana,* where the somatic hybrid products of a chemical fusion of protoplasts have been used to modify the alkaloid and disease-resistant traits of commercial tobacco cultivars (Pandeya et al. 1986).Somatic hybrids were produced by fusing protoplasts, using a calcium-polyethylene glycol treatment, from a cell suspension of chlorophyll-deficient *N.* rusfica with an albinomutant of *N. tabacum* (Douglas et al. 1981). The wild *N. rusfica* parent possessed the desirable traits of high alkaloid levels and resistance to black root rot. Fusion products were selected as bright green cell colonies, the colour being due to the genetic complementation for chlorophyll synthesis in the hybrid cells. Plants recovered by shoot organogenesis showed a wide range of leaf alkaloid content but had a high level of sterility. However, after three backcross generations to the cultivated N. tabacum parent, plant fertility was restored in the hybrid lines, although their alkaloid content and resistance to blue mould and black root rot were highly variable. Interestingly, neither parent was known topossess significant resistance to blue mould. (Pandeya and White 1994).

2. Plant genetic engineering

The selective deliberate transfer of beneficial gene(s) from one organism to another to create new genetically improved crops, for higher crop yield, disease and pest, abiotic stress resistance varieties is a powerful technique for crop improvement programme using tissue culture methods. *Agrobacterium* mediated genetic transformation and particle bombardment are few important methods for crop improvement programme are discussed in the following sections.

2.1 *Agrobacterium* mediated genetic transformation and crop improvement

Phytopathogenic gram-negative soil bacteria, *Agrobacterium tumefaciens* and *A. rhizogenes* cause neoplastic diseases (crown gall or hairy root, respectively) in many dicotyledonous plant species. The *A. tumefaciens* infects plants at the crown, usually through a wound site, causing cancerous growths of proliferating plant cells known as crown gall tumors. This disease in itself is ergonomically important and effects most dicotyledonous plants causing millions of dollars' worth of damage to plants. In the 1940s, from experimental observations, it was concluded that a factor is transmitted from the invading bacteria to the host plant cell. Further studies demonstrated that the disease is actually the direct result of the transfer of a particular DNA fragment (genes) from the bacterium to the plant cell. In addition to its chromosomal DNA, *Agrobacterium* contains a much smaller circular DNA molecule called a Ti (tumor-inducing) plasmodia, of which a small piece, called the T-DNA (Transferred-DNA), is the factor transferred into plant cells. The T-DNA becomes stably integrated into the plant's chromosomes, from where it is able to perturb the natural functions of the plant. The T-DNA encodes genes, which, when expressed, bring about the production of new enzymes that are able to alter the hormone balance within the infected cell. This brings about de-differentiation and cell division, leading to proliferation of cells and the formation of tumors. This appears to be of little benefit to the bacterium. However, other genes are also present in the T-DNA which, when expressed, are able to synthesize novel compounds from naturally occurring plant precursors. These novel compounds cannot be metabolized by the plant but are a good source of nutrients for the bacterium.

Crown gall disease affects hundreds of species, particularly fruits, nuts and ornamental plants such as roses. The disease recently caused severe damage and economic losses among walnut tree growers in California. Once *A. tumefaciens* infects a plant, the bacterium travels throughout the root system, and wipe out entire crop. The only option for farmers is to destroy the plants. The practice of replacing the tumor-inducing genes with other DNA began in the 1970s and led to the widespread use of the bacterium in research. The microbe is particularly well suited for testing the functions of individual plant genes because bacterial DNA disrupts the genome at the point of insertion (Fig. 2).

Agrobacterium is used to insert a piece of DNA in the middle of a plant gene, thus inactivating the gene," says Derek W. Wood, who led part of the University of Washington project. "We can then analyze the mutation to see what it does to the plant." This experiment has been done on a large scale. Many thousands of mutant potato and alfalfa plants are available to plant researchers from institutions like the University of Wisconsin in Madison. The microbe has also been used to create transgenic crops, including new strains of corn and soybeans. *Agrobacterium*-mediated gene transfer into a wide variety of crop species has already become a realizable technique for plant improvement by transferring agronomically useful genes such as insect resistance, virus resistance (Beachy et al., 1990), resistance against bacterium and herbicide resistance.

The *A. tumefaciens* genome has a very unusual structure. Some 5,400 genes reside on four DNA elements—a circular chromosome, a linear chromosome, and two smaller circular structures called plasmids. Many bacteria have circular chromosomes and some have linear chromosomes, but *Agrobacteria* are the only species known to have both structures together (Table 4).

2.2 Hairy root culture for pharmaceuticals

The ability of *Agrobacterium rhizogenes* to induce hairy roots in a range of host plants has lead to studies on it as a source of root-derived pharmaceuticls (Flores et al. 1999). Tepfer (1990) summarized 116 plants belonging to 30 dicotyledonous families wherein hairy roots have been induced. Hairy roots re induced by transfer of T_DNA from the plasmid of *A. rhizogenes* (Ambros et al. 1986) to host tissue, resulting in root formation.

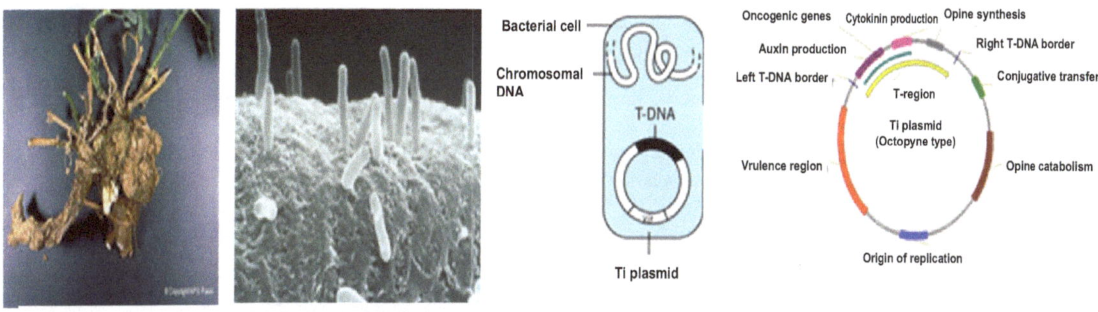

Crowngall on alfalfa

Agrobacterium tumefaciens
Attached to a plant cell

Figure 2 :

Table 4 : Genetic transformation of tree species using *Agrobacterium tumefaciens*

Species	Strain	Construct	Explant	Infection and co-cultivation details	Kanamycin conc. for selection (μg/ml)	Gene expression	Detection method	Reference
Actinida deliciosa (Kiwi fruit)	EHA101	pLAN411, pLAN421 (*npt II, gus*)	Hypocotyl	Hypocotyles were shaken in bacterial suspensin for about 20 min, blotted dry and co-cultivated for 1 day	50	*npt II, gus*	NPT II Enzyme assay, Histo chemical enzyme assay	Uematsu et al 1991
Actinida deliciosa (Kiwi fruit)	LBA4404	pRokla-EG *npt II, soybean* β-*1,3 endonuclease*	Leaf disc petiole, stem segment	Infected with bacteris	50	*npt II, Soybean β-1,3 endonuclease*	PCR and Southern blot analysis, Northern blot and Western blot analysis	Nakamura et al 1999
Carica papaya (Papaya)	LBA4404	pB1121 (*npt II , gus)*	Petioles from multishoot	Just dipped and co-cultivated for 2 days	50	*gus, gus*	PCR and Southern blot analysis, Northern blot and Western blot analysis	Yang et al. 1996
castanea sativa (European chesnut)	LBA4404	p35SGUSINT (*npt II, uid A)*	Hypocotyl	Hypocotyles were gently shaken in bacterial suspension for about 10 min, bltted dry and co-cultivated for 2 days	50	*npt, uid A*	PCR, Histochemical enzyme assay	Seabra and Pais 1998
Citrus aurantifolia (Lime oragne)	EHA 105	p35SGUSINT (*npt II, gus*)	Internodal segment	Infected with bacterium for 15-30 mins and then co-cultivated for 3 days	10	*npt II, gus, npt, gus*	PRC and Southern blot analysis, ELISA and histochemical enzyme assay	Pna et al 1997
Citrus aurantium (Sour organge)	EHA 101	pGA482GG-CTVCP (*gus, ctv-cp*)	Internode	Explants co-cultivated for 1-3 days	100	*ctv-cp, gus, ctv-cp*	PCR and Southern bot analysis, Histochemical enzyme assay, Western blot analysis	Guttierrez et al 1997
Citrus sinensis - Osbeck × *Poncirus trifolia* (Citrange)	EHA 105	p35SGUSINT (*npt II, gus*)	Epicotyl	Explants incubated 15 min in 15 ml bacterial suspension, co-cultivated for 1,3,5 days	100	*npt II, gus, gus*	PCR and Southern blot analysis, Northern blot and Western blot analysis, Histochemical enzyme assay	Cervera et al. 1998

Citrus sinensis (Washington novel orange)	LBA4404, EHA 101, EHA 105, C58	p35SGUSINT (*npt II, gus*)	Epicotyl	Epicotyl segments infected for 10 min in bacterial suspension (5 × 10^8 cfu/ml) blotted dry, co-cultivated for 2-3 days.	100	*npt ii, gus, gus*	PCR and Southern blot analysis, Northern blot and Western blot analysis, Histochemical enzyme assay	Bond and Roose 1998
Diospyros kaki (Japanese persimmon)	LBA 4404, EHA 101	pB1121, pSMAK251 (*npt II, gus*)	Hypocotyl	Segments of hypocotyl immersed in bacterial suspension approximately (5 × 10^8 cfu/ml for 15 min and then co-cultivated for 3 days.	100	*npt II, gus, gus*	PCR and Southern blot analysis, Northern blot and Western blot analysis, Histochemical enzyme assay	Nakamura et al 1998
Eucalyptus camaldulensis (Eucalyptus)	A6, LBA 4404, GV 3111, AGL1, GV3850	each containing *npt II, gus*	Leaf disc petiole, stem segment	Co-cultivated for 2 days in bacterial suspension 10^9 cells/ml on callusing medium.	9	*npt II, gus, gus*	Southern blot analysis, NPT II enzyme assay, Histochemical enzyme assay	Mullins et al. 1997
Hevea brasiliensis (Rubber)	GV 2260	p35SGUSINT (*npt II, gus*)	Callus	Calli submerged in an overnight grown *Agrobacterium* suspension for 1 min and then co-cultivated for 2 days.	100	*npt II, gus, npt II, gus, gus*	Southern blot analysis, Histochemical enzyme asay by Sudan III Staining (Jensen 1962)	Arokiaraj et al 1998
Larix kaempferi × *L. decidua* (Hybrid Larix)	C58	pPMP90, pMRKE70 Km (2 copies ofnpt II)	Somatic embryo	Bacterial suspension at OD 0.3 (approximately 10^8 cfu/ml) applied to embryonal masses kept on proliferation medium in dark for 48 h.	50	*npt II*	PCR and southern blot analysis	Leeve et al. 1997
Malus domestica (Royal Gala apple)	EHA 105	p35SGUSINT (*npt II, gus*)	Internode	Epicotyl segments dipped in overnight grown bacterial suspension culture, a density of (2	100	*gus, gus*	Histochemical enzyme assay	Liu et al 1998

				$\times 10^8$ cfu/ml), blotted dry, co-cultivated for 48 h				
Malus domestica (Marshal MacIntosh apple)	EHA 105	p35SGUSINT (*npt II, gus*)	Leaf disc petiole, stem segment	Wounded leaves co-cultivated in dark for 48h after infecting with bacterial suspension	25	*npt II, gus, npt II, gus, gus*	PCR and Southern blot analysis, ELISA and NPT enzyme assay by histochemical enzyme assay	Bolar et al. 1999
P. alba × grandidentata (Populus hybrid NG 5339)	C58	pPMG85/587 (*npt ii*, EPG syntnase \aro A)	Leaf segment	1-5 ml bacterial bropth culture infected for 30 min and then co-cultivated for 48-96h	50	*npt II, aro A, npt Ii, aro A*	Southern blot analysis	
Prunus domestica (Plum)	EHA 101, EHA 101	pCGN7001, pCGN7314 (*npt ii, gus)*	Hypocotyl slices	Explant soaked in 10 ml bacterial suspensions and co-cultivated for 48-72h (*Agrobacterium* 10^5 celles/ml)	75	*npt II, gus, npt II, gus*	NPT II enzyme assay, EPSP synthase protein activity by Western blot, Nopaline assay by Electrophoresis	Mante et al. 1991
Prunus dulcis (Almond)	LBA 4404, EHA 101, EHA 105, C58	p35SGUSINT (*npt ii*, *gus*)	Leaf	Leaf wouned from mid-rib dipped in bacterial suspension and co-cultivated for 3-4 days	15	*npt II, gus, gus*	Southern blot analysis, NPT II enzyme assay and ELISA, Histochemical enzyme assay	Miguel et al. 1999
Pyrus communis (Pear)	EHA 101	pFM3002,pFAJ300 (*npt II, uid A, attacin E*)	Leaf	Infected with bacterial suspension 5 $\times 10^7$ cuf/ml	100	*npt II, uid A, attacin E, attacin E*	PCR and Southern blot analysis, Nothern blot and Western blot analysis	Reynoird et al. 1999
Santalum album (Sandal wood)	LBA4404,EHA 101	pKIW 105, pIG121-Hm (*npt Ii, uid A*)	Somatic embryo	Infected with bacterial suspension and co-cultivated for 24 h	50	*npt II, uid A, npt II, uid A*	PCR and Southern blot analysis, NPT II enzyme assay, Histochemical enzyme assay	Shiri and Rao 1998

3. Immobilization of plant cells for the production of secondary metabolites

Improvement in the secondary metabolite production of cell cultures is often associated with the organization and differentiation of plant cells. The concept of organization and differentiation led to the use of immobilization technology, which has long been used for microbes and enzymes. Immobilization is defined as a technique, which confines a catalytically active enzyme or cells on a fixed support and prevents its entry into liquid phase (Lindsey and Yeoman 1983). Immobilized plant cells have been used for single and multi-step biotransformation of precursors to desired products as well as for the de-novo biosynthesis of secondary metabolites. Immobilization can have a dramatic impact on cellular physiology and secondary product formation.

4. Plant molecular marker assisted breeding

Increasingly plant scientists exploit the characteristic feature of better yielding 'hybrids' in plant. Hybrid vigour, or heterosis as it is scientifically known, exploits the fact that some offspring from the progeny of a cross between two known parents would be better than the parents themselves. Many hybrid varieties of several crop species are being grown all over the world today. An example of this is the hybrid tomatoes that we eat commonly.

A technique that uses molecular marker to select for a particular trait of interest, such as yield. A molecular marker is a short sequence of DNA that is tightly linked to the desirable trait (such as disease resistance) that selection for its presence ends up selecting for the desirable trait. E.g. maize that is tolerant to drought and maize streak virus. Some time the embryo rescue is one of the potential tissue culture technique f tissue culture which can be use to develop hybrid plants where successful crossing is not possible due to abortive fruit set.

5. Plant tissue culture by biotransformation

Biotransformations are chemical reactions catalyzed by cells, organs or enzymes. Biotransformations explore the unique properties of biocatalysts, namely their stereo-and regiospecificity and their ability to carry out reactions at no extreme pH values and temperatures. The biotransformation is another technique of tissue culture in which chemical conversion of an exogenously supplied substance can be done by the living cell culture. It is for commercial exploitation of secondary metabolites. Biotransformation is defined as chemical transformations that are catalyzed by microorganism or their enzymes (Vaniserce et al.2004). The unlimited enzymatic potential of plant cell culture can in principle is used for bioconversion purpose. Biotransformation is an area of biotechnology that has achieved a considerable attention. It is the process in which pharmaceutically less important precursor is converted into the compound of interest. Biotransformation of digitoxin to gitoxin purpurea glycoside-A & purpurea glycoside-B from suspension culture of

D. Purpurea (Srivastava and Srivastava,2007) Such modification can takes place at several positions in the molecules.Cell suspension cultures, immobilized cells, enzyme preparations and hairy root cultures can be applied for the production of food additives or pharmaceuticals by biotransformation process.

6. Biotechnology - Means to achieve food security

Agricultural productivity is important for food security in that it has an impact on food supplies, prices, and the incomes and purchasing power of farmers. Improving food security at the national level requires an increase in the availability of food through increased agricultural production.

Historically, increased food production in the developing countries can be attributed to the cultivation of more land rather than to the deployment of improved farming practices or to the application of new technologies. By its very nature, agriculture threatens other ecosystems, a situation that can be exacerbated by over-cultivation, overgrazing, deforestation and bad irrigation practices. However, increased demands for food in Asia, Europe and North Africa have to be met by increasing yields because most land in these areas is already used for agriculture. It is in this scenario that various biotechnology techniques can come handy to be employed to enhance the yield and productivity.

Different types of crops have been produced using the molecular tools of biotechnology and are beginning to be utilized in agricultural systems all over the world. Biotechnology has the potential to assist farmers in reducing on-farm chemical inputs and produce value-added commodities. Conversely, there are concerns about the use of biotechnology in agricultural systems including the possibility that it may lead to greater farmer dependence on the providers of the new technology. Biotechnology is technology based on biology, especially when used in agriculture, food science, and medicine. The UN Convention on Biological Diversity has come up with one of many definitions of biotechnology. “Biotechnology means any technological application that uses biological systems, living organisms, or derivatives thereof, to make or modify products or processes for specific use.”

6.1 Attaining food security

Global food productivity is undergoing a process of rapid transformation as a result of technological progress in the fields of communication, information, transport and modern biotechnology. A general observation is that technologies tend to be developed in response to market pressures, and not to the needs of the poor who have no purchasing power. As agriculture is the main economic activity of rural communities, optimizing the levels of production will generate employment and income, and thus uplift the wealth and well-being of the community. Improving agricultural production in developing countries is fundamental to reducing poverty and increasing food security.

Investment to raise agricultural productivity can be achieved through the introduction of superior technologies like meristem culture for virus free plants, micropropagtion for obtaining elite disease free planting material. Other such measures would include using techniques which are: (i) environmentally sound, preserving resources and maintaining production potential (ii) profitable for farmers and workable on a long-term basis (iii) providing food quality and sufficiency for all people (iv) socially acceptable (v) socially equitable, between different countries and within each country

The production problems experienced by farmers vary between countries and communities, and technological solutions need to be relevant to those circumstances, i.e. one solution will not be suitable everywhere. Indeed, such programs are now widely accepted as being at the core of sustainable agriculture. Producing nutritionally enhanced properties in staple crops eaten by the poor could reduce the burden of disease in many developing countries. For example scientists at the International Crops Research Institute for the Semi-Arid Tropics (ICRISAT, India) have developed a pearl millet variety enhanced with beta-carotene. This has not only resulted in producing a crop which is widely used by the poor with increases productivity but has also added value to the nutrition content of the crop. There is a need to direct research efforts to areas which are able to generate sustainable and long term solutions to food problems and which are not governed by the considerations of only pure commercial interests.

6.2 Biotechnology and food processing

Biotechnology includes a wide range of diverse technologies and they may be applied in each of the different food and agriculture sectors. It includes technologies such as gene modification (manipulation) and transfer; the use of molecular markers; development of recombinant vaccines and DNA-based methods of disease characterisation/diagnosis; in-vitro vegetative propagation of plants; embryo transfer and other reproductive technologies. It also includes a range of technologies used to process the raw food materials produced by the crop and forestay. It is an area that receives relatively little attention from the media, but which is very important for food security in many developing countries. Biotechnology has been associated with general improvement of health and life styles of human beings for quite some time. In fact a lot of research being conducted in the field of biotechnology is concentrated in the area of genetic engineering, DNA fingerprinting, recombitant DNA (rDNA) analysis and other related areas are directed to unravel the mysteries of life and diseases to enhance the quality of life and improve living standards. Needless to say that this exciting and yet unexplored (unexplored in the sense that still a vast ocean is still lying there to be conquered as far as human conditions and diseases are concerned that offers tremendous opportunities to drug manufacturing companies and academic institutions to undertake and sponsor research and pilot projects for giving shape to ideas and concepts.

The Department of Biotechnology (DBT) was set up under the Ministry of Science and Technology in 1986. This gave a new boost to the development of biotechnology in India. The DBT has set up many Centres of Excellence in the country. These centers are responsible for generating skilled manpower, developing research initiatives and opportunities as well as supporting R&D efforts of private industries and providing platform to them to out source their research activities to these centres. This has promoted interactions between the academics and the industry which has resulted in several industry houses and entrepreneur initiatives to take root and grow biotechnology in India.

The Indian government has laid down a decent regulatory framework to approve GM crops and r-DNA products for human health. A proactive government policy allows stem cell research in the country while having in place sound ethical guidelines. The product patent regime which has come into force since the year 2005 and resulted in giving a message to the world as well as Indian Industry that India supports world regulatory framework and rewards new research and initiatives. The second amendment of the Indian Patents Bill include a 20-year patent term, emergency provisions and commencement of R&D immediately after the filing of patents. The bill is compatible with the provisions of WTO and TRIPS and make Indian laws compatible with what has been agreed within the framework of these multi lateral negotiations.

Several states have taken their own initiatives in terms of defining their own biotech policies to give an impetus to the industry in this sector and to biotechnology in India as a whole. States like Andhra Pradesh, Karnataka, Gujrat, Maharashtra, Kerala, Tamil Nadu and Himachal Pradesh are developing biotech parks. They are encouraging research activities, establishing links between their research institutions and industry. In the future, biotechnology can improve health by using and understanding genetic code to fight diseases, and it gives doctors the tools to not only treat disease, but prevent and cure diseases with a personalized approach. Stem cells are building block cells for every organ, tissue and cell in the human body. Preservation of one's own stem cells could one day help restore parts of one's body that become damaged due to disease or injury.

6.3 Biotechnology in Indian context

Biotechnology is the sunrise industry and perhaps one of the fastest emerging sectors in the new generation Indian economy, resulting in tremendous growth in research initiatives and jobs creation. The vast base of intellectual prowess and well developed skill sets, R&D facilities and cost advantage all tend to make India as one of the most promising markets in the world and offer great opportunities to business and industry for making investments and reaping the benefits thereof. The government is also trying hard to frame and modify the policy framework so as to make India a vibrant hub of bio technology research and infrastructure development centre.

Biotechnology in India can be divided into three broad areas – Medicinal sector relating to improved human and animal health care and drug research, agricultural sector leading to introduction of new crops and improved plant produce and industrial arena leading to establishment of new economy companies and industries.

The core competence of Indian biotech industry exists in the following areas and sectors are (1) Growth of fermentation-based products (2) Use of plant and animal parts for extracting value added products of high purity (3) Use of cell and microbial culture techniques (4) Plant breeding techniques and animal breeding technology based on molecular methodology (5) Plant cell/tissue culture, etc. (6) DNA technology of plants and animals (7) Isolation of plants and animal products (8) Bioprocess engineering (9) Gene manipulation of microbes and animal cells.

India is a success story as far as development in the Bio technology sector is concerned for betterment of health care and animal produce and agricultural growth and plant produce. According to a NASSCOM-KPMG Study, the biotech industry's R&D and services will reach a turnover of US$3 billion by 2010 and the bioinformatics market will touch US$ 2 billion. Indian companies have become providers of biotech information to clients around the world. Sequencing genes and delivering genomic information for big pharmaceuticals companies is the next boom industry. India has the potential to become one of the prime forces in the development and manufacture of genomic drugs with right climate and well directed research efforts and initiatives. At present, there are about 190 biotech companies in India. The Indian Biotech Scenario India comprises a large market for biotech based products. However most of these products are not developed here but are imported by domestic and foreign firms. As per the analysis done by capitalmarket.com, the Indian biotech industry is expected to reach Rs. 4,40,000 Crores in 2020.

6.4 Biotechnology in food- risk assessment

Probably no discovery in green biotechnology arena has had, in so short a time, such far reaching consequences on agriculture as the method reported in 1983 for the genetic modification of plants using gene technology. In 2005, such genetically modified varieties comprised 60% of global soy bean cultivation, 14% of maize, 28% of cotton and 18% of rape seed; between 2003 and 2005 the overall increase of the area worldwide given over to GM crops was 33%. This clearly demonstrates that the application of gene technology in agriculture has economically been very successful.

Genetic modifications in crop plants have focused primarily on the production of varieties for cutting harvest losses due to insects and weeds. More recent developments are directed to protection against viral and fungal infections, the enhancement of tolerance towards drought and salinity, the formation of male sterile plants for the generation of productive hybrids, and the improvement of the nutritional quality of crop plants and increasing the shelf life of many perishable items such as tomatoes by identifying gene responsible for early ripening of the same.

While the benefits from genetically modified organisms are many and more and more research is getting directed towards the field, the use of GMOs may also involve potential risks for human health and development. Many genes used in GMOs have not been in the food supply before. While new types of conventional food crops are not usually subject to safety assessment before marketing, assessments of GM foods are generally undertaken before the first crops are commercialized. There is a need for risk assessments to consider both the intended and unintended effects of such foods in the food supply. GM foods currently traded on the international market have passed risk assessments in several countries and are not likely, nor have been shown, to present risks for human health. However still a lot of misgivings are there which need to be addressed painstakingly.

The application of modern biotechnology to food production presents new opportunities and challenges for human health and development. Recombinant gene technology, the most well-known modern biotechnology, enables plants, animals and microorganisms to be genetically modified (GM) with novel traits beyond what is possible through traditional breeding and selection technologies. It is recognized that techniques such as cloning, tissue culture and marker-assisted breeding are often regarded as modern biotechnologies, in addition to genetic modification.

The challenge of producing more food grains to feed the ever increasing population of world and reducing the agricultural waste and increasing productivity with less resources has brought companies to invest in GM crops and undertake new research in the field. The benefits are many and results encouraging. What is needed is a concerted efforts to dispel fears.

6.4.1 Assessment of the impact of GM foods on human health

Introduction of GM crops for human consumption has been fraught with controversies and conflicts of interests of various parties such as governments, corporations and research institutions and farmers and consumers. In order to balance the stakes of all the stake holders and carry out adequate risk assessment procedure, it is imperative that the first and most important assessment- impact on human health is given due consideration by policy makers around the world. It is not that the world governing bodies are not aware of the issue. In July 2003, the Codex Commission adopted the following principles, which though are not binding on national governments but are considered while carrying out the assessment:

(a) Principles for the risk analysis of foods derived from modern biotechnology;

(b) Guideline for the conduct of food safety assessment of foods derived from recombinant-DNA plants;

(c) Guideline for the conduct of food safety assessment of foods produced using recombinant-DNA microorganisms.

These principles and guidelines presuppose carrying out a pre market assessment, performed on a case-by-case basis and including an evaluation of both direct effects (from the inserted gene) and unintended effects (that may arise as a consequence of insertion of the new gene). These principles and guidelines in order to assess the quality and impact of GM foods require investigation of the following points:

(a) direct health effects (toxicity);
(b) tendency to provoke allergic reactions (allergenicity);
(c) specific components thought to have nutritional or toxic properties;
(d) stability of the inserted gene;
(e) nutritional effects associated with the specific genetic modification; and
(f) any unintended effects which could result from the gene insertion.

Potential direct effects on human health

The potential direct health effects of GM foods are generally comparable to the known risks associated with conventional foods, and include, for example, the potential for causing allergy and toxicity and its impact on the nutritional quality and microbiological safety of the food. While many of these points have not been traditionally assessed for conventional food, the safety assessment of GM food follows a stepwise process aided by a series of structured questions. Factors taken into account in the safety assessment include:-

- Identity of gene of interest, including sequence analysis.
- Source of gene of interest;
- Composition of GMO;
- Protein expression product of the novel DNA;
- Potential toxicity;
- Potential allergy cause

It is important that due consideration is given by all the policy decision makers to these areas so as to instill a sense of confidence among those who are skeptical about the use of GM crops for human consumption. This will have a positive impact on the growth of food security among nations.

6.4.2 Economic cost of adopting GM crops

Numerous reports from organizations either in favour or critical of GM foods have been published, and numerous claims for increased or decreased profitability of agricultural practices including GMOs can be found in world literature.

A review of the United States National Centre for Food and Agricultural Policy concludes that biotechnology is having, and will continue to have, a significant impact on improved yields, reduced grower costs and reduced pesticide use. GM Bt cotton seems to have relevant benefits for smallholder farmers in many areas around the world. On the other hand, some report lower yields, continuing dependency on chemical sprays, loss of exports and critically reduced profits for farmers as a consequence of using biotechnology.

A United States Department of Agriculture report on the economic consequences of GM crops summarized a positive impact of the adoption of Bt cotton on net farm returns, but a negative impact in the case of Bt maize. An improvement of returns has also been seen with herbicide-tolerant maize, whereas no significant impacts were observed with herbicide-resistant soybean. A very detailed study by the European Commission on the economic impact of GM crops on agriculture found that a quick adoption by farmers in the USA was the result of strong profitability expectations. However, there was no conclusive evidence on the farm-level profitability of GM crops.

The most immediate and tangible ground for farmer utility of GM crops appears to be the combined effect of performance and convenience of GM crops — in particular, herbicide-tolerant varieties. These crops allow for greater flexibility in growing practices and, in given cases, for reduced or more-flexible labour requirements. For insect-resistant crops like Bt maize, yield losses are more limited than for conventional maize. However, the cost-efficiency of Bt maize depends on a number of factors, especially growing conditions.

Profitability of GM crops should be analyzed within a long-term time frame. Firstly, there are important yearly fluctuations in yields and prices and it is difficult to isolate the possible effects of biotechnology. Secondly, developments on the supply and on the demand sides of the food chain have to be jointly considered. A recent study analyzing international diffusion of gains in the use of GMOs shows the need for differentiation between crops and regions. In China, a region with a typically high baseline of pesticide use and cases of pesticide poisoning in farmers, a report showed that the use of Bt cotton substantially reduced the use of pesticides without reducing the output per hectare or the quality of cotton. This resulted in substantial economic and health benefits for small farmers.

There seems to be evidence of profitability of certain GM crops under specific situations, especially growth conditions which are significantly dependent on regional agro-ecological factors, especially the baseline of pest pressure and pesticide use. On the other hand, there seem to be situations where these factors would not provide for profitability of growing GM crops, or where other practices for planting may be more valuable because of various regional or market-related reasons.

Impact of GM Crops on World Agriculture Scenario

Probably no discovery in plant sciences has had, in so short a time, such far reaching consequences on agriculture as the method reported in 1983 for the genetic modification of plants using gene technology. In 2005, such genetically modified varieties comprised 60% of global soy bean cultivation, 14% of maize, 28% of cotton and 18% of rape seed; between 2003 and 2005 the overall increase of the area worldwide given over to GM crops was 33%. This clearly demonstrates that the application of gene technology in agriculture has economically been very successful.

Genetic modifications in crop plants have so far focused primarily on the production of varieties for minimising harvest losses due to weeds, and the generation of insect-resistant varieties to decrease losses from insect damage. More recent developments are directed to protection against viral and fungal infections, the enhancement of tolerance towards drought and salinity, the formation of male sterile plants for the generation of productive hybrids, and the improvement of the nutritional quality of crop plants, for example by modifying the fatty acid composition in oil seeds.

The advent of the genetically modified seeds and plant for increasing productivity and reducing crop losses is a boon for countries like India from the point of view of food security and fighting hunger and poverty. Government also needs to frame suitable and transparent policy framework for bringing out comprehensive legislation on the industry and suitable safeguards to avoid the possible pitfalls.

Rapid strides in this field of agricultural biotechnology has allowed and opened new doors to scientists, companies and policy makers to explore the possibilities of use of the technology in farming. Today even in developing countries, more and more land is being planted with genetically modified varieties of an ever-expanding number of crops. Research efforts are being made to genetically modify most plants with a high economic value such as cereals, fruits, vegetables etc. Rapid strides in this field of biotechnology has allowed and opened new vistas of opportunities to scientists and companies to explore the possibilities of use of the technology in farming. Today even in developing countries, more and more land is being planted with genetically modified varieties of an ever-expanding number of crops. Research efforts are being made to genetically modify most plants with a high economic value such as cereals, fruits, vegetables etc. Challenge for countries like India is to reap the benefits of the new technology and safeguard its interests through various measures. The challenge of producing more food grains to feed the ever increasing population of India that has already crossed one billion mark with less resources has bought companies to invest in GM crops.

A lot of awareness campaigns have to be conducted to reach out to the farmers to brief them about the benefits of using seeds that are resistant to pests, diseases, herbicides, and crops which are tolerant to drought, cold, salinity and other harsh environments. This will bring in confidence among the farmers as well as policy makers.

Initiatives to promote biotechnology in India

Keeping in view the tremendous opportunities offered by various facets of biotechnology sector for the development of nation and addressing issues of food security, agricultural production and yield, industrial pollution and application and medicinal benefits, the Indian Government has taken a number of initiatives to promote the growth of the biotechnology sector in India.

An Empowered Group of Ministers (EGOM) while considering the provisions of setting up of Special Economic Zone (SEZ) and the SEZ Act and Rules has relaxed provisions in terms of both the land area and built-up area requirement for biotechnology sector to the tune of 10 hectares and 40,000 square metres. This would encourage entrepreneurship, innovation and greater participation of small investors in the areas of biotechnology in general and health and agriculture in particular.

The Government has taken special measures to promote the Biotechnology industry and to facilitate increase in turnover of biotech sector. Initiatives have been taken to provide an enabling environment for industrial growth such as exemption of biotech sector from compulsory licensing; permitting 100 per cent FDI in the sector, reducing the area of SEZ to bring it at par with IT sector; providing fiscal incentives to in-house R&D recognized industries in terms of exemption of custom duty on capital goods, reduction in import duty and 150 per cent weighted deduction against expenditure incurred on in-house R&D.

Department of Biotechnology is also supporting R&D and technology development, both in the public and private sector, for developing products and processes which provide affordable biotechnology solutions to the growing food insecurity and health problems and issues. R&D in agriculture biotechnology is being supported to develop improved crop varieties resistant to abiotic and biotic stress, specially drought. Salt resistant transgenic rice has been developed and is currently under field trial. Bio fortification of important staple crops such as rice, wheat and maize for enhancing macronutrients such as iron and zinc has been initiated. The country also participated in the International Rice Genome Programme with ten other countries. Full sequence of rice genome has been completed. It is expected that the decoded rice genome will help in the discovery of gene and DNA makers for development of improved varieties.

In the health sector, research is continuing for developing low cost and affordable solutions for public health. A number of vaccines and diagnostics have been developed which are under different stages of trial. Diagnostics for HIV, Japanese Encephalitis have been transferred to industry. Rota virus vaccine is in the advanced stages of trial along with the industry.

Indian biotech investment

India, the darling of the world as far as bio technology sector is concerned offers tremendous opportunities to companies to make investment in the sector in India. Growth of all three areas of bio technology- medicinal, agricultural and industrial and conducive climate for the same make India as one of the ideal destination for world investment flow in India. Following are a few areas where opportunities exist for India:

Agriculture sector

Hybrid seeds, including genetically modified seeds such as Bt cotton represent new business opportunities based on yield improvement, and development of a production base in biopesticides and biofertilisers would facilitate India's entry into the growing organic or natural foods market. The Genetically Modified crops like corn, cotton, millet, mustard and other nutritionally improved vegetables also provide good potential in the agriculture sector and also leads to improvement in farm produce and productivity per hectare.

Vaccines

India's huge and growing population makes it among the world's largest markets for vaccines of all types. India faces a growing demand for new-generation and 'combination' vaccines, such as DPT with Hepatitis B, Hepatitis A and injectable polio vaccine, besides several veterinary and poultry vaccines. Apart from conventional vaccines, the rDNA have further market potentials and offer great opportunities to companies in the economy.

Medicinal discovery

Opportunities exist for enhancing production facilities and economies of scale based on licensing, joint ventures, setting up of new production bases and establishing royalty sharing arrangements for all therapeutic and medicinal products approved for marketing in India, namely Insulin, Alpha, Interferon, Hepatitis B surface antigen based vaccine, Erythropoietin, Streptokinase, Chymotrypsin, and others.

Medicinal research

New research and developments in the field make India as a hub of cutting edge technology for development of new products and medicines having ready and developed market for the same. Indian pharmaceutical companies possess competitive skills in chemical synthesis and process engineering and extraction technologies, which they can leverage to develop new drugs and formulae. New investment into research and successful defending of patents by a number of Indian companies in ten world markets has opened new vistas of opportunities for Indian companies.

Clinical research and trials

With clinical trials in India costing less than a fraction of what it costs in developed markets, clinical research organizations can seek research and trial projects in India from international companies, provided they are able to demonstrate best international practices and follow up procedures.

Bioinformatics

Indian bioinformatics companies can play a significant role in critical areas such as data mining, lexican, mapping and DNA sequencing and extraction, molecule design simulation in world market for bilinformatics services. Complex algorithm writing and the use of computational capacities to study the 3D structures of proteins are the main skills required in this arena and India offers good investment opportunities for the same.

Plant production biotechnology

There is scarcely any aspect of plant production that will not undergo profound changes as a result of the application of biotechnology. Commercial applications of plant genetic engineering have not yet occurred. At the present time, more traditional aspects of biotechnology such as tissue culture have had an important impact, especially in the acceleration of the breeding process for new varieties and in the multiplication of disease-free seed material.

Realizing the potential of plant tissue culture technology in revolutionizing the commercial agriculture sector by enabling mass propagation of elite, high yielding and disease free plants throughout the year, the Department of Biotechnology (DBT) has identified it as a priority area and initiated a number of programmes aimed at development and commercialization of the technology in an integrated manner. A number of research and development projects in various research institutes and universities have been supported for perfecting protocols of important plant species namely forest trees, horticulture crops and plantation crops both Govt. and private partnership (Table-5).

Table 5 : List Of Plant Species Which Have Been Perfected For Large Scale Propagation

Plant category	**Plants**
Fruits	Banana, Grapes, Pineapple, Strawberry, Sapota
Cash crops	Sugarcane, Potato
Spices	Turmeric, Ginger, Vanilla, Large Cardamom, Small Cardamom
Medicinal plants	Aloe vera, Geranium, Stevia, Patchouli, Neem
Ornamentals	Gerbera, Carnation, Anthurium, Lily, Syngonium, Cymbidium
Trees	Teak, White teak, Bamboo, Eucalyptus, populus

Govil and Gupta (1994) reported 21 Indian commercial units in production in 1994 with targeted production capacity of 80 m plants and 14 units in the pipeline with annual production capacity of 60 m plants. Four units had annual targeted production capacity of over 10 m, eight with capacity of 1-5 m plants while nine units with targeted production of 5 m plants each were in the pipeline. Of a total of 75 units covered in this article, details of production capacity of 30 Indian commercial units are available. An analysis shows that there are 4 major units with production capacity varying from 15 to 25 m, 3 major units with capacity of 10 m, 12 units with production capacity of 5 to 6.25m, 4 units with production ranging from 1.5 to 3 m, 7 units including 2 pilot plants with production capacity of 1 m, totalling to 190 m plants per annum.A perusal of plants micropropagated by Indian commercial units (Table.2,6) indicates that plants belonging to the horticulture segment of the ornamental industry are the major items being produced This is also the international trend as shown earlier (Govil and Gupta, 1994). It is clear that of 14 units engaged in micropropagation of fruit crops, 12 are producing banana, 5 are aseptically multiplying strawberry, while 3 each are engaged in papaya and pineapple production. Only 6 units are producing 7 forest tree species 72 leaving enough scope for expansion. Twenty four units are producing a large number of ornamentals. A few vegetable crops and some plantation crops like cardamom, vanilla, sugarcane and ginger are also being multiplied (Table. 6). Indian units need to strike an intelligent balance between product diversity as well as quantity and should try to introduce novelty and exploit commercial potential of endemic ornamentals as new entrants. In fact, production by micropropagation units is market-driven and based on advance orders. The production of large number of ornamental and horticultural plants reflects their demand by consumers as well as their willingness to pay the higher cost. Globally, the floriculture market alone is supposed to be of Rs. 750 billion level. While considering costbenefit ratio for identification of products, social and environmental aspects whose monetary value cannot be estimated, should also be given due consideration. With the liberalised industrial policy, incentives provided by the Government and demonstration efforts made by the commercial units, tissue culture products of our units are fast capturing domestic market as well as increasingly becoming popular in international markets. Looking at the brighter prospects of the industry, Government of India had fixed export target to Rs. 1300 million for cut-flowers, seeds, live plants and tissue culture products in 1995-96 (Kumar, 1994).

Table 6 : List of Current Commercial Micropropagated Tissue Culture Units in India.

Sl.No.	**Name of the Unit**	**Fruit Crops**
1	Harrisons Malayalam Ltd.	Banana, *Citrus, Grapes, Papaya, Pomegranate, Strawberry
2	Godrej Plant Biotech Ltd.	Banana, Pineapple, Strawberry
3	IAHS	Banana, Cherryapple, Peach, Pineapple, Strawberry
4	Nath Seeds	Banana
	Bio Tissue Labs	Banana
	Microplantae,	Banana
	Kumar Gene Tech,	Banana
	SIV Industries Ltd.	Banana
	Maharashtra State Seed Corporation,	Banana
	National Agricultural & Scientific Research Foundation	Banana
5	A.V. Thomas & Co	Banana, Pineapple

Sl.No.	Name of the Unit	Fruit Crops
6	SPIC	Banana, Datepalm, Mango, Papaya (Female)
7	Enzo-chem	Papaya
8	Agrigene International	Apple
9	Microplantae	Strawberry
	Indrayani Biotech	Strawberry
	Saubhagya Kalpataru Pvt. Ltd.	Banana, Strawberry
	FOREST TREES	
1	Harrisons Malayalam Ltd.,.,.	Eucalyptus
2	Tata Tea Ltd	Eucalyptus
3	SIV Industries Ltd	Eucalyptus
4	Thapar Corp. R&D Centre,	Eucalyptus
5	IAHS	Teak, Kadamb
6	SPIC	*Calistemon*, *Eucalyptus*, Jackfruit, *Simaruba*
7	Godrej Plant Biotech	*Salavadora*
8	Saubhagya Kalpataru Pvt. Ltd.	Bamboo
ORNAMENTALS		
	NAME OF THE UNIT	Foliage plants
1	Harrisons Malayalam Ltd	*Aglaonema*, *Calathea*, *Cordyline*, *Diffenbachia*, *Ficus*, *Philodendron*, *Spathiphyllum*, *Syngonium*
	NAME OF THE UNIT	Flowering plants
1		*Alstroemeria*, *Anthurium*, *Callalily*, Carnation, *Cestrum*, *Chrysanthemum*, *Gerbera*, *Impatiens*, *Limonium*, Rose (Miniature)
2	IAHS	*Alpinia*, *Anthurium*, *Calethea*, *Callalily*, *Chrysanthemum*, *Cordyline*, *Croton*, *Diffenbachia*, *Dracaena*, *Eryngium*, ***Ficus***, *Freesia*, *Gardenia*, *Geranium*, *Gerbera*, *Gladiolus*, *Hydrangea*, *Iris*, *Impatiens*, *Lilies*, *Maranta*, Orchids, *Philodendron*, Rose, *Spathiphyllum*, *Syngonium*
3	SPIC,	*Anthurium*, SPIC *Anthurium*, *Calathea*, *Cordyline*, *Gerbera*, *Maranta*, Orchids, Rose, *Syngonium*
4	Godrej Plant Biotek	*Anthurium*, *Calathea*, *Ficus*, Lily, Ornamental Bamboo, *Spathiphyllum*, *Syngonium*
5	Agrigene International	*Ficus*, *Spathiphyllum*
	A.V. Thomas & Co.	*Chrysanthemum*, *Gerbera*, Lily, Orchids,

Sl.No.	Name of the Unit	Fruit Crops
		Rose
6	Cadilla Laboratories Ltd	*Calathea*, Lily, *Spathiphyllum, Syngonium*
7	Kalpataru Irrigation Systems Ltd.	Lily
8	Rallis India	Carnation, Orchids
	COSTFORD promoted Unit	*Dendrobium* (Orchid), *Phaelenopsis*
1	Phytoclone	*Dendrobium*
2	Pocha seed,	Ornamentals
3	Khoday Biotek	Ornamentals
4	Padamjee Paper Mills, ,	Foliage plants
5	Florida Plant Cultures	Foliage plants
	Green Laboratories	Foliage plants
6	National Agricultural & Scientific Research Foundation	*Anthurium, Spathiphyllum*
7	Raj Agrobase Pvt. Ltd.,	Orchids
8	SIV Industries Ltd.,	Orchids
9	Beena Nursery	Orchids
10	Indrayani Biotech	Rose
11	Kumar Gene Tech,	Flowering plants
12	Microplantae	Flowering plants
13	Khushboo Kenibreed Plants	Flowering plants
14	Kenibreed Plants	*Cymbidium*, *Oncidium*, *Phalaenopsis*, *Vanda*
15	Saubhagya Kalpataru Pvt. Ltd	Rose
16	SIV Industries Ltd.	*Chrysanthemum*, *Philodendron*
	NAME OF THE UNIT	VEGETABLE CROPS
1	Harrisons Malayalam Ltd.	Potato, *Asparagus*
2	SPIC	Potato
3	Proagro	*Brassica napus, Brassica juncea, Lycopersicon esculentum* (tomato)
4	Frontier Biotech	Microtubers of Potato (*Solanum tuberosum*)
5	Hindustan Lever Ltd.	Microtubers of Potato (*Solanum tuberosum*
	AgroIndia Biotechnologies	Microtubers of Potato (*Solanum tuberosum*
6	Kalyani Agrotech	Microtubers of Potato (*Solanum tuberosum*
7	IAHS	*Asparagus*, *Cauliflower, Cabbage*, Tomato
8	Agrigene International	*Asparagus officinalis* (exotic vegetable)
1	NAME OF THE UNIT	PLANTATION CROPS
2	Harrisons Malayalam Ltd.	Cardamom, Sugarcane, Vanilla
3	IAHS	Cardamom, Coffee, Ginger, *Miscanthus*, Sugarcane
4	Tata Tea Ltd.	Coffee, Tea
5	SPIC	Cardamom, Vanilla
	A.V. Thomas & Co	Cardamom, Ginger, Turmeric, Vanilla, Black Pepper, Tea
6	Bush Boake Allen	Vanilla

Contd. ...

Sl.No.	Name of the Unit	Fruit Crops
7	Dhampur Sugar Mills,., ,	Sugarcane
8	Frontier Biotech	Sugarcane
9	Saubhagya Kalpataru Pvt. Lt	Sugarcane
10	Western India Tissue Tech Ltd.,	Sugarcane
11	Agri India Biotechnologies	Sugarcane
12	EID Parry Ltd.	Sugarcane
13	Rallis India Ltd	Sugarcane
14	Vasant Dada Sugar Institute	Sugarcane
15	Kalpataru Irrigation Systems Ltd.	Coffee, Tamarind
16	National Agricultural & Scientific Research Foundation	Cardamom

Recommendations

The widespread application of in vitro propagation depends on its cost-competitiveness and is profitable only when there is an associated advantage over conventional propagation methods. We should set up modest, commercially viable units at appropriate sites taking into account cost of land, construction, availability of infrastructure such as water, power, sanitation, labour and distance from airport. We should encourage horizontal technology transfer from viable operating indigenous units rather than purchasing technology from abroad. Import of equipment needs to be minimised and costly consumables like agar-agar should be substituted by 'ragi' or 'maida' and sucrose by Indian table sugar. In-house R&D units and interaction with nearby R&D institutes need to be strengthened to optimise production and solve contamination, hardening and related problems. The units should carefully select plant material taking into consideration availability of stocks, reproducible protocol with rapid multiplication rates, demand in domestic and international markets as well as build good market reputation by delivering quality products on time. The micropropagation industry is particularly well suited for a country like India as it is environment friendly and labour-intensive. It can really solve our unemployment problem to a reasonable extent, only if planned judiciously.

Blessed with diverse agro climatic zones and availability of both skilled manpower as well as labour at cost effective rates, combined with liberalised Government policies and incentives provided, Indian tissue culture industry, though a little late starter, is all set to have its presence felt on the international scenario and make a dent in the national economy by earning considerable foreign exchange on a recurring basis.

Provision of seeds

Plant breeding has been enhanced considerably by in vitro development of improved varieties which are better adapted to a specific environment. The application of tissue culture has several advantages, including rapid reproduction and multiplication, availability of seed material throughout the year etc. Since the application of tissue culture does not require very expensive equipment, this technology can be applied easily in developing countries and can help to improve local varieties of food-crops. For example, using traditional methods for propagating potatoes.

Reduced use of agrochemicals

Biotechnology can help reduce the need for agrochemicals which small farmers in developing countries often cannot afford. A reduction in the use of agrochemicals implies fewer residues in the final product. This is expected to enhance the productivity and land fertility as well as reduction in toxic elements in crops.

Increased production

Biotechnology can be used in many ways to achieve higher yields; for example by improving flowering capacity and increasing photosynthesis or the intake of nutritive elements. Productivity increases may lead to lower prices, which is a vital policy objective in many a developing nation.

Improved harvesting

The cloning of plants can help to reduce the work necessary for harvesting. When individual plants show more uniform characteristics, grow at the same speed and ripen at the same time, harvesting will be less laborious. A reduction in the workload is not only an objective in highly industrialized countries, it can also be very important for small farmers in developing countries.

Improved storage

Food shortages would not exist in many countries if the problem of post-harvest losses could be solved. In the future, genetic engineering may be used to remove plant components that cause early deterioration of the harvest. For instance, a technique to reduce the presence of a normal tomato enzyme involved in the softening of ripe tomato fruit has been patented and would be found very useful for enhancing the shelf life of crops of various varieties.

Bioinformatics

Bioinformatics and computational biology involve the use of techniques and methods including applied mathematics, informatics, statistics, computer science, artificial intelligence, chemistry and biochemistry to solve biological problems. Major research efforts being carried out in the arena include sequence alignment, gene finding, genetic engineering, DNA finger printing, genome assembly, protein structure alignment, protein structure prediction, prediction of gene expression and protein-protein interactions, and the modeling of evolution. This is the sunrise and multi billion dollar industry and millions of research hours are being spent in the field in developed countries to explore better and more effective results based on data.

The terms bioinformatics and computational biology are often used interchangeably. However bioinformatics more properly refers to the creation and advancement of algorithms, computational and statistical techniques, and theory to solve formal and practical problems posed by or inspired from the management and analysis of biological data. Computational biology, on the other hand, refers to hypothesis-driven investigation of a specific biological problem using computers, carried out with experimental and simulated data, with the primary goal of discovery and the advancement of biological knowledge.

In the last few decades, advances in molecular biology and the equipment available for research in this field have allowed the increasingly rapid sequencing of large portions of the genes. This deluge of information has necessitated the careful storage, organization and indexing of sequence information. Information science has been applied to biology to produce the field called Bioinformatics.

The simplest tasks used in bioinformatics concern the creation and maintenance of databases of biological information. Nucleic acid sequences (and the protein sequences derived from them) comprise the majority of such databases. While the storage and or organization of millions of nucleotides is far from trivial, designing a database and developing an interface whereby researchers can both access existing information and submit new entries is only the beginning. These are some of the concepts associated with the field of bioinformatics:

(a) Finding the genes in the DNA sequences of various organisms.
(b) Developing methods to predict the structure and/or function of newly discovered proteins and structural RNA sequences.
(c) Clustering protein sequences into families of related sequences and the development of protein models.
(d) Aligning similar proteins and generating phylogenetic trees to examine evolutionary relationships.

Conclusion

Since last two decades there have been considerable efforts was made in the use of plant cell cultures in bioproduction, bioconversion or biotransformation and biosynthetic studies. The potential commercial production of pharmaceuticals by cell culture techniques depends upon detailed investigations into the biosynthetic sequence. The great enthusiasms of biotechnologists have been seen in the potential use of cell culture in the production of valuable secondary products. Plant tissue culture is a noble approach to obtain their substances in large scale. Many companies in India and abroad are showing interest in this direction. Tissue culture is an alternative way for the production of phytochemical of therapeutic importance.

Other techniques like hairy root culture, biotransformation, immobilization and elicitations are used for the increased production of secondary metabolites. By this approach an increase production of secondary metabolites are obtained. In present scenario it is an effective and efficient procedure for converting less medicinally important plant metabolites to a valuable product.

Blessed with diverse agro-climatic zones and availability of both skilled manpower as well as labour at cost-effective rates, combined with liberalised Government policies and incentives provided, Indian tissue culture industry, though a little late starter, is all set to have its presence felt on the international scenario and make a dent in the national economy by earning considerable foreign exchange on a recurring basis.

References

Ajithkumar, D. and Seeni, S. 1998. Rapid clonal multiplication through *in vitro* axillary shoot proliferation of *Aegle marmelos* (L) Corr., a medicinal tree. *Plant Cell Reports*, 17: 422-426.

Ambros, P.F., Matzke, A.J.M. and Matzyke, M.A.1986.Localization ofAgrobacterium rhizogenes T-DNA in plant chromosomes by in situhybridization. *The EMBO. J.*, 5: 2073-2077.

Amin, M. N. and Jaiswal, V. S. 1988. Micropropagation as an aid to rapid cloning of a guava cultivar. *Scientia Horticulture*, 36: 89-95.

Amin, M. N. and Jaiswal, V. S. 1993. In vitro res'ponse of apical bud explants from mature tree of Jack fruit (*Artocarpus heterophyllus*). *Plant Cell Tissue Organ Culture*, 33: 59-65.

Amin, M.N. and Jaiswal, V.S. 1987. Rapid clonal propagation of guava through *in vitro* shoot proliferation on nodal explants of mature trees. *Plant Cell Tisue Organ Culture*, 9: 235-241.

Anon. 1994. General Questions and Answers on Biotechnology. Ottawa: Food Protection and Inspection Branch, Agriculture and Agri-Food Canada.

Arokiaraj, P., Yeang, H. Y., Cheong, K. F., Hamzah, S., Jones, H., Coomber, S. and Charlwood, B. V. 1998. CaMV *355* promoter directs ß-glucuronidase expression in the laticiferous system of transgenic *Hevea brasiliensis* (rubber tree). *Plant Cell Reports*, 17: 621-625.

Bhojwani, S. S., Mullins, K. and Cohen, D. 1984. *In vitro* propagation of *Pyrus pyrifolia*. *Scientia Horticulture*, 23: 247-254.

Bodde, T. 1982 Genetic engineering in agriculture: another green revolution. *Bioscience* 32, 572-575.

Bond, J. E. and Roose, M. L., 1998. *Agrobacterium*-mediated transformation of commercially important citrus cultivar Washington novel orange. *Plant Cell Reports*, 18: 229-234.

Brown, D.C.W. and Thorpe, T.A.1995. Crop improvement through tissue culture. World *Journal of Microbiology & Biotechnology*. 11: 409-415.

Cervera, M., Pina, J. A., Juarez, J., Navarro, L. and Pena, L. 1998. *Agrobacterium* - mediated transformation of citrange: factors affecting transformation and regeneration. *Plant Cell Reports*, 18: 271-278.

Chalupa, V. 1984. *In vitro* propagation of Oak *(Quercus robur* L.*)* and Linden *(Tilia cordata* Mill.). *Biologia Plantarum*, 26(5): 374-377.

Chalupa, V. 1988. Large sale micropropagation of *Quercus robur* L. using adenine-type cytokinins and thidiazuron to stimulate shoot proliferation. *Biologia Plantarum*, 30(6): 414-421.

Chand, P. K., Sahoo, Y., Pattnaik, S. K. and Patnaik, S. N. 1995. *In vitro* meristem culture-An efficient *ex situ* conservation strategy for elite mulbery germplasm. In: Mohanty R.C. (Ed) Environment: Change and Management, Kamala Raj Enterprises, New Delhi, India, pp. 127-123.

Collins, G.B. & Grosser. J.W. 1984. Culture of Embryos. In Cell Culture and Somatic Cell Gnetics of Plants,Vol. 1. ed Vasil. I:K.pp. 241-257. New York: Academic Press.

Cox, P.A., Balick, M.J., 1994, *The ethanobotanical approach to drug discovery,* Sci Am , 82–7.

Das, A. B. 1999. The effect of different physiochemical factors on the regeneration of *Cymbopogan polyneuros* Stapf via callus culture and subsequent chromosomal stability. *Israel Journal of Plant sciences* 47:195-198.

Das, A. B., Mallick, R., Chatterjee, A.1998. High frequency regeneration of *Cymbopogon martinii* var. motia as controlled by different physio-chemical factors and their chromosomal stability. *Perspective in Cytology and Genetics* 9: 503-511.

Das, A. B., Das, P., 1991. Shoot multiplication and regeneration related isozyme markers in meristem cultures of *Musa*. *Orissa Journal of Horticulture* 19: 107-117.

Das, A. B., Das, P., 1992. Rapid multiplication of *Musa acuminata* 'Dwarf Cavendish' through meristem culture and evaluation of chromosomal stability. *Orissa Journal of Horticulture* 20: 22-27.

Das, A. B., Mallick, R., Chatterjee, A., 1997. High frequency regeneration of *Cymbopogon polyneuros* Stapf as controlled by different physiochemical factors and their chromosomal stability. *Plant Tissue Culture* 7(1): 1-11.

Das, A. B., Rout, G. R., Das, P., 1995. Somatic embryogenesis from callus culture of a timber yielding tree *Hardwickia binata* Roxb. *Plant Cell Reports* 15(1-2): 147-149.

Das, P., Das, A. B. 1998. Esterase marker during callus mediated regeneration of *Typhonium trilobatum* Schott, a medicinal plant. *Orissa J. Hort.* 26(2): 50-55.

Das, P., Rout, G. R., Das, A. B., 1993. Somatic embryogenesis in callus culture of *Mussaenda erythrophylla* L. Cv. Queen Sirikit and Rosea. *Plant Cell Tissue Organ Culture* 35: 199-201.

Das, T. and Mitra, G. C. 1990. Micropropagation of *Eucalyptus tereticornis* Smith. Plant *Cell Tissue and Organ Culture* 22: 1053.

Datta, S. K., Datta, K. and Paramanik, T. 1983. *In vitro* clonal multiplication of mature trees of *Dalbergia sissoo* Roxb. *Plant Cell Tissue Organ Culture* 2: 15-20.

Davi, Y.S., Mukherjee, B. B. and Gupta. S. 1994. Rapid cloning of elite Teak (*Tectona grandis* L.) by *in vitro* multiple shoot production. *Indian Journal of Experimental Biology*, 32: 668-671.

Douglas, G.C, Keller, W.A & Setterfield, G. 1981. Somatic hybridization between *Nicotiana ruslica* and *N.tabacum.* L Isolation and culture of protoplasts and regeneration of plants fromcell cultures of wild-type and chlorophyll-deficient strains. *Canadian Journal of Botany* 59, 208--219.

Encina, C. L., Barcelo-Munoz, A., Herrero-castano, A. and Pliego Alfaro, F. 1994. *In vitro* morphogenesis of *Annona cherimola* Mill, bud explants. *Journal of Horticultural Science*, 69(6): 1053-1059.

Evans, D. A., Sharp, W. R., Ammirato, P. V. and Yamada, Y. 1983. *Handbook of plant cell culture* ", Macmillan Publishing Co., New York , 1, 4.

Evans, D.A & Bravo, j.E. 1988 Agricultural applications of protoplastfusion. In Pla/1t Biotechnology, ed Marby, T.j. pp. 51-91.Austin: IC Institute, University of Texas.

Farnsworth, N. R., 1985. *The role of medicinal plants in drug development. In*: Kroogsgard-Larsen, P., Brogger Christenses, S., (Eds), *Natural products and drug development* , Denmark: Munsgaard Copenhagen, 17–30.

Feher, A & Dudits, D. 1994 Plant protoplasts and cell fusion and direct DNA uptake: culture and regeneration systems. In : *Plant Cell and Tissue Culture*, eds Vasil. I.K. & Thorpe, TA pp. 71-118. Dordrecht: Kluwer Academic.

Gamborg, O.L., Shylak, j.P. & Shahin, E.A 1981 Isolation, fusion,and culture of plant protoplasts. In : *Plant Tissue Culture: Methodsand Applications in Agriculture*, ed Thorpe. T.A pp. 115-153.New York: Academic Press.

Gautheret, R. J. 1939. *C.R. Hebd. Seances Acad. Sc .,* 208 : 118-120.

Ghosh, N., Smith, D. W., Chatterjee, A., Das A. B. 2010. Regeneration and clonal propagation of the fastest growing leguminous tree *Albizia falcataria* (L.) Fosberg. Botanical Society of America (Abstract) in Developmental and Structural Section.

Glaser, V., 1999, *Billion-dollar market blossoms as botanicals take root,* Nat Biotechnol , 17: 17–8.

Govil, S., and Gupta, S.C. 1994. Plantmicropropagation industry in India retrospects and prospects. *Jour. Sci. Ind.* Res., 53: 768–777.

Goyal, Y. and Arya, H. C. 1984. Tissue culture of desert trees: I clonal multiplication of *Prosopis cineraria* by bud culture. *Journal of Plant Physiology*, 115: 183-189.

Goyal, Y. and Arya, H. C. 1985. Tissue culture of desert trees: II. clonal multiplication of *Zizyphus in vitro. Journal of Plant physiology*, 119:399-404.

Goyal, Y., Bingham, R. L. and Felker, P. 1985. Propagation of the tropical tree, *Leucaena leucocephala* K 67, by *in vitro* bud culture. *Plant Cell Tissue and Organ Culture* 4: 3-10.

Gulati, A. and Jaiwal, P. K. 1996. Micropropagation of *Dalbergia sissoo* from nodal explants of mature trees. *Biologia Plantarum*, 38(2): 169-175.

Gutierrez-E, M. A., Luth, D. and Moore, G. A. 1997. Factors affecting *Agrobacterium*-mediated transformation in citrus and production of sour orange (*Citrus aurantium* L) plants expressing the coat protein gene of citrus tristeza virus. *Plant Cell Reports*, 16: 745-753.

Hammerschlag, F. A. Bauchan, G. R. and Scorza, R. 1987. Factors influencing *in vitro* multiplication and rooting of peach cultivars. *Plant Cell Tissue and Organ Culture*, 8: 235-242.

Hinchee, M.AW., Corbin, D.K, Armstrong, CL., Fry, j.E., Sato,S.s., Deboer, D.L., Petersen, W.L.. Armstrong, TA, Connor-Wand, D.Y., Layton, j.G. & Horsch, R.B. 1994 Plant transformation. In : *Plant Cell and Tissue Culture,* eds Vasil. LK. & Thorpe,TA pp. 231-270. Dordrecht: Kluwer Academic.

Hossain, M., Rahaman, S. M., Zaman, A., Joarder, O. L. and Islam, R. 1992. Micropropagation of *Morus laevigata* Wall, from mature trees. *Plant Cell Reports*, 11: 522-524.

Huang, F. H., Al-Khayri, J. M. and Gubur, E. E. 1994. Micropropagation of *Acacia mearnsii.* In vitro *Cell Developmental Biology*, 30: 70-74.

Ivanicka, J. 1987. *In vitro* micropropagation of mulberry, *Morus nigra L Scientia Horticulturae*, 32: 33-39.

Jaiwal, P. K., Bala. S., Dahiya. S. and Gulati A. 1995. Plant regeneration from c.otyledons of *Dalbergia sissoo* Roxb. - *a* leguminous timber tree. *Physiol and Molecular Biology of Plants*, 1: 37-44.

Kaur. K., Verma. B. and Kant, V. 1998. Plants obtained from the khair tree *(Acacia catechu* Willd.) using mature nodal segments. *Plant Cell Reports*, 17: 427-429.

Kaveriappa, K. M., Phillips, L. M. and Trigiano, R. N. 1997. Micropropatation of flowering dog wood *(Cornus florida)* from seedlings. *Plant Cell Reports*, 16: 485-489.

Kleese, R.A & Duvick, ON. 1980 In *Genetic Improvemerzt of Crops:Emergent Techniques,* eds Rubenstein, I., Gengenbach, B., Phillips,R.L., Green, CE. pp. 24-43. Minne-apolis: University of Minnesota.

Kuckuck, H., Kobabe, G. & WenzeL G. 1991 *Fundamentals of PlantBreeding.* Berlin: Springer Verlag.

Kumar K B (1994) Commercialization of micropropagation. In: Mukhopadhyay S (ed) Commercialization of Micropropagated Plants in India (pp 25-29). Naya Prakash, Calcutta.

Kumar, V., Radha, A. and Chitta, S. K. 1998. *In vitro* plant regeneration of fig *(Ficus carica L cv.* gular) using apical buds from mature trees. *Plant Cell Reports*, 17: 717-720.

Kumar. K.B.1994.Commercialization of micropropagation. In: Mukhopadhyay S (ed) Commercialization of Micropropagated Plants in India (pp 25-29). Naya Prakash, Calcutta

Lal R. & LaL S. 1990 Crop Improvement Using Biotechnology. Boca Raton, FL: CRC Press.

Leeve, V., Lelu, M. A., Jouanin, L, Cornu, D. and Pilate, G. 1997. *Agrobacterium tumifaciens* - mediated transformation of hybrid Larch (*Larix* Kaempferi × L. *decidua*) and transgenic plant regeneration. *Plant Cell Reports*, 16: 680-685.

Lindsey, K., Yeoman, M. M., 1983. Novel experimental systems for studying the production of secondary metabolites by plant tissue cultures. In: Mantell, S. H., Smith, H., (Eds). *Plant Biotechnology* . London : Cambridge Univ. Press, 39–66.

Linington, I. M. 1991. *In vitro* propagation of *Dipterocarpus altus* and *Dipterocarpus intricatus. Plant Cell Tissue and Organ Culture*, 27: 81-88.

Liu, Q., Salih, S. and Hummerschlag, F. 1998. Etiolation of 'Royal Gola' apple *(Malus × domestica* Borkh.) shoots promotes high-frequency shoot organogenesis and enhanced ß-glucuronidase expression from stem internodes. *Plant Cell Reports*, 18 : 32-36.

Loh, C. S. and Rao, A. M. 1989. Clonal Propagation of guajava (*Psidium guajava* L.) from seedlings and grafted plants and adventitious shoot formation *in vitro. Scientia Horticulturae*, 39: 31-39.

Maheshwari, P. & Rangaswamy, N.5. 1965 Embryology in relation to physiology and genetics. *Advances in Botanical Research* 2 : 218-321.

Mante, S., Morgens, P. M., Scorza, R., Cordts, J. M. and Callahan, A. M. 1991. *Agrobacterium*-mediated transformation of Plum (*Prunus domestica* L.) hypocotyl slices and regeneration of transgenic plants. *Biotechnology*, 9 : 853-857.

Mantell, S.H. and Smith, H., 1983. *Plant Biotechnology* , Cambridge University Press, Cambridge , 3.

Mathur, J. and Mukunthakumar, S. 1992. Micropropagation of *Bauhinia variegata* and *Parkinsonia aculeata* from nodal explants of mature tress. *Plant Cell Tissue and Organ Culture, 28*: 119-121.

Miller, C. O., Skoog, F., Von Saltza, M. H. and Strong, F. M., 1955. J . *Am. Chem. Soc.,* 77:1392.

Miller, G. A., Coston, D. C., Denny, E. G. and Rameo, M. E. 1982. *In vitro* propagation of 'Nemaguard' peach root stock. *Horticultural Science*, 17(2): 194.

Mridula, M. K., Gupta, P. K. and Mascarenhas, A. F. 1983. Rapid multiplication of *Sapiurn sebiferum* Roxb. by tissue culture. *Plant Cell Tissue Organ Culture*, 2: 133-139.

Mullins, K. V., Llewellyn, D. J., Hartney, V. J., Strauss, S. and Dennis, E.S. 1997. Regeneration and transformation *of Eucalyptus camaldulensis. Plant Cell Reports*, 16: 787-791.

Naik, G. R. 1998. Micropropagation studies in medicinal and aromatic plants. *In:* Khan, I. A,, Khanun, A. (Eds), Role of biotechnology in medicinal and aromatic plants. Hyderabad : Ukaz Publications, 50–56.

Nakamura, Y., Sawada, H., Kobayashi, S., Nakajima, I. and Yoshikawa, M. 1999. Expression of soyabean M, ß-endogluconase cDNA and effect on disease tolerance in Kiwifruit Plants. *Plant Cell Reports*, 18: 527-532.

Oka, S. and Ohyama, K. 1986. Mulberry (*Morus alba* L.) In: Bajaj YPS (Ed) Biotechnology in Agriculture and Forestry, Vol. 1, Trees 1, Springer-verlag, Berlin, pp. 384-392.

Paily, J. and D' Souza, L. 1986. *In vitro* clonal propagation of *Lagerstroemia flos-reginae* Retz.. *Plant Cell Tissue Organ Culture*, 6: 41-45.

Palai, S. K., Das, A. B. 2002. Large scale propagation of *Musa balbisiana* cv. Muguni through *in vitro* techniques. *In:* Proceedings of the State level Seminar of Advances in production of quality planting materials of Horticultural crops. *Orissa Horticultural Society*, OUAT, BBSR, pp. 133-136.

Palmer, CE. & Keller, W.A 1994 In vitro culture of oilseeds. InPlant Cell and Tissue Culture, eds Vasil, LK. & Thorpe, TApp. 413-455. Dordrecht: Kluwer Academic.

Pandeya, R. & White, F.H. 1994 Delgold: a new Rue-curedtobacco. *Canadian Journal of Plant Science* 64 : 233-236.

Pandeya, R., Douglas, G.C, Keller, W.A, Setterfield, G. &Patrick, Z.A 1986. Somatic hybridization between *Nicotiana mstica* and *N. tabacum*: development of tobacco breeding strains with disease resistance and elevated nicotine content. Zeilschriftfur Pflanzenzuchtung 96: 346-352.

Papadatou, P., Pontikis, C. A., Ephtimiadou, E. and Lydaki, M. 1990. Rapid multiplication of guajava seedlings by *in vitro* shoot tip culture. *Scientia Horticulturae*, 45: 99-103.

Park, Y. G. and Son, S. H. 1988. Regeneration of plantlets from cell suspension culture derived callus of white poplar *(Populus alba* L.). *Plant Cell Reports*, 7: 567-570.

Pattanaik, S. K. and Chand, P. K. 1997. .Rapid clonal propagation of three mulberries *Morus cathayana* Hems; *M. ihou* Koiz. and *M. serrata* Roxb. through *in vitro* culture of apical shoot buds and nodal explants from mature trees. *Plant Cell Reports*, 16: 503-508.

Pattnaik, S. K., Sahoo, Y. and Chand, P. K. 1996. Micropropagation of a fruit tree, *Morus australis* Poir. syn *M. acidosa* Griff. *Plant Cell Reports*, 15: 841-845.

Pattnaik, S., Pradhan, C., Naik, S. K. and Chand, P. K. 2000. Shoot organogenesis and plantlet regeneration from hypocotyls-derived cell suspension of tree legume, *Dalbergia sissoo* Roxb. *In vitro Cell Development Biology Plant* 36: 407-411.

Payne, G. F., Bringi, V., Prince, C., Shuler, M. L., 1991. Plant cell and tissue culture in liquid systems. Munich : Hanser Publ ., 1–10.

Pena, L., Cervera, M., Juarez, J., Navarro, A., Pina, J. A. and Navarro, L. 1997. Genetic transformation of lime (*Citrus aurantifolia* Swing.): factors affecting transformation and regeneration. *Plant Cell Reports*, 16 : 731-737.

Pezzuto, J. M., 1995. Natural product cancer chemoprotective agents. *In*: Arnason, J. T,, Mata, R., Romeo, J. T., (Eds). Recent advances in phytochemistry, *Phytochemistry of medicinal plants* , 29 : 19–45.

Plucknett, D.L & Smith, N.J.H. 1986 Sustaining agricultural yields.*BioScience*36, 40-45.

Pradhan, C., Kar, S., Pattnaik, S. and Chand, P.K. 1998b. Propagation of *Dalbergia sissoo* Roxb. through *in vitro* shoot proliferation from cotyledonery nodes. *Plant Cell Reports* 18 : 122-126.

Pradhan, C., Pattanaik, S., and Chand, P. K. 1998c. Rapid *in vitro* proliferation of East Indian Rose wood (*Dalbergia letifolia* Roxb.) through high frequency shoot proliferation from cotyledonary nodes. *J.Plant Biotechemistry and Biotechnology* 7 : 61-64.

Pradhan, C., Pattanaik, S., Dwari, M., Pattnaik, S. N. and Chand, P. K. 1998a. Efficient plant regeneration from cell suspension –derived callus of East India rose wood (*Dalbergia latifolia* Roxb.) *Plant Cell Reports* 18 : 138-142.

Purohit, S. D. and Ashish, D. 1996. Micropropagaton of *Sterculia urens* Roxb. and endangered tree species. *Plant Cell Reports* 15 : 704-706.

Purohit, S. D. and Kukda, G. 1994. *In vitro* propagation of *Wrightia tinctoria. Biologia Plantarum* 36(4): 519-526.

Purohit, S. D. and Tak, K. 1992. *In vitro* propagation of an adult tree *Feronia limonia L.* through axillary branching. *Indian Journal of Experimental Biology*, 30: 377-379.

Purohit, S. D., Kukda, G., Sharma, P. and Tak, K. 1994. *In vitro* propagation of an adult tree *wrightia tomentosa* through enhanced axillary branching. *Plant Science* 103: 67-72.

Quraishi, A. and Mishra, S. K. 1998. Micropropagation of nodal explants from adult tree of *Cleistanthus collinus*. *Plant Cell Reports,* 17: 430-433.

Raghavan, V.1994. *In vitro* methods for the control of fertilizationand embryo development. In Plant Cell and Tissue Culture, edsVasiL I.K. & Thorpe, T.A pp. 173-194. Dordrecht: KluwerAcademic.

Raghavaswamy, B. V., Himabindu, K. and Lakshmi Sita, G. 1992. *In vitro* micropropagation of elite rose wood *Dalbergia latifolia* Roxb.). *Plant Cell Reports,* 11: 126-131.

Rahman, M. A. and Blake, J. 1988. Factors affecting *in vitro* proliferation and rooting of shoots of jack fruit (*Artocarpus heterophyllus* Lam.). *Plant Cell Tissue Organ Culture*, 13: 179-187.

Rama, P. and Pontikis, C. A. 1990. *In vitro* propagation of olive (*Olea europaea sativa* L.) 'Kalamon'. *Journal of Horticultural Sciences*, 65(3): 347-353.

Randolph, LF. 1945. Embryo culture of Iris seeds. *Bulletin of the. American Iris Society* 97: 33-45.

Rathore, T. S., Deora, N. S. and Sekhawat, N. S. 1992. Cloning of *Maytenus emarginata* (Willd.) a tree of the Indian Desert through tissue culture. *Plant Cell Reports*, 11: 449-451.

Rathore, T. S., Singh, R. P. and Shekhawat, N. S. 1991. Clonal propagation of desert teak (*Tecomella undulata*) through tissue culture. *Plant Science*, 79: 217-222.

Ravishankar, G.A., Rao, R.S., 2000. Biotechnological production of phyto-pharmaceuticals. *J Biochem, Mol Biol Biophys,* 4 : 73 –102.

Rossi, L, Hohn, B. and Tinland, B. 1993. The vir D2 protein of *Agrobacterium tumefaciens* carries nuclear localization signals important for transfer of T-DNA to plants. *Molecular and General Genetics*, 239*:* 345-353.

Roy, S. K., Raham, S. L and Dutta, P. C. 1988. *In vitro* propagation of *Mitragyna parvifolia* north. *Plant Cell Tissue Organ Culture*, 12*:* 75-80.

Roy, S. K., Rahman, S. L. and Majumdar, R. 1990. *In vitro* propagation of jack fruit (*Artocarpus heterophyllus* Lam.). Journal of Horticultural Science, 65(3): 355-358.

Rugini , E. and Verma, D. C. 1982. Micropropagation of difficult-to-propagate almond (*Prunus amygadalus,* Batsch) cultivar. *Plant Science Letters.* 28: 273-281.

Rugini, E. 1984. *In vitro* propagation of some live (*Olea europaea sativa* L.) cultivars with different rooting-ability and medium development using analytical data from developing shoots and embryos. *Scientia Horticulturae*, 24*:* 123-134.

Sasson, A.1993 Biotechnologies in Developing Countries: Present amd Future, Vol. 1. Paris: United Nations Educational, Scientific and Cultural Organization.

Schieder, O. & Kohn, H. 1986. Protoplast fusion and generation ofsomatic hybrids. In: *Cell Culture and Somatic Cell Genetics of Plants.* Vol. 3, ed Vasil, I.K. pp. 569-588. New York: Academic Press.

Sekhawat, N. S., Rathore, T. S., Singh, R. P., Deora, N. S. and Rao, S. R. 1993. Factors affecting *in vitro* clonal propagation of *Prosopis cineraria. Plant Growth Regulation*, 12 : 273-280.

Shen, X. S. and Mullins, M. G. 1984. *In vitro* propagation of pear, *Pyrus communis* L. cultivars 'Williams Bon Chretien', 'Packham's Triumph' and 'Beurre Bose'. *Scientia Horticulturae* 25*:* 51-57.

Singha, S. 1982. *In vitro* propagation of crabapple cultivars. *Horticultural Science*, 17(2):191-192.

Singha, S. 1984. Influence of two commercial agars on *in vitro* shoot proliferation of 'Almery' crabapple and 'seckel' pear. *Horticultural Science*, 19(2): 227-228.

Skoog F., Miller C.O. 1957. Chemical regulation of growth and organ formation in plant tissues cultured in vitro. *Symposia of the Society for Experimental Biology*. 11:118-131

Snir, I. 1982. *In vitro* propagation of sweet cherry cultivars. *Horticultural Science*, 17:192-193.

Snir, I. 1984. *In vitro* propagation of 'canino' apricot. Horticultural Science, 19(2): 229-230.

Sriskandarajah, S., Mullins, M. G. and Nair, Y. 1982. Induction of adventitious rooting *in vitro* in difficult-to-propagate cultivars of apple. *Plant Science Letters*, 24: 1-9.

Srivastava, S. and Srivastava, A.K.2007. Hairy root culture for mass-production of high-value secondary metabolites. *Crit. Rev. Biotechnol.*, 27: 29-43.

Tilton, V.R. & Russel, S.H. 1984 Applications of *in vitro* pollination/fertilization technology. *Bio Science* 34, 239-241.

Vaniserce, M., Lee, C., Nalawade, S.M., Lin, C.Y and Tasy, H. 2004. Studies on the production of some important secondary metabolite from medicinal plants by Plant tissue culture. *Bot. Bull. Acad. Sin.*, 45: 1-22.

Villalobos, V.M. & Engelmann, F. 1995 *Ex situ* conservation ofplant gerrnplasm using biotechnology. *World Journal of Microbiologyand Biotechnology* 11: 375-382.

Wang, P. and Hu, C. 1984. *In vitro* cloning of the deciduous timber tree *Sassafras randaiense.* *Journal of Plant Physiology*, 133: 331-335.

White, P.R. (1939). Potentially unlimited growth of excised plant callusin an artificial nutrient. *Am J Bot* 26:59–64.

Yadav, V., Lal, M. and Jaiswal, V. S. 1990. Micropropagation of *Morus nigra* L. from shoot tip and nodal explants of mature trees. *Scientia Horticulturae*, 44*:* 61-67.

Yang, Jiu-Sherng, Yu, Tsong-Ann., Cheng. Ying-Huey and Yeh. Shyi-Dong. 1996. Transgenic Papaya plants from *Agrobacterium*-mediated transformation of petioles of *in vitro* propagated multi-shoots. *Plant Cell Reports*, 15*:* 459-464.

Yeung, E.C, Thorpe, TA. & Jensen, CL 1981 *In vitro* fertilization and embryo rescue. In : *Plant Tissue Culture: Methods and Applicationsin Agriculture*, ed Thorpe, T.A. pp. 253-271. New York:Academic Press.

Zapata-Arias, F.]., Torrizo, LB. & Ando, A. Current developments in plant biotechnology for genetic improvement: the case ofrice (*Oryza sativa* L). *World Journal of Microbiology and Biotechnology* 11 : 393-399.

Zenkteler, M. 1990 In vitro fertilization and wide hybridization in higher plants. CRC *Critical Reviews in Plant Science* 9: 267-279.

Zhong, J.-J., Yu, J.-T & Yoshida, T. 1995 Recent advances in plantcell cultures in bioreactors. *World Journal of Microbiology &Biotechnology* 11: 461-467.

□□□

Green Agriculture : Newer Technologies, 2012
© Kambaska Kumar Behera (ed.), pp. 353-389
New India Publishing Agency, New Delhi (India)
e-mail : info@nipabooks.com; website : www.nipabooks.com

Chapter-14

Fertigation Techniques in Fruit and Vegetable Crops

[1]R. P. Sharma and K.K. Behera[2]
[1]Division of Crop Production, Indian Institute of Vegetable Research, P.Box. No. 01, Jakhini,Varanasi-221305.
Email: rpsharma64@yahoo.com
[2]Dept.of Bio-Science and Biotechnology, Banasthali University, Banasthali, Rajasthan – 304022, India.
E-mail : kambaska@yahoo.co.in

SUMMARY

Fertigation is the injection of concentrated soluble nutrients into irrigation water to enhance crop production. It is used extensively in commercial agriculture and horticulture. In combination with micro-irrigation (drip irrigation), this technique forms an efficient method for precisely applying nutrients close to the crop root zone, especially when a polyethylene mulch is used. Fertigation is used to spoon feed additional nutrients or correct nutrient deficiencies detected in plant tissue analysis and usually practiced with high value crops such as vegetables, turf, fruit trees, and ornamentals. Water supply for fertigation kept separate from domestic water supply to avoid contamination. Fertilizing fruit and vegetable crops requires a delicate balance between yield, quality, and environmental impact. Fertigation increases response to nitrogen (N) and provides greater opportunity to control rates to optimum levels to the root zone. Providing nutrients through the irrigation system enables more flexibility in a fertilizer program. Several types of irrigation systems are available for use in crop production. For any system used, approved backflow control valves and interlock devices are necessary to prevent accidental contamination of the water source due to irrigation system failure or shutdown.

1.0 Introduction

Both the amount and availability of irrigation water is becoming inadequate and scarcer over time. This problem could be addressed by a switchover from traditional surface irrigation to improved and efficient irrigation techniques such as micro irrigation (MI). The area irrigated through MI is although increasing temporally but still covers quite smaller area as compared to traditional irrigation methods. It could be due to socio-economic and techno- know-how factors of users. Its use in vegetables and horticultural crops has shown a promising proposition.

Fertilizer application in present day scenario is inevitable for a profitable agriculture. The fertilizers being a costly input, its efficient use can be ensured through fertigation. Research conducted at various locations indicated that fertigation is one of the efficient methods of application of fertilizers, which resulted in increased yield of various crops and reduction in amount of use of fertilizer. Very meager information on fertigation or on automated fertigation is available and that too limited for a few crops. This information base also needs to be expanded if we need to make a real headway in convincing the farmers to adopt horticulture with drip fertigation. The problems of non-availability of human labour as well as its increased cost could only be answered to a large extent through proper fertigation under automated drip irrigation system.

2.0. Present status of the technology at national and international levels

a. International Status

2.1. General

Today rainfed agriculture accounts for about 80% of global cultivated area and produces 60% of world's food, whereas irrigated agriculture which is more intensively managed produces 40% of world's food from just 18% of global cropland (Rockstrom et al.2007). Irrigated agriculture has significant impacts on water resources, for instance, it accounts for 67% of global freshwater withdrawals and 87% of consumptive water use (Döll and Siebert, 2002). Irrigated agriculture has expanded by 48% during the past century and is projected to expand by 20% by 2030 (FAO, 2003). Ecosystem needs and environmental flows requirements are unlikely to be met in large parts of the world (Smakhtin et al. 2004). The FAO (2007) data show that for the 90 developing countries, the water requirement ratio (irrigation water requirements in km^3/ total agricultural water withdrawals in km^3) was around 38% in 2000, varying from 25% in areas of abundant water resources (Latin America) to 40% in Near East/North Africa and 44% in South Asia where water scarcity results in higher ratio. At the country level, variations are even higher: 10 countries used more than 40% of their water resources for irrigation in 2000, a situation which FAO (2007) considers critical based on its comprehensive global water balance approach calculated in a uniform way and comparable across countries. An additional nine countries used more than 20% of their water resources, a threshold that could be used to indicate impending water scarcity (Fig. 1). By the year 2000, irrigation water use had been several times higher than their annual renewable water resources for three countries not mentioned above viz. Libya (712%), Saudi Arabia (643%) and Yemen (154%). None of these countries is a major agricultural producer at the same time these all are quite arid. Libya relies largely on fossil groundwater withdrawn from the desert and transported a substantial distance to water users. These countries desalinize water to meet their water needs. Groundwater mining also occurs at the local level in several other countries (Giordano and Vilholth, 2007).

2.2 Drip irrigation and its adaptation in sub-surface drip irrigation management

Drip irrigation systems allow water to be applied uniformly and slowly at the plant location so that essentially all the water is placed in the root zone (Johnson *et al.*, 1991). Drip systems are categorized according to their placement in the field:

- Surface drip irrigation: Water is applied directly to the soil surface.
- Subsurface drip irrigation (SDI): Water is applied below the soil surface through perforated pipes.

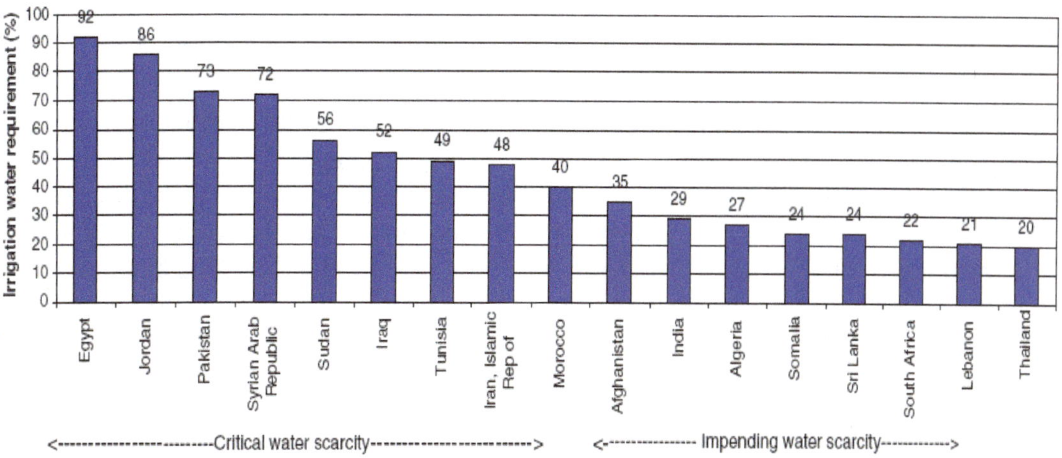

Figure 1 : Water scarcity: Irrigation water withdrawal as percentage of renewable water resources (*Source :* FAO, 2007; and Khan and Hanjara, 2009).

Both the surface and subsurface drip has been used worldwide for crops including citrus, cotton, sugarcane, some vegetables, sweet corn, ornamentals, lucerne and potato (Raine et al. 2000; Alejandro and Eduardo, 2001; Thorburn et al. 2003; Bhattari et al. 2004; Shock et al.2004; Lamm and Trooien, 2005, Saxena and Gupta, 2004). Subsurface drip has proven to be an efficient irrigation method with potential advantages of high water use efficiency, fewer weed and disease problems, less soil erosion, efficient fertilizer application, maintenance of dry areas for tractor movement at any time, flexibility in design, and lower labour costs than in a conventional drip irrigation system. However, there are also potential disadvantages with SDI, which mainly relate to poor or uneven surface wetting and risky crop establishment (Camp et al. 2000; Lamm, 2002; Raine and Foley, 2001).

2.2.1. Lateral drip line

Tapes and tubes are available for use as laterals. Tape products are thinner than tubes (Neufeld et al.1993). Commonly, tube wall thickness ranges from 0.4 mm to 1.5 mm (Hanson et al. 2000). Camp et al. (2000) identified two classes of tape wall thickness. Flexible thin-walled (0.15 mm to 0.30 mm) tapes are typically used for shallow installation, whilst thicker-walled (0.38 mm to 0.50 mm) tapes are installed deeper or where the soil does not provide sufficient support to prevent collapse by equipment or soil weight. O'Neill et al. (2002) used 0.38 mm thickness of tape for potato (*Solanum tuberosum* L.), corn (*Zea mays* L.), alfalfa (*Madicago sativa*) and pinto bean (*Phaseolus vulgaris* L.) production in sandy loam soils. Successful production of lucerne with sub surface drip irrigation was recorded by Thompson (2005) in Victoria, using 0.38 mm tape.

2.2.2. Tape installation depth

The use of surface *versus* subsurface drip irrigation varies by region and by crop, and is often based on perceived constraints on the vertical placement of the drip tape/tube or laterals (Clark and Smajstrla, 1996). With SDI, the choice of drip tape depth is influenced by crop, soil, climate characteristics and anticipated cultural practices, but it generally ranges from 0.02 to 0.7 m (Camp,1998). It is often in the range of 0.05 to 0.2 m for shallow rooted horticultural crops. From the literature, a depth of 0.15 m for lettuce would be appropriate on the sandy soils at UWS, used for experiments later in this status report. Although installation depth is generally decided for horticultural reasons, another consideration for determining depth is that deeper placement (0.45 m) will be required if the primary aim is to reduce soil evaporation and capture the potential benefit of improved water use efficiency (yield and quality) that is possible with SDI (Bryla et al. 2003).

With the shallow systems, relatively deeper installation should reduce soil evaporation and also allow for a wider range of cultural practices. However, as noted above, deeper installation may limit the effectiveness of the SDI system for seed germination/crop establishment. Deeply placed drip lines may require an excessive amount of irrigation for germination/crop establishment. This practice can result in off-site environmental effects (Camp, 1998), and it reduces water-use efficiency. Deeper placement may restrict the availability of surface applied nutrients and other chemicals (Camp and Lamm, 2003).

Relatively shallow tape placement has been tried for many years to assist germination (Burt and Styles, 1994). Recent examples include broccoli on sandy loam soil (Roberts et al.2008) and corn on a silt loam (Lamm and Trooien, 2005). Germination of tomato (*Lycopersicon esculentum* Mill.) under SDI was better with drip line depths of 0.15 and 0.23 m than at 0.3 m on clay loam soil (Schwankl et al. 1990). It can be assumed that shallow placement is especially important for establishment if there is no supplementary source of surface irrigation.

Shallow placement of drip tape is generally required also for satisfactory growth of shallow rooted crops in sandy soils, which have limited capillary water movement (Broner and Alam, 1996); although this is not always the case, as Rubeiz et al. (1989) found higher zucchini (*Cucurbita pepo*) yield at 0.15 m depth than 0.04 m depth on a coarse loam soil. In Australia, tape depth of 0.25 to 0.30 m is used in the Queensland cotton (*Gossypium spp.*) industry on cracking clay soils (Raine et al. 2000). There are regional differences in the tape placement, with growers in NSW generally installing more deeply than in Queensland (Raine et al.2000).

2.2.3. Lateral spacing

An overview of published studies shows that lateral spacing ranges from 0.25 to 5 m for SDI, as determined by crop behaviour, cultural practices soil and properties. Wider lateral spacing is practiced in heavy textured soil (Camp, 1998). Closer spacing is recommended for sandy soil (Phene and Sanders, 1976).Lateral spacing is generally one drip line per row/bed or an alternative row/ bed with one drip line per bed or between two rows (Lamm and Camp, 2007).With row crops such as tomatoes, laterals are often spaced 1-2 m apart. Lateral spacing of 1.5 m in sub-surface drip-irrigated corn was successful in a silt loam soil (Darusman et al.1997).Lateral placement of 0.3 m is recommended for subsurface systems in the loamy sand soil of South Carolina for vegetable crops; cowpea (*Vigina unguiculata*), green bean (*Phaseolus vulgaris*), yellow squash (*Cucurbita pepo*), muskmelon (*Cucumis melo*) and broccoli (*Brassica oleracea*) (Camp et al.1993). Lateral spacing of 2 m intervals on a 1:2 drip tape:crop row has been successful in Queensland for cotton (Raine et al. 2000).

2.2.4. Installation

Lateral lines should be laid following the contour of the land as closely as practicable to avoid pressure variations within the line due to elevation change (Haman and Smajstrla, 2003). The first step in installing a successful drip system is maintaining proper hydraulic design. This allows the system to deal with constraints related to soil characteristics, field size, shape, topography, and water supply.

Lateral diameter and length influence water application uniformity (Kang et al. 1999). Lamont et al. (2002) observed in vegetables in the USA that a tape diameter of 125– 200 mm was the industry standard and common for subsurface drip irrigation where rows range from 90 m to180 m. In Greece, 17 mm polyethylene pipe was used at the shorter row length of 30 m for sugar beet (*Beta vulgaris* L.)research using subsurface drip (Sakellariou-Makrantonaki et al.2002).

2.2.5. Emitters

Emitters are plastic devices which precisely deliver small amounts of water. Hla and Scherer (2003) described two types of emitter. Point-source emitters discharge water from individual or multiple outlets. Line-source emitters have perforations, holes, porous walls, or emitters extruded into the plastic lateral lines (Ayars, et al. 2007). Line-source emitters are generally used for widely spaced crops such as vines, ornamentals, shrubs and trees. Point source emitters are used for small fruits, vegetables and closely spaced row crops (Bucks and Davis,1986).The emitters used for SDI are much the same as those used for surface drip, but the emitter is fixed internally in the drip line (Harris, 2005c).

2.2.6. Emitter spacing

Soil characteristics and plant spacing determine emitter spacing. Spacing used in Queensland are mostly between 0.3 m and 0.75 m for row crops (Harris, 2005d). Kamara et al. (1991) used 0.3 m emitter spacing for drip-irrigated cotton grown in sandy loam soil in the USA. Similarly, an emitter spacing of 0.3 m was suitable for corn production for deep silt loam soils under subsurface drip (Lamm and Aiken, 2005). In a semi-arid environment, 0.45 m emitter spacing was used in clay loam soils for drip-irrigated corn (Howell et al.1995). In general, emitter spacing should normally be less than the drip lateral spacing and closely related to crop plant spacing (Lamm and Camp, 2007).

2.3. Flushing Capacity

A critical area of design that impacts on system performance is the flushing capacity. Many drip systems appear to have been installed with inadequate flushing capacity, resulting in sediment deposition, decreases in flow volumes and blockages (Pitts et al. 1996). This will produce higher backpressures in the mains, which may also affect system performance (Lamm and Camp, 2007). Retrofitting large valves or increasing the number of valves may solve some flushing problems (Raine et al. 2000).

2.4. Water Application Uniformity

Water application uniformity in micro-irrigation depends on system uniformity and spatial uniformity in the field (Wu et al.2007).

2.4.1. System uniformity

The system uniformity is affected by system design factors such as lateral diameter and emitter spacing (Wu et al.1986), and manufacturing variation (Bralts et al. 1981a). It is also considered to include emitter clogging (Bralts et al.1981b). The parameters used to evaluate microirrigation system application uniformity are: the Uniformity Coefficient (UC); emitter flow variation (qvar); and Coefficient of Variation (CV) of emitter flow (Bralts and Kensar, 1983; Wu et al. 1986). Using these parameters, Ayars et al. (1999) discussed various drip tape products and determined the values of these uniformity parameters. System uniformity values predicted by design or evaluation models are similar for both surface and subsurface drip (Camp et al.1997).

2.4.2. Spatial uniformity in the field

The spatial uniformity in the field refers to variation in soil water. In addition to system design factors noted above (Wu et al. 2007), it includes variation due to field topography and soil hydraulic properties (Burt and Styles, 1994; Burt et al.1997).

2.4.3. Causes and consequences of Non-uniformity

The causes of non-uniformity include unequal drainage and unequal application rates (Burt, 2004). Even where system uniformity is high, variation in soil properties, such as hydraulic conductivity, can affect drainage and lead to variation in water content. Application uniformity may be directly related to yield (Solomon,1984b;Letey, 1985). Non-uniformity in one field (45%) was estimated to be mainly due to pressure differences, with only 1% due to unequal drainage and 2% due to unequal application rate (Burt, 2004).Burt (2004) considered the typical manufacturing coefficient of variation in tube today is only 0.02 to 0.06, which will be negligible. Soil 'excavating' by subsurface emitters was shown to increase flow rate by 2.8% to 4%, but not sufficiently to affect uniformity calculations (Sadler et al. 1995).

One consequence of non-uniform application is increased drainage (Ben-Asher and Phene, 1993; Phene and Phene, 1987), assuming irrigation for uniformly good crop growth. Drainage may also occur if the application is uniform but the soil water holding capacity or hydraulic properties are not uniform.

Obtaining sufficiently moist soil for germination and crop establishment by applying uniform irrigation to soils which are inherently variable is a challenging issue for SDI (Patel and Rajput, 2007). They found that to provide adequate irrigation water for potato plants in the early growth period, they had to be over-irrigated, leading to more downward movement of water on sandy loam soil than upward capillary movement of water.

2.4.4. Minimising Non-uniformity

Overall, minimising non-uniformity of the drip system requires: a design which considers the topography of the field (Wu et al. 2007) periodic checking of the system (Clark and Phene, 1992), and irrigation scheduling (volume and frequency) (Burt et al.1997). Greater irrigation uniformity can be achieved by using pressure-compensating emitters in surface and subsurface drip (Schwankl and Hanson, 2007). Flow meters are widely recommended to check the system performance in sub surface drip irrigation (Alam et al. 2002). They are used to determine the rate and volume of water applied in an automated irrigation control system (Ayars and Phene, 2007).

2.4.5. Comparison of uniformity in surface and subsurface drip

In SDI, emitter clogging and accumulation of salt caused by evaporation is less than in surface drip (Hills et al.1989a). More uniform water content was observed in the root zone with SDI than surface drip (Ghali and Svehlik, 1988). In an SDI system more uniform water content in root zone was observed than surface drip, and thus drainage would be less with SDI (Ben-Asher and Phene, 1993; Phene and Phene, 1987).

2.5 Discharge rate and irrigation frequency in relation to crop and soil type

Most surface and subsurface drip irrigation systems generally consist of emitters that have discharge rates less than 8 L/h (ASAE, 2001). A discharge rate of 0.25 L/h gave high yield of corn in sandy loam soils of Israel (Assouline, 2002), although the difference in yields between discharge rates was not statistically significant. In a silt loam soil a discharge rate of 0.5 L/h gave the highest onion (*Allium cepa*) yield (Shock et al.2005, a). In a drip system, frequency and emitter discharge rate determine the soil water availability and plant water uptake pattern (Coelho and Or, 1996; 1999) and consequently yield (Bucks et al.1981; El-Gindy and El-Araby, 1996).

Illustrating the importance of matching irrigation frequency to soil type, Ruskin (2005) reported that a coarse textured sandy soil required drip lines with higher flow rates and shorter irrigation cycles than clay soil. Similarly, shallow rooted vegetable crops on fine sandy soils in Florida required frequent (once or more per day) water application (Haman and Smajstrla, 2002). Conversely, in a clay loam soil, drip irrigation applied every second day achieved maximum tomato yield (Dalvi et al.1999). High frequency irrigation seems to be especially important for coarser-textured soils, high frequency SDI gave best yields of processing tomato in a sandy loam soil (Ayars et al.1999) and of potato in loamy soils in China (Wang et al. 2006). High frequency water application under drip enables maintenance of salts at reasonable levels within the rooting zone (Mmolawa and Or, 2000 b).

The main reported benefit of increased irrigation frequency with SDI is the increased yield. A less commonly reported benefit of increased irrigation frequency is improved crop establishment (Phene and Beale, 1976). As crop establishment is a common problem in SDI, it is surprising that there seem to be relatively few studies of irrigation frequency in relation to establishment. More frequent or pulsing irrigation, which involves applying small increments of water multiple times per day rather than applying large amount for long duration, has been advocated to improve surface and near surface soil moisture wetting for crop establishment

(Lamm and Camp, 2007). However, there is a lack of operational guidelines for SDI (Lamm and Camp, 2007). Other potential benefits of high frequency SDI are reduced deep drainage of water (Ayars et al. 1999), although for this it will be important to have both uniform application and uniform soil and crop growth. High frequency SDI may have lower water requirement, as shown by Wendt et al. (1977).

The flow rate of the drip line has to match the particular soil type. When soil hydraulic conductivity decreases, the pressure head of the soil next to the emitter will increase, which reduces the flow rate of emitters (Warrick and Shani, 1996). On the other hand, emitter discharge decreases due to backpressure, which depends on the soil type, possible cavities near the dripper outlet, and the drip system hydraulic properties (Shani et al. 1996). When the pressure in the emitter increases this may significantly reduce the source discharge rate (Lazarovitch et al.2005). Crop type also influences optimum irrigation frequency, even amongst vegetable crops. For example, on loam soil, cantaloupe (*Cucumis melo*) yield was higher with weekly irrigations compared to daily irrigations, whilst onion yield was higher for daily irrigation compared with weekly irrigation (Bucks et al. 1981).

In most cases, supplementary irrigation has been used in establishment (Schwankl et al. 1993; Howell et al. 1997). Of the many papers dealing with irrigation management with SDI, few appear to have independently varied management for the establishment and growth periods other than adjust the crop factor. It appears that crops are often over-watered in the establishment period (Enciso et al. 2007; Patel and Rajput, 2007) to ensure establishment. This has been reported to increase drainage (Howell et al. 1997). One topic which appears to have received no study is the need to vary irrigation frequency through the life of a crop to meet different requirements. Frequent irrigation may be needed for good establishment, but frequent irrigation subsequently should reduce deep drainage, and increase water use efficiency. This approach is analogous to securing establishment by increasing irrigation rate above the crop requirement determined by K_c and ET_o (Howell and Meron, 2007), but with less risk of increased drainage.

2.6. Fertigation via drip irrigation

The application of nutrients together with the irrigation water, there are some considerations directly relevant to drip systems. Fertigation is a sophisticated and efficient method of applying fertilizers with irrigation water (Magen, 1995). All soluble nutrient sources are suitable for fertigation, and selection is generally according to the cost and the other element in the salt. Fertigation contributes to the achievement of higher yields and better quality by increasing fertilizer efficiency (Haynes, 1985; Imas, 1999), regardless of whether DI or SDI is being used. In addition, minimization of leaching below the root zone may be achieved by fertigation (Hagin and Lowengart, 1996; Hanson, 1996).

Although fertigation can be used with any drip irrigation system, a major potential advantage of subsurface drip is that water and nutrients are potentially used more efficiently when compared to surface installation (Phene et al.1987).

Frequency of fertilizer injection can range from once a week to daily for drip irrigated vegetable crops (Marr, 1993). Combined SDI and nutrient management schemes have been developed for several vegetable crops, including collard, mustard, spinach, and romaine lettuce (Thompson and Doerge, 1995 a & b, 1996) and corn (Lamm et al. 2001). Subsurface drip irrigation and fertilizer management together has been found to increase yield on tomato, sweet corn and cantaloupe (Ayars et al. 1999), sweet corn (Bar-Yosef, 1989), cabbage and zucchini (Rubeiz et al. 1989). Water and fertigation requirements need to be established for each crop, as significant differences occur. For example, watermelon yield may be increased by maximising the interactive effects of water and nitrogen applied through SDI on sandy loam soil (Pier and Doerge, 1995), whereas for broccoli production with SDI on sandy loam soils, fertigation frequency had no effect on yield (Thompson et al. 2002).Vazquez et al. (2005) observed substantial drainage during the crop establishment period of processing tomato under drip irrigation, when the roots explore only a small volume of soil and water absorption capacity is small (Jackson and Bloom, 1990). The excessive irrigation and associated drainage of tomatoes during establishment caused large N losses (Vazquez et al. 2006). So, if extra irrigation is required to ensure establishment and this creates a risk of drainage, the fertigation regime needs to be varied to minimize the risk of N leaching.

2.6.1. Frequency of fertigation

As plants grow, their demand for nutrients change with time and as such, some salts that are easily taken up by plants may get depleted sooner than the excluded ones. This preferential uptake of solutes can lead to high concentrations of the excluded salts in the rhizosphere that could prove to be detrimental for optimum plant growth. Thus fertigation is often necessary to augment nutrient fertilizers. A fertigation scheduling plan is often compounded by the changing climatic conditions and the changing demands of fertilizer requirements of growing plants. Fertigation can be applied with each irrigation or on a scheduled basis to prevent nutrient stress. Nevertheless, fertigation should be carried out, not to adversely alter the solute dynamics in the root zone, but should provide tolerant and optimum concentrations of nutrients and salts in the rhizosphere. Hence, accurate prediction of when and how much fertilizer to apply is critical for fertigation management. The amount of fertilizer to be applied depends on the plant requirement at the time of application. The frequency of application for fertilizers depends among other factors on the soil type, system design constraints, and the length of the growing season.

As such, we know that the nutrient uptake increases with plant growth, some schedules ensure that the fertigation rate increases according to the crop growth curve. According to Hochmuth (1992), the frequency of fertigation is usually not as critical as achieving the right rate of application at a given crop stage. However, Locascio et al. (1997b) Locascio and Smajstrla (1989) found that with 40% pre-plant N and K application, similar yields were obtained with six 2-weekly or 12 weekly applications, either all equal or scheduled with initially small amounts that increased progressively with plant growth, and with daily or weekly fertigation. The frequency of fertigation – daily, weekly or 2- weekly – and whether the applications were uniform or increased progressively to match the plant growth were not critical, so that fertigation can be planned to suit the equipment available and the grower's convenience. Application of 100% of the N and K either pre-plant or by fertigation resulted in lower production than the split application (Locascio et al. 1989). On finer-textured soils, response to fertigation was not as consistent as on coarse ones, although N, and sometimes K, are most usually applied through fertigation to increase nutrient use efficiency. With subsurface drip-irrigation, broccoli (*Brassica oleracea* var. *italica*) yields were similar with fertigation at 1, 7, 14 and 28-day intervals (Thompson et al. 2003). It is apparent that to maximize crop yield on coarse-textured soils, 30 to 40% of the N and K must be applied preplant and the remainder by fertigation and that the actual schedule for fertigation is not critical.

Subsurface drip irrigation provides incremental application of nitrogen and water. With good management, this has been reported to reduce NO_3 - leaching and contamination of groundwater in lettuce production (Thompson and Doerge, 1996). For crops such as broccoli, celery and lettuce, N uptake is low in the first half of the season and higher before harvest. Fruiting crops such as tomatoes, pepper and melons require little N until flowering, then increase N uptake, reaching peak uptake during fruit set. These factors need consideration for drip irrigation with fertigation (Hartz, 1996). Yabaji et. al. (2009) conducted a study of application timing in a SDI cotton in 2005-2006. The conclusion of this study was that of the timing of termination of fertilizer injection did not affect lint yield. Two application strategies, a 5-week (early bloom) and 8-week (peak bloom), periods were compared to provide evidence of this.

2.6.2. Nitrogen fertilization

Sources of N that perform similarly to one another in the fertigation of vegetables include ammonium nitrate, calcium nitrate, ammonium sulfate and potassium nitrate (Hartz and Hochmuth, 1996; Locascio et al.1982; Locascio et al. 1984, Locascio and Martin, 1985). Also, urea can be applied via fertigation, but, studies have shown that nitrification of urea may be slow in fumigated soils (Fiskell and Locascio, 1983), and that some nitrate-N should be applied after soil fumigation, especially in cooler soils.Manufactured N fertilizers are the most important sources of N to cultivated crops, and consumption of N fertilizer has increased from 22 to 85 million metric tons in the past 30 years in the United States (Havlin et al. 2005). Between 1972 and 2001, N fertilizer use increased about 45% in the United States (Martin et al., 2006). Nitrogen fertilizer use in agricultural production has been a key in the overall increase in food and fiber, but it also has been identified as a major contributor to elevated concentration levels of nitrates in groundwater and surface water in some states (Huang et al.1999). High concentrations of nitrates in drinking water supplied by ground and surface water are a concern because of the potential cause of human health problems. Nitrates in drinking water are linked to potentially fatal infant methemglobinemia (blue baby syndrome), and nitrosamine, a potent carcinogen affecting a wide range of organs in many animal species (Huang et. al.1999). Nitrate levels in wells south of Lubbock were found to be increasing, such that 20% have NO_3-N levels > 10 μg mL $^{-1}$ (Hopkins et al. 1993; Chua et al.2003).

About 76% of cotton production acreage receives N fertilizer (Martin et al.2006). Recommendations for N fertilization rates for cotton the western United States is spring soil NO_3-N test (Zhang et al.1998). Nitrogen management in irrigated cotton has a strong correlation to lint yield. Response of irrigated cotton to N fertilizer is likely when 0 to 60 cm of soil NO_3-N is below 70 kg NO_3-N ha^{-1} (Hutmacher et al. 2004; Bronson et al. 2006).

2.6.3. Nitrogen application timing

The capability to utilize N for cotton plant's seed and lint production is influenced by both the quantity and timing of N and water application (Guinn and Mauney, 1984; Mullins and Burmester, 1990). Two strategies of N management have been created on the basis of timing of N, the first being N fertilizer efficiency is influenced by leaching, volatilization, or denitrification losses, the second is the N supply to the plant affects the performance of the plant at different physiological stages of development (Tucker, 1974). The proper timing of N application is essential to avoid early season deficiencies or excesses during the growing season (Tucker, 1974). Nitrogen uptake rate ranging from 0.18 kg ha^{-1} d^{-1} rate up to 0.29 kg ha^{-1} d^{-1} in the 168 kg ha^{-1} N from planting to 28 DAP suggesting that fertilizer N application can be delayed in cotton until 28 DAP (Boquet and Breitenbeck et al. 2000). Analysis in that the maximum N uptake occurred between 49 and 71 DAP and was 2.9 and 4.3 kg ha^{-1} d^{-1} for cotton receiving 84 and 168 kg N ha^{-1} (Boquet and Breitenbeck et al. 2000).

The scheduling of nutrient application with drip irrigation is critical to the efficient use of nutrients, especially on coarse-textured soils, and requires some change in the way fertilizer is applied. When all nutrients were applied preplant in the bed, as with overhead- and surface-irrigated, polyethylene-mulched crops, both sprinkler and drip irrigation resulted in similar yields of tomato (Doss et al. 1980; Locascio and Myers, 1974) and of watermelon (*Citrullus lanatus*) (Elmstrom et al. 1981). When part of the N and K was applied preplant and part by fertigation with drip irrigation, yields were higher than with overhead irrigation for tomato (Locascio and Myers, 1974), muskmelon *(Cucumis melo)* (Shmueli and Goldberg, 1971), and strawberry (Locascio and Myers, 1975). With 100% preplant application of N and K, tomato yields were lower than when 50% was applied by fertigation (Dangler and Locascio, 1990). On a coarse-textured soil preplant application of all the P and of 40% of the N and K, with 60% of the N and K fertigated with drip irrigation tomato yields were greater than when all nutrients were applied preplant (Locascio and Smajstrla, 1989; Locascio et al. 1997b).

With drip irrigation on a coarse-textured soil, it is essential to supply only part of the N–K requirement via fertigation and to avoid over-irrigation. With part of the nutrients applied at planting, nutrient leaching is reduced, nutrient use efficiency is increased, and this generally results in higher yields than if all the nutrients were applied either preplant or through the drip system (Locascio et al. 1997b). In a 2-year study on fine-textured soils, however, yields were higher when 100% of the nutrients were applied before planting than when all or part of them were applied by fertigation (Locascio et al. 1997b). Split applications of nutrients were reported to maximize production of pepper (Hartz et al. 1993) and muskmelon (Bogle and Hartz, 1986). Preplant incorporation of N and K in the root zone provides nutrients for early growth during a period when irrigation may not be required, and before fertigation begins to supply nutrients throughout the bed as crop growth continues.

2.6.4 Nitrogen uptake by crops

Nitrogen accumulation in cotton is probably near the maximum for the season at the first open boll, when leaf shedding begins (Halevy, 1976; Li et al. 2001; Chua et al.2003). Exceeding the optimum rate of applied N fertilizer rate increases the total N uptake, but decreases the lint yields and fertilizer efficiency (Boquet and Breitenbeck, 2000).Management decisions that can result in improved NUE in crops include the timing of N application. Pre-plant applications of N are subject to loss from leaching, volatilization, or denitrification, thus decreasing the efficiency of N fertilizer. Norton and Silvertooth (1998) reported a reduction in the N fertilizer needed and increases in NUE, when pre-plant N was avoided in irrigated cotton in Arizona. Silvertooth (2001) reported the rate of N uptake is at the maximum at peak bloom in cotton. Improvement of the application timing of N fertilizer injections in SDI systems in crops could save up to 60 lbs N per acre, without affecting the yields (Bronson et al. 2003; Chua et al. 2003).

2.6.5 Fertigation of other nutrients

All soluble nutrients can be applied effectively by fertigation with drip irrigation, but N and K are the main nutrients applied in this way, because they move readily with the irrigation water. All other needed nutrients generally can be applied most efficiently preplant. Fertigation P and most micro-nutrients move very poorly in the soil and do not reach the root zone. Needed P, secondary elements, and micro-nutrients are most efficiently applied preplant in the root zone. Use of fertigation to apply P and micro-nutrients together with Ca and Mg may cause precipitation and blockage of the emitters (Imas, 1999), and therefore should be minimized. When conditions require that P be applied by fertigation, it should be applied alone and the irrigation water should be acidified, to prevent clogging of the emitters (Rolston et al. 1991). Were micronutrient deficiencies occurring and applications are made via fetigation, completely soluble sources or chelates can be used.

Potassium is also easily soluble in water and applied through drip irrigation. Phene and Beale, (1976) have shown that daily low rate application of nitrogen and potassium with a high frequency drip irrigation system improved nutrient uptake efficiency of sweet corn in sandy soils and reduced leaching loss.

DI may also manage the placement and availability of immobile nutrients (eg. P). The restricted mobility of the phosphate ion implies that pre-irrigation mixing of P in both clay and sandy soils is necessary, supplemented by addition to the irrigation solution, to obtain a uniform P concentration in the soil volume (Bar-Yosef and Sheikholslami, 1976). Immobile nutrients are delivered at the centre of the soil root volume rather than on top of the soil in subsurface drip (Martinez et al. 1991). Fertigation with P in SDI has improved yield, root growth and environmental performance in tomato (Ayars et al. 1999) and sweet corn (Phene et al. 1991).

Suitable K sources for fertigation include potassium chloride, potassium sulfate, and potassium nitrate, which generally perform very similarly to one another (Locascio et al. 1997a). Potassium (K) fertilizers, when added to irrigation waters, do not generally cause any adverse chemical reactions that plug irrigation pipes and emitters. But they may well cause precipitation of insoluble salts if mixed with other fertilizers. For example, if calcium nitrate is mixed with K sulphate insoluble calcium sulphate will result (Rolston et al. 1986). The soil solution K is usually too low for adequate plant nutrition during any crop season. As a result, K has to be replenished through release of fixed K, exchangeable and structural K. But these K-replenishing processes may not guarantee sufficient K in the soil for optimum plant growth (Sparks and Huang, 1985).

On soils low in organic matter, S deficiencies may occur if some S-containing fertilizer is not applied either before planting or through fertigation; the required S can be supplied by applying part of the fertigated N or K in the form of S-containing fertilizers such as ammonium sulfate, ammonium thiosulphate, or potassium sulfate. On a low-S soil cabbage yields were higher when S was applied by fertigation than when applied preplant (Susila and Locascio, 2001); this indicates the importance of S fertigation on coarse-textured soils.

2.7. Growth and Yield of Vegetables in Surface and Sub-surface Drip Irrigation

As a general guide, crops which are suitable for surface drip irrigation are also suited to SDI (Lamm and Camp, 2007). With good agronomic practices, increased yields have been reported for a wide range of crops. These include lettuce (Hanson et al. 1997); sugarbeet (Sharmasarkar et al. 2001; Sakellariou-Makrantonaki et al. 2002); soluble solid content in transplanted muskmelon (*Cucumis melo* L.) (Hartz, 1997); onion (Hanson and May, 2004; Shock et al.2004); and green bean (*Phaselous vulgaris* L.) (Metin-Sezen et al. 2005).

The crop response to SDI differs with crop growth characteristics and rooting pattern (Lamm and Camp, 2007). In lettuce, little yield difference was found between SDI and furrow irrigation in a sandy loam soil (Hanson et al. 1997).

Potato yield was increased 27% with SDI over sprinkler irrigation, while reducing irrigation needs by 29%, provided there were drip lines in each crop row (DeTar et al. 1996). SDI had greater yield and higher water use efficiency than surface drip, furrow and sprinkler irrigation with cantaloupe, zucchini and oranges when irrigation was close to consumptive use (Davis and Pugh, 1974).

Information on root distribution is useful to understand crop responses to irrigation and fertigation, especially with the limited wetted soil volume that develops under subsurface drip (Phene et al.1991). Phene and Beale (1976) showed that root length and rooted soil volume of sweet corn could be improved by frequent irrigation with shallow SDI. They revealed that frequent irrigation maintained a portion of the root zone within the optimal matric potential range. In high-frequency irrigated corn, root length density and water uptake patterns are determined primarily by the soil water distribution under the drippers, whether the drippers are placed on, or beneath the crop row (Coelho and Or, 1999). Most of the root system is concentrated in the top 40 cm of the soil profile in drip irrigated processing tomatoes (Machado and Oliveira, 2003). Unfavourable results obtained with drip irrigation have often resulted from inadequate root growth and distribution (Brown and Don Scott, 1984), especially in heavy textured soil (Meek et al.1983). Supply of aerated water with subsurface drip system can maintain aeration of the root zone in heavy clay soils and significantly increase yield of vegetable soyabean and zucchini (Bhattarai et al. 2004).

Subsurface drip irrigation can minimise the period between crops, especially with reduced tillage, and facilitate more intensive cropping. Multiple cropping with SDI has several practical advantages. The subsurface system does not require staking of the drip tubing during initial plant development, does not interfere with machine or manual thinning, weeding, spraying and harvesting of crops as does surface drip irrigation of vegetable crops (Bucks et al. 1981).A continuous cropping system of head lettuce and cabbage by using no tillage could be a potential advantage with subsurface drip (Chase, 1985). Minimal tillage on semi-permanent beds has been widely adopted in the Sydney region, although not with SDI (Senn and Cornish, 2000). Multiple cropping of vegetables such as cowpea, green bean, squash, and muskmelon in the spring season and broccoli in the autumn season were possible without yield reduction in a humid area (Camp et al.1993).

2.8. Soil properties and drip system's performance in the vegetables

2.8.1 Role of soil texture and structure

Hanson et al. (1997) compared furrow, drip and subsurface drip irrigation for lettuce on sandy loam soils. There was more sand and less silt under furrow irrigated plots in the top layer of soil (0-0.3 m) due to greater infiltration than drip plots. Sand, silt and clay contents of the 0-0.3 m depth interval were quite constant with distance in subsurface drip. Changes in clay content, cation levels and the pore space around emitters were observed in long term subsurface drip irrigation with processing tomato, rockmelons and onions (Barber et al.2001).These authors concluded that these changes could have inhibited the movement of water by altering soil hydraulic properties and reducing the spread of the irrigation wetting-front in clay soils. In one study in heavy textured soil in a region where secondary salinity is a problem, subsurface drip irrigation increased the rate of salinization compared with furrow irrigation because of improved structure and reduced slaking and dispersion in subsoil which led to increased solute movement through the soil profile (Hulugalle et al. 2002). Slaking and dispersion are used to measure the structural stability of soil (Daniells et al.2002). Gypsum improves soil structural stability and economic use of gypsum depends on soil properties and seasonal condition (Greene and Ford, 1980; Ford et al. 1980). Soil conditioners applied by drip irrigation have also increased water stable aggregation in the wetting zone around the drippers (Shaviv et al. 1987).

Drip irrigation can improve plant water availability in medium and low permeability fine-textured soil, and in highly permeable coarse-textured soil in which water and nutrients move quickly downward from the emitter (Cote et al. 2003). Continuous irrigation at a rate equal to evapotranspiration was optimal for medium textured soils whilst greater application rate was required for coarse textured soils to minimise deep percolation losses (Ghali and Svehilk, 1988). Many experiments have been conducted in both modelling and field research to investigate plant water availability and root uptake pattern in different soil types (Or, 1996; Or and Coelho, 1996; Mmolawa and Or, 2000 a and b; Thourban et al. 2003, b).

2.8.2. Role of soil hydraulic properties

Knowledge of soil hydraulic properties assists design of irrigation systems (Mehta and Wang, 2004). Non-uniformities in hydraulic properties and infiltration rates are considered to be major reasons for inefficiencies in drip irrigation and may cause non-uniformities in soil water content and could potentially affect plant growth. Soil hydraulic conductivity is a limiting factor for water uptake by plants under drip irrigation, particularly in sandy soils (Li et al. 2002). However, in clay loam soils, subsurface drip irrigation resulted in very non-uniform soil water contents above the depth of emitters (Amali et al. 1997), which may be corrected by using a membrane under the drip tube.

2.8.3. Soil chemical responses to drip and sub-surface drip irrigation

For row crops, the drip emitters are often placed at the centre of row beds, below which most salt loading or leaching would probably occur. In one study, soil electrical conductivity, pH and soluble cations were lower under subsurface drip than surface drip (Nightingale, 1985), suggesting increased leaching. Haynes (1990) observed that the conversion of fertigated ammonium sulphate and urea into nitrate-N caused acidification in the wetted soil volume to the surface (0-20 cm) of silt loam soils, also suggesting an increase in leaching. Similarly, acidification throughout the soil profile was observed in vegetable beds in tomato crops (Stork et al. 2003), again suggesting leaching of NO_3. This hypothesis finds support in an investigation of commercial production of processing tomato where subsurface drip irrigation, combined with excessive fertilizer application, was thought to cause the leaching of nitrogen (and phosphate) to groundwater depths (Stork et al.2003). Under drip irrigation of tomato crops on sandy loam soils, Vazquez et al.(2005) found that greater drainage occurred during the crop establishment period, which increased the leaching of nitrates previously stored in the soil profile.

2.8.4 Soil moisture around point source - Effect of discharge, applied volume and frequency of irrigation

Water movement under the point source of drip irrigation is very complex, as it is applied through emitter the water spreads in all the directions, above the soil surface, entry into the soil as well as its movement within the soil. Therefore, the resulting water infiltration process is three dimensional with respect to three space co-ordinates. Many researchers in India and abroad have tried to evaluate and expressed the movement of water and moisture distribution over time and space. Drip irrigation systems consist of small emitters, either buried or placed on the soil surface, which discharge water at a controlled rate. Water infiltration takes place in the region directly around the emitter, which is small compared with the total soil volume of the irrigated field. As a result, three-dimensional transient infiltration occurs. This differs from more traditional techniques of flood or sprinkler irrigation, where water infiltrates through most or all of the soil surface area, and water infiltration can usually be adequately simulated by one-dimensional vertical movement (Brandt *et al.* 1971; Bresler, 1977). As the frequency of irrigation increases, the infiltration period becomes a more important part of the irrigation cycle. The hydraulic properties are critical because they control the infiltration phase (Bresler, 1978). Irrigation management under high-frequency water applications involves controlling the quantity of water passing through the root zone by regulating the drip discharge rate according to the soil hydraulic properties (Hanson et al. 1996). Goldberg et al. (1976) developed emitter discharge and wetted diameter as an empirical power relationship. Schwartzman and Zur (1986) described the relationships between the emitter discharge, total wetted volume, width and depth of wetted front and saturated hydraulic conductivity of the soil with a set of empirical formulae. Zozueta et al. (1995) had derived a relation for wetted diameter with emitter discharge and the basic infiltration rate, root depth and dimensionless constant.

Several procedures and approaches have been adopted to mathematically explain the flow situation in the soil beneath the drip source. Richards' equation, which combines Darcy's law with conservation of mass, is usually used to describe the three-dimensional infiltration and subsequent redistribution of water in the soil (Molz, 1981; Coelho and Or, 1996; and Clothier and Green, 1997). The major difficulty to solve the Richards equation is due to non-linear dependency of the hydraulic conductivity (K) on the unknown water content (θ).

National Status

Fertilizer in the agricultural is an important area of concern. India's green revolution gave a positive increase to the sector. The sector experienced a faster growth rate and presently India is the third largest fertilizer producer in the world. By the end of 20th century, developing countries had increased their utilization to 60 % of the world's fertilizer use (Beard, 2000).Developing countries, account for production of 55 % of total nitrogenous fertilizers. In India, total capacity of the industry has reached beyond a level of 121.10 lakh MT of nitrogen (inclusive of an installed capacity of 208.42 lakh MT of urea after reassessment of capacity) and 53.60 lakh MT of phosphatic nutrient. The Fertilizer Industry in India started its first manufacturing unit of Single Super Phosphate in Ranipet near Chennai with a capacity of 6000 MT a year (Fertilizer Industry, 2009). Presently there are 57 large fertilizers plants in the country producing urea, Di-Ammonium Phosphate, Complex fertilizer, Ammonium Sulphate and Calcium Ammonium Nitrate.

Several researchers in India have published empirical relationships for spatial distribution of soil moisture. One of the foremost of these were conducted by Kaul (1979), who studied the hydraulics of soil moisture front under drip irrigation source. He reported that the soil moisture in the wetted zone, resulting from the point source of water application, manifested itself by a rapid increase in the soil moisture content in the soil layer close to the point of water application. Later, Rema Devi (1983) evaluated the soil moisture distribution as a function of elapsed time at different rates of water application and reported that the vertical and the horizontal moisture fronts could be represented by exponential and second order polynomial equations respectively. Nehra et al. (1991) and Singh et al. (2001) studied the movement of moisture front advance from a surface point drip source and established the empirical relationships between the emitter discharge, time of application and the vertical or the horizontal distance of the advance front. Jaiswal and Lal (2001) fitted several mathematical models for the wetted front advance under line source of drip. They found that Quadratic, Hoerl, Weibull, Vapour pressure, Power fit, Exponential association, Sinusoidal, Heat-capacity, Modified-geometric and Modified-exponential models were best fitted for wetted width and time.

2.9. Fertigation and crop response in India

Fertigation has been reported as the most efficient method of fertilizer application, as it ensures application of the fertilizers directly to the plant roots (Patel & Rajput, 2001a). In fertigation, fertilizer application is made in small and frequent doses that fit within scheduled irrigation intervals matching the plant water use to avoid leaching. Fertilizer use efficiency upto 95 % can be achieved by drip fertigation (Table 1).

Table 1 : Fertilizer use efficiency

Nutrient	Fertilizer use efficiency, %		
	Soil application	Drip	Drip and fertigation
N	30-50	65	95
P	20	30	45
K	50	60	80

Significant savings in the use of fertilizers and increase in yield of various crops have been reported by researchers (Table 2) (Anonymous, 2001, Patel and Rajput, 2001a, 2001b, and 2002). Although liquid fertilizers are most appropriate for use in fertigation, but in India the lack of availability and high cost of liquid fertilizers restricts their use. Studies with granular fertilizers also have revealed their feasibility.

Table 2 : Response of crops under fertigation

Sl. No.	Crop	Saving in fertilizer, %	Increase in yield,%
1.	Okra	40	18
2.	Onion	40	16
3.	Banana	20	11
4.	Castor	60	32
5.	Cotton	30	20
6.	Potato	40	30
7.	Tomato	40	33
8.	Sugarcane	50	40

(Rajput and Patel, *2002*)

Study conducted at IIHR, Banglore indicated marginally higher yield of mango with 80% evaporation replenishment rate than 40%. Fruit number and yield was on par in treatments with 100 and 75 % of recommended dose of fertilizer, A study revealed that fertigation of guava with NPK at the time of fruit setting resulted in more yield (76.3 kg/tree) followed by fertigation at time of flowering (67.10 kg/ tree). The research conducted at various places on fertigation indicated that crop yields were substantially increased from 26 - 40 % in pomegranate, 11 - 41% in straw berry and 8 - 31% in grape (Table 3).

Table 3 : Comparative yields due to fertigation

Crop	Yield (t/ha)	
	Fertigation	**Conventional**
Pomegranate	57.0-76.3	40.3-56.4
Strawberry	19.3- 23.8	10.5-14.0
Grape	24.3-41.0	13.0-37.0

(Kumar, *2001*)

An automated drip fertigation system installed in guava and mango orchards at CIAE, Bhopal could perform excellent with uniformity coefficient, distribution uniformity and statistical uniformity in the range of 96-98%; and no emitters were clogged. The system could be used for 5-20 hectares of mango and 3-20 hectares of guava to become techno-economically feasible with sell price of fruits at rate of Rs.15 per kg. The benefit cost ratio was 1.58 and 2.10 for 20-hectare area of mango and guava orchard, respectively with pay back period of nine and seven years (Singh et al. 2009).

Future of fertigation in India

Use of drip irrigation is growing fast in India. About 1.3 million-hectare areas under vegetables and high value crops were being irrigated through drip irrigation in India during 2008 (Table 4). Fertigation is the essence of drip irrigation. Drip irrigation should be viewed as a method of growing crops and not only as a method of irrigation. Traditional irrigation methods are used mainly for irrigation, while fertigation is an integral part of drip irrigation system.

Table 4 : Area under drip irrigation in India

State	Area, ha
Rajasthan	15248
Maharashtra	462240
Haryana	6243
Andhra Pradesh	317935
Karnataka	169795
Gujarat	158727
Tamil Nadu	124951
West Bengal	123
Madhya Pradesh	12518
Chattishgarh	2627
Orissa	3361
Uttar Pradesh	10577
Punjab	10427
Kerala	14119

Contd. ...

Sikkim	80
Nagaland	0
Goa	762
Himachal Pradesh	116
Arunachal Pradesh	613
Jharkhand	133
Bihar	107
Grand Total	1310956

(NCPAH, 2008)

Fertigation is a must in order to realize the full potential and benefits of drip system. In many cases, depending on the year and location, the drip system is used mainly as a fertigation system. Many times irrigation is required in situation of abundant rainfall, but there is need to apply fertilizer because of leaching conditions. Applying the fertilizers in small doses at high frequency will ensure a continuous and stable supply of nutrients to meets the growth requirements of plants without leaching the fertilizers below the root zone.The Government has planned to bring 14 Mha area under micro irrigation during XI Plan. The net potential area for drip irrigation is estimated to be 21.3 Mha for the country (Narayanamoorthy, 2004). Therefore, with increasing area under drip irrigation there is great prospect of using fertigation system for many crops to enhance yield and fertilizer saving.

2.10. Drip irrigation in salt affected environment

Under the Indian conditions, Agrawal and Khanna (1983) reported the results of an experimental trial conducted at CCS HAU, Hisar, for growing radish crop with saline tube well water (EC = 6.5 dS m^{-1}) and good quality canal water (EC = 0.25 dS m^{-1}). The results revealed the utility of drip irrigation in two ways. The yield was higher with drip being maximum in subsurface than surface drip (Table 5). The yield reduction was much less in case of saline water in drip as compared to surface irrigation. However, it may be mentioned that benefits of subsurface drip are not always forthcoming and the usefulness has to be investigated considering the soil, crop, salinity of the soil and water etc.

Field experiments were also conducted at CCS HAU farm for tomato, cauliflower, cabbage, brinjal, watermelon, grapes, cotton, and sugarcane under deep water table conditions (> 5m) and shallow water table conditions (<1.5m) in sandy loam soils to study the comparative performance of drip irrigation with saline tube well water (EC=6.5 dS m^{-1}) and good quality canal water (EC = 0.28 dS m^{-1}) for different irrigation schedules based on ratio of the depth of irrigation to the potential evapotranspiration (IW/PET) varying from 0.3 to 1.0. Drip irrigation performed better under deep water table conditions but the performance of drip irrigation under shallow water table conditions was mixed (Singh and Kumar, 1989, Singh et al. 1990 and Singh and Kumar, 1994).

Table 5 : Water use efficiency under different methods of irrigation with saline and good quality water

Method of Irrigation	Good quality water (EC= 0.25 dS m^{-1})		Saline water (EC= 6.5 dS m^{-1})	
	Yield (t ha^{-1})	**WUE (kg ha^{-1} cm^{-1})**	**Yield (t ha^{-1})**	**WUE (kg ha^{-1} cm^{-1})**
Subsurface drip	2.68	3000	2.36	2600
Surface drip	1.75	1900	1.57	1800
Surface irrigation at 35 mm CPE	1.64	1400	0.99	900
Surface irrigation at 60 mm CPE	1.39	1200	0.67	600

Kumar and Sivanappan (1983) concluded that drip irrigation gave higher crop yield than any other irrigation method when irrigating with saline water. They developed iso-soil salinity curves for the root-zone at different durations from the day of application of 5 levels of saline water (EC 0.85, 2.5, 5.0, 7.5, and 10.0 dS m^{-1}) in equal amount by micro-tube, nozzle and orifice type emitters. They prescribed that saline water having an EC of 7.5 dS m^{-1} is safe for growing crop with the drip irrigation. Jain (1984) reported that drip lines installed at 90 and 150 cm distance in paired rows of tomato and cotton increased the yields almost 3 times over the conventional flood irrigation method and attained 30-50 per cent economy in water use. He also reported that salt concentration in root zone under drip was minimum at the drip points. Compared to the drip points, salt concentration was twice at 10-15 cm distance and thrice at 30-60 cm. Similarly moisture content at drip points was 20 and 40 per cent higher respectively as compared to those at 30 and 60 cm distance. The major drawback of irrigation with drippers is the high salt concentration that develops at the wetting front. Accumulated salt cause difficulties in the planting of subsequent crops because effective leaching of salts require flooding. Another problem reported is the clogging of drippers due to precipitation of salts.

Subba Rao et al.(1987) observed up to 50 per cent decrease in yield of tomato when EC of irrigation water exceeded 6 dS m^{-1}. Singh and Kumar (1988) studied the comparative performance of trickle and subsurface irrigation systems on tomato at different EC and IW/PET ratios. Results reported in Table 6 explain the effect of irrigation scheduling on yield and salt build-up. Apparently, a low IW/PET ratio for irrigation scheduling seems to be preferable to get high yields, save water and to minimize salt build-up.

Table 6 : Yield, irrigation depth, water use efficiency and soil EC for tomato

Year	IW/PET	Yield (t ha^{-1})	Irrigation depth (cm)	WUE (kg ha^{-1} cm^{-1})	EC (dS m^{-1})
1986	0.7	5.47	40.4	135	2.03
	0.5	15.14	29.4	515	0.82
	0.3	14.22	21.2	610	0.70
1987	0.7	13.06	38.1	343	0.83
	0.5	12.23	30.0	397	0.32
	0.3	7.83	23.5	333	0.55

Jain and Pareek (1989) observed that salt accumulation was minimal in drip irrigation when saline waters of EC ranging from 2.7 to 9.0 were used to irrigate date palm trees. Similar results were reported by Singh *et al.* (1990) when sodic waters containing RSC 2.1, 8.45 and 12.45 meq l^{-1} were applied to grow the kinnow (*Citrus reticulata*) plantation. Drip irrigation system was also found more effective in the establishment of fruit garden on salt affected soils (Dwivedi et al.1990) whereas Pampattiwar et al. (1993) reported higher water use efficiency with drip method of irrigation over the conventional method.

Irrigating tomato and brinjal crops through drip using canal water, and waters of 4 and 8 dS m^{-1} at three IW/CPE levels (0.75, 1.00 and 1.25) at different irrigation intervals of 2, 3 and 4 days gave better yield at higher IW/CPE ratios (CSSRI, 2000). It was also observed that when total amount of water application remained constant, 13 and 33 per cent higher yield was observed at irrigation intervals of 3 and 4 days in comparison to the interval of 2 days. In another study on tomato under drip, the yield decreased from 38.7 to 29.8 t ha^{-1} as salinity of irrigation water increased from 0.21 to 5 dS m^{-1}, which was about 24 per cent less over the normal water application (Kadam and Patel, 2001).

2.10.1. Drip irrigation studies at CSSRI, Karnal/AICRP centers

Studies on drip irrigation have been conducted at CSSRI, Karnal since its inception. Three levels of RSC 0.6, 4 and 8 meq l^{-1} (at a fixed value of EC of 3 dS m^{-1}) were given through drip irrigation in kinnow orchards at 3 l hr^{-1} at an irrigation interval of 3 days. The distribution of moisture, chlorides, SAR and the nutrient status were monitored in the root zone (CSSRI, 1986). Since rainfall during monsoon season leached down the salts, no build-up of salts was observed over the years.

A study conducted on sugarcane at Trichurapalli centre of AICRP on Salt Affected Soils and Use of Saline Water in Agriculture revealed that irrigation scheduling under drip with alkali waters (pH 8.8, EC 2.2 dS m^{-1}, RSC 12.9 meq l^{-1} and SAR 18.2) at 60 per cent of pan evaporation (PE) gave higher water use efficiency than 80 and 100 per cent of PE and farmers' practice (surface irrigation), under both the sub-treatments of no gypsum or 50 per cent application of gypsum requirement (Table 7) (CSSRI, 2000).

Table 7 : Effect of irrigation schedules on growth and yield of sugarcane under drip irrigation

Main treatments of irriga-tion at percentage of PE	Yield (t ha^{-1})		Water applied		Water use efficiency (t ha^{-1} cm^{-1})	
	50 per cent of gypsum requirement	No gypsum applica-tion	Depth (cm)	Reduction in water applied (%)	50 per cent of gypsum requirement	No gypsum application
100	99.3	95.6	44.8	7.14	2.21	2.13
80	107.4	98.5	33.8	42.01	3.18	2.91
60	96.6	91.1	23.0	108.69	4.20	3.96
Farmers' practice	99.5	93.8	48.0	-	2.07	1.95

Singh et al. (2000) compared the plant performance and the soil salinity before and after three years of application of 0.4, 4.0, 8.0 and 12.0 dS m^{-1} saline water through drip and basin irrigation in sapota crop at Khanpur farm, CSSRI-RRS, Anand, Gujarat. The plant performance and the soil salinity after the experiment showed that drip irrigation had performed better for growth of plant and less salinity build- up was observed compared to the basin methods in all the treatments.

Table 8 : Interaction effect between EC and IW/CPE ratio on fruit yield (t/ha) of tomato and chilli in drip and surface irrigation

IW/CPE ratio	EC_{iw} levels (dS/m)			Mean	EC_{iw} levels (dS/m)			Mean
	Control	4	8		Control	4	8	
	Drip Irrigation				Surface Irrigation			
Tomato				2003-2004				
0.75	26.47	21.74	18.66	22.29	26.45	22.48	17.31	22.08
1.00	28.63	23.46	17.07	23.05	26.39	22.54	16.95	21.96
1.25	28.52	23.72	16.98	23.07	27.79	21.99	15.00	21.61
Mean	27.87	22.97	17.57		26.87	22.34	16.42	
2004-2005								
0.75	53.02	37.95	27.51	39.49	50.03	40.83	25.05	38.64
1.00	53.58	37.78	26.31	39.22	44.64	33.43	24.6	34.22
1.25	50.59	37.88	25.78	38.08	44.44	32.93	22.55	33.31
Mean	52.40	37.87	26.53		46.37	35.73	24.07	
Chilli				2004				
0.75	1.71	0.14		0.62	2.40	0.08		0.83
1.00	2.05	0.50		1.02	2.74	0.24		0.99
1.25	2.44	0.70		1.05	2.76	0.41		1.06
Mean	2.07	0.45			2.63	0.24		
2005								
0.75	4.80	3.25	0.10	2.72	4.92	2.85		2.59
1.00	5.01	2.76	0.10	2.62	5.11	1.74		2.28
1.25	5.45	2.88	0.10	2.81	5.01	1.62		2.21
Mean	5.09	2.96	0.10		5.01	4.32		

An experiment was conducted at AICRP, Agra to assess the tolerance of tomato-chilli rotation with treatment combinations of saline irrigation water (Canal, EC_{iw} 4 and 8 dS/m) and irrigation schedule (IW/CPE ratio 0.75, 1.00 and 1.25) (CSSRI, 2007). Irrigation interval for drip irrigation was 4 days and depth of water application in each irrigation was 4 cm. The fruit yield of tomato decreased significantly with increasing EC_{iw} in both drip and surface irrigation system. With EC_{iw} 4 and 8 (dS/m), the tomato fruit yield reduced by 18 and 37 in 2003-04 and 28 and 49 percent in 2004-05 in drip irrigation and 17 and 39 and 23 and 48 percent in surface irrigation system respectively (Table 8). Since the yield in surface and drip methods of irrigation did not vary much, it could be inferred that drip irrigation method could not play a significant role in case of tomato at these salinity levels. Although data are not available to confirm the observation, irrigation scheduling being same in the two methods, advantage of drip irrigation to facilitate frequent application of water was not utilized in the present set-up. However, equivalent yield of tomato could be obtained with 25-30 percent less water in drip than with surface method. Amongst the IW/CPE ratios 0.75 to 1.00 ratio could be used. After the harvest of winter tomato, chilli was transplanted during summer season. The saline irrigation affected chilli more in 2003 compared to 2004. The yield with EC_{iw} 4 (dS/m) declined by 78 percent in 2004 and 42 percent in 2005 over BAW. At higher EC_{iw} 8 dS/m crop failed completely in both the years. Overall, growing of the chilli crop is not advisable with saline water during summer season.

The fruit yield of chilli in 2005-06 (Winter season) significantly decreased with increasing EC_{iw} levels in both drip and surface irrigation system (Table 9). The EC_{iw} 4 and 8 dS/m reduced the fruit yield by 36 and 40 percent in drip irrigation and 40 and 54 percent in surface irrigation system, respectively. The IW/CPE ratio treatments were found non-significant.

Table 9 : Effect of EC and IW/CPE ratio on yield (t/ha) of chilli in drip and surface irrigation

IW/CPE ratio	**EC_{iw} levels (dS/m)**			**Mean**	**EC_{iw} levels (dS/m)**			**Mean**
	Control	**4**	**8**		**Control**	**4**	**8**	
	Drip Irrigation				**Surface Irrigation**			
0.75	15.40	9.82	9.43	11.55	10.24	6.38	5.60	7.41
1.00	15.60	9.87	9.26	11.58	10.34	6.36	4.34	7.01
1.25	14.74	9.52	8.64	10.97	10.30	5.63	4.28	6.74
Average	15.21	9.74	9.10		10.30	6.13	4.74	
CD 5%		6.8				10.7		
IW/CPE ratio		NS				NS		
EC x IW/CPE		NS				NS		

Technological gaps

- Many of the technological inputs related to duration and time schedule for fertigation, the quantity and doses of the fertilizer inputs are not available as a fertigation package for various crops in the region.
- Through the fertigation/ chemigation how much enhancement in fertilizer use efficiency as well as higher water productivity can easily be achieved as compared to the direct field application of fertilizers/ chemicals is not known.
- Regular uptake of nutrients by the plants as well as its rate is not known under drip irrigation, which can provide full potential/ maximum plant growth and yield under drip irrigation.
- Technological information synthesized after properly conducted experimentations is not available on crops' response to various fertigation schedules for the region.
- Information on fertigation uniformity through evaluation of the system performance particularly in automated drip fertigation systems are missing in the region.

Conclusion

For all fruit and vegetable crops, the major pathway for mineral nutrient uptake is via the root system. Through the traditional system only 60% of the applied fertilizer is effective to the crop and rest 40% is lost due to contamination, leaching and other menace**.** In Fertigation more efficient use of fertilizers and probably lessens fertilizer contamination through leaching below the plant root zone. In its broadest sense, fertigation is "a spoon feeding system" to the crop by injecting soluble fertilizers in to water in the irrigation system. It is an extremely effective and efficient method of applying fertilizers and other chemicals via the drip irrigation system. However, it does require more management and attention to details than other methods of fertilizer application. Success in using this system will depend on a sound fertility program based on soil testing and a drip irrigation system that is designed and operated properly.

References

Aggarwal, M.C. and Khanna, S.S. 1983. *Efficient Soil and Water Management in Haryana*. Bulletin HAU, Hissar. 118 p.

Ajdary, Khalil; Singh, D.K., Singh, A.K. and Khanna, Manoj. 2007. Modelling of Nitrogen Leaching from Experimental Onion Field under Drip Fertigation. *Agricultural Water Management*. 89 (1-2):15-28.

Alam, M., Trooien, T.P., Lamm, F.R., and Rogers, D.H. (2002) Filtration and maintenance subsurface drip irrigation systems. Publication No: MF-2361. Kansas State University.

Alejandro, P., and Eduardo, P. (2001) Managing water with high concentration of sediments for drip irrigation purposes. Proc. 2nd Int'l Symp. Preferential flow, water movement and chemical transport in the environment. Eds. Bosch, D.D., and King, K.W. ASAE publication, Pp: 290-292.

Amali, S., Rolston, D.E., Fulton, A.E., Hanson, B.R., Phene, C.J., and Oster, J.D. (1997) Soil water variability under subsurface drip and furrow irrigation. *Irri. Sci.* 17(4): 151-155.

Anonymous,2001. National committee on plasticulture applications in horticulture, progress report 2001, Dept. of agriculture and co operation, MOA, GOI, New Delhi, PP:44

ASAE, (2001) ASAE Standard S526.2, JAN01, Soil and Water Terminology, ASAE, St.Joseph, Michigan.

Assouline, S. (2002) The effects of microdrip and conventional drip irrigation on water distribution and uptake. *Soil Sci. Soc. Am. J.* 66: 1630-1636.

Ayars J.E., and Phene, C.J. (2007) Automation, Microirrigation for crop production design, operation and management, Eds. Lamm, F.R., Ayars, J.E., and Nakayama, F.S. Elsevier, Pp: 259-284.

Ayars, J.E, Bucks, D.A., Lamm, F.R., and Nakayama, F.S. (2007) Introduction, Microirrigation for crop production design, operation and management, Eds. Lamm, F.R., Ayars, J.E., and Nakayama, F.S. Elsevier, Pp: 473-551.

Ayars, J.E., Phene, C.J., Hutmacher, R.B., Davis, K.R., Schoneman, R.A., Vail, S.S., and Mead, R.M. (1999) Subsurface drip irrigation for row crops: A review of 15 years research at the Water Management Research Laboratory. *Agric. Water Manag.* 42: 1-27.

Barber, S.A., Katupitiya, A., and Hickey, M. (2001) Effects of long-term subsurface drip irrigation on soil structure. *Proc. of the 10th Australian Agronomy Conference.*

Bar-Yosef, B. (1989) Sweet corn response to surface and sub-surface trickle P fertigation. *Agron. J.* 81(3): 443-447.

Bar-Yosef, B., and Sheikholslami, M.R. (1976) Distribution of water and Ions in soils irrigated and fertilized from a trickle source. *Soil Sci. Soc. Am. J.* 40: 575-581.

Beard,J .2000. Educational catalogue, Land scape contractors association(San Diego chapter), California.

Ben-Asher, J., and Phene, C.J. (1993) The effect of surface drip irrigation on soil water regime evaporation and transpiration. In : *Proc. 6th Int'l Conf. Irrig.*, May 3-4, Tel-Aviv, Israel. Pp: 35-42.

Bhattari, S.P., Huber, S., and Midmore, D.J. (2004) Aerated subsurface irrigation gives growth and yield benefits to zucchini, vegetable soybean and cotton in heavy clay soils. *Ann. Appl. Biol.* 144: 285-298.

Bogle, C.R., and T.K. Hartz. 1986. Comparison of drip and furrow irrigation for muskmelon production. *HortScience* 21:242-244.

Boquet, D. J., and G. A. Breitenbeck. 2000. Nitrogen rate effect on partitioning of nitrogen and dry matter by cotton. *Crop Sci.* 40:1685-1693.

Bralts, V.F., and Kesner, C.D. (1983) Drip irrigation uniformity estimation. *Transactions of the ASAE* 24: 1369-1372.

Bralts, V.F., Wu, I.P., and Gitlin, H.M. (1981a) Manufacturing variation and drip irrigation uniformity. *Transactions of the ASAE* 24(1): 113-119.

Bralts, V.F., Wu, I.P., and Gitlin, H.M. (1981b) Drip irrigation uniformity considering emitter plugging. *Transactions of the ASAE* 24(5): 1234-1240.

Brandt, A., Bresler, E., Diner, N., Ben-Asher, I., Heller, J. and Godelberg, D. 1971. Infiltration from a trickle source: I. Mathematical models. *Soil Science Society of America Proceedings* 35: 675–682.

Bresler, E. (1978) Analysis of trickle irrigation with application to design problems. *Irri. Sci.* 1:3-17.

Bresler, E. 1975. Two-dimensional transport of solutes during non-steady infiltration from a trickle source. *Soil Science Society of America Proceedings.* 39: 604–613.

Broner, I., and Alam, M. (1996) Irrigation Subsurface Drip, Fact Sheet No: 4.716 Colorado State University Cooperative Extension.

Bronson, K. F., J. D. Booker, J. P. Bordovsky, J. W. Keeling, T. A. Wheeler, R. K. Boman, M. N. Parajulee, E. Segarra and R. L. Nichols. 2006. Site specific irrigation and nitrogen management for cotton production. *Agron. J.* 98:212-219.

Bronson, K. F., T. Y. Chua, J. D. Booker, J. W. Keeling, and R. J. Lascano. 2003. Inseason nitrogen status sensing in irrigated cotton: II. leaf nitrogen and biomass. *Soil Sci. Soc. Am. J.* 67:1439-1448.

Brown, D.A., and Don Scott, H. (1984) Dependence of crop growth and yield on root development and activity. Roots, Nutrients, Water flux, and Plant Growth. Eds. Barber, S.A., and Boulding, D.R. *ASA Special Publication* No: 49.

Bryla, D.R., Banuelos, G.S., and Mitchell, J.P. (2003) water requirements of subsurface drip-irrigated faba bean in California. *Irri. Sci.* 22: 31-37.

Bucks, D.A., and Davis, S. (1986) Historical Development. Trickle irrigation for crop production design, operation and management. Eds. Nakayama, F.S., and Bucks, D.A. Elsevier, Pp: 1-26.

Bucks, D.A., Erie, L.J., French, O.F., Nakayama, F.S., and Pew, W.D. (1981) Subsurface trickle irrigation management with multiple cropping. *Transactions of the ASAE* 24(6): 1482-1489.

Burt, C. M., Clemmens, A. J., Strelkoff, T. S., Solomon, K. H., Bliesner, R. D., Hardy, L. A., Howell, T. A., and Eisenhauer, D.E. (1997) Irrigation performance measures: efficiency and uniformity. *J. of Irri. And Drain. Engi.* 123(6): 423-442.

Burt, C.M. (2004) Rapid field evaluation of drip and microspray distribution uniformity. *Irrigation and drainage system* 18: 275-297.

Burt, C.M., and Styles, S.W. (1994) Drip and micro-irrigation for trees, vines, and row crops: Design and management, California Polytechnic State University, San Luis Obispo, California.

Camp, C. R. (1998) Subsurface Drip Irrigation: A review. *Transactions of the ASAE* 41(5): 1353-1367.

Camp, C.R., and Lamm, F.R. (2003) Irrigation systems: Subsurface drip. *Encyclopaedia of Water Science* Pp: 560-564.

Camp, C.R., Garrett, J.T., Sadler, E.J., and Busscher, W.J. (1993) Micrirrigation management for double-cropped vegetables in a humid area. *Transactions of the ASAE* 36(6): 1639-1644.

Camp, C.R., Lamm, F.R., Evans, R.G., and Phene, C.J. (2000) Sub surface drip irrigation – past, present, and future. *Proceedings of the 4th Decennial National Irrigation Symposium,* Nov. 14-16.

Chase, R.G. (1985) Subsurface trickle irrigation in a continuous cropping system. *Proc. 3rd Int'l Drip Irrigation Congress, Fresno*, CA, USA. Pp: 909-914.

Chua, T. T., K. F. Bronson, J. D. Booker, J. W. Keeling, A. R. Mosier, J. P. Bordovsky, R. J. Lascano, C. J. Green, and E. Segarra. 2003. In-season nitrogen status sensing in irrigated cotton: I. Yield and nitrogen recovery. *Soil Sci. Soc. Am. J.* 67:1428- 1438.

Clark, G.A., and Phene, C.J. (1992) Automated centralised data acquisition and control of irrigation management system. *ASAE Paper* No: 92-3021,11pp.

Clark, G.A., and Smajstrla, A.G (1996) Design considerations for vegetable crop drip irrigation systems. *HortTechnology* 6(3): 155-159.

Clothier, B.E. and Green. S.R.1997. Roots: The big movers of water and chemical in soil. *Soil Science*. 162:534–543.

Coelho, E.F., and Or, D. (1996) A parametric model for two-dimensional water uptake by corn roots under drip irrigation. *Soil Sci. Soc. Am. J. 60*: 1039-1049.

Coelho, E.F., and Or, D. (1999) Root distribution and water uptake patterns of corn under surface and subsurface drip irrigation. *Plant and Soil* 206: 123-136.

Coelho, F.E. and O., Dani. 1996. A parametric model for two-dimensional water uptake intensity by corn roots under drip irrigation. *Soil Science Society of America Journal* 60: 1039–1049.

Cote, C.M., Bristow, K.L., Charlesworth, P.B., Cook, F.J., and Thorburn, P.J. (2003) Analysis of soil wetting and solute transport in subsurface trickle irrigation. *Irri. Sci.* 22(3/4): 143-156.

CSSRI, 2000. Evaluation of irrigation schedules for sugarcane using alkali water through drip irrigation. Biennial Report (1998-2000). AICRP on Management of Salt Affected Soils and Use of Saline Water in Agriculture. Coordinating Unit, CSSRI, Karnal : 72.

Dalvi, V.B., Tiwari, K.N., Pawade, M.N., and Phirke, P.S. (1999) Response surface analysis of tomato production under micro-irrigation. *Agric. Water Manag.* 41(1): 11- 19.

Dangler, J.M., and S.J. Locascio. 1990. Yield of trickle-irrigated tomatoes as affected by time of N and K application. *Journal of the American Society for Horticultural Science* 115:585-589.

Davis, S., and Pugh, W.J. (1974) Drip irrigation: Surface and subsurface compared with sprinkler and furrow. In: *Proc. 2nd Int'l Drip Irrigation Cong.*, July, 7-14. San Diego, California, USA. Pp. 109-114.

DeTar, W.R., Browne, G.T., Phene, C.J., and Sanden, B.L. (1996) Real-time irrigation scheduling of potatoes with sprinkler and subsurface drip systems. *In Proc. Int'l Conf. on Evapotranspiration and Irrigation Scheduling*, Nov 3-6, San Antonio, Texas, USA, Pp: 812-824.

Döll, P., Siebert, S., 2002. Global modeling of irrigation water requirements. *Water Resources Research* 38 (4), 1037.

Doss, B.D., J.L. Turner, and C.E. Evans. 1980. Irrigation methods and in-row chiseling for tomato production. *Journal of the American Society for Horticultural Science* 105:611-614.

Dwivedi, P.N., Verma, L.P., Ali, A., Singh, V.K., Dixit, R.S. and Singh, R.K. 1990. Studies on combined effect of drip irrigation system and soil conservation techniques on plant performance and runoff water quality in salt effected soils. *Proc. of 14th International Congress on Irrigation and Drainage*, Rio-de-Janerio. Brazil. No/.1-B: 431-443.

El-Gindy, A.M., and El-Araby, A.M. (1996) Vegetable crop response to surface and subsurface drip under calcareous soil. Evaporation and irrigation scheduling. *Proc. of the Inter.* Conf. Nov. 3-6 San, Antonio, Texas.

Elmstrom, G.W., S.J. Locascio, and J.M. Myers. 1981. Watermelon response to drip and sprinkler irrigation. *Proceedings of the Florida State Horticultural Society* 94:161-163.

FAO, 2003. World Agriculture: Towards 2015/2030. An FAO Perspective. Food and Agriculture Organization of the United Nations/Earthscan, Rome, Italy/USA.

FAO, 2007. Review of agricultural water use per country. AQUAST, FAO, Rome. <http://www.fao.org/nr/water/aquastat/water_use/index.stm>. (Accessed in Jan 2010)

Fertilizer Industry, 2009. http://www.economywatch.com/business-and-economy/fertilizer-industry.html.

Fiskell, J.G.A., and S.J. Locascio. 1983. Changes in available N for drip irrigated tomatoes from preplant and fertigation N sources. *Soil and Crop Science Society of Florida Proceedings* 42:180-184.

Ford, G.W., Smith, I.S., Badawy, N.S., and Roulston, A. (1980) Effects of gypsum of two red duplex wheat soils in Northern Victoria. *National Soils Conference*, 19 -23 May, Sydney.

Ghali, G.S., and Svehlik, Z.J. (1988) Soil-water dynamics and optimum operating regime in trickle irrigated fields. *Agric. Water Manag.* 13: 127-143.

Giordano, M., Vilholth, K. (Eds.), 2007. The Agricultural Groundwater Revolution: Opportunities and Threats to Development. CABI Publication, Wallingford UK and Cambridge MA, USA.

Goldberg, D., Gornat, B. and Rimon, D. 1976. *Drip Irrigation Principles, Design and Agricultural Practices*. Drip Irrigation Scientific, Kfar Shmariahu, Israel.

Greene, R.S.B., and Ford, G.W. (1980) The effect of gypsum on the structural stability of two Victorian soils. *National Soils Conference*, 19 -23 May, Sydney.

Guinn, G., and J. R. Mauney. 1984. Fruiting of cotton. II. Effects of moisture status on flowering. *Agron. J.* 68:701-705.

Hagin, J., and Lowengart, A. (1996) Fertigation for minimizing environmental pollution by fertilizers. *Fertilizer research* 43: 5-7.

Halevy, J., 1976. Growth rate and nutrient uptake of two cotton cultivars grown under irrigation. *Agron J.* 68:701-705.

Haman, D.Z., and Smajstrla, A.G. (2002) Scheduling tips for drip irrigation of vegetables. Publication No: AE259. Florida extension service, University of Florida, USA.

Haman, D.Z., and Smajstrla, A.G. (2003) Design tips for drip irrigation of vegetables. Publication No: AE093, Florida Cooperative Extension Service, University of Florida.

Hanson, B.R. (1996) Fertilising row crops with drip irrigation. Irrigation Journal. Available at: www.greenmediaonline.com/ij/1996/1296.

Hanson, B.R., and May, D.M. (2004) Response of processing and fresh-market onions to drip irrigation. International Symposium on Irrigation of Horticultural crops. Eds. Snyder, R.L. Davis, CA, USA. *Acta Horticulturae* 664: 399-405.

Hanson, B.R., Fipps, G., and Martin, E.C. (2000) Drip irrigation of row crops: what is the state of the art? *Proc. 4th decennial Irrigation Symposium. ASAE publication.*

Hanson, B.R., Schwankl, L.J., Schulbach, K.F., and Pettygrove, G.S. (1997) A comparison of furrow, surface drip, and subsurface drip irrigation on lettuce yield and applied water. *Agric.Water Manag.* 33: 139-157.

Harris, G.A. (2005c) Sub-surface drip irrigation - System components , DPI&F Note, Brisbane.

Harris, G.A. (2005d) Sub-surface drip irrigation - System design , DPI&F Note, Brisbane.

Hartz, T.K. (1996) Water management in drip-irrigated vegetable production. *HortTechnology* 6(3): 165-168.

Hartz, T.K. (1997) Effects of drip irrigation scheduling on muskmelon yield and quality. *Scientia Horticulturae* 69: 117-122.

Hartz, T.K., and G.J. Hochmuth. 1996. Fertility management of drip-irrigated vegetables. *HortTechnology* 6:168-172.

Hartz, T.K., M. LeStrange, and D.M. May. 1993. Nitrogen requirements of drip irrigated pepper. *HortScience* 28:1097-1099.

Havlin, J. L. Tisdale, S. L., W. L. Nelson, J. D. Beaton, 2005. Soil fertility and fertilizers. 7th ed. Pearson Education Inc., Upper Saddle River, NJ.

Haynes, R.J. (1985) Principles of fertilizer use for trickle irrigated crops. *Fertilizer Research*, 6:235-255.

Haynes, R.J. (1990) Movement and transformations of fertigated nitrogen below trickle emitters and their effects on pH in the wetted soil volume. *Fertilizer Research* 23: 105- 112.

Hills, D.J., Tajrishy, M.A.M., and Gu, Y. (1989a) Effects of chemical clogging on drip tape irrigation uniformity. *Transactions of the ASAE* 32(4): 1202-1206.

Hla, A.K., and Scherer, T.F. (2003) Introduction to micro-irrigation. Fact sheet No: AE- 1243. North Dakota State University.

Hochmuth G J 1992 Fertilizer management for drip-irrigated vegetables in Florida. *HortTechnology* 2, 27–31.

Hopkins, J. 1993. Water-quality evaluation of the ogallala aquifer, Texas. Texas Water Development Board Report 342, Austin, TX.

Howell, T.A., and Meron, M. (2007) Irrigation scheduling. Microirrigation for crop production design, operation and management. Eds. Lamm, L.R., Ayars, J.E., and Nakayama, F.S. Elsevier, Pp: 61-130.

Howell, T.A., Schneider, A.D., and Evett, S.R. (1997) Subsurface and surface microirrigation of corn-Southern High Plains. *Transactions of the ASAE* 40(3): 635-641.

Howell, T.A., Schneider, A.D., and Stewart B.A. (1995) Subsurface and surface irrigation of corn – U.S. southern High Plains Proc. Of Fifth Int'l Microirrigation Congress, April 2-6 Orlando, Florida, *ASAE*, Pp 375-381.

Huang, Wen-Yuan, and Uri, Noel D., 1999. The Economic and Environmental Consequences of Nutrient Management in Agriculture. Nova Science. Commack, NY.

Hulugalle, N.R., Friend, J.J., and Kelly, R. (2002) Some physical and chemical properties of hardsetting Alfisols can be affected by trickle irrigation. *Irri. Sci.* 21(3): 103-113.

Hutmacher, R.B., R.L. Travis, D.W. Rains, R.N. Vargas, B.A. Roberts, B.L. Weir, S.D. Wright, D.S. Munk, B.H. Marsh, M.P. Keeley, F.B. Fritschi, D.J. Munier, R.L. Nichols, and R. Delgado. 2004. Response of recent Acala cotton varieties to variable nitrogen rates in the San Joaquin Valley of California. *Agron. J.* 96:48–62.

Imas, P. (1999) Recent techniques in fertigation horticultural crops in Israel. *Proc. of Recent trends in nutrition management in horticultural crop.* 11-12th Feb. Dapoli, Maharashtra, India.

Jackson, L.E., and Bloom, A.J. (1990) Root distribution in relation to soil nitrogen availability in field grown tomatoes. *Plant Soil* 128: 115-126.

Jain, B.L. 1984. Saline water management for higher productivity. *Indian Farming*. 34(7): 13-15, 43.

Jain, B.L. and Pareek, O.P. 1989. Effect of drip irrigation and mulch on soil and performance of date palm under saline water irrigation. *Annals of Arid Zones*. 28:245-248.

Jaiswal, C.S. and Lal, Chhdei. 2001. Wetting front advancement under surface source line condition. *In*: Singh, H.P., Kaushik, S.P. and Kumar, Ashwini, Murthy, T.S., and Samual, Jose C. (*Eds.*) *Microirrigation, CBIP*. Proceedings of the International conference on Micro and Sprinkler Irrigation Systems, held during February8-10, 2000 at Jain Irrigation Hills, Jalgaon, Maharashtra, India : 142-148.

Kadam, J.R. and Patel, K.B.2001. Effect of saline water through drip irrigation system on yield and quality of tomato. *J. Maharashtra Agric. Univ*. 26(1): 8 - 9.

Kamara, L., Zartamn, R., and Ramsey, R.H. (1991) Cotton root distribution as a function of trickle irrigation emitter depth. *Irri. Sci.* 12: 141-144.

Kang, Y., Yuan, B., and Nishiyama, S. (1999) Design of micro-irrigation laterals at minimum cost. *Irri. Sci.* 18: 125-133.

Kaul, R.K.(1979).Hydraulics of moisture from advance in drip irrigation. Ph.D.Thesis, IARI, New Delhi.

Khan, S. and Hanjra, M.A. 2009. Footprints of water and energy inputs in food

production – Global perspectives. In: *Food Policy* 34: 130-140.

Kumar, Ashwani, 2001. Status and issues of fertigation in India. Microirrigation, CBIP, publication: 418-427.

Kumar, V. and Sivanappan R.K. 1983. Utilization of salt water by the drip system. Proceedings Second National Seminar on Drip Irrigation. March 5-6, 1983, Tamil Nadu Agricultural University, Coimbatore : 47-55.

Lamm, F.R. (2002) Advantages and disadvantages of subsurface drip irrigation. Proc. Int'l Meeting on Advances in Drip/Micro Irrigation, Puerto de La Cruz, Tenerife, Canary Islands, December 2-5.

Lamm, F.R., and Aiken, R.M. (2005) Effect of irrigation frequency for limited subsurface drip irrigation of corn. *Proc. Irrig. Assoc. Int'l Tech. Conf.* Nov. 6-8, 2005, Phoenix, Arizona.

Lamm, F.R., and Camp, C.R. (2007) Subsurface drip irrigation. Microirrigation for crop production design, operation and management, Eds. Lamm, F.R., Ayars, J.E., and Nakayama, F.S. Elsevier, Pp: 473-551.

Lamm, F.R., and Trooien, T.P. (2005) Dripline depth effects on corn production when crop establishment is non-limiting. *Applied Engineering in Agriculture* 21(5): 835-840.

Lamm, F.R., Trooien, T.P., Manages, H.L., and Sunderman, H.D. (2001) Nitrogen fertilization for subsurface drip irrigated corn. *Transactions of the ASAE* 44: 533-542.

Lamont, J.W., Orzolek, D.M., Harper, K.J., Jarrett, R.A., and Greaser, L.G. (2002) Drip Irrigation for Vegetable Production, Penn State Agricultural Research and Cooperative Extension Factsheet.

Lazarovitch, N., Simunek, J., and Shani U. (2005) System dependent boundary condition for water flow subsurface source. *Soil Sci. Soc. Am. J.* 69:46-50.

Letey, J. (1985) Irrigation uniformity as related to optimum crop production-additional research is needed. *Irri. Sci.* 6: 253-263.

Li, H., R. J. Lascano, E. M. Barnes, J. Booker, L. T. Wilson, K. F. Bronson, and E. Segarra. 2001. Multispectral reflectance of cotton related to plant growth, soil water and texture and site elevation. *Agron.* J. 93:1327-1337.

Li, Y., Wallach, R., and Cohen, Y. (2002) The role of soil hydraulic conductivity on the spatial and temporal variation of root water uptake in drip-irrigated corn. *Plant and Soil* 243: 131-142.

Locascio, J.S. (2005) Management of irrigation for vegetables: past, present, future, *Hort Technology* 15(3): 482–485.

Locascio, S.J., and A.G. Smajstrla. 1989. Drip irrigated tomato as affected by water quantity and N and K application timing. *Proceedings of the Florida State Horticultural Society* 102:307-309.

Locascio, S.J., and F.G. Martin. 1985. Nitrogen source and application timing for trickle-irrigated strawberries. *J. Amer. Soc. Hort. Sci.* 110:820-823.

Locascio, S.J., and J.M. Myers. 1974. Tomato response to plug-mix mulch and irrigation method. *Proceedings of the Florida State Horticultural Society* 87:126-130.

Locascio, S.J., and J.M. Myers. 1975. Trickle irrigation and fertilization method for strawberry. *Proceedings of the Florida State Horticultural Society* 88:185-189.

Locascio, S.J., G.J. Hochmuth, F.M. Rhoads, S.M. Olson, A.G. Smajstrla, and E.A. Hanlon. 1997b. Nitrogen and potassium application scheduling effects on drip-irrigated tomato yield and leaf tissue analysis. *HortScience* 32:230-235.

Locascio, S.J., J.G.A. Fiskell, and F.G. Martin. 1984. Nitrogen sources and combinations for polyethylene mulched tomatoes. *Proceedings of the Florida State Horticultural Society* 97:148-150.

Locascio, S.J., J.M. Myers, and J.G.A. Fiskell. 1982. Nitrogen application timing and source for drip irrigated tomatoes. p. 323-328. *In* A. Scaife (ed.) *Proceedings of the Ninth International Plant Nutrition Colloquium*, Warwick Univ., Warwick, England, 1982.

Locascio, S.J., S.M. Olson, R.C. Hochmuth, A.A. Csizinszky, and K.D. Shuler. 1997a. Potassium source and rate for polyethylene-mulched tomatoes. *HortScience* 32:1204-1207.

Machado, R.M.A., and Oliveira, M.R. (2003) Comparison of tomato root distributions by minirhizotron and destructive sampling. *Plant Soil* 255(1): 375-385.

Magen, H. (1995) Fertigation: An overview of some practical aspects. *Fertiliser News of India* 40(12): 97-100.

Marr, C.W. (1993) Fertigation of vegetable crops Publication No: MF-1092. Kansas State University.

Martin, J. H., Waldren, R. P., Stamp, D. L. 2006. Principles of field crop production. 4th ed. Pearson Prentice Hall., Upper Saddle River, NJ.

Martinez, H.J.J., Bar-Yosef, B., and Kafkafi, U. (1991) Effect of subsurface and subsurface fertigation on sweet corn rooting, uptake, dry matter production and yield. *Irri. Sci.* 12: 153-159.

Mehta B.K., and Wang Q.J. (2004) Irrigation in a variable landscape: Matching irrigation systems and enterprises to soil hydraulic characteristics, Final Report, Department of Primary Industries, Victoria.

Metin-Sezen, S., Yazar, A., Canbolat, M., Eker, S., and Celikel, G. (2005) Effect of drip irrigation management on yield and quality of field grown green beans. *Agric. Water Manag.* 71(3): 243-255.

Miller, M.L., Charlesworth, P.B., Katupitiya, A., and Muirhead, W.A. (2000) A comparison of new and conventional sub surface drip irrigation systems using pulsed and continuous irrigation management. In: *Proc. Nat'l Conf. Irrig. Assoc. Australia*. May 23-25 Melbourne, Australia.

Mmolawa, K., and Or, D. (2000a) Water and solute dynamics under drip-irrigated crop: experiment and analytical model. *Transactions of the ASAE* 43(6): 1597-1608.

Mmolawa, K., and Or, D. (2000b) Root zone solute dynamics under drip irrigation: A review. *Plant and soil* 222: 163-190.

Molz, F. J.1981. Models of Water Transport in the soil–plant system. *Water Resources Ressearch.* 17(5):1245–1260.

Mullins, G. L and C. H. Burmester. 1990. Dry matter , Nitrogen, phosphorous, and potassium accumulation by four cotton varieties. *Agron. J.* 82:729-736.

Narayanamoorthy, A. 2004. Drip irrigation in India: can it solve water scarcity. *Water Policy.* 6(2): 117-130.

NCPAH (2008). Area under Micro Irrigation. National Committee on Plasticulture Application in Horticulture, New Delhi.

Nehra, V.S., Singh, Jaspal and Tyagi, N.K. 1991. Simulation of soil moisture distribution pattern under trickle source through analytical approach, *Indian Jouranl of Power and River valley Development.* 41(1); 12-16.

Neufeld, J., Davison, J., and Stevenson, T. (1993) Sub-surface drip irrigation. *Fact sheet No*: 97-13. University of Nevada.

Nightingale, H.I., Phene, C.J., and Patton, S.H. (1985) Trickle irrigation effects on soil –chemical properties. *3rd Int'l Drip Irrigation Congress, Fresno,* CA, USA. Pp: 730-735.

Norton, E. R. and J. C. Silvertooth. 1998. Field validation of soil solute profiles in irrigated cotton. *Agron. J.* 82:729-736.

O'Neill, M., Pablo, R., and Begay, T. (2002). Development and evaluation of drip irrigation for North Western New Mexico. *36th Annual Progress Report*, New Mexico State University.

Or, D. (1996) Drip irrigation in heterogeneous soils: steady state field experiments for stochastic model evaluation. *Soil Sci. Soc. Am. J.* 60: 1339-1349.

Or, D., and Coelho, F.E. (1996) Soil water dynamics under drip irrigation: Transient flow and uptake model. *Transactions of the ASAE* 39(6): 2017-2025.

Pampattiwar, P.S., Suryawanshi, S.N., Gorantiwar, S.D., Pingole, L.V. 1993. Drip irrigation for pomegranate. *Maharashtra J. Horti*. 7(1):46-50.

Patel, N., and Rajput, T.B.S. (2007) Effect of drip tape placement depth and irrigation level on yield of potato. *Agric. Water Manag* 88: 209-223.

Patel, Neelam and Rajput,T.B.S., 2001 b. Fertigation of okra using commercially available granular fertilizers. *Proce. Of International symposium on importance of potassium in nutrient management for sustainable crop production in India* held at New Delhi from Dec. 3-5,pp: 270-273.

Patel, Nelam and Rajput,T.B.S., 2001 a. Effect of fertigation on growth and yield of onion. CB Publication. 282. pp: 451-454.

Phene, C. J., and D. C. Sanders. 1976. High-frequency trickle irrigation and row spacing effects on yield and quality of potatoes. *Agron. J.* 68(4):602-607.

Phene, C.J., and Beale, O.W. (1976) High frequency irrigation for water nutrient management in humid regions. *Soil Sci. Soc. Am. J.* 40: 430-436.

Phene, C.J., and Phene, R.C. (1987) Drip irrigation systems and management. *ASPAC Food and Fertilizer Tech. Ctr., Taiwan, Ext. Bull.* No: 244.

Phene, C.J., Davis, K.R., Hutmacher, R.B., and McCormick, R.L. (1987) Advantages of subsurface irrigation for processing tomato. Eds. Sims, W.L. Davis, California. *Acta Horticulturae* 200:101-114.

Phene, C.J., Davis, K.R., Hutmacher, R.B., Bar-Yosef, B., Meek, D.W., and Misaki, J. (1991) Effect of high frequency surface and subsurface drip irrigation on root distribution of sweet corn. *Irri. Sci.* 12: 135-140.

Pier, J.W., and Doerge, T.A. (1995) Concurrent evaluation of agronomic, economic and environmental aspects of trickle irrigated watermelon production. *J. Environ. Qual.* 24: 79-86.

Pitts, D., Peterson, K., Gilbert, G., and Fastenau, R. (1996) Field assessment of irrigation system performance. *Appl. Engr. Agric.* 12(3): 307-313.

Raine, S.R., and Foley, J.P. (2001) Application systems for cotton irrigation-Are you asking the right questions and getting the answer right? In: *Proc. Nat'l Conf. Irrig. Assoc*. Australia. May 23-25 Melbourne.

Raine, S.R., Foley, J.P., and Henkel, C.R. (2000) Drip irrigation in the Australian cotton industry: A scoping study. NECA publication no: 179757/1. National Centre for Engineering in Agriculture, University of Southern Queensland, Toowoomba.

Rajput, T.B.S. and Patel, Neelam. 2002. Fertigation: theory and practice. Publication No. IARI/WTC/2002/2.

Rema Devi, A.N. 1984, Soil moisture distribution pattern under drip irrigation system. *Proceedings Second National Seminar on Drip Irrigation*. March 5-6, 1983, Tamil Nadu Agricultural University, Coimbatore : 86-93.

Roberts, T.L., White, S.A., Warrick, A.W., and Thompson, T.L. (2008) Tape depth and germination method influence patterns of salt accumulation with subsurface drip irrigation. *Agric. Water Manag*. 95(5): 669-677.

Rockstrom, J., Lannerstad, M., Falkenmark, M., 2007. Assessing the water challenge of a new green revolution in developing countries. *PNAS* 104 (15), 6253–6260.

Rolston D E, Miller R J and Schulbach 1986 Management principles: Fertilization. *In* Trickle Irrigation for Crop Production, Design, Operation and Management. Eds F S Nakayaina and D A Bucks. pp 317–344. Elsevier.

Rolston, D.E., Biggar, J.W., and Nightingale, H.I. 1991. Temporal persistence of spatial soil-water pattern under trickle irrigation. *Irrigation Science.* 12:181-186.

Rubeiz, I.G., Oebker, N.F., and Stroehlein, J.L. (1989) Subsurface drip irrigation and urea phosphate fertigation for vegetables on calcareous soils. *J. Plant Nutrition* 12(12): 1457-1465.

Ruskin, R. (2005) Subsurface drip irrigation and yields. Available at: www.geoflow.com/agriculture.

Sadler, E.J., Camp, C.R., and Busscher, W.J. (1995) Emitter flow rate changes by excavating subsurface drip irrigation tubing. *Proc. Fifth Int'l Microirrigation congress*, Eds. Lamm, F.R. ASAE, St. Joseph, Michigan, 2-6 April, Orlando, FL, USA. Pp: 763-768.

Sakellariou-Makrantonaki, M., Kalfountzos, D., and Vyrlas, P. (2002) Water saving and yield increase of sugar beet with subsurface drip irrigation. *The Int. J.* 4(2-3): 85-91.

Saxena, C. K. and Gupta, S.K. 2006. Effect of soil pH on the establishment of litchi (*Litchi chinensis* Sonn.) plants in an alkali environment. *The Indian Journal of Agricultural Sciences*. 76:547-549.

Saxena, C.K., Gupta, S.K. 2004. Drip Irrigation for Water Conservation and Saline / Sodic Environments in India: A Review. Natural Resources Engineering and Management and Agro-Environmental Engineering. *Proceedings of International Conference on Emerging Technologies in Agricultural and Food Engineering* (2004). Indian Institute of Technology, Kharagpur, India December 14-17, 2004. Anamaya Publishers, New Delhi, 110 030, India. Pp 234-241.

Schwankl, L.J., and Hanson, B.R. (2007) Surface drip irrigation. Microirrigation for crop production, Eds. Lamm, F.R., Ayars, J.E., and Nakayama, F.S. Elsevier, Pp: 431- 472.

Schwankl, L.J., Grattan, S.R., and Miyao, E.M. (1990) Drip irrigation burial depth and seed planting depth effects on tomato germination. *Proc. Third International Irrigation Symp.* Oct. 28- Nov.1 Phoenix, Arizona. ASAE, St. Joseph, Michigan. Pp: 682-687.

Schwankl, L.J., Schulback, K.F., Hanson, B.R., and Pettygrove, S. (1993) Irrigating lettuce with buried drip, surface drip and furrow irrigation system: Overall performance. Management of irrigation and drainage system. Eds. Allen, R.G. 21-23 July, Utah, USA.

Schwartzman, M., and B. Zur 1986. Emitter spacing and geometry of wetted soil volume. *Journal of Irrigation and Drainage Engineering,* 112: 242-253.

Senn, A.A., and Cornish, P.S. (2000) An example of adoption of reduced cultivation by Sydney's vegetable growers. In : *Soil 2000: New Horizons for a New Century.*

Shani, U., Xue, S., Gordin-Katz, and Warrik, A.W. (1996) Soil-limiting flow from subsurface emitters - Pressure measurements. *J. Irrig. and Drain. Engin.* 122(5): 291- 295.

Sharmasarkar, F.C., Sharmasarkar, S., Held, L.J., Miller, S.D., Vance, G.F., and Zhang, R. (2001) Agroeconomic analysis of drip irrigation for sugarbeet production. *Agron. J.* 93: 517-523.

Shaviv, A., Ravina, I., and Zaslavsky, D. (1987) Field evaluation of methods of incorporating soil conditioners. *Soil and Tillage Research* 9: 151-160.

Shmueli, M., and D. Goldberg. 1971. Sprinkle, furrow and trickle irrigation of muskmelon in an arid zone. *HortScience* 6:557-559.

Shock, C.C., Feibert, E.B.G., and Saunders, L.D. (2004) Plant population and nitrogen fertilization for subsurface drip-irrigated onion. *HortScience* 39(7): 1722-1727.

Shock, C.C., Feibert, E.B.G., and Saunders, L.D. (2005,a) Onion response to drip irrigation intensity and emitter flow rate. *HortTechnology* 15 (3): 652-659.

Silvertooth, J.C. 2001. Nitrogen management for cotton. University of Arizona Cooperative Extension pub. AZ1200. (Online). Available at http://ag.arizona.edu/crops/cotton/ soilsmgt/ nitrogen_managemant.html (verified 20 July 2009).

Singh, DK, Singh, RM and Rao, KVR. 2009. Development/ adoption and evaluation of automatic fertigation system for mango and guava. Final report of research project No. 505. Central Institute of Agricultural Engineering, Bhopal, M.P., India: 1-42.

Singh, Pratap and Kumar, R. 1989. Comparative performance of trickle irrigation for tomato. *J. Agril. Engg.,* ISAE. 26(1):39-48.

Singh, Pratap and Kumar, R.1994. Effect of irrigation methods for cauliflower grown in heavy soils with shallow watertable. *J. Agril. Engg*., ISAE. 31:36-43.

Singh, R.V., Chauhan, H.S., Tafera, Abu. 2001. Wetting front advance for varying rates of discharge from a trickle source. *In*: Singh, H.P., Kaushik, S.P. and Kumar, Ashwini, Murthy, T.S., and Samual, Jose C. (*Eds.*) *Microirrigation, CBIP. Proceedings of the International conference on Micro and Sprinkler Irrigation Systems*, held during February8-10, 2000 at Jain Irrigation Hills, Jalgaon, Maharashtra, India : 125-128.

Singh, V.P., Samra, J.S. , Gill, H.S. 1990. Use of poor quality water through drip system in kinnow orchards. In; *Proc. XI International Congress on the use of Plastics in Agriculture.* Oxford & IBH Pub. Co. Pvt. Ltd., N.Delhi. 165-175.

Smakhtin, V., Revenga, C., Döll, P., 2004. A pilot global assessment of environmental water requirements and scarcity. *Water International* 29.

Solomon, K.H. (1984b) Yield related interpretation of irrigation uniformity and efficiency measures. *Irri. Sci.* 5: 161-172.

Sparks D L and Huang P M 1985 Physical chemistry of soil Potassium. *In* Potassium in Agriculture. Ed. RD Munson. Pp 201–265. ASA-CSSA-SSSA, Madison, Wisconsin, USA.

Storke, P.R., Jerie, P.H., and Callinan, A.P.L. (2003) Subsurface drip irrigation in raised bed tomato production – Soil acidification under current commercial practice. *Australian J. of Soil Research* 41: 1305-1315.

Subba Rao, N.Subbiah, G.V. and Ramaiah, B. 1987. Effect of saline water on tomato yield. And soil properties. *J.Indian Soc. Soil Sci. Research*. 5(2):407-409.

Susila, A.D., and S.J. Locascio. 2001. Sulfur fertilization for polyethylenemulched cabbage. *Proceedings of the Florida State Horticultural Society* 114:318-322.

Thompson, L. (2005) Adoption of permanent subsurface drip irrigation for forage production to enhance the economical and environmental sustainability of Victoria's dairy industry. Project No: GF3/036.

Thompson, L.T., S.A. White, J. Walworth, and G.J. Sower. 2003. Fertigation frequency for subsurface drip-irrigated broccoli. *Soil Science Society of America Journal* 67:910-918.

Thompson, T.L. and Doerge, T.A. (1995a) Nitrogen and water rates for subsurface trickle irrigated romaine lettuce. *HortScience* 30(6): 1233-1237.

Thompson, T.L., and Doerge, T.A. (1995b) Nitrogen and water rates for subsurface trickle –irrigated collard, mustard and spinach. *HortScience* 30(7): 1382-1387.

Thompson, T.L., and Doerge, T.A. (1996) Nitrogen and water interactions in subsurface trickle – irrigated leaf lettuce: Plant response. *Soil Sci. Soc. Am. J.* 60: 163-168.

Thompson, T.L., Doerge, T.A., and Godin, R.E. (2002) Subsurface drip irrigation and fertigation of broccoli: Yield, quality and nitrogen uptake. *Soil Sci. Soc. Am. J. 66*: 186-192.

Thorburn, P., Biggs, J., Bristow, K., Horan, H., and Huth, N. (2003) Benefits of subsurface application of nitrogen and water to trickle irrigated sugarcane. 11th Australian Agronomy Conference paper.

Thorburn, P.J., Cook, F.J., and Bristow, K.L. (2003, b) Soil dependant wetting from trickle emitters: implications for system design and management. *Irri. Sci.* 22: 121-127.

Tucker, B. B. 1974. Efficient use of fertilizer for maximizing profits in cotton production. *In : Proceedings Beltwide cotton production research conferences.* 158-162. National Cotton Council. Memphis. TN.

Vazquez, N., Pardo, A., Suso, M.L., and Quemada, M. (2005) A methodology formeasuring drainage and nitrate leaching of mineral nitrogen from arable land. *Plant Soil* 269: 297-308.

Vazquez, N., Pardo, A., Suso, M.L., and Quemada, M. (2006) Drainage and nitrate leaching under processing tomato growth with drip irrigation and plastic mulching. *Agriculture ecosystem and environment* 112: 313-323.

Wang, F., Kang, Y., and Liu, S. (2006) Effects of drip irrigation frequency on soil wetting pattern and potato growth in North China Plain. *Agric. Water Manag.* 79 (3): 248-264.

Warrick, A.W., and Shani, U. (1996) Soil-limiting flow from subsurface emitters-Effect on uniformity. *J. Irrig. And Drain.* 122(5): 296-300.

Wendt, C.W., Onken, A.B., Wilke, O.C., Hargrove, R.S., Bausch, W., and Barnes, L. (1977) Effect of irrigation systems on the water requirement of sweet corn. *Soil Sci. Soc. Am. J.* 41: 785-788

Wu, I.P., Barragan, J., and Bralts, V.F. (2007) Field performance and evaluation. Microirrigation for crop production design, operation and management, Eds. Lamm, F.R., Ayars, J.E., and Nakayama, F.S. Elsevier, Pp: 357-387.

Wu, P., Gitlin, H.M., Solomon, K.H., and Saruwatari, C.A. (1986) Design principles- Trickle irrigation for crop production. Eds. Nakayama, F.S., and Bucks, D.A. Elsevier, Pp: 53-92.

Yabaji, R., Nusz, J. W., Bronson, K. F., Malapati, A., Booker J. D., Nichols, R. L., Thompson, T. L. 2009. Nitrogen Management for Subsurface Drip Irrigated cotton: Ammonium Thiosulfate, Timing, Canopy Reflectance. *Soil Sci. Soc. Am. J.* 73: 589-597.

Zazueta, F.S., Clark, G.A., Smajstrla, A.G. and Carrilo, M. 1995. A simple equation to estimate soil–water movement from a drip irrigation source. *In*: F.R. Lamm (ed.) *Microirrigation for a changing world: Conserving resources/ preserving the environment. Proceedings of the Fifth International Microirrigation Congress* (American Society of Agricultual Engineers) held during 2-6 April 1995 at Orlando, Florida, USA. pp. 851–856.

Zhang, H., G. Johnson, B. Raun, N. Basta and J. Hattey. 1998. OSU soil test interpretations. Okhlahoma Cooperative Extension Service Fact Sheet No. 2225.Oklahoma State Univ., Stillwater. USA.

□□□

Green Agriculture : Newer Technologies, 2012
© Kambaska Kumar Behera (ed.) pp. 391-396
New India Publishing Agency, New Delhi (India)
e-mail : info@nipabooks.com; website : www.nipabooks.com

Chapter-15

Basics of Livelihood in Rural Perspective

Shubhadeep Roy [1], Neeraj Singh [1], R. Roy Burman [2] and Baldeo Singh [3]
[1] *Scientist, IIVR, Varanasi. Email: shubhadeepiari@gmail.com*
[2] *Senior Scientist, Division of Agril. Extension, IARI, New Delhi.*
[3] *Former Jt. Director (Extension), IARI, New Delhi.*

SUMMARY

In the context of increasing population pressure, diminishing per capita land holding, and employment crisis, livelihood security of Indian rural population has become important concern today. The livelihood of the people is dynamic, diverse and complex in nature. The livelihood comprises the people themselves, the assets they posses, the activities they do and the output they gain from their activities. According to the need, availability of the resources and capacity, people used to select their livelihood strategies like agricultural intensification, livelihood diversification or they migrate to the other places to find alternative livelihoods.

1. Livelihood Security

Livelihood security is a buzz word today. The country like India, where population is more than 117 crore (July, 2010) and where nearly 25 percent of the total population are living below poverty line, certainly it is crucial to think about their livelihood security. Though 'Livelihood' appears as a small word, but it is a very complex process. Different people access to different forms of livelihood assets, and the type of livelihood strategies they take up are different and their livelihood strategies are always influenced and mediated by the socio- cultural, political and institutional environment from the very household level to the global level. So, as the socio- cultural, political and institutional environment changes, people also change their livelihood strategies. Thus the livelihoods of the people are diverse, dynamic and complex in nature.

1.1. So, what is livelihood? According to the definition given by the Institute of Development Studies, A livelihood comprises the capabilities, assets (including both material and social resources) and activities required for means of living. And a livelihood is sustainable when it can cope with and recover from stresses and shocks maintain or enhance its capabilities and assets, while not undermining the natural resource base.

2. Now, let us discuss about the nature of human livelihoods. The analysis of livelihood starts from the household level. It is about who the people are, what they do, what assets they are having and ultimately the living scenario they gain from what they do. Let us discuss the things more precisely:

2.1. The people: Peoples are different in their livelihood capacities. It is the attitude of the people and other environmental factors which make them different in livelihood capacities to support for enhancement and exercise of livelihood strategies.

2.2. The activities: Activities means what people do. According to Chambers and Conway (1991) many livelihoods are largely predetermined by accident of birth. Livelihoods of this sort may be ascriptive: in village India, children may be born into a caste with an assigned role of potters, shepherds or washer people. Gender as socially defined is also a persistent ascriptive determinant of livelihood activities. And not necessary ascriptively, a person may be born, socialized and apprenticed into an inherited livelihood- as a cultivator with land and tools, a pastoralist with animal, a forest dweller with trees, a fisherperson with boat and tackle or a shopkeeper with shop and stocks, and each of these may in turn create a new household in the same occupation. A person or household may also choose a livelihood, especially through education and migration.

2.3. The assets : The assets of livelihood may be categorized into two categories:

2.3.1. Tangible assets : Mainly stores and resources of household may come under tangible assets category. Stores may include food stocks for consumption and sale, stocks of animal feed, gold, jewellery, textiles, savings in bank etc.

The examples of livelihood resources are land, water, trees, livestock, farm machineries and tools, household utensils etc. A farm family should have optimum level of stores and resources in its procession to maintain a sustainable livelihood. It depends on the motivation or more precisely the wisdom of the individuals how well they can maintain their stores and resources.

2.3.2. Intangible assets : According to Chambers and Conway (1991), claims and access to the household may come under the intangible assets category.

Claims are demands and appeals which can be made for material, moral or other support or access. The support may be in terms of food, implements, loans, gifts or work. Claims may be made on individuals or agencies, on government or the international community.

Access is the opportunity in practice to use a resource, store or service or to obtain information, material, technology, employment, food or income.

Service may include transport, education, health and market.

Information may include extension services, radio, television and news paper.

Technology includes improved techniques of cultivation in rural perspective.

Employment and other income earning activities include rights to common property resources such as fuel wood or animal grazing on communal land. Government sponsored employment generation programme or poverty eradication programme like Mahatma Gandhi National Rural Employment Guarantee Act (MNREGA), National Livelihood Security Mission and many more are also the example of the livelihood access to the population.

So, here comes the responsibility of the government, that how well the claims and accesses may be distributed among the people.

Livelihood Flow Chart

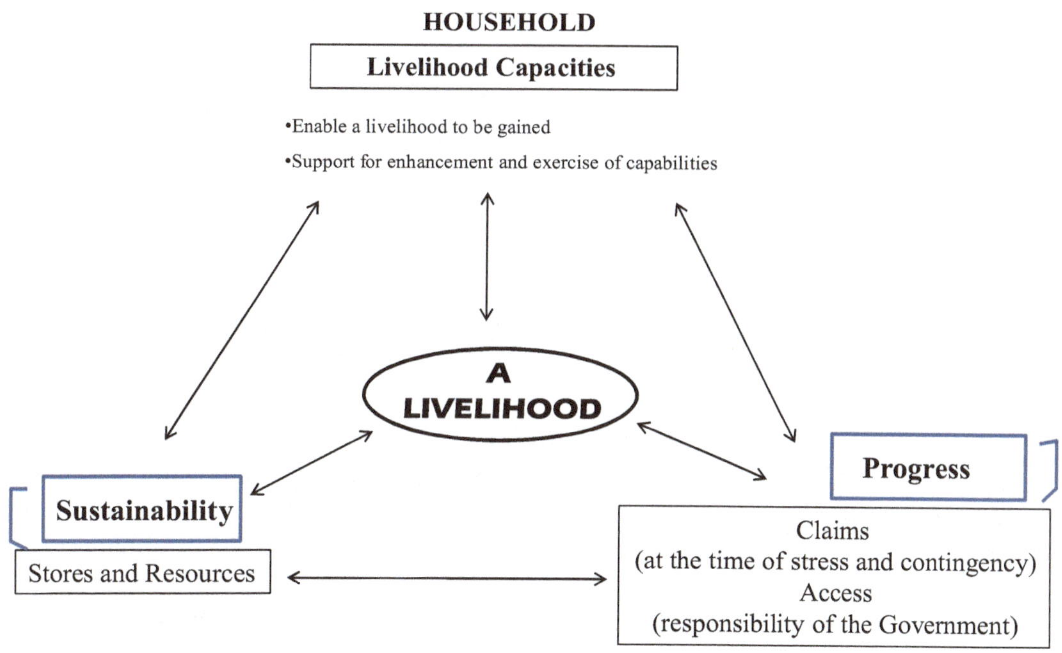

2.4. Gains or output : Whatever has been discussed earlier, cumulatively they deliver a living to the population, and they gain according to their activities.

3. Livelihood Strategiesc

According to the DFID's sustainable livelihood glossary, the term "Livelihood Strategies" denotes, "the range and combination of activities and choices that people make in order to achieve their livelihood goals. Livelihood strategies include how people combine their income generating activities, the way in which they use their assets, which assets they choose to invest in and how they manage to preserve existing assets and income. Livelihoods are diverse at every level, for example, members of a household may live and work in different places engaging in different activities, either temporarily or permanently. Individuals themselves may rely on a range of different income generating activities at the same time" (DFID, 2001). For the farming based household livelihood strategies may be divided into three sub categories.

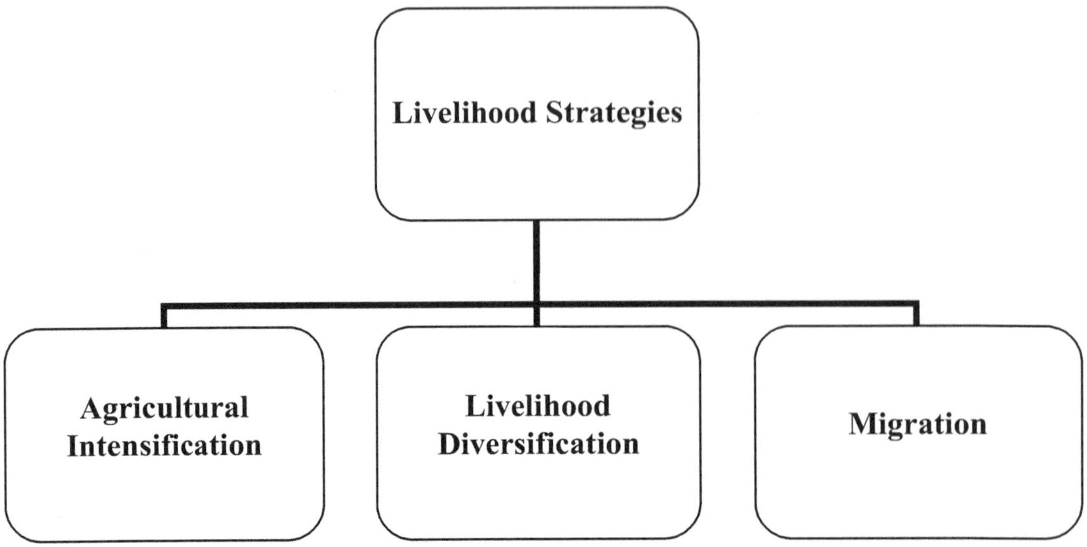

3.1. Agricultural intensification : Agricultural Intensification refers to the use of a greater amount of non- agricultural recourses (labour, input etc.) for a given land area, so that higher output is produced (Hussain and Nelson, 1999). It generally focuses on the increased productivity of agricultural commodities best suiting the agro- ecological conditions of the region and the farm and existing market outlet. The growth rate of population in India is about 1.37 percent per year and due to urbanization cultivable land is decreasing day by day. So, agricultural intensification is the only option to maintain the food security of the country.

3.2. Livelihood diversification : Livelihood Diversification is defined as "the process by which households construct a diverse portfolio of activities and social support capabilities for survival and also to improve their standard of living." (Ellis,

2000). There is a greater need of livelihood diversification with the increasing inability of agriculture to deliver in the post- liberalization era, climate change, poverty and other uncertainties over which rural people have no control. Livelihood diversification can strengthen livelihoods, helps household to cope with the adversity. It may guarantee smooth consumption, stable income and meet adequate labour requirement.

According to the nature and strategy to diversify livelihood of the people, the livelihood diversification may be classified into two categories :

3.2.1. Choice diversification : People diversify for the better off of their living standard. In this case their basic needs are already fulfilled and they strive for the higher strata of the living standard. They start saving more and they are happy with their activities.

3.2.2. No-Choice Diversification : It may be called as necessity response or defensive response to cope with or survive in the adverse situation. When people stumble to meet the basic need for their families, they have to shift to some other activities with which they might not be happy.

3.3. Migration : People migrate to other places from their native to ensure a better livelihood. It may be by choice: people migrate for education, job or better business and may be by compulsion: the wage labourers used to migrate to cities for getting works in off season. It is obvious that choice migration is acceptable and people should go for this.

4. Conclusion

Diversification is most important in situation of high vulnerability, absent market and poor infrastructure. In India 52% of the total labour force engaged in agriculture and allied activities, whereas 13.7% of the total Gross Domestic Product (GDP) comes from agriculture. Sometimes labourers do uneconomic works in their small piece of land, what is termed as disguised employment, which is one of the major causes of poverty in rural India. Few remedies have been discussed here:

Strengthening the rural- urban linkages: Urban area is expanding everywhere in India. The rate of urbanization is 2.4% per year and 29% of the total population of India live in urban areas. They are mostly engaged in service sector (54.6% of total GDP) or industrial sector (28.2% of total GDP). The people engaged in service and industrial sectors have money in their hand and this money have to be channelized to the rural sector. Strengthening rural- urban linkages may be the correct step here. Farming should be need based and market oriented. There is a demand of high value vegetables and fruits in urban areas. So, farmers have to shift for such kind of crops. But vegetables and fruits are highly perishable in nature as well as specific knowledge and skills are also required for its cultivation.

Indian Institute of Vegetable Research (IIVR), Varanasi, IIHR, Bangalore, IARI, New Delhi are doing rigorous research and extension work at national level to promote high value vegetable and fruit crops as well as providing training to the farmers and officials to enhance their knowledge and skill in horticultural crop cultivation.

Next important is access to the urban food market and for this physical infrastructure, road network and affordable transport are required. About 40% of the habitations in India are not connected by good roads. Pradhan Mantri Gram Sadak Yojana (2000) has been launched to enhance connectivity all over the country.

Another very important point is market information. Farmers should have proper information on how market operates, including price fluctuation and consumer preferences so that they can decide about their correct marketing channel.

Moreover the rural agriculture has to be transformed to agribusiness. There is a need to make the farmers entrepreneurial through systematic motivational training to take a shift from subsistence farming to commercially and economically viable agribusiness.

References

Chambers, R. and Conway, G.R. (1991). Sustainable rural livelihoods: Practical concepts for the 21st century. IDS discussion paper 296.

DFID (Department For International Development). 2001. Sustainable Livelihoods guidance Sheets.

Ellis, F. (2000) Rural livelihoods and diversity in developing countries, Oxford University Press.

Scoons, Ian. (1998). Sustainable Rural Livelihoods: A framework for analysis. IDS working paper 72, 1998.

□□□

Green Agriculture : Newer Technologies, 2012
© Kambaska Kumar Behera (ed.), pp. 397-414
New India Publishing Agency, New Delhi (India)
e-mail : info@nipabooks.com; website : www.nipabooks.com

Chapter-16

Abiotic Stresses in Pulse Crops – Dimensions and Solutions

Om Gupta and Anita Babbar
AICRP on Chickpea (Lead Centre),
Deptt. of Plant Breeding & Genetics, JNKVV, Jabalpur (M.P.)
E-mail : omgupta_jnkvv@rediffmail.com

SUMMARY

Abiotic stress can be reduced by choosing the most appropriate pulse species and adjusting agronomy (sowing time, plant density, soil management) to ensure sensitive crop stages occur at the most favourable time in the season. For example choosing the optimum sowing time, species and cultivars with appropriate phenology can reduce the effect of frost and drought in pulse crops in dry land environments, which termed stress.

Most appropriate approach of dealing with stresses caused by extremes in the abiotic environment is to develop cultivars resistant to specific stresses. However, breeding and selection for resistance to these stresses is often considerable difficult because of the unpredictability of climatic conditions. Drought and heat escape through earliness in flowering and maturity is probably the characteristic most widely used by breeders for pulses and other crops to escape drought especially in low rainfall, terminal drought environment. Lack of simple and accurate screening procedures to screen parental genotypes and breeding population for various abiotic stresses is the major bottlenecks in the development of stress tolerant cool season pulse crops.

Germplasm for resistance to various abiotic stresses are not widely available to breeders. Recent studies suggest that wild species of cool season pulses posses desired traits for a number of abiotic stresses. Future effort is required to identify desirable genes from this germplasm for transfer to adapted cultivars by conventional and / or biotechnological approaches to develop abiotic stresses resistant cultivars.

1. Introduction

Pulses are widely recognized as having an important role and an integral part of Indian agriculture, supplementing higher amount of quality protein in Indian diet than that in Asia and World as a whole. Out of several pulse crops grown during winter season in water deficit environment the chickpea is the most preferred by the dry land farmers of Indian subcontinent. Besides chickpea, lentil and pea are also sown on conserved soil moisture from monsoon rains. Although some available moisture is always left even the crops are challenged by several stresses and therefore produces mean seed yield much below those of the competing cereals. Moreover, the farmers of rainfed areas hardly use any fertilizer.

Considering the limited breeding efforts and development of management packages for pulses, compared with cereals, there is considerable scope for further improvement of pulse yields in India. However, in recent years fluctuations in pulse yield and frequent crop failure in several regional environments caused by biotic and abiotic stresses are threatening the future expansion. The relative importance of abiotic stresses affecting pulse production in India is poorly understood.

This paper is aimed to discuss the important abiotic stresses in pulses and to suggest agronomical, physiological and breeding strategies to overcome these stresses and improve the productivity in prone areas.

2. Present status

Globally pulses are second in importance only to cereals, the global pulse production stands at 60.7 million tonnes from an area of 69.7 m ha with an average yield of 817 Kg/ha^{-1}. India contributes 25 percent to the global basket from an area of about 32 percent (22-23 million ha) under pulse crops (Fig.1, 2). The annual production being 13.8 million tonnes. Major crops grown in the country are – Chickpea (7.71 m ha), Pigeonpea (3.63 m ha), Mungbean (3.34 m ha), Urdbean (3.17 m ha), Lentil (1.47 m ha) and Field pea (0.79 m ha). Pulses like Chickpea and Pigeonpea, the country's share is 66 percent and 90 percent in the global production. The country has produced around 13.8 million tonnes of pulses in 2007-08. Nationally the most important states for pulses are Madhya Pradesh, Uttar Pradesh Maharashtra, Rajasthan, Karnataka and Andhra Pradesh, which together accounts 75 percent production. The most important pulse crop in the country is chickpea, which accounts for 42 percent of the production from 32 percent area and the four states together contribute 87 percent of the chickpea production.

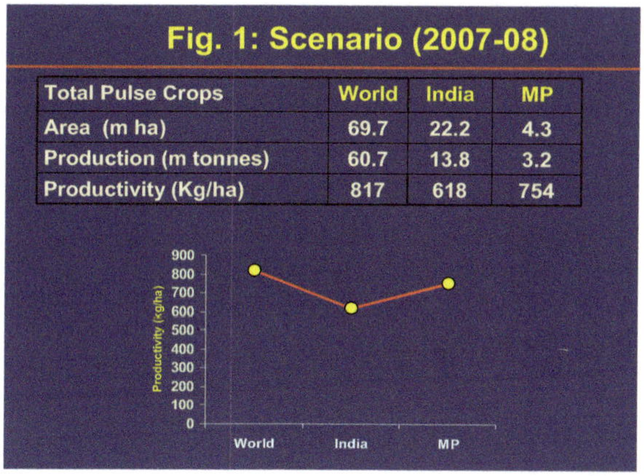

Figure 2 : Production scenario of pulses in India (2007-08)

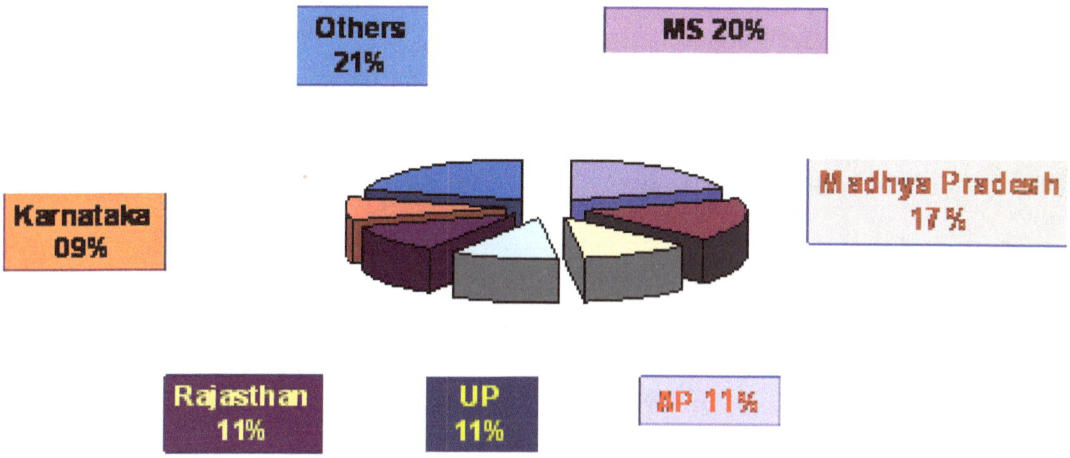

Contribution of Pulses in Nutrients supply in India, Asia and world

Nutrition supply through pulses	World	Asia	India
Pulse supply/capita/year (Kg)	5.90	5.60	12.80
Calorie/Capita/day (number)	55.90	52.50	122.20
Protein/capita/day (g)	3.50	3.20	7.20
Fat/capita/day (g)	0.40	0.40	1.10

Source : Directorate of Eco. & Stat, New Delhi

3. Constraints

The major constraints that limit the full realization of potential yield of pulses include biotic and abiotic stresses in the pulses growing areas besides socioeconomic factors. About 90 percent of the rabi pulses are grown as rainfed in marginal lands often poor in nutrients and do not receive monetary inputs limiting high yield potential.

Among abiotic stresses, terminal drought, high temperature during reproductive stage, cold sensitivity during vegetative and flowering stages water lodging, heat, boron toxicity and salinity/alkalinity throughout the crop period inflict major yield losses and instability in production. In the semi arid areas where chickpea is the important crop, saline under ground water is causing a great threat to its cultivation. A critical review on drought management in pulse crop is being discussed (Serraj *et. al.*, 2003, Dua *et. al.*, 2003)

Other constraints to productivity are

1. Excessively wet soil condition during Kharif and water stress during rabi
2. Low seed replacement rate
3. Inadequate plant stand
4. Delayed planting of rabi crops
5. Diversification of varieties
6. Inadequate and imbalanced use of nutrients
7. Non availability of Rhizobium resistant to moisture stress conditions
8. Biotic stresses
9. Technological constraints
10. Socio economic constraints

4. Phenology

The extent of damage caused by abiotic stresses depends on the pulse species, the prevailing environmental condition and stage of crop growth.

The phenology of pulse crop varies with species, time of sowing and location. For a particular stress, the time at which it occurs in the plants life cycle will affect plant production through its yield components (total biomass, number of pods, number of seeds, seed weight). Life cycle of the pulse crop may be divided into the following five phases:-

Phase I - From planting to seedling emergence.
Phase II - From seedling emergence to first flowering
Phase III - From first flowering to pod appearance.
Phase IV - From pod appearance to the beginning of seed formation.
Phase V - From seed formation to physiological maturity.

5. Abiotic stresses identified in major pulse crops

Crop	Season/ Nitches	Abiotic stress
Chickpea	Timely sown	Low temperature Terminal drought Salt stress
Lentil	Late sown	Terminal drought Cold

		Moisture/Temperature
Pea	Late sown	Shattering
Pigeonpea	Kharif- Early Medium late Pre-rabi	Water logging Cold, terminal drought/ water logging Cold, terminal drought
Mungbean & Urdbean	Kharif Zaid Rabi	Pre harvest sprouting, terminal drought Pre harvest sprouting Temperature stress, drought, terminal drought.

5.1 Drought stress

Drought is the most important constraint to yield in all pulse growing regions, particularly in the low rainfall areas. Practically, it is difficult to separate drought and stress and they generally occur simultaneously. The type of drought stress which affect pulse production are: lack of or low opening rainfall at planting, inter mittant drought stress by breaks in winter rainfall and the terminal drought stress, resulting from receding soil moisture and increasing high temperature in spring.

Terminal drought is the most common form of drought stress depends upon the amount and distribution of rainfall capacity of the soil to store moisture and the evaporating demand of the atmosphere.

Symptoms

There are some major effects of drought on pulse productivity:

- Failure to establish the desired plant stands that means reduced plant population in a scattered manner.
- Stunted growth with less number of branches and yellowing of lower leaves
- Quick withering of leaves and reduction in yield due to sub optimal moisture availability

Chickpea has the greatest ability to tolerate intermittent drought and respond to subsequent rainfall due to its more indeterminate growth habit as compared to other cool season pulses. Seed yield loss in chickpea due to terminal drought can vary between 30-60 per cent depending upon the location and climatic conditions during the crop season.

Lentil is considered relatively tolerant to drought, potential losses in yield can range between 6-54 per cent.

Fieldpea is also sensitive to drought and yield losses varies from 21-54 per cent. However, fieldpea with the correct phenology may escape drought and produce respectable yield when compared with other pulses.

5.2 Heat stress

The optimal temperature for cool season pulses range between 10-30^{0}C. Temperature falling outside the optimum range cause stress. Daily maximum temperature above 25^{0}C is considered as the thresh hold level for heat stress. The critical temperature for heat tolerance seems to be higher in chickpea then lentil and fieldpea.

Symptoms

Pulses are particularly sensitive to heat at full bloom stage. A few days of exposure to temperature (30-35^{0}C), caused heavy yield losses through flower drop and pod abortion. Marked reduction in seed yields after brief episodes of high temperatures during seed filling can diminish seed set, seed weight and accelerate senescence and reduce yield.

Management practices suggest sowing of drought resistant varieties- save moisture from rain water before sowing, proper management of weeds to maintain loss of moisture.

5.3 Cold

The major symptoms of cold in winter pulses are-

- Reduced leaf production
- Area expansion
- Poor dry matter production
- Chlorosis
- Necrosis of the older leaves
- Reduced ovule fertilization and poor seed set

These symptoms are generally observed in northern India. A daily average temperature between 0-10^{0}C is considered a threshold level for cold or chilling stress in cool season pulses. Under such circumstances reproductive organs mainly flower buds and flower are most susceptible part to cold or chilling injury. Chickpea is more sensitive to chilling injury than fieldpea or lentil. Breeding for cold tolerance in chickpea has been reviewed by Chaturvedi *et al.* 2009 .

5.4 Frost

The threshold level for frost injury in winter pulses occurs when daily minimal temperature is below 0^{0}C without snow cover. Frost damage in spring is a major cause of yield loss in crops including pulses. The most important stress in the freezing process is ice formation and the associated mechanical damage of the tissue. Flowering, early pod formation and seed filling are the most sensitive stages. Critical temperatures for frost injury appear to be higher for chickpea than fieldpea, lentil and faba bean. Among the pulses fieldpea seems to be the most susceptible to frost injury during the reproductive stage. Chickpea is a more indeterminate species than the other pulses and will continue to produce fresh flowers and pods provided there is adequate moisture in the soil profile and the temperature conditions during this period remain favourable. This allows chickpeas to compensate for loss of early flowers due to frost or cold injury. The important symptoms are

- Drooping and fading of flowers
- No seed setting
- Prolonged vegetative growth with less number of pods.

5.5 Water logging

Seed germination is very susceptible to water logging. Poor crop establishment is common problem when water logging occurs at seedling emergence. Among cool season pulses, faba bean is relatively tolerant to water logging at germination compared with lentil, fieldpea or chickpea. Water logging six days after germination of fieldpea can delay the emergence by up to five days and reduce the final plant density by 80 per cent. Water logging depresses vegetative growth of plants but affects root growth more than shoot growth. Once seedlings are established, the sensitivity of plants to water logging stress is generally increases with advancing age. Thus the ability to survive and recover following water logging is dependant on the timing of the water logging event relative to the stage of growth. Ability to recover generally declines sharply as reproductive growth (Flowering and pod filing stage) approaches. Seed yield loss resulting from water logging can be substantial. Limited data for lentil, fieldpea and chickpea suggest that reduction in yield can vary from negligible to almost 100 per cent, depending upon the species, stage of growth when waterlogged, duration, and the extent of root zone affected by water logging. Management practices to reduce the effects of water logging include the species choice (Faba bean tolerant; lentil highly susceptible), paddock selection, time of sowing, seedling rate and drainage.

5.6 Excessive moisture and high input conditions

Symptoms

- Due to more availability of nutrients and less nitrogenous compounds, the leaves turn yellow.
- Falling of lower leaves and decrease in number of branches occurs.

Management

- Avoid water stagnation in field.
- Rouging of every 3rd row, would be done in case of excessive vegetative growth.

It avoids the favourable condition for the diseases which occurs under high humidity and high moisture conditions.

5.7 Excessive vegetative growth

It occurs mostly in North and North eastern part of India where the conditions are favourable for growth of chickpea upto long duration .Due to excessive vegetative growth the crop canopy changes which effect the microclimate. Thus increase the positively of more leafy diseases, reduction in pods and lodging.

Management

1. Excessive vegetative growth may be managed by delay in sowing the crop by 2-3 weeks.
2. It can also minimized by distant plots

5.8 Soil pH and chlorosis due to Fe deficiency

Cool season pulses are sensitive to acid soil conditions and require a neutral to alkaline soil pH for their optimum growth and yield. Among the cool season pulses, lentil is the most sensitive to low pH followed by chickpea, faba bean and fieldpea. A drop of one pH unit below the threshold value (pH 5.5) can cause more than 86 per cent reduction in seed yield of lentil. Cool season pulses generally appear to be more sensitive to acidity than cereals.

Nutrient availability can change dramatically with small changes in soil pH, particularly in mineral soils low in organic carbon. On such soils, toxicity of Al, Fe and Mn occur when the pH falls from 5.5 to 4.5. Root growth of pulses is severely restricted on acid soils. Symptoms of nutrient deficiency and water stress, therefore commonly occur in pulse growing on unsuitability acid soils.

Deficiencies of Fe, Zn and Mn occur above pH 8.0. There are many reports of Fe deficiency in chickpea and lentil and Zn deficiency in field pea, chickpea and lentil. Fe deficiency is by far the most commonly occurring nutrient disorder on high pH and calcareous soils across cool season pulses. Losses in yield of highly susceptible genotypes range from 22 to 50 per cent, both in chickpea and lentil. Appearance of Fe deficient symptoms is highly transient across cool season pulse crops appearing under wet (above field capacity) soil moisture conditions and low temperatures. Screening germplasm grown on high pH (>8.1) calcareous soils has been very effective in identifying genotypic differences in susceptibility to Fe deficiency in cool season pulse crops.

5.9 Salinity and Sodicity

Cool season pulses are relatively sensitive to salinity. Salinity damage to pulse crops is shown by characteristic symptoms of excess ion accumulation. Necrosis of the outer margins and yellowing of the older leaves are the first signs of salinity. As salinity intensifies, these symptoms progress to younger leaves and older leaf die and abscise. Salinity intensifies anthocyanin pigmentation in leaves and stems in desi chickpea but in kabuli chickpea these tissues become yellow. Crop response to salinity also changes with crop age. For example, lentil, faba bean and fieldpea are more sensitive at germination than at subsequent growth stages and the converse is true for chickpea. To quantify the effect salinity on plant growth, it is necessary to establish critical values, relating salt concentration and their effect on growth and yield reduction. Glasshouse studies and field observations suggest that lentil and fieldpea may have greater salinity tolerance than faba bean and chickpea. The latter stages of chickpea growth, particularly flowering are more sensitive to salinity than the early vegetative stages. Scope for genetic enhancement of salinity resistance in cool season pulses is limited by the lack of variability in germplasm with the desired levels of resistance. Sensitive pulse crops (e.g. chickpea and faba bean) should not be grown on salt affected soils.

Sodicity adversely affects cool season pulses by reducing germplasm and seedling establishment with increasing exchangeable sodium percentage (ESP; 15-20). Glasshouse studies and field observations suggest that chickpea and lentil are more sensitive to sodicity than faba bean and fieldpea. The level at which 50 per cent reduction of seed yield occur is about 10 ESP for chickpea and 15 ESP for lentil. As with salinity, seed yield appears more sensitive to sodicity than does vegetative growth. In lentil and chickpea, the number of pods per plant and mean seed weight were severely decreased, whereas the number of seeds per pod was little affected. In comparison, canola a moderately sodic-tolerant species, can withstand 35 ESP without significant reduction of growth or yield.

6.0 Strategies for improving abiotic stresses management

Abiotic stress can be reduced by choosing the most appropriate pulse species and adjusting agronomy (sowing time, plant density, soil management) to ensure sensitive crop stages occur at the most favourable time in the season. For example choosing the optimum sowing time, species and cultivars with appropriate phenology can reduce the effect of frost and drought in pulse crops in dry land environments, which termed stress.

Most appropriate approach of dealing with stresses caused by extremes in the abiotic environment is to develop cultivars resistant to specific stresses. However, breeding and selection for resistance to these stresses is often considerable difficult because of the unpredictability of climatic conditions. Drought and heat escape through earliness in flowering and maturity is probably the characteristic most widely used by breeders for pulses and other crops to escape drought especially in low rainfall, terminal drought environment.

Lack of simple and accurate screening procedures to screen parental genotypes and breeding population for various abiotic stresses is the major bottlenecks in the development of stress tolerant cool season pulse crops.

Germplasm for resistance to various abiotic stresses are not widely available to breeders. Recent studies suggest that wild species of cool season pulses posses desired traits for a number of abiotic stresses. Future effort is required to identify desirable genes from this germplasm for transfer to adapted cultivars by conventional and/ or biotechnological approaches to develop abiotic stresses resistant cultivars.

Attempts have been made to define chickpea ideotypes for different growing conditions. In the drought prone environments, traits that help plants to escape or tolerate drought should be considered in the ideotype. Saxena and Johansen (1990) suggested that an ideotype for drought- stress environments should have early maturity, a deep root system, and a smaller leaf size. In the Mediterranean climate, the crop experiences a cool and wet winter followed by rapid warming in spring, leading to terminal drought. Thus, the ideotype for the Mediterranean environments should include early flowering and tolerance to cold during flowering (Sedgley *et al.* 1990). It has also been suggested that a compact plant type with erect growth habit and short internodes could help resist excessive growth in high input conditions (Dahiya and Lather, 1990). A spontaneous mutant with short internodes and compact growth habit, E 100 YM, has been identified (Dahiya *et al.*,1984) and used in ideotype breeding. Promising progenies with compact growth habit have been obtained, which can be grown at high plant density (Lather, 2000). Some of the aspects pertaining to management are being discussed.

6.1 Agronomical aspects

6.1.1 Soil water conservation

1. One deep tillage using mould board plough helps in improving the infiltration rate and soil moisture storage
2. Repeated tillage should be carried out in fallow lands after each effective rainfall
3. Removal of weeds and creation of soil mulch is effective in conserving soil water

Table 1 : Effect of tillage on soil water storage in dry land research area, Hissar (Oswal and Khanna 1983)

Tillage Operation	Depth of soil (cm)	Infiltration rate (cm/ha)	Soil water storage (mm/120 cm soil depth)
Mould board ploughing	30.0	7.6	184
Disc ploughing	12.5	5.7	178
Ploughing with local plough	12.0	5.2	168

6.1.2 Seeding technology

To obtain higher productivity in dry land crops is to attain good germination and uniform plant population under field condition.

- Sowing with ridger seeder at the bottom of furrows in 30+60 cm system (one pair of row in each furrow at 30 cm spacing within pair, and 60 cm spacing between two paired rows) could provide a good crop stand in dryland chickpea.

- In extreme conditions, seeding of only one row at the centre of furrow could be feasible.

Table 2 : Seed yield of dry land chickpea with different seeding methods (Malik & Agrawal, 1990)

Treatment	Seed yield (Kg/ha)
Local plough	1525
Tractor drawn ridger seeder	1969
Bullock drawn ridger seeder	1879

6.1.3 Fertility Management

- Deep placement of 20 Kg N + 40 to 60 Kg P_205 ha^{-1} has been found economical to boost the grain yield of chickpea.
- Advance application of fertilizer by the end of monsoon season could also be practiced profitably in chickpea.
- Integrated nutrient management (organics, inorganic fertilizers and bio-fertilizers) increase water use efficiency.

6.1.4 Choice of crop /variety

Mothbean, Mungbean more adopted in light textures soil of arid Rajasthan.

- Urdbean more production heavy textured vertisols of decean pleateau
- Lathyrus and Horsegram-insensitive to moisture.

6.1.5 Efficient cropping systems

The common intercropping systems for different dryland areas:

- Sorghum+ Pigeonpea
- Maize/Peart millet + Mungbean /Urdbean
- Groundnut + Pigeonpea
- Cotton + Mungbean/ Urdbean
- Pigeonpea+ Mungbean /Urdbean/Cowpea
- Rice+ Pigeonpea
- Chickpea + Mustard/Linseed/Barley/Coriander
- Lentil+ Linseed

6.1.6 Mulching

Mulching is an important agronomic measure that prevents soil erosion and facilitates infiltration and reduced runoff and evaporation losses. Crop residues viz , sorghum and maize stubbles, dry grains, wheat straw and Pigeonpea stalk can be used as surface mulch

6.2 Physiological Indices for higher productivity in Pulses under drought conditions

Among abiotic stresses constraining yield of rainfed pulses, drought stress ranks high on a global basis, because of the multitude of environments, farming, cropping systems in which pulses are grown and the wide diversity among the genotypes in the expression of an array of agriculturally significant characteristics.

Conceptually to a physiological approach for improving drought resistance should be –

1. Identification of various physiological traits that complement each other in a pyramidic manner by selective incorporation of these traits into a single cultivar.
2. Characterization of growing environments for drought stress in terms of intensity and time of occurrence of plant life cycle
3. Identification of the mechanism of resistance (at cell, organ or plant level.) It would assist breeders to selectively incorporate few of these traits into the best agronomic background of the cultivars needed for present day production systems in drought prone environments. Early maturity and the phonological adjustment are the dynamic nature of adaptation to drought escape.
4. To achieve dehydration avoidance and tolerance it is necessary that emphasis should be given to these characteristics in the adapted pulse varieties
    - Root attributes : depth, length density and hydraulic conductivity
    - Shoot attributes : Canopy structure leaf movements, leaf reflectance and stomatal conductance
    - Seedling establishment : at low soil water potential
    - Early growth vigour
    - Leaf area maintenance
    - Higher water use efficiency (WUE)
    - Developmental plasticity – Adjustment of phenology to intermittent stress pattern, recovery and remobilization of pre- anthesis assimilates

In chickpea, water stress during 60-100 days of crop growth appeared to be the most critical and severely affects the yield. The leaf air temperature differential (LATD) can be used as an effective indicator of water stress effects in chickpea

- Rooting depth and density are the main drought avoidance traits identified to confer seed yield under terminal drought environments (Turner *et. al.* 2001)
- For higher productivity a genotype should have more plant height, higher TDM, early flowering, pod formation and maturity, higher chlorophyll in leaves and pods walls, higher sugar and protein contents in the leaves.

6.3 Role of plant growth regulators and agrochemicals to overcome drought

Heremath et. al. (2002). Found that soaking Pigeonpea, Chickpea seeds with $Cacl_2$ (20%) for 2 hrs in 1:1 ratio (seed : solution) is very effective in giving higher seedling emergence, dry weight of seedling, shoot length, root length and leaf expansion. Seed hardening techniques are useful for induction of drought tolerance in Pulses (Sabbian et al. 2000).

6.3.1 Seed hardening techniques

Seed hardening techniques for induction of drought tolerance in pulses
(Sabbian et al. 2000)

Crop	Techniques
Chickpea	Seed soaking for 4 hrs in 10% Potassium di hydrozen phosphate solution in ⅓ volume and air dried
Pigeonpea	Seed soaking for 4 hrs in $Znso_4$ (1000 ppm solution in ⅓ volume (100 Kg of seeds in 33 lit solution) and air dried
Greengram	Seed soaking for 4 hrs in $MgSo_4$ (100 ppm) solution in ⅓ volume of and air dried
Blackgram	Seed soaking for 4 hrs in the following equie-volume solution
(i)	Succinic acid 20 ppm
(ii)	Ascorbic acid 20 ppm
(iii)	Potassium sulphate/Manganese sulphate/Zinc sulphate-100 ppm

6.4 Conceptual framework and breeding strategies

The various steps before particular traits can be recommended for use in breeding program aimed at improving drought tolerance are

- Validate the potential contribution of specific traits
- Search for genotypes with high levels of expressions of the desired traits
- Mode of inheritance of the trait
- Develop rapid and efficient screening method for segregating population
- Incorporate the traits inter agronomically superior genotype to give insight for manipulating the management strategies and directing suitable breeding program to improve drought tolerance in pulses
- Characterization of environment in terms of soil type, water, status and climate variation like temperature regime, rainfall and potential evaporation will guide the relevant screening methodology for evaluating germplasm lines.
- Computerized mapping and ready visualization of drought environment through Geographic information system (GIS) technique.

6.4.1 Screening methods

The utility of particular characters as selection criterion in a drought program will depend upon the assessment of plant performance at a critical developmental stage in short time.

- Screening for drought resistance involves
- Evaluating germplasm lines under field conditions with and without irrigation
- Estimation of yield, yield components and total dry matter
- Assessment of morphological characters early growth vigour, height of fertility zone, biological yield and harvest index could be employed as rapid screening methodology.

6.4.2 Varietal improvement

Dought and heat stress

Chickpea genotypes identified for abiotic stresses

- Drought tolerance, high root vigour
- ICC 4958, Phule G5, Katila, Phule G 81-1, Annigiri, K 850, BGD 86, BG 256, KPG 59, KAK 2
- Drought escape due to earliness
- ICCV 2, ICCV 96029, ICC 96030,
- Cold tolerance (set pod at 4%)
- FLIP 82 -85 C, 82-131 C , K 1189 , ICCV 88501, ICCV 88502, ICCV 88503, ICCV 88506, ICCV 88510, GNG 469, ILC 8262
- Drought resistance
- Vijay, K 850, Pusa 362
- Salt tolerance
- Karnal Channa 1 (CSG 8962), CSG 8893, E 1004, H 89-84, H 81-69
- High temperature tolerance
- JG 14, ICC x 830107

The Sources of drought resistance

1. Cultivated varieties- First priority
2. Land races
3. Related wild species or introduced by genetic engineering.

Salinity tolerance

The salt tolerant lines identified by Dua and Sharma (1995) has lower Na^+ in root than the sensitive genotypes. A salt tolerant variety Karnal Chana 1 (CSG-8963) has been released in India that can be grown in saline soils with electrical conductivity upto 6 ds/m (Dua *et. al.* 2001).

Suitable varieties for salt stressed environments

Crop	Varieties
Chickpea	Desi – CSG 8977, CSG 8962 (Karnal chana 1), CSG 8943 Kabuli – CSG 88101
Pigeonpea	CSPPS 89261,89258,89256,89406,89314,89313,(icpl 227)
Pea	Pant P 109, DDR 4, KFPD 7
Greengram	K 1148,PS 16, PDM 85-199, LAM 88-4 (For summer) ML 395, MH 83 20, Pusa 105 (Rainy Season)
Blackgram	KU 311, , T9, KU 313, KU 309
Cowpea	C 152, RC 80, GC 2, RC 77
Fababean	VH 82

Cold tolerance

Over 6000 germplasm and breeding lines were screened for cold tolerance at ICARDA and 11 kabuli germplasm lines and 121 breeding lines were found tolerant. The best sources of tolerance in the cultigen were ILC 8262, ILC 8617 and FLIP 8-82 C with a consistent score of 3 (Singh *et. al.,* 1995). ICRISAT has developed a number of cold tolerant lines *viz.* 88502, ICCV 88503, ICCV 88506, ICCV 88510 and ICCV 88516 which are able to set pod at low temperatures (Gaur et al.. 2007), A pollen selection method has been successfully used. Clarke and Siddique (2004) have studied response of chickpea genotypes to chilling tolerance during reproductive development .Besides cultivated species large number of accessions of eight wild species were evaluated were evaluated and only *C.reticulatum* and *C. Echinospermum* were adjudjed as cold tolerant (Malhotra, 1998).

Chilling tolerance in chickpea –Novel methods for crop improvement

- Selection of desirable alleles during the haploid phase of growth by pollen selection.
- Screening for molecular markers closely associated with chilling tolerance

6.5 Genetic option in managing abiotic stresses

Efforts on breeding for resistance to abiotic stress has limited success in the past as these were difficult to screen. It is now realized that they causes more losses to productivity than those attributed to the biotic constraints.

- Terminal drought and heat stress which result into forced maturity and may reduce seed yield by 50% can be manipulated in increasing yield through exploring drought escape by shortening crop duration.
- Possible genetic improvement strategies are outlines ranging from empirical selections for yield in drought environments to a physiological genetic approach.
- Indirect selection based upon tightly linked genetic markers seems to be more promising in pyramiding the resistance genes. This relies an exploitation of the tightly linked markers and establishment of a convenient and low cost detection procedure.
- Salinity is increasing at an alarming rate in arid/semiarid areas under irrigated conditions. Apart from costly and energy requiring techniques, approach for genetic improvement in pulses will help in keeping the water table low and maintain the salt balance within tolerable limits and ultimately improve the yield potential under salt stress environments.
- Tissue culture technique has been successful in screening chickpea germplasm against salinity. In vitro transformation through *Agrobactericum* in Pulses is expected to incorporate salt tolerance
- Availability of better salinity tolerance in wild relatives *Atylosia albicans, C. yamashitaee C. reticulatum C. echinospermum and C. bigugam* holds promise for enhancing salt tolerance in cultivated species of Pigeonpea and chickpea respectively.

Conclusion

From the overall discussions, it is evident that the adverse effects of abiotic stresses on productivity of pulses, though cannot be totally avoided but can be minimized to a reasonable extent through the application and transfer of appropriate technology directed to improve soil moisture conservation and water use efficiency. By knowing the amount of available water in the soil profile at the time of planting, the farmers can be guided to adopt the suitable management strategies and choose suitable cultivar.

Breeding for stress environments is possible provided it is conducted with strategies and methodologies that little have common with those used in breeding for favourable environments. Adaptation overtime can be improved by breeding for specific adaptation to a given type of stress environments.

Losses from the abiotic stresses can be achieved by taking advantage of the agronomical options physiological as well as genetic aspects. Temporal variability of stress environment permits exposure of the some breeding material to variable combinations of stresses over a short period.

- This is need to develop genotypes having reduced dark respiration rate during vegetative and reproductive phase of growth as it enhances the yield.
- In future care should be taken to retain the desirable characters while developing short to medium duration varieties for dry land areas.

- Sincere efforts are needed to strengthen the utilization of molecular biotechnology in terms of qtls for rooting depth, osmotic adjustment, mobilization of assimilates from vegetative parts to seeds etc. would be helpful in improving the different abiotic stresses and finally productivity of pulses.
- There is clear need to develop methods for maintenance and creation of *ex situ* collections of perennial species to enable their evaluation for important agronomic traits.

References

Abbo,S., Lev-Yadun S. and Galwey N. (2002) Breeding for osmotic adjustment in chickpea (*Cicer arietinum* L.) In : *Proceedings of the 12th Australian Plant Breeding Conference*, McComb, J.A., Ed.,Australian Plant Breeding Association, Perth, Australia, 463.

Ahmed,F., Gaur, P.M. and Croser, J.S. (2005) Chickpea (*Cicer arietinum* L) In a Book on " Genetic resources, chromosome engineering and crop improvement series- Grain Legumes (Vol. 1) Edited by Ram J. Singh and Prem P. , Jauhar. Published by CRC Press Tayler & Francis Group 6000 Broken sound Parkway NW, Scule 300, Boca Ratan, FL 33487-2742.

Chaturvedi, S.K.,D.K.Mishra, P.Vyas and Neelu Mishra (2009) Breeding for cold tolerance in chickpea. (Mini review) *Trends in Biosciences* 2 (2) :1-6

Clarke, H.J. and Siddique, K.H.M., (2004) Response of chickpea genotypes to chilling tolerance during reproductive development, *Field Crops Res.*, 90:23.

Croser, J.S. et al. (2003) Low temperature stress: Implications for chickpea (*Cicer arietinum* L.) improvement, *Critical Reviews in Plant Sciences*, 22:185.

Dahiya, B.S. et al. (1984) Useful spontaneous mutants in chickpea , International Chickpea *Newsl.*,11:5.

Dahiya, B.S. and Lather, V.S. (1990) A breakthrough in chickpea yields – paper 1, *International Chickpea Newslt*, 22:6.

Dua, R.P. and Sharma, P.C. (1995) Salinity tolerance to kabuli and desi chickpea genotypes. *Int.Chickpea and Pigeonpea Newsl.*, 2:19.

Dua, R.P., Chaturvedi, S.K., Shiv Sewak (2001) Reference Varieties of Chickpea for IPR Regime, Indian Institute of Pulses Research, Kanpur.

Gaur, P.M., Tripathi,S. Gowda,C.L.L. Pande,S.,Sharma,H.C. ,Sharma, K.K., Kashiwagi,V.,Vadej V.,Krishnamurthy, L. ,Varshney, R.K.Malikaarjuna,N. and Hoisington, D.A. (2007). International efforts in chickpea improvement. In Legumes for Ecological Sustainability :Emerging Challenges and opportunities. Indian Society of Pulses research and Development (eds. Ali et al.) IIPR Kanpur (India) pp.359-378

Heremath, S.M., Chetti, M.B. and Kalpana, M. (2002) Role of plant growth regulators and Agrochemicals to overcome drought. In: *Physiological approaches for enhancing productivity potential under drought conditions* (M.B. Chetti, S.M. Heremath and M. Kalpana eds.) published at M/S Saraswati Printers Dharwad, pp 75-82.

Krishnamurthy,L. et al.(2003) Genetic diversity of drought avoidance root traits in the mini-core germplasm collection of chickpea, Int. Chickpea and Pigeonpea Newsl., 10:21.

Leport, L. et al., (1999) Physiological responses of chickpea genotypes to terminal drought in a Mediterranean type environment, *European J. Agron.*, 11:279.

Malhotra, R.S. (1998) Breeding chickpea for cold tolerance. 3rd *European Conference on Grain Legumes*. Opportunities for high quality,healthy and Added- Value Crops to meet European demands,Valldolid, Spain,14-19 Nov.pp.152

Morgan, J.M., Rodriguez, M.B., and Knights, E.J. (1991) Adaptation to water deficit in chickpea breeding lines by osmoregulation: Relationship to grain yields in the field, *Field Crop Res.*, 27:61.

Saxena, N.P. (2003) Management of drought in chickpea- a holistic approach. In Management of Agricultural Drought- Agronomic and Genetic options, Saxena, N.P. Ed. Oxford & IBH Publishing Co. Pvt. Ltd., New Delhi, p. 103.

Saxena, N.P. and Johansen ,C. (1990)Chickpea ideotypes for genetic enhancement of yield and yield stability in South Asia. In Chickpea in the Nineties: *Proceedings of the Second International Workshop on Chickpea Improvement,* ICRISAT, Patancheru, India, p. 81.

Sedgley, R.H., Siddique, K.H.M., and Walton, G.H.(1990) Chickpea ideotypes for Mediterranean environments. In : Chickpea in the Nineties: *Proceedings of the Second International Workshop on Chickpea Improvement*, ICRISAT , Patancheru, India, 87.

Serraj, R., Gaur, P.M., Krishnamurthy,L. Kahsiwagi, J. and Crouch, J.H. (2003) Pulses in New perspective. *Proceedings of the National Symposium on crop diversification and natural resource management*, 22-22 Dec.,2003 IIPR, Kanpur (Editors- Masood Ali, B.B. Singh, Shiv Kumar, Vishwadhar pp. 472-479)

Siddique,K.H.M., Loss,S.P., Regan, K.L. and Jettner, R.L.(1990). Adaptation and seed yield of cool season grain legumes in Mediterranean environments of south- western Australia. *Australian Journal of Agril.Research.*50:375-387

Singh, K.B. (1990)Winter chickpea: Problems and potential in the Mediterranean region. In Present Status and Future Prospect of Chickpea Crop Production and Improvement in the Mediterranean Countries, saxena, M.C., Cubero, J.I., and Wery, J.,Eds., Options Mediterraneennes, Serie A: Seminaires Mediterraneenes: No. 9, CIHEAM, Zaragoza, Spain.

Singh, K.B., Malhotra,R.S. and Saxena,M.C. (1990) Sources for tolerance to cold in Cicer species, *Crops Sci.*, 30:1136.

Singh, K.B., Malhotra,R.S. and Saxena,M.C.(1995) Additional sources of tolerance to cold in cultivated and wild Cicer species, *Crops Sci.*, 35:1491.

Singh, N.B. and Singh, R.A. (1984) Screening for alkalinity and salinity tolerance in chickpea. *Int. Chickpea Newsl.*, 11:25.

Subbian, P., Annadurai, Kan Palaniappan, S.P. (2000) Agriculture facts and figures, Kalyani Publishers, Ludhiana, India.

□□□

Green Agriculture : Newer Technologies, 2012
© Kambaska Kumar Behera (ed.), 415-438
New India Publishing Agency, New Delhi (India)
e-mail : info@nipabooks.com; website : www.nipabooks.com

Chapter-17

Concept and Relevance of Organic Farming in Present Context

R. K. Mathukia
Department of Agronomy, College of Agriculture,
Junagadh Agricultural University, Junagadh – 362 001 (Gujarat)
E-mail : rkmathukra@gmail.com

SUMMARY

Organic farming is not new to Indian farming community. Several forms of organic farming are being successfully practiced in diverse climate, particularly in rainfed, tribal, mountains and hill areas of the country. Among all farming systems, organic farming is gaining wide attention among farmers, entrepreneurs, policy makers and agricultural scientists for varied reasons such as it minimizes the dependence on chemical inputs (fertilizers; pesticides; herbicides and other agro-chemicals) thus safeguards/improves quality of resources, and it is labour intensive and provides an opportunity to increase rural employment and achieve long term improvements in the quality of resource base. The popularity of organic farming is gradually increasing and now organic agriculture is practiced in almost all countries of the world, and its share of agricultural land and farms is growing.

1. Introduction

Green revolution technologies such as greater use of synthetic agro chemicals like fertilizers and pesticides, adoption of nutrient responsive, high-yielding varieties of crops, greater exploitation of irrigation potentials etc. has boosted the production out put in most of cases. Without proper choice and continuous use of these high energy inputs is leading to decline in production and productivity of various crops as well as deterioration of soil health and environments. The most unfortunate impact on Green Revaluation Technology (GRT) on Indian Agriculture is as follows:

1. Change in soil reaction
2. Development of nutrient imbalance /deficiencies
3. Damage the soil flora and fauna
4. Reduce the earth worm activity
5. Reduction in soil humus / organic matter
6. Change in atmospheric composition
7. Reduction in productivity
8. Reduction in quality of the produce
9. Destruction of soil structure, aeration and water holding capacity
10. Breeding more powerful and resistant pests and diseases

All these problems of GRT lead to not only reduction in productivity but also deterioration of soil health as well as natural eco-system. Moreover, to day the rural economy is now facing a challenge of over dependence on synthetic inputs and day by day it change in price of these inputs. Further, Indian Agriculture will face the market competition due to globalization of trade as per World Trade Organization (WTO). Thus apart from quantity, quality will be the important factor. Agriculture gave birth to various new concepts of farming such as organic farming, natural farming, bio-dynamic Agriculture, do-nothing agriculture, eco-farming etc.

The essential concept of these practices is "Give back to nature", where the philosophy is to feed the soil rather them the crop to maintain the soil health. Therefore, for sustaining healthy ecosystem, there is need for adoption of an alternative farming system like organic farming.

1.1 Concept of organic farming

The basic concepts behind organic farming are:

1. It concentrates on building up the biological fertility of the soil so that the crops take the nutrients they need from steady turnover within the soil nutrients produced in this way and are released in harmony with the need of the plants.
2. Control of pests, diseases and weeds is achieved largely by the development of an ecological balance within the system and by the use of bio-pesticides and various cultural techniques such as crop rotation, mixed cropping and cultivation.
3. Organic farmers recycle all wastes and manures within a farm, but the export of the products from the farm results in a steady drain of nutrients.
4. Enhancement of the environment in such a way that wild life flourishes. In a situation where conservation of energy and resources is considered to be important community or country would make every effort to recycles to all urban and industrial wastes back to agriculture and thus the system would be requiring only a small inputs of new resources to "Top Up" soil fertility.

1.2 Definition of organic farming

Many scientists at different levels have elaborated the concept of organic farming; the important descriptions are as follows:

According to US Department of Agriculture (USDA), organic farming is defined as 'a system that is designed and maintained to produce agricultural products by the use of methods and substances that maintain the integrity of organic agricultural products until they reach the consumer. This is accomplished by using substances, to fulfill any specific fluctuation within the system so as to maintain long term soil biological activity, ensure effective peak management, recycle wastes to return nutrients to the land, provide attentive care for farm animals and handle the agricultural products without the use of extraneous synthetic additives or processing in accordance with the act and the regulations in this part'.

Organic agriculture has been defined differently, but the description offered by Lampkin (1990) appears to be the most comprehensive one covering all essential features. As per this description, organic agriculture is a production system, which avoids or largely excludes the use of synthetic compounded fertilizers, pesticides, growth regulators and livestock feed additives. To the maximum extent feasible, organic farming system relies on crop rotations, crop residues, animal manures, legumes, green manures, off-farming organic wastes and aspect of biological pest control to maintain soil productivity and tilth, to supply plant nutrients and to control insects, weeds and other pests. The concept of soil as living system that develops the activities of beneficial organisms is central to this definition.

The term "organic" is best thought of as referring not to the type of inputs used, but to the concept of the farm as an organism, in which all the components - the soil minerals, organic matter, microorganisms, insects, plants, animal and humans - interact to create coherent, self-regulating and stable whole. Reliance on external inputs, whether chemical or organic, is reduced as far as possible. Organic farming is holistic production system (Lampkin *et al.,* 1999).

National Organic Standards Board of the U.S. defines organic farming as an ecological production management system that promotes and enhances biodiversity, biological cycles and soil biological activity (Lieberhardt, 2003).

Organic farming refers to organically grown crops which are not exposed to any chemicals right from the stage of seed treatments to the final post harvest handling and processing (Pathak and Ram, 2003).

There are also different opinions on nomenclature of organic farming. Some call it as ecofarming i.e., farming in relation to ecosystem. Others prefer the term biological farming (farming in relation to biological diversity); yet others prefer the term biodynamic farming (biologically dynamic and ecologically sound and sustainable farming) or macrobiotic agriculture (agriculture in relation to macro-fauna). Whatever be the name, the basic point is that organic farming is the farming based on natural principles, which alone are sustainable. Organic farming is a matter of giving back to nature what we take from it. It is safe, inexpensive, profitable and sensible. Organic farming is not mere non-chemicalism in agriculture; it is a system of farming based on integral relationship. So, one should known the relationships among soil, water, plants, and microflora and the overall relationship between plants and animal kingdom, of which, man is the apex animal.

It is the totality of these relationships, which is the backbone of organic farming (NPCS, 2008).

1.3 Area under organic farming

Table : Area under organic farming in % of total agricultural area in important countries (2003)

Country	% of Cultivated Area	Country	% of Cultivated Area
Austria	11.30	Australia	2.31
Switzerland	9.70	France	1.40
Italy	7.94	USA	0.23
Denmark	6.51	Japan	0.10
Sweden	6.30	China	0.06
United Kingdom	3.96	India	0.03
Germany	3.70		

Bhattacharya and Gehlot *(2003)*

India represents only 0.03% area (43000 ha) out of total cultivated (143 million ha) area.

1.4 Importance of organic farming

The agriculture today in the country is hampered by erosion of natural resources viz., land, water, biodiversity, fast declining soil fertility and use efficiency of inputs, such as water, fertilizer and energy. Demographic pressure accelerates the former and the faulty agronomic practices account for the latter problems. The modern agriculture with its potential takes the country out of the food trap and to reach an era of self sufficiency in food grain production.

The present day for self sufficiency in food grain production may not last longer unless we develop a sustainable agricultural system which maintains and /or improves soil fertility and productivity with greater acceptance of biological principles so as to assure adequate/more food production in future. Besides plants are more prone to pest and diseases in intensive agriculture, use of chemicals can have residues on the produce, in the soil and in ground water. With more of purchased inputs cost of production is also mounting up. Pesticides use in paddy, cotton and vegetables which occupy less than 30 per cent of total area account for more than 80 per cent of the chemicals used.

Organic farming practices that reduces the pressure on land, water and biodiversity without adverse effects on agricultural production and nutritive value of food comprise judicious use of organic manure, viz. farm yard manure, compost, crop resides, vermicompost etc. integrated use of efficient nutrient management practices, cropping systems, conjunctive use of rain, tank and under ground water, integrated pest management and conservation of genetic resources. Among them, soil fertility is give top attention due to its dynamic action with various physical, chemical and biological properties. Besides this, following advantages derived from organic farming:

1.5 Advantages of organic farming

1. Organic manures produce optimal conditions in the soil for high yields and good quality crops.
2. They supply all the nutrients required by the plant (NPK, secondary and micronutrients).
3. They improve plant growth and physiological activities of plants.
4. They improve the soil physical properties such as granulation and tilth, giving good aeration, easy root penetration and improved water holding capacity. The fibrous portion of the organic matter with its high carbon content promotes soil aggregation to improve the permeability and aeration of clay soils while its ability to absorb moisture helps in the granulation of sandy soils and improves their water holding capacity. The carbon in the organic matter is the source of energy for microbes which helps in aggregation.
5. They improve the soil chemical properties such as supply and retention of soil nutrients and promote favourable chemical reactions.
6. They reduce the need for purchased inputs.
7. Most of the organic manures are wastes or by-products which on accumulation may lead to pollution. By way of utilizing them for organic farming, pollution is minimized.
8. Organic fertilizers are considered as complete plant food. Organic matter restores the pH of the soil which may become acidic due to continuous application of chemical fertilizers.
9. Organically grown crops are believed to provide healthier and nutritionally superior food for man and animals than those grown with commercial fertilizers.
10. Organically grown plants are more resistant to disease and insects and hence only a few chemical sprays or other protective treatments are required.
11. There is an increasing consumer demand for agricultural produces which are free of toxic chemical residues. In developed countries, consumers are willing to pay more for organic foods.
12. Organic farming helps to avoid chain reaction in the environment from chemical sprays and dusts.
13. Organic farming helps to prevent environmental degradation and can be used to regenerate degraded areas.
14. Since the basic aim is diversification of crops, much more secure income can be obtained than to rely on only one crop or enterprise.

1.6 Objectives of organic farming

The objectives of organic agriculture have been expressed in the standard document of the International Federation of Organic Agriculture Movement (IFOAM) as follows:

- To produce food of high nutritional quality in sufficient quantity.
- To work with natural systems rather than seeking to dominate them.
- To encourage and enhance the biological cycles within farming system involving micro-organisms, soil flora and fauna, plants and animals.
- To maintain and increase the long term fertility of soils.
- To use, as far as possible, renewable resources in locally organized agricultural systems.
- To work as much as possible, within a closed system with regard to organic matter and nutrient elements.
- To give all livestock, conditions of life that allow them to perform all aspects of their innate behaviour.
- To avoid all forms of pollution that result from agricultural techniques.
- To maintain the genetic diversity of the agricultural system and its surroundings, including the protection of plant and wildlife habitats.
- To allow agricultural producers for adequate return and satisfaction from their work including a safe working environment.
- To consider the wider, social and ecological impact of the farming system.

1.7 Essential characteristics of organic farming

The most important characteristics are as follows:

- Maximal but sustainable use of local resources.
- Minimal use of purchased inputs, only as complementary to local resources.
- Ensuring the basic biological functions of soil-water-nutrients-human continuum.
- Maintaining a diversity of plant and animal species as a basis for ecological balance and economic stability.
- Creating an attractive overall landscape which given satisfaction to the local people.
- Increasing crop and animal intensity in the form of polycultures, agroforestry systems, integrated crop/livestock systems etc. to minimize risks.

1.8 Key principles in organic farming systems

Organic agriculture systems are based on three strongly interrelated principles under autonomous ecosystems management: mixed farming, crop rotation and organic cycle optimization. The common understanding of agricultural production in all types of organic agriculture is managing the production capacity of an agro-ecosystem. The process of extreme specialization propagated by the green revolution led to the destruction of mixed and diversified farming and ecological buffer systems. The function of this autonomous ecosystem management is to meet the need for food and fibres on the local ecological carrying capacity.

1.8.1 Mixed farming

In organic agriculture systems, one strives for appropriate diversification, which ideally means mixed farming or the integration of crop and livestock production on the farm. In this way, cyclic processes and interactions in the agro-ecosystem can be optimized like using crop residues in animal husbandry and manure for crop production. Diversification of species biotypes and land use as a means to optimize the stability of the agro-ecosystem is another way to indicate the mixed farming concept. The synergistic concept among plants, animals, soil and biosphere support this idea (Sharma, 2008).

1.8.2 Crop rotation

Within the mixed farm setting, crop rotation takes place as the second principle of organic agriculture. Besides the classical rotation involving one crop per field per season, inter cropping, mixed cropping and relay cropping are other options to optimize interactions. In addition to plant functions, other important advantages such as weed suppression, reduction in soil-borne insects and diseases, complimentary nutrient supply, nutrient catching and soil covering can be mentioned (Sharma, 2008).

1.8.3 Organic cycle optimization

Each field, farm or region contains a given quantity of nutrients. Management should be used in such a way that optimal use is made of this finite amount. This means that the nutrients should be recycled and used a number of times in different forms. Care should be taken that only a minimum amount of nutrients actually leave the system so that “import” of nutrients can be restricted. The quantity of nutrients available to plants and animals can be increased within the system by activating the **edaphon** (aggregate of organisms that live in the soil), resulting in increased weathering of parent material (Sharma, 2008).

1.9 Organic farming in India: relevance in present context

- In India, only 30% of total cultivable area is covered with fertilizers, where irrigation facilities are available and in the remaining 70% of arable land, which is mainly rainfed, negligible amount of fertilizers is being used. Farmers in these areas often use organic manure as a source of nutrients is readily available either in their own farm or in their locality.

- The North Eastern Hills of India provides considerable opportunity (18 million hectare) for organic farming due to least utilization of chemical inputs, which can be exploited for organic production.

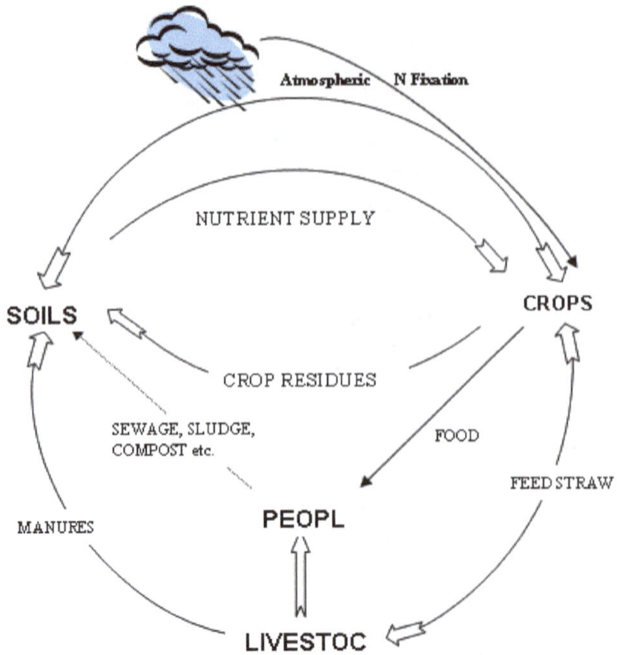

Fig.: Organic cycle of an organic farming

- India is an exporting country and does not import any organic products. The main market for exported products is the European Union. Recently India has applied to be included on the "EU-Third-Country-List", another growing market is USA.
- There has been plenty of policy emphasis on organic farming and trade in the recent years in India.
- The 10th five-year plan emphasizes promotion and encouragement to organic farming in India with the use of organic waste, IPM and INM.
- Even 9th five-year plan had emphasized the promotion of organic produce in plantation crops, spices and condiments with the use of organic and bio-inputs.
- There are many states and private agencies involved in promotion of organic farming in India. These include various ministries and departments of the government at the central and state levels such as:
    - Universities and Research Centres
    - Non Govt. Organizations (NGO)
    - Eco Farms
    - Certification Bodies like INDOCERT, ECOCERT, SKAL and APOF etc.

The central and state governments have also identified *Agri-Export Zones* for agricultural exports in general and organic products in some states:

- In Uttar Pradesh and Uttaranchal, the Diversified Agriculture Support Project (DASP) is promoted for organic farming.
- In Bangalore and Nilgiris, with 50 outlets in south India helps for supply the organic products from small growers.
- IRFT (International Recourses for Fairer Trade) based in Mumbai, procures organic cotton and agro products to sell them to Indian and foreign buyers to help the rural poors.
- Ion Exchange, Mumbai, a private company is engaged for export and domestic marketing of organic products in India.
- In Himachal Pradesh, the net incomes per hectare from organic farming was found to be 2-3 times higher both in case of maize and wheat due to higher production and also for higher price were obtained by organic produce.
- In Haryana, net returns were higher (2-3 times) in basmati rice, soybeans, arhar and wheat because of 25 to 30 % price premium on organic produce and lower cost of production and marketing.
- In Maharashtra, popularization of organic cotton production was due to high cost benefit ratio of organic cotton 1:1.63 as against 1:1.47 for conventional cotton.
- In Gujarat, organic production of chickoo, banana and coconut had higher profitability (Naik, 2001).
- In Karnataka, groundnut, jowar, cotton, coconut and banana were grown as organic. The major problems faced by organic farmers were found to be initial lower yields, no price incentives, no separate markets for organic produce, besides lack of and high costs of certification (Singh, 2003).

1.10 Constraints / Limitations of organic farming in India

- **Small holding**: The average size of an operational holding is 1.57 ha and further decreasing gradually due to population pressure.
- **Poor infrastructure facilities**: i.e. lack of sufficient soil testing laboratories.
- Lack of technological knowledge, lack of knowledge for use of bio-fertilizers, bio-pesticides, bio-control, IPM and INM etc.
- Organic farming takes four years for a farmer to free his land completely stopping the use of chemical as nutrients and crop savers.
- The neighbouring farmers do not well co-operate regarding use of fertilizers, pesticides, weedicides etc.
- Decrease in production of high yielding crops like rice, wheat which needs high fertility status to get potential yield.
- The competitive uses of organic materials such as dung-cakes for domestic cook fuel in villages and bagasses as fuel in sugar factories and villages.
- Wheat and rice straws are disposed off by burning, instead of return to the soil.
- Dung, slurry and pig manure and other waste used directly in the field (without compositing), which damage the crop and pollute the ground water.
- Most of organic materials are bulky in nature, hence very difficult to store, carry and use.

- Sewage, sludge contains pathogens and, some of them survive more than six months, which may hazard the human life and prove fatal for the animal.
- City garbage contains undecomposed materials such as metal, plastic, glass, stones, needles etc. which causes many problems,
- Biocontrol agents are available only for few selected insect pests.
- Complicated organic certification process and also high cost of certification.
- High price expectations, delayed delivery, quality restrictions, lack of certification and marketing network are the major problems for organic producers.
- Major Indian and multinational companies are not interested in biopesticides, also dealer's interest in chemical pesticides.

Conventional Farming v/s Organic Farming

Conventional Farming	*Organic Farming*
i. It is based on *economical* orientation, heavy mechanization, specialization and misappropriate development of enterprises with unstable market oriented programme.	i. It is based on *ecological* orientation, efficient input use efficiency, diversification and balanced enterprise combination with stability.
ii. Supplementing nutrients through fertilizers, weed control by herbicides, plant protection measures by chemicals and rarely combination with livestock.	ii. Cycle of nutrients within the farm, weed control by crop rotation and cultural practices, plant protection by non-polluting substances and better combination of livestock.
iii. Based on philosophy of to feed the crop/ plants.	iii. 'Feed the soil not to the plant' is the watch word and slogan of organic farming.
iv. Production is not integrated into environment but extract more through technical manipulation, excessive fertilization and no correction of nutrient imbalances.	iv. Production is integrated into environment, balanced conditions for plants and animals and deficiencies need to be corrected.
v. Low input : output ratio with considerable pollution.	v. High input : output ratio with no pollution.
vi. Economic motivation of natural resources without considering principles of natural up gradation.	vi. Maximum consideration of all natural resources through adopting holistic approaches.

2. Nutrient Management in Organic Farming

2.1 Introduction

In order to realize the potential of production systems on a sustained basis, efficient management of resources is crucial (essential). A successful farming system relies on the management of organic matter to enhance physico-chemical and biological properties of the soil. The effects of soil organic matter are dynamic as it is a source of gradual release of essential plant nutrients; improves soil structure, its drainage, aeration and water holding capacity (WHC); improves soil buffer capacity; influence the solubility of minerals and serves as a source of energy for the development of micro-organisms.

The use of biological inputs such as N-fixing bacteria, mycorrhiza or soil fauna as a means of enhancing the endemic biological activities are the means of biological soil management. Direct management is also achieved by the use of organic matter inputs, for the purpose of providing feeding materials to biological populations. Management techniques such as tillage and fertilization also influence the activity of the biota by improving the physical and chemical environment of the soil.

According to a conservative estimate, around 600 to 700 mt of agricultural waste is available in the country but it is not managed properly. We must convert waste into wealth by converting this biomass into energy, nutrient to starved soil and fuel to farmers. India produces about 1800 mt of animal dung per annum. Even if ⅔ of the dung is used for biogas generation, it is expected to yield about 440 mt/ annum of manure, which is equivalent to 2.90 mt N, 2.75 mt P_2O_5 and 1.89 mt K_2O.

2.2 Concept of biological INM

The concept of biological INM is the continuous improvement of soil productivity on long-term basis through appropriate use of organic manures, green manures, BGA, biofertilizers and other biological derived materials and their scientific management for optimum growth, yield and quality of crops and intensive cropping systems in specific agro-ecological situations.

2.3 Definition of biological INM

According to Sanchez (1994), we should rely on biological processes by adapting germplasm to adverse soil conditions, enhancing soil biological activity and optimizing nutrient, cycling to minimize external inputs and maximize the efficiency of their use.

It can also be defined as "a system for approaching of soil nutrient management which maintain soil health, soil fertility, sustaining agricultural productivity and improving farmers' profitability through effective, judicious and intensive use of biological based nutrient management resources". The resources are biofertilizers, organic manures green manuring crop rotation, N-fixing organisms, mycorrhizae, PSM etc.

2.4 Role of different sources for biological INM

2.4.1 Organic manures

Term 'manure' was used originally for denoting materials like cattle manure and other bulky natural substances that were applied to land, with the object of increasing the production of crops. Therefore, manures are defined as the plant and animal wastes which are used as sources of plant nutrients (Reddy, 2005).

Urine is normally low in phosphorus and high in potash, where as about equal parts of nitrogen may be excreted in faeces and urine of the cattle. Hence the manure in which the proportion of the urine was allowed to drain away would be relatively low in N and K. Poultry manure is very important for organic farming due to there will be no loss of urine, since both liquid and solid portions are excreted together.

Fresh poultry manure creates local alkalinity; it may hamper the standing crop. Therefore, it is recommended to preserve the excreta at least for six months with suitable amendments and appropriate microbes.

Advantages of Manuring

- Manures supply plant nutrients including micro nutrients
- They improve soil physical properties
- Increase nutrient availability
- Provide food for soil micro organisms
- Provide buffering action in soil reaction
- Improve soil tilth, aeration and WHC of the soil

On the basis of concentration of nutrients, manures can be grouped into two categories:

(A) Bulky Organic Manures

Contain small percentage of nutrients and they applied in large quantities like FYM, compost, green manure, biogas slurry, night soil, sewage and sludge, poultry manure, sheep and goat manure, animal waste, crop residue etc.

A.1 Farm yard manure (FYM)

Most commonly used organic manure in India. It refers to the decomposed mixture of dung and urine of farm animals along with litter and left over materials from roughages or fodder fed to the animals. It contains 0.5% N, 0.2% P_2O_5 and 0.5% K_2O. Urine contains 1% N and 1.35% K_2O. Litter is the straw, peat, sawdust and dry leaves used as bedding material for farm animals and birds. The N present in urine is mostly in the form of urea which is subjected to volatilization losses. Chemical preservatives are used to reduce losses and enrich FYM e.g. gypsum, kaolinite and super phosphate. These preservatives absorb urine and prevent volatilization loss of urea and also add nutrients (Reddy, 2005).

A.2 Compost

Compost means 'a product obtained by the controlled decomposition of organic wastes (*composting*), finally used as organic manure'. Composting is the process of reducing animal and vegetable refuse (except dung) to a quickly utilizable condition for improving and maintaining soil fertility. The final well decomposed manure having lower C:N ratio is termed as 'compost'. The recycling of organic materials by biological decomposition as manure is very important for organic farming as it kills weed seeds, pathogenic organisms, and dispose off agricultural / industrial wastes to produce a uniform, slow release organic fertilizer which stimulates soil's life, improve soil structure and control insect-pests and diseases. Compost contains 0.5-0.15-0.5% N-P-K, respectively (Reddy, 2005).

A.3 Biogas slurry

Instead of directly using the animal dung for composting it can be used for production of biogas by feeding through Biogas Plants. It contains (1–1.8% N, 0.4–0.9% P_2O_5 and 0.6-1% K_2O) due to low volatilization losses of ammonia (Reddy, 2005).

A.4 Night soil (Poudrette)

Night soil is human excreta, both solid and liquid. It contains 5.5% N, 4% P_2O_5 and 2% K_2O. The dehydration of night soil, as such or after admixture with absorbing materials like soil, ash, charcoal and sawdust produces a poudrette that can be used easily as manure. Poudrette contains about 1.32% N, 2.8% P_2O_5 and 4.1% K_2O (Reddy, 2005).

A.5 Sewage and sludge

The solid portion in the sewage (human excreta + water) is called *sludge* and liquid portion is *sewage water* (Reddy, 2005). It can be recycled for crop fertilization, irrigation to the crop, aquaculture production, application to forest land, biogas production and land reclamation. It was estimated that total waste generated by 217 million people in urban areas is 39 mt/ year (2001). The total NPK content of this would be 2.5 lac tonnes of N, 2.6 lac tonnes of P and 2.6 lac tonnes of K. Both the components are separated and are given a preliminary fermentation and oxidation treatments to reduce bacterial contamination and offensive smell, otherwise soil quickly becomes **"sewage sick"** owing to the mechanical clogging by colloidal matter in the sewage and the development of anaerobic organisms which not only reduce the nitrate already present in the soil but also produce alkalinity. These defects can be removed by thoroughly aerating the sewage in the settling tank by blowing air through it. The sludge that settles at the bottom in this process is called **"activated sludge"** (3.6% N, 2% P_2O_5 and 1% K_2O).

A.6 Sheep and goat manure

The droppings of sheep and goat contain higher nutrients than FYM and compost. On an average, the manure contains 3% N, 1% P_2O_5 and 2% K_2O). It is applied to the field in two ways- i) Sweeping of sheep and goat sheds are placed in pits for decomposition and it is applied later to the field. ii) Sheep penning- wherein sheep and goats are allowed to stay over night in the field and urine and faecal matter is added to soil (Reddy, 2005).

A.7 Poultry manure

Poultry manure can supply higher N and P to the soil than other bulky organic manures. The average nutrient content is 2.87% N, 2.93% P_2O_5 and 2.35% K_2O (Reddy, 2005).

A.8 Green manuring

Green undecomposed plant material used as manure is called green manure. By growing green manure crops (usually leguminous crops) are grown in the field and incorporating it in its green stage in the same field is called green manuring. It adds organic matter and nitrogen to the soil. On an average green manuring gives 60-80 kg N/ha (Reddy, 2005).

(B) Concentrated Organic Manures

These have required in small quantities and contain higher nutrients as compared to bulky organic manures. The most commonly used are oilcakes, fish meal, meat meal, blood meal, horn and hoof meal, bird guano, raw bone meal etc. which act a good source of organic manures for organic farming system.

B.1 Oilcakes

Oilcakes are generally grouped into two groups, viz., *edible* oilcakes suitable for feeding the cattle and other domestic animals and *non-edible* oilcakes exclusively used as manure due to their higher content of plant nutrients (Reddy, 2005). It has been estimated that India produced about 2.5 million tonnes of oilcakes annually.

Table : Average nutrient content of different oilcakes (Reddy, 2005)

Oilcakes	Per cent composition		
	N%	P%	K%
Edible oilcakes (feed for livestock)			
Safflower (decorticated)	7.9	2.20	1.9
Groundnut	7.3	1.5	1.3
Cotton seed (decorticated)	6.5	2.9	2.2
Non-edible oilcakes (not fed to livestock)			
Safflower (un-decorticated)	4.9	1.4	1.2
Cotton seed (un-decorticated)	3.9	1.8	1.6
Caster	4.3	1.8	1.3
Neem	5.2	1.0	1.4

Non-edible oilcakes are used as manure especially for horticultural crops. Nutrient present in oilcakes, after mineralization, are made available to crops 7-10 days after application. Oilseed cakes need to be well powdered before application for even distribution and quicker decomposition. **Neem cake** acts as **Nitrification Inhibitor**.

B.2 Fish meal

Sea food canning industries are present in almost all coastal states of India. Fishes which is not preferred for table purposes due to their small size, bonny nature and poor taste can be converted into very good organic manure. The fish is dried, powdered and filled in bags. It contains average nutrients are 4-10, 3-9 and 0.3-1.5% NPK. These manures are highly suitable for fruit orchards and plantation crops (Reddy, 2005).

B.3 Meat meal

An adult animal can provide 35 to 45 kg of meat after slaughter or death. It contains 8-9% N and 7% P_2O_5 (Reddy, 2005).

B.4 Blood meal

Blood manure contain about 13-20%N, rich in iron and its application gives a deep rich colour to foliage (Reddy, 2005).

B.5 Horn and Hoof meal

A healthy animal can give about 3 to 4 kg of horn and hoof. These materials are dried, powdered, bagged and marketed as manure. It contains 13% N (Reddy, 2005).

B.6 Guano (Bird / Fish)

The excreta and dead remains of the bird is called *bird guano* (11-14% N and 2-3% P_2O_5) and the refuse left over after the extraction of oil from the fish in factories, dried in cemented yards and used as manure is called as *fish guano* (7% N and 8% P_2O_5) (Reddy, 2005).

B.7 Raw bone meal

An excellent source of organic phosphorus. It contains 3 to 4% N and 20 to 25% P_2O_5 (Reddy, 2005).

2.4.2 Vermicomposting

(A) Definition

The process of composting organic wastes through domesticated earthworms under controlled conditions is vermicomposting (Reddy, 2005).

Earthworms have tremendous ability to compost all biodegradable materials. Wastes subjected to earthworm consumption decompose 2 to 5 times faster than in conventional composting. During composting the wastes are deodorized, pathogenic micro-organisms are destroyed and 40 to 60% volume reduction in organic wastes take place. It is estimated that the earthworms feed about 4 to 5 times their own weight of material daily.

Earthworm bears both male and female reproductive organs. However, two worms are needed for successful copulation. The self fertilization does not occur generally in the earthworms. Fertilization takes place in the egg case or cocoon.

Earthworm species such as *Eisenia foetida, Eudrilus eugeniae, Lumricus rubellus, Lampito mauritii* and *Perionix excavatus* have been recommended for vermiculture technology. Vermicompost is the compost which is prepared by earthworms. It is a mixture of worm casting (faecal excretions) organic materials including humus, live earthworms, their cocoons and other micro organisms.

(B) Vermiculture

It is the process of rearing and breeding of earthworms in controlled condition and presently it is known as earthworm biotechnology. It is estimated that 1800 worms which is an ideal population for one sq. meter can feed on 80 tonnes of humus per year. Faecal matter or excretions of earthworms is known as vermin cast. Vermiwash is a liquid fertilizer collected after the passage of water through a column of worm activation, which is useful for foliar spray. It may be diluted with water before use. It can also be diluted with 10% urine of cow. The average nutrient content of vermicompost is about 0.5 to 0.9 - 0.1 to 0.2 - 0.67 % N-P-K, respectively.

(C) Types of earthworms

There are about 3000 species of earthworms reported in the world. Among them 509 species are available in India. Some of the important species used for vermicomposting are *Eisenia foetida, Eudrilus eugeniae, Lampito mauritii, Perionyx excavatus, Octochaetona serrata,* etc. These earthworms are mainly divided into two groups viz., i) ***Epigeic*** (Surface feeder) - which feeds at or near the soil surface, mainly on plant litter, dead roots and other plant debris. For example *Eisenia foetida, Eudrilus eugeniae, Lampito mauritii, Perionyx excavatus, Octochaetona serrata* etc. are very important for vermicomposting. ii) ***Endogeic***

(Geophagus/humus feeder) – which feeds deeper beneath the soil surface, ingesting large quantity of organically rich soil. Not suitable for vermicomposting.

(D) Characteristics of compost worms

The following are the basic characteristics of earthworm species suitable for vermicomposting:

i) The worms should have feeding preference and adaptability to wide range of organic materials.
ii) It should be efficient converter of plant/animal biomass to body proteins.
iii) It should be tolerant to diseases, wide adaptability to environmental factors and have least inactivity period.
iv) It should have high consumption, digestion and assimilation rates
v) The worms should produce large number of cocoons.
vi) Growth rate, maturity from young one to adult stage should be fast.
vii) The worms should feed near the surface of organic matter.

(E) Vermiwash - A Liquid manure

It is a transparent pale yellow coloured fluid collected after the passage of water through a column of worm action **or** it a collection of excretory products and mucus secretions of earthworm along with nutrients from the soil organic molecules. It is very useful as a foliar spray to enhance the plant growth and yield and to check development of diseases.

(F) Benefits of vermicompost

1. When added to clay soil loosens the soil and provides the passage for the entry of air.
2. The mucus associated with it being hygroscopic, absorbs water and prevents water logging and improves water holding capacity.
3. In the vermicompost, some of the secretions of worms and the associated microbes act as growth promoter along with other nutrients.
4. It improves physical, chemical and biological properties of soil in the long run on repeated application
5. The organic carbon in vermicompost releases the nutrients slowly and steadily into the system and enables the plant to absorb these nutrients.
6. The multifarious effects of vermicompost influence the growth and yield of crops.
7. Earthworm can minimize the pollution hazards caused by organic waste by enhancing waste degradation.

(G) Application of vermicompost

In orchards the dose depends on the age of the tree. It can be used @ 500 g in small fruit plants and 3-4 kg/tree, whereas for vegetable crops @ 3 kg/10 m^2 area. For general use in agriculture, vermicompost should be applied @ 5 t/ha. Vermicompost is mixed with equal quantity of dried cow dung and used as broadcast when seedlings are 12-15 cm height and water should be sprinkled.

2.4.3 Green manuring

(A) Definition

Crops grown for the purpose of restoring or increasing the organic matter content in the soil are called *green manure crops* while their green undecomposed plant material used as manure is called *green manure*. Their use in cropping system is generally referred as *green manuring*. It is obtained in two ways-either by grown *in situ* or brought from out site. In both ways, the organic material should be worked into the soil while they are fairly young for easy and rapid decomposition (Reddy, 2005).

i) *In situ green manuring* : Growing of green manure crops in the field and incorporating it in its green stage in the same field (i.e. *in situ*) is termed as *green manuring*.
ii) *Green leaf manuring*: is the application of green leaves and twigs of trees, shrubs and herbs collected from nearby location and adding to the soil. Forest tree leaves are the main source of green leaf manuring. Legumes are usually utilized as green manure crops as they fix atmospheric nitrogen in the root/stem nodules through symbiotic association.

(B) Advantages of green manuring

1. It adds organic matter to the soil. This stimulates the activity of soil micro organisms.
2. Green manuring concentrates plant nutrient in the surface layer of the soil.
3. It improves the structure of soil by deep rooting system.
4. It facilitates the penetration of rain water, thus decreasing run off and soil erosion.
5. It holds plant nutrients that would other wise be lost by leaching (e.g. N).
6. It increases the availability of certain plant nutrients like P, Ca, K, Mg and Fe.
7. It checks weed growth by quick initial growth.
8. It aid in reclamation of sodic soils by release of organic acids.

(C) Desirable Characteristics for Green Manure Crops

The criteria for which green manure crops are selected should have following characters,

- It should be high biomass production
- It should be deep rooting system
- It should be leguminous family
- It should be fast initial growth
- It should be more leafy than woody
- It should be low C/N ratio
- It should be non-host for crop related pathogens
- It should be easy and abundant seed producer
- It should be useful for 'by-products'.

2.4.4 Recycling of organic residues

(A) Organic Residues

A variety of organic residues include crop residues in the form of straw, husk, forest litter; animal wastes like dung urine, bones etc., guano, city or household residues, oilcakes, by-products of food and sugar industries, pond silt, marine wastes, sea weeds and human habitation wastes. There are two major components of crop residues available, i.e. *harvest refuse* (straw, stubbles, haulm of different crops) and *process wastes* (nut shell, oilcakes and cobs of maize, bajra and sorghum). *Crop residues* are defined as 'the non-economic plant parts that are left in the field after harvest and remains that are generated from packing sheds or that are discarded during crop processing'. The *benefits* of proper organic residue recycling are that they supply essential plant nutrients, improve soil properties, protect the soil from erosion hazards, reducing residue accumulation at the sites they produced, providing employment as well as income to many, enhancing environmental qualities and illustrate that man is not a waste generator but also its wise utilizer/ manager (Reddy, 2005).

(B) Methods of Recycling

Organic residues can be recycled in soil by different methods like incorporation, burning, surface mulching, composting etc.

(i) Incorporation: The crop residues like maize, rice, sorghum, wheat straw can be directly applied to the field and ploughed in the soil before the rainy season has beneficial effect on soil properties. Farm wastes can be ploughed in the soil (0-20 cm layer). After harvesting of cotton, sugarcane, sorghum etc. can be incorporated into the soil by use of rotavator implements which directly adding small pieces of crop residues in the soil.

***(ii) Burning**:* A large quantity of sugarcane trash, cotton stalks, caster stalks etc are available and many farmers burn them in the field. It is not advisable practice as burning kills the soil fauna and flora, increases losses of N, C, S and possibly some other nutrients in volatilization and results in unfavourable soil conditions. Although burning releases Ca, Mg and k from crop residues but increases the potential loss due to leaching and erosion.

***(iii) Surface Mulching**:* One unique and simple way of profitable recycling the crop residues is their use as surface mulching materials. Mulches are thermo insulators, have smother effect on weeds, protect the soil from rain drop impact, reduce salinization and barriers to vapour transfer thus conserve soil moisture. It is also beneficial for soil micro organisms and on degradation adds organic matter to the soil.

***(iv) Composting**:* As discussed earlier, compost is the stabilized and sanitized product of composting which is beneficial to soil health and plant growth. A huge quantity of crop wastes/residues and animal wastes are always available on a farm. Properly recycled, these residues form excellent compost in one to six months, depending upon the composting process used. The important methods of composting are NADEP compost, vermicompost, sugarcane trash compost, obnoxious weed compost and recycling of pond silt accumulation alone or by enriching composting units.

2.4.5 Biofertilizers (Microbial inoculants)

The atmosphere over an hectare of land consists of 80,000 tonnes of N. Though atmospheric N is present in sufficient quantity (78%), it is not available to plants since it exists in inert form. Biological nitrogen fixation is the conversion of atmospheric N by living organisms into forms that plants can use. This process is carried out by a group of bacteria and algae, which fix atmospheric nitrogen (N_2) into assimilable forms of nitrogen (NH_3)

It can be defined as biofertilizers or microbial inoculants are preparations containing live or latent cell of efficient strain of N-fixing or P-solubilizing micro organisms used for seed or soil application with the objectives of increasing the numbers of such micro organisms in the soil or rhizosphere and consequently improve the extent of microbiologically fixed N for plant growth.

(A) Use of Biofertilizers

Azospirillum is applied as seed treatment or soil application in crop like rice, sugarcane, pulses, soybean and vegetables. It increases root length, top dry weight, root dry weight, total leaf area and yield. The inoculants like nitroplus (legume inoculants) and VAM (*Vesicular Arbuscular Mycorrhizae*) are also effective for crop yield improvement. The *Bacillus* sp. and *Pseudomonas* sp. are helpful in synthesizing the insoluble form of phosphorus. The combined application of phosphobacteria, rock phosphate and FYM to commercial crops has greatly enhanced biomass production, uptake of nutrients and yield.

(B) Enrichment of Compost with Microbial Inoculants

Compost prepared by traditional method is usually low in nutrients and there is need to improve its quality. Enrichment of compost using low cost nitrogen fixing and phosphate solubilizing microbes is one of the possible ways of improving nutrient status of the soil. It could be achieved by introducing microbial inoculants, which are more efficient than the native strains associated with substrate materials. Both the nitrogen fixing and phosphate solubilizing microbes are more exacting in their physiological and ecological requirements. The only alternative is to enhance their inoculum potential in the composting mass.

(C) Benefits of Biofertilizers in Organic Farming

- Biofertilizers are eco-friendly and do not have any ill effect on soil health and environment.
- They reduce the pressure on non-renewable nutrient sources/fertilizer.
- Their formulations are cheap and have easy application methods.
- They also stimulate plant growth due to excretion of various growth hormones.
- They reduce the incidence of certain disease, pathogen and increase disease resistance.
- The economic benefits to cost ratio of biofertilizers is always higher.
- They improve the productivity of waste land and low land by enriching the soil.

(D) Types of Biofertilizers

a. Biological N fixing micro-organisms
b. Phosphate solubilizing and mobilizing micro-organisms
c. Potash solubilizing micro-organisms
d. Sulphur mobilizing micro-organisms
e. Arbuscular mycorrhizal fungi
f. Growth promoting substance excreting micro-organisms

(i) Biological N-fixing micro-organisms

Biological N-fixing micro-organisms help in reduction of atmospheric N_2 to NH_3. The N-fixing organisms such as *Rhizobium* spp. which live in symbiotic association with roots of leguminous vegetables, forming nodules and free living fixers

Azotobacter spp. and *Azospirillum* spp. which live in association with root system of crop plants. There are two types of rhizobia: (i) the slow growing *Bradyrhizobium* and (ii) the fast growing *Rhizobium*. *Azospirillum* fix N from 10 to 40 kg/ha and saves N fertilizer inputs by 25 to 30%. Azotobacter inoculation saves N fertilizer by 10 to 20%.

a) *Rhizobium* and *Bradyrhizobium*

They symbiotically fix N with leguminous plants increasing the amount of available N for uptake by plants. The quantum of N fixation ranges from 50-300 kg N/ha/crop under most optimum conditions i.e. Cow pea 80-85 kg/ha, Red gram 168-200 kg/ha, Groundnut 50-60 kg/ha and Lucerne 100-300 kg/ha can fix symbiotically N by legume crop root nodules. An increase in yield about 10-20 % has been observed in pulses treated with *Rhizobium*.

b) Azola

Azola symbiotically can fix 30-100 kg N/ha and increase yield up to 10-25% and also survive at high temperature in flooded rice crop.

c) *Azotobacter*

Azotobacter is a free living aerobic N-fixing bacteria can fix 10-25 kg N/ha/season in cereals. 50% of N requirement of crop can be reduced through *Azotobacter* inoculation along with FYM. *A. chroococcum* is the dominant species in arable soils. Vegetable crops such as tomato, brinjal and cabbage responded better to *Azotobacter* inoculation than other crops.

d) *Azospirillum*

Azospirillum inoculation helps to fix nitrogen from 15 to 40 kg/ha. It is useful in cereals for better vegetative growth and also saving inputs of nitrogenous fertilizers by 25-30%.

e) *Beijerinckia*

Its production is high in acidic soils. *B. indica* is a common species. It is generally present in the rhizosphere of plantation crops such as coconut, arecanut, cashewnut, cocoa and pepper.

(ii) Phosphate solubilizing and mobilizing micro-organisms

Several soil bacteria particularly *Pseudomonas striata* and *Bacillus polymixa* and fungi *Aspergillus awamori* and *Penicilium* spp. possess the ability to bring insoluble phosphates into soluble forms by secreting organic acids. Arbuscular mycorrhizal fungi (AMF) are also responsible for converting fixed phosphorus into available phosphorus through inoculation of efficient strains of AMF, 25 to 50% of P fertilizer can be saved.

(iii) Potash solubilizing micro-organisms

The bacterium, *Frateuria aurantia* was isolated from banana plant from Orissa soil. These bacteria have solubilizing power of 90% within 22 days when the mineral source of K is in fixed form. These bacteria were tested on banana and paddy which increased the yield by 20 and 25%, respectively. It can be used as soil application for all types of crops @ 2.5 kg/ha. It can be mixed with @ 200-500 kg FYM in furrows before sowing. The bacterium can save up to 50-60 % of cost of K fertilizer.

(iv) Sulphur mobilizing micro organisms

Sulphur present as insoluble sulphur form at 30-35 cm deep in soil and are associated with oxides of iron and aluminium. *Acetobacter pasteurianus* helps in converting this non-usable form to usable form. The use of 625 g/ha of *A. pasteurianus* influenced the levels of sulphur in crops like vegetables, cabbage, turnip, onion etc.

(v) Arbuscular mycorrhizal fungi (AMF)

AMF improve plant growth through better uptake of nutrient like P, Zn, Cu etc. and make the plant root more resistant to pathogens, improve soil texture, WHC, disease resistance and better plant growth. AMF saves 25-50 kg P/ha in addition increase the yield up to 10-12%.

(vi) Growth promoting substance erecting micro-organisms

The specific strain of plant growth promoting rhizobacteria (PGPR) could colonize roots of crops like potato, beet root, apple and legumes. They enhance plant growth indirectly by depriving the harmful micro-organisms. PGPR belong to many genera including *Agrobacterium, Arthrobacter, Azotobacter, Bacillus, Pseudomonas, Cellulomonas, Rhizobium* etc.

References

Bhattacharya, P. and Gehlot, D. (2003). Current status of regulatory mechanism in organic farming. *Fertilizer News,* 49(11): 33-38.

Lampkin, N. (1990). "Organic Farming". Farming Press Book, Ipswich, UK. p. 715.

Lampkin, N.; Foster, C.; Padel, S. and Midmore, P. (1999). The policy and Regulatory Environment for organic Farming in Europe. In : *Organic Farming in Europe: Economics and Policy*, Vol. 2, (eds: Dabbert, Haring, Zanoli), Zed Book, London.

Lieberhardt, B. (2003). What is organic agriculture? What I learned from my transition. In: *Organic Agriculture, Sustainability, Markets and Policies, Organization for Economic Cooperation and Development* (OECD) and CABI, Wallingford, UK. pp. 31-44.

Naik, G. (2001). Organic Agriculture. In: Implications of WTO Agreements for Indian Agriculture(eds: S.K. Datta and S.Y. Deodhar). *CMA Monograph No. 191*, IIM, Ahmedabad.

NPCS (2008). "The Complete Book on Organic Farming and Production of Organic Compost". NIIR Project Consultancy Services, Asia Pacific Business Press Inc., New Delhi.

Pathak, R.K. and Ram, R.A. (2003). Approaches for green food production in horticulture. In: Precision Farming in Horticulture (Eds: H.P. Singh, Gorakh Singh, J.C. Samuel and R.K. Pathak), Central Institute for Subtropical Horticulture, Lucknow.

Reddy, S.R. (2005). "Principles of Agronomy". Kalyani Publishers, Ludhiana.

Sanchez, P.A. (1994). Tropical soil fertility research: Towards the second paradigm. In: Inaugural and State of the Art Conferences. *Transactions 15th World Congress of Soil Science*, Acapulco, Mexico. pp. 65-88.

Sharma, A.K. (2008). "A Handbook of Organic Farming". Agrobios (India), Jodhpur.

Singh, S. (2003). Marketing of Organic Produce and Minor Forest Produce. *Indian Journal of Agricultural Marketing*, 17(3): 77-83.

□□□

Green Agriculture : Newer Technologies, 2012
© Kambaska Kumar Behera (ed.), 439-456
New India Publishing Agency, New Delhi (India)
e-mail : info@nipabooks.com; website : www.nipabooks.com

Chapter-18

Predominant Farming Systems and Alternatives in Madhya Pradesh

A.K.Jha,V.B.Upadhyay and S.L.Vishwakarma
Department of Agronomy,
JNKVV, Jabalpur-482004 (M.P.)
E-mail : amitagcrewa@rediffmail.com

SUMMARY

Multiplicity of cropping systems has been one of the main features of Indian agriculture and it is attributed to rain-fed agriculture and prevailing socio-economic situations of farming community. It has been estimated that more than 250 double cropping systems are followed throughout the country and based on rationale of spread of crops in each district in the country, 30 important cropping systems have been identified. Cropping systems of a region are decided by and large, by a number of soil and climatic parameters which deter mine overall agro ecological setting for nourishment and appropriateness of a crop or set of crops for cultivation. Much of the Madhya Pradesh area also coincides with the rainfed semi-arid agro-ecological region and sub-regions which are upland, forested with low agricultural productivity and predominately tribal population. It would appear that over the centuries – including in recent history tribals have been inexorably moved off the better lands, into areas which from an agricultural perspective have low productive capacity and little potential for growth. Low agricultural productivity is one of the main causes of persistently high poverty rates in the state and the poor masses not only depends upon the farming system but also chosen different alternatives for their sustanability

Introduction

Madhya Pradesh covers 30.8 million hectare geographical area with a population of 60.3 million in 2001. Of the total lands, nearly 65% area is arable lands, where several kinds of food grain, oilseed, pulse, fiber, vegetable, fruit and fodder crops are grown depending on the suitability to various heterogeneous agro-climatic conditions. Besides crop husbandry, other farming systems viz. dairy, poultry, goatry, fish farming and horticulture also exist in pockets depending on the socio-economic status of the farmers, infra structural facilities (marketing, transport, processing and storage etc.), irrigation facilities and technological development etc. These farming systems are mostly fallowed to fulfill the domestic needs of the farmers. Thus, these farming systems are not followed in commercial scale to enhance the income and employment. There will be tremendous presser on available agricultural lands and other natural resources to ensure the food security in future for ever increasing human and livestock population. Therefore, integrated farming system may be one of the best alternatives for enhancing the farm income and employment generation to small and marginal farmers by efficient utilization of all available natural resources.

Madhya Pradesh is agriculturally divided into 11 agroclimatic zones by considering the existing soil types, rainfall and cropping systems as per norms of National Agricultural Development Programme (Table 1). Crop husbandry is predominant farming system in almost entire state. The cropping intensity varies from 104% in zone III to 140% in VI and XII zones. Cropping intensity (140%) in zone XII is appreciable, where annual precipitation is low (600-800 mm) with less (13.2%) irrigation facilities. On the other hand, cropping intensities in I (125%), VI (140%), VII (111%), VIII (123%) zones are not much so efficient looking to the irrigated areas from 31.4 to 41.7% coupled with high rainfall (1000-1500 mm). Thus, there is an urgent need to identify the ways and means for improving the land and water use efficiency in these zone to enhance the income and employment opportunities. Almost all agro-climatic zones have sufficient area under dense forest with slopy topography, and thus, there are enough scope to introduce agro-forestry base farming system in the state.

Holding Size of Farmers

Different major groups of farmers consist with 84.01 lakh operational holdings. They operate 221.11 lakh hectares of land in the state (Table 2). Marginal, small, semi medium, medium and large groups of farmers held 37.33, 22.82, 20.69, 15.32 and 3.84 % of total number of operational holdings, respectively in the state, while respective groups of farmers operate 6.37,12.59,21.88,35.15 and 24.0% of total lands. The large holding groups of farmers operate maximum lands (16.44 ha) without paying much attention in agricultural operations. Farming operations of farmers having large size holding are done by the labourers and timely availability of adequate labourers during farm operation peaks causes serious problem for successful farming. Marginal and small groups of farmers have shortage of working capital to invest on progressive farming.

Table 1 : Characteristics features of agro-climatic zones of Madhya Pradesh

	Agro-climatic zone	Geographical area (lac ha)	Forest area (%)	Irrigated area (%)	Annual rainfall (mm)	Annual PET (mm)	Soil type	Important crops	Cropping intensity (%)
I	Chhattisgarh plains	9.24	52.59	41.7	1300-1600	1400-1500	Shallow to medium, light to medium loamy, red + yellow, medium AWC slopy	Rice, maize, blackgram, pigeonpea, wheat, linseed, lathyrus	125
III	Northern hills of Chhattisgarh	28.16	41.61	6.5	1200-1600	1400-1500	Shallow, light texture, red + yellow undulated topography, hilly, medium AWC	Rice, small millets, maize, pigeonpea, wheat, gram, niger	104
IV	Kymore plateau and Satpura Hills	49.96	28.45	12.4	1200-1500	1300-1500	Medium to deep, clayey, black red + yellow, slopy, low AWC	Rice, soybean, maize, pigeonpea, wheat, gram, lentil, linseed	122
V	Vindhyan plateau	42.58	28.16	16.2	1000-1200	1300-1500	Medium to deep, clayey, black, medium to high AWC	Soybean, sorghum, kharif pulses, wheat, gram, linseed	118
VI	Central Narmada Valley	15.11	32.45	33.6	1000-1200	1300-1500	Deep, clayey, black, high AWC, ill-drainage	Soybean, kharif pulses, rice wheat, gram, linseed, sugarcane,	140
VII	Gird	37.40	20.32	31.4	800-900	1400-1900	Shallow to medium, alluvial, grey colour, low to medium AWC, ravenous topography	Pearl millet, soybean, groundnut, sorghum, kharif pulses, wheat, mustard	111
VIII	Bundelkhand	20.80	16.34	32.1	800-1000	1300-1500	Shallow to medium, black + red, medium AWC	Soybean, jowar, bajra,sesame, groundnut,rice,pulses,wheat, gram,linseed,vegetable	123
IX	Satpura plateau	21.97	38.35	13.1	1000-12000	1300-1500	Shallow to medium, black, clayey, low to medium AWC	Soybean, maize ,small millets, wheat, gram, linseed, vegetables, chillies	116
X	Malwa plateau	51.76	10.84	22.0	1000-1000	1600-2000	Medium to deep, black, clayey, medium to deep AWC	Soybean, maize, sorghum, cotton, wheat, gram	136
XI	Nimar valley	24.70	38.86	21.7	800-1000	1600-2000	As above	Cotton, sorghum, soybean, groundnut, pulses, gram, wheat	110
XII	Jhabua hills	6.75	19.10	13.2	600-800	1600-2000	Medium to deep, black, loamy slopy, hilly, medium to high AWC	Soybean, maize, jowar, wheat, chickpea, caster	140

AWC – Available water retention capacity

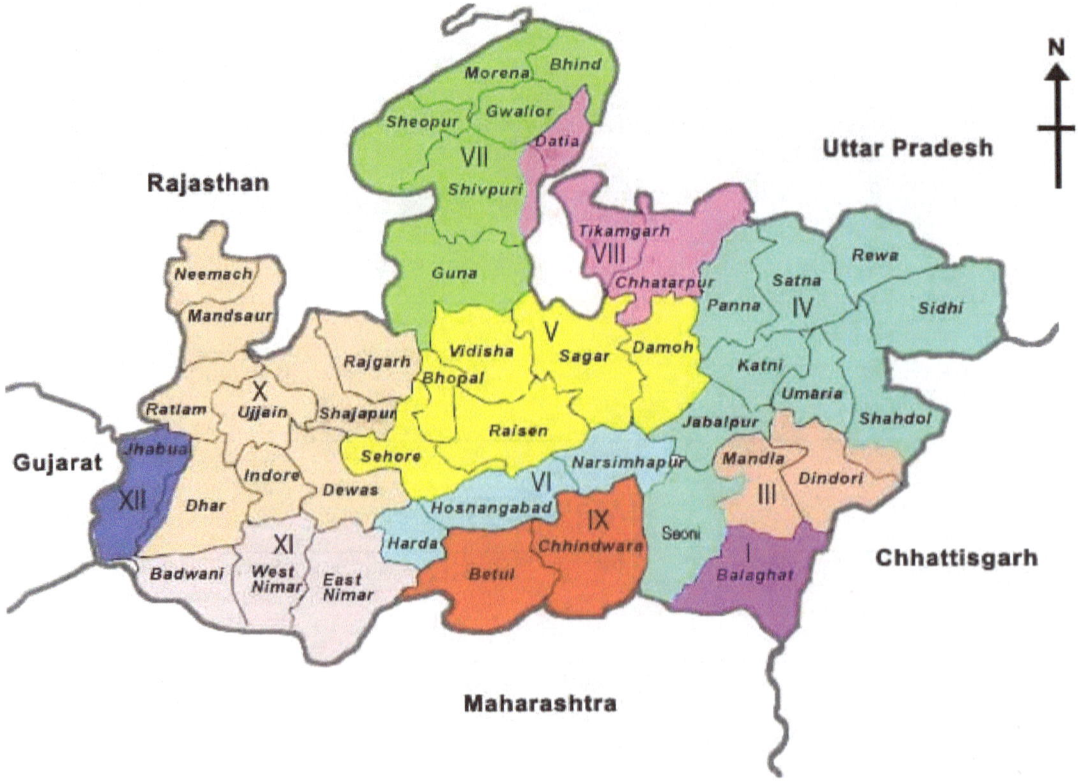

Table 2 : Number of operational holding area operated and operational holding size under major size group in M.P.

Major size group	Number of operational holding (in, 000)	Area operated by major size group (in, 000 ha)	Operational holding size (ha)
Marginal (less than 1.0 ha)	3136	1409	0.45
Small (1.0 to 2.0 ha)	1917	2783	1.45
Semi medium (2.0 to 4.0 ha)	1738	4838	2.78
Medium (4.0 to 10.0 ha)	1287	7772	6.04
Large (Above 10.0 ha)	323	5309	16.44
All holdings	8401	22111	2.63

Socio-economic Status

Major social categories viz. general, others backward classes, scheduled caste, and scheduled tribes respectively, contribute nearly 32, 31, 20, and 17 % of total population (60.3 million) of the State. Generally, schedule caste and scheduled tribe farmers have very small land holding and most of them are land less or working on the farms as a labour (Table 3). Though some of the schedule tribe farmers have considerable size of land holdings, their lands are undulating and degraded. Poor investment capacity for farming and illiteracy are major constraints associated with Scheduled tribe farmers. General and other backward categories of the farmers have their fragmented holdings and thus, they face several problems in efficient utilization of their available managerial resources. Farmers of each category in the society are not evenly linked with effective agricultural extension services.

Table 3 : Number of operational holding area operated and operational holding size under major group of the society of M.P.

Major group of society	Number of operational holdings (000)	Area operated by major size group (000 ha)	Operational holding size (ha)
General	1815	7127	3.92
OBC	4453	11496	2.58
SC	814	1370	1.68
ST	1319	2118	1.60
	8401	21111	2.67

Zone-wise Existing Farming Systems

Though existing farming systems of the state are not systematically documented, the important farming systems which exist under different agro-climatic zones of Madhya Pradesh are described in Table 4. This information is based on the status report of the various agro-climatic zones of the state.

Table 4 : Existing of important farming systems in different agro-climatic zones

	Identified farming systems	Agroclimatic Zone	Leading district
(1)	Crop production (a)	All agroclimatic Zone	All districts
(2)	a + dairying (b)	All agroclimatic Zone	All districts
(3)	a + vegetable production (c)	I, IV, VI,VIII,IX, X, XI, XII	All districts
(4)	a + b + c	All Zone except III	All districts
(5)	a + b + poultry (d)	V, VI, IX, X, XI, XII	All districts
(6)	a + d	IV, VI, IX, X, XI, XII	All districts
(7)	b + d	V, VI, VII, IX, X, XI, XII	All districts
(8)	a + b + agro-forestry	III, IX, XI, XII	Mandla, Dindori, Betul, Chhindwara, Khandwa
(9)	a + c + honey collection	III	Mandla, Dindori,
(10)	a + b + c + mushroom production	IV, VI	Jabalpur, Seoni, Narsinhpur

Identified farming systems		Agroclimatic Zone	Leading district
(11)	a + fisheries	IV, VIII	Tikamgarh, Seoni, Jabalpur
(12)	a + singhara cultivation + fisheries	IV	Tikamgarh, Rewa, Jabalpur, Satna
(13)	a + beetle vine cultivation (e)	IV, VIII	Tikamgarh, Jabalpur, Satna, Rewa
(14)	a + b + c + e	VIII, IV	Tikamgarh, Jabalpur, Satna, Rewa
(15)	b + c + e	VIII, IV	Tikamgarh, Jabalpur, Satna, Rewa

Zone-wise Farming Systems with Major Constraints

I. Chhattisgarh plains (Balaghat district only) zone

Crop production is predominant farming system of this zone. Though farmers are raising bullocks and he-buffalows for agricultural operations viz. draft and transportation etc., they rear cows or she buffalows also to produce the milk just to fulfill the domestic needs. Only a few farmers perform dairy farming on commercial scale either alone or in combination with crop production. Mejority of the farmers grow vegetable crops like cucurbits, *Suran, Banda,* okra, *Popat* and *Ambadi* during rainy season and potato, bringal and tomato during winter season only for domestic uses. Only a few farmers grow vegetable crops as a farm business particularly, in Balaghat, Kirnapur, and Katangi blocks of Balaghat district. Nearly 200 farm holdings of Katangi block in this district grow sugarcane and perform "*Gur*" making on commercial scale.

Constraints

(1) Rice is predominant crop in Balaghat district of the zone. Though rainfall is good, its ill-distribution resulting the frequent and long dry spells which pose dominance of monocropping and sometimes failure of rice also.

(2) Enough forest lands (41.7%) exist in the zone, but farmers are not well aware about the concept of agro-forestry

(3) Farmers are traditionally growing vegetables like *Banda* and *Suran* on the boundary of paddy fields and *Popat* and *okra* on the bunds of paddy fields for their domestic usages, but vegetable cultivation is not well organised on commercial scale.

(4) Farmers have very poor knowledge about growing of green fodder for milch animal owing to very low milk production from poor local breeds

(5) Poultry business is not well known to the farmers
(6) Socio-economic problems like Nexallitism, poor literacy, low investment capacity and cost groupism etc. are hurdles in adoptation of modern farming technology
(7) Poor extension services

II. Northern hills of Chhattisgarh zone

Crop husbandry is major agricultural business of the farmers. The soils are slopy and hilly resulting in excessive runoff causing soil-erosion. The soils very shallow depth (5 to 20 cm) poor available water retention capacity (AWC) and very low fertility status. Thus, low yielding crops viz. small millets and niger are predominantly grown. The farmers raise local and poor breeds of cattle like cow and goats alongwith bullocks and buffalloes for draft work. Mostly goats are raised for milk production and rarely for meet production. They follow still wild grazing practice of cattle jointly in herds of all farmers. Some of the farmers rear poultry for domestic purpose only. Farmers believe to earn money by working as a labourers elsewhere rather than giving attention on farming enterprises. Occasionally, they perform collection of forest products viz. *tendu patta, char, amla, Mahua* flower & fruits, honey and bamboo etc. to earn the money.

Constraints

(1) Soils with poor AWC do not permit post-rainy season cropping under rainfed conditions
(2) Poor knowledge about improved breeds of animals
(3) Inadequate supply of green fodder to milch and other animals throughout the year
(4) Lack of knowledge of MPTS (Multi purpose tree species) suitable for agro-forestry purpose
(5) Farmers do not know the importance of bee keeping and processing of forest by products
(6) Poor linkage of extension services

III. Kymore Plateau and Satpura Hills Zone

This zone has diversified of farming situations due to variation in soil conditions, (topography, fertility, AWC, colour, depth) and socio-economic status of farmers. Therefore, several crops are grown. Nearly 12.4% area of arable land is irrigated. Normally, rainfall is good but erratic rainfall-distribution with long soil moisture stress spells during crop season results into dominance of monocropping. During kharif rice, soybean, maize, pulses (pigeonpea, black gram), oilseeds (sesame and niger) and in rabi wheat, chickpea, linseed and lentil are the predominant crops. Several farming systems viz. animal husbandry, poultry, fisheries, piggery and production of beetle vine, singhara; vegetable and fruit (Guava, Mango, Citrus, Papaya) etc. are associated with crop husbandry depending on the edaphic and environmental conditions, socio-economic status of the farmers, technological advancement and development of irrigation resources. But these farming systems are not followed in commercial scale.

Constraints

(1) Dominance of rainfed agriculture
(2) Low workability of land during rainy season due to heavy texture
(3) Inadequate availability of green fodder throughout the year
(4) Poor marketing facilities for vegetable and fruit crops
(5) Poor knowledge of improved breeds of cattle and their management
(6) Knowledge of supporting business like bee keeping and mushroom cultivation for farmers belonging to different categories of society
(7) Sighara cultivation, beetle vine cultivation, fish farming and piggery are still cost related agri-business of the society
(8) Poor infra-structural facilities for post harvest processing, marketing and storage of farm produce
(9) Socio-economic status of the farmers restricts the adoption of modern farming systems
(10) Poor extension services for popularization of farming system diversification

IV. Vindhyan Plateau Zone

Most of the area is rainfed and subjects to soil erosion losses. The soil is sticky when wets and becomes too hard when dries. Thus, tillage operations are difficult for crop production. Though rainfall is good (> 1000 mm) in the zone, the crops mostly suffer due to erratic rains under rainfed conditions. Mostly lands are kept fallow during rainy season. Like other zones, rearing of cattle for draft and milch purposes and growing of some seasonal vegetables for household uses are practiced besides the crop production. Dairying, poultry and vegetable production are another farming practices neighbouring to the towns to fulfill the location specific needs. These farming practices are not commercialized in other pockets of zone.

Constraints

(1) Difficulty in workability of lands in agricultural operations.
(2) Frequent failure of kharif soybean due to early termination of rains at reproductive phase
(3) Unavailability of green fodder throughout the year due to less irrigation facilities
(4) Poor knowledge of farmers for animal husbandry, poultry and production vegetable and fruit crops

(5) Inadequate infrastructural facilities (cold storage, marketing, transport and processing of farm produces)
(6) Socio-economic problems are hurdles in the adoptation of improved farming systems
(7) Poor extension services

V. Central Narmada Valley Zone

Like other zones, crop production is main farming system in this zone also. The existence of other farming systems viz. animal husbandry, vegetable production, poultry is also very limited. The irrigation facility is available in about 44% area, besides a good rainfall (>1000 mm) in the zone. Thus, fisheries are also associated in the areas having water reservoirs mostly by a particular caste of the society. In slopy lands, plantation of forest trees is also prevalent among the rich farmers.

Constraints

(1) Poor workability of land for agricultural operations due to stickiness and hardness of soils under wet and dry conditions, respectively.
(2) Problem of ill-drainage and water logging due to poor permeability of water.
(3) Mostly lands are kept fallow during rainy season resulting in low land use efficiency
(4) Non-availability of improved breeds of cattle, poultry and fishes
(5) Inadequate knowledge about the proper nutritional management of cattle, poultry and fishes
(6) Less attention in vegetable and fruit production
(7) Poor socio-economic status of farmers
(8) Less technological intervention
(9) Poor infra structural facilities
(10) Poor knowledge about the management of heavy soils and undulating lands by introducing agro-forestry

VI. Gird Zone

The prevailing farming systems of the zone are crop production (a), a + dairying (b), a + b + vegetable production (c), a + b +poultry (d), a + d and b + d. Farmers mostly pay attention on crop production. Though animal production is practiced on large scale, no proper care is made for it. Poultry and vegetable production are practiced by caste basis in the society.

Constraints

(1) Salinity and alkalinity hazards are common in agricultural lands

(2) High soil erosion losses from the alluvial soils have resulted in the development of ravine in Bhind and Morena districts
(3) About ⅔ area of the cultivable area of zone is still rainfed and the annual rainfall is less with very high ET
(4) Problems for control of insects, pests and diseases in dominating pulse and oilseed crops
(5) Low productivity of milch animals
(6) Poor socio-economic status of farmers
(7) Poor attention of vegetable and fruit production
(8) Poor extension services
(9) Poor knowledge about integrated farming systems
(10) Wild grazing of sheep, goats and cattle
(11) Poor infra structural facilities

VII. Bundelkhand Zone

This zone consists with wide diversity in farming systems, but crop production is one. Good number of farmers practice animal husbandry and vegetable production without commercialized views. Because of good sources of irrigation water, fish farming, beetle vine cultivation and *singhara* cultivation are practiced in pockets by a particular community of the societies.

Constraints

(1) Problems of salinity and alkalinity in soils in pockets
(2) Soils are deficient in Zn, Mn besides low status of N with medium P and K contents
(3) Poor knowledge about improved breed of cattle and fishes
(4) Very poor infra structural facilities (market, cold storage, transportation, banking and processing etc.)
(5) Unavailability of seeds of vegetables, spices and fruit crops
(6) Poor socio-economic status of the farmers
(7) Weak extension services

VIII. Satpura Plateau Zone

This agroclimatic zone has also much diversity in prevailing farming systems. Crop production is predominant farming systems and majority of the farmers are rearing cattle to meet the domestic needs alongwith crop production. Some of the farmers are growing vegetable crops like potato, bringal, tomato, cauliflower, cucurbits, okra and spices like chillies and coriander. Saunsar block of Chhindwara is very popular for commercial cultivation of citrus. Cultivation of commercial crop sugarcane is followed by the farmers of this zone in the pockets. Though rainfall is good (>1000 mm) in this zone, the crops suffers with severe drought due to long dry spells during crop season.

Constraints

(1) High erodibility of soil due to steapy topography with shallow to medium depth and good AWC
(2) Less awareness about modern advancement in agriculture
(3) Poor socio-economic status of the farmers
(4) Lack of extension services
(5) Poor infra-structural services
(6) Non-availability of improved cattle breeds

IX. Malwa Plateau Zone

This zone has moderate (900-1000 mm) rainfall with long dry spells. The soils have medium to high depth and good AWC. Limited irrigation facilities are available. Crop production is predominant farming system. Animal husbandry, vegetable production and cultivation of spices viz. chillies, onion, and few medicinal crops are other farm enterprises associated with crop production.

Constraints

(1) Soil experience drought ness during the intermittent dry spell periods
(2) The length of growing period ranges from 90-150 days in a year
(3) 70-90% area of kharif is fallow and rainfed
(4) Inadequate supply of improved varieties of vegetable crops and spices
(5) Socio-economic problem for mechanization of farming systems
(6) Imperfect drainage conditions limits the root growth
(7) Saline and alkaline soils are scatterly distributed in the zone
(8) Poor knowledge of improved breeds of cattle and poultry

X. Nimar Valley Zone

Crop production is the pre-dominant farming system of this zone. Though farmers are rearing animals for draft and transportation purposes, they rear cow or buffaloes also to produce milk for domestic also. Avery limited number of farmers perform dairy as a farm business either alone or in combinations with crop production. Some farmers also grow vegetables like cucurbits, *banda*, cabbage, and cauliflower during rainy season and potato, onion, peas and tomato during winter season in very limited areas only for domestic usages. A variety of farming systems like poultry, agro-forestry and fruit cultivation etc. are also associated with crop production depending on the edaphic and environmental conditions and socio-economic status of the farmers. However, none of these farming enterprises has comfortable position except crop husbandry.

Constraints

(1) Imperfect drainage restricts the plant growth in kharif
(2) The length of growing period ranges from 90-150 days in a year
(3) Insufficient water supply in rabi season
(4) Crop cultivation is totally rainfed
(5) Inadequate supply of breeds/varieties
(6) Lack of fodder supply throughout the year to draft and milch animals

XI. Jhabua Hills Zone

A number of farming enterprises like crop production, animal husbandry, poultry (Kadaknath breed), fisheries and fruits (Guava, mango, banana, citrus, papaya) production etc. are associated with crop husbandry depending an the edaphic and environmental, socio-economic status of the farmers, technological advancement and development of irrigation resources.

Constraints

(1) Undulating topography
(2) Gross annual deficit of 800 to 900 mm of water between rainfall and ET
(3) Frequent dry spells during crop growing period lead to failure of crops
(4) About 60% of the soils are prone to erosion
(5) Nearly 65% soils of the area have severe limitations of soil depth, topography and water retention

Potential Alternatives

(I) For Balaghat district in Chhattisgarh plain zone

According to suitability of farming situations following alternatives appeared to be feasible for improving the monetary gain and employment:
In low and semi deep land conditions:

(1) The fields are bunded into small pieces of about 1000-2000 m^2 size and water is impounded easily. Generally, no herbicides are applied to control the weeds. Thus, practice of fish farming by introducing quick developing fish breeds may be practiced.
(2) Dairy enterprise can be easily popularized by introducing rice bean (*Phaseolus calcaratus*) and grass pea (*Lathyrus sativa*) during rainy and winter seasons, respectively as nutritious fodder crop under rainfed conditions in these areas.

(II) In intermediate low land (bunded upland) conditions:

(1) Cultivation of vegetable crops viz. *Banda* and *Suran*, Spices viz. turmeric, ginger and medicinal crops viz. *Safed Musli* to be popularized

(2) Mushroom cultivation may also be beneficially introduced in these areas

(3) Dairy can be popularized by introducing improved method of feeding paddy straw, and growing of maize/sorghum + cowpea in upland fields during rainy season and grass pea during winter season as fodder crops under rainfed condition.

(III) For northern hills of Chhattisgarh zones

(1) Niger is widely grown from July to August months and its flowers attract honey bees. Secondly; honey collection from local wild honies is also common practice of tribal farmers and niger it can be grown throughout the year due to its photo-natural habitat. Thus, there is a great scope to introduce bee keeping in the areas as a supporting farming in this zone, where niger is predominantly grown.

(2) This zone consists with slopy and hilly lands, where bamboo plantation may be quite successful. The practice of bamboo plantation may be helpful to prevent the soil degradation by addition of organic matter and checking of soil losses by erosion. The use of bamboo for preparation of baskets can be popularized as a cottage industry besides the sale of bamboo to paper industry located at Amalai (Shahdol) near by the area. Thus, crop husbandry + Agro forestry with bamboo plantation can be popularized.

(3) There are sufficient grasslands for cattle grazing, but these grasslands are full of natural wild grasses. These grasslands can be improved by seeding of improved grasses and controlled grazing. Simultaneously, plantation of perennial trees which can supply fodder to the cattle besides fire fuel can also be introduced to encourage animal husbandry such as dairying and goat rearing alongwith crop production.

(4) Poultry may be other most profitable enterprise alongwith cropping by introducing it on commercial scale.

(5) The processing of forest by-products (*Dona pattal* and bamboo baskets) and cultivation of medicinal crops alongwith crop production can be one of the possible promising alternatives.

(6) Management and recycling of rain water for crop production and other enterprises in stress condition.

(IV) For Kymore plateau zones

(1) Popularization of bee keeping alongwith crop husbandry in Kundam and Dheemarkheda blocks of Jabalpur district and Lakhanadaun block of Seoni

(2) Introduction of *Lakh* production in Sidhi, Shahdol, Katni and Seoni districts alongwith crop husbandry
(3) Improved piggery among the SC community of the zone
(4) Bamboo cultivation and preparation of baskets alongwith other may be introduced
(5) Cultivation of medicinal crops in Sidhi, Shahdol, Umaria, Panna and Seoni districts alongwith crop husbandry
(6) Popularization of beetle vine cultivation with vegetable production among a site specific community in Jabalpur, Satna and Rewa districts
(7) Dairying and poultry in all districts of the zone under both irrigated and rainfall ecosystems

(V) For Vindhyan plateau zones

(1) Production of medicinal crops, spices, flowers and vegetables crops alongwith crop husbandry may be popularized
(2) Introduction of mushroom production alongwith animal husbandry and crop production
(3) Introduction of agro-forestry and animal husbandry in Sagar, Damoh, Bhopal, Sehore and Raisen districts
(4) Commercialization of dairying and poultry in all districts of the zone

(VI) For Central Narmada valley zones

(1) Dairying alongwith crop production in commercial scale
(2) Agro-forestry + dairying + crop husbandry
(3) Dairying and vegetable production
(4) Cultivation of medicinal crops alongwith crop production
(5) "*Gur*" making by sugarcane production alongwith crop production
(6) Fish farming in water reservoirs may be popularized

(VII) For Gird zones

(1) Commercial dairying with crop production + agro-forestry
(2) Commercial goatry with crop production + agro-forestry
(3) Cultivation of fruit and vegetable crops with crop production
(4) Fish farming and bee keeping alongwith crop husbandry in Chambal Command areas
(5) Plantation of subabool and other forages in ravenous lands to popularize animal husbandry and goatry

(VIII) For Bundelkhand zones

(1) Cultivation of fruit and vegetable crops with crop husbandry or animal husbandry
(2) Commercial goatry with crop production and agro-forestry
(3) Commercialization of dairy with crop production
(4) Mushroom cultivation or bee keeping alongwith crop husbandry + vegetable production
(5) Cultivation of spices like ginger or Coriander and medicinal crops alongwith agro-forestry for MPTS

(IX) For Satpura plateau

(1) Cultivation of fruit and vegetable crops with crop husbandry or animal husbandry
(2) Cultivation of spices like chillies, ginger and coriander and medicinal crops alongwith agro-forestry
(3) Bee keeping alongwith crop husbandry + vegetable production
(4) Sugarcane cultivation and "*Gur*" making alongwith crop husbandry
(5) Growing of intercrops in orange orchards and bee keeping

(X) For Malwa plateau

(1) Dairying with crop production
(2) Poultry and dairying
(3) Medicinal crops and spices with crop husbandry
(4) Crop husbandry and mushroom cultivation

(XI) For Nimar valley

(1) Growing of fruit and vegetable crops with crop husbandry
(2) Crop production and agro-forestry
(3) Dairying with crop production including fodder crop production
(4) Crop production and sericulture
(5) Dairying + poultry + mushroom cultivation

(XII) For Jhabua hills zones

(1) Dairying + poultry alongwith crop production
(2) Crop production + agro-forestry
(3) Crop production + bee keeping + mushroom cultivation
(4) Dairying + agro-forestry

Technological Interventions for Improving the Productivity of Farming Systems

Several technologies have been evolved which can be popularized among the farmers for improving the soil productivity for enhanced income besides the generated of employment opportunities.

(I) The farmers of the state use very limited quantity (32kg NPK/ha) of fertilizers. Thus, there is an excellent opportunity to grow high value crops (medicinal crop, spices, flower crops and fruit crops) which require less fertilizers to increase the productivity of farms. (Chhinnasamy et al 1994, Bheemaiah et al, 1995, and Chhinnasamy et al 1997).

(II) According to Awasthi and Upadhyaya (1999), a multiple cropping of rice-potato-spinch-spinch sequence was successful under guava-based agri-horticultural system utilizing the waste water of surrounding dairies for irrigation. An additional guava yields of 41.6 q/ha was recorded with this system besides the rice yield of 19.76 q/ha. Jain and Tiwari (2000) found that growing of lentil + linseed (3:1 rows) in between the rows of guava orchard proved to be remunerative even upto 10 years under rainfed condition in clay loam soils of Jabalpur, Madhya Pradesh. Similarly, cultivation of turmeric or zinger under shaded condition of 12 years old guava orchard gave additional yields of associated crops with the guava fruit yields. (Khare et al, 2001 and Koshta et al, 2001).

(III) Most of the marginal, small and medium farmers can not use fertilizers to realize the potential crop yields. Integrated use of organic manures including recycling of farm wastes, crop residues and animal residues with fertilizers in agriculture may be rationalized by the judicious combination of different farming systems viz. crop production, animal production, poultry and mushroom cultivation etc. (Bheemaiah et al, 1995, Jana and Getaham, 1999; and Oman and Rao 1999).

(IV) Plantation of *Leucanea leucocephala, Albizia Lebhack, Terminalia anjuana and Eucalyptus teretiorinis* in undulating lands reduced losses of soil by splash erosion. Dairy and goatry could be successfully introduced in such areas (Awasthi et al, 1998 and Solanki and Ram Newaj,1999)

(V) Bee keeping is good practice to be followed by the formers for generation of additional profit on the farms. This practice could be successfully adopted where water is available to honey bees and flowering blooms are available for honey bee visits to collect the honey. Niger being a photo neutral crops can be grown through out the year and the blooms of niger are altraction of honey bees. Therefore, bee keeping can be introduced in such areas where niger cultivation is practiced traditionally (Project Report 2003)

(VI) Crop residues of rice, wheat, berseem and soybean could be used for commercial of production of mushroom provided water is available almost during thoughout the year. This practice will be helpful to generate additional income by the farmers without investment (Manjunath and Itnal, 2003)

(VII) Piggery of local breed is maintained almost in all zones and SC farmers mostly perform this farming system. Therefore, this farming system can be popularize by providing trainings and improved breeds to improve the productivity and profits.

(VIII) Popularization of bamboo plantation to improve the degraded lands and enhance the income by developing cottage industries like prepration of baskets and others goods from bamboo in zone I, III and IV.

Future Thrust

(1) Research on generation improved technology for site specific farming system

(2) Identification of site specific availability of various farm resources and existing indigenous practices of farming systems

(3) To ensure the regular fodder supply through out the year

(4) Development of proper infra structural facilities viz. marketing, storage, processing and transport etc.

(5) To coordinate the facilities of financial supports in adoption of new farming system by the farmers

(6) Strengthening of extension services.

References

Awasthi, M.K.; Upadhya, S.D.; Jain, K.K. and Sahu, M.L. (1998). Interception loss in agro-silvicultural systems under rainfed conditions. *1st Int. Agron. Congress*; IARI, New Delhi, Nov. 23-27, pp 485

Awasth, M.K. and Upadhyaya, S.D. (1999). Watershed approach for wasteland development in central India. *Int. Conf. Managing Natural Resources for sustainable Agriculture in 21st Century*, IARI, New Delhi, Feb. 14-18, pp. 1199-1200

Bheemaiah, G; Subramanyam, M.V.R.; Syed Ismail (1995) Intercropping in Faidherbia albida with arable crops under different tree spacing in alfisol. *Indian Journal of Dryland Agriculture Research and Development*, 10: 0-19.

Chinnasamy, C; Jayanathi, C; Rangasamy, A (1994) Farming system research methodology for problem identification and research prioritization. *Proc. Integrated farming system research and management for sustainable agriculture,* ICAR Summer Institute, 6-15, June pp 54-64.

Chinnasamy, D; Jayanathi; C. Rangasamy, A. (1997) Sustainable productivity of cropping and soil fertility management through integrated farming system. *3rd IFOAM-Asia Scientfic Conf. & Gen. Assembly. "Food security in harmony with nature"* UAS Banglore 1-4th Dec. P-20.

Jain, K.K. and Tiwari, N.P. (2000) Crop compatibility for sustained productivity under Guava based agri-horticultural systems. *National Seminar on Biodiversity Development*, April 19-20, JNKVV. Jabalpur pp. 28-29

Jana, B. and Getahum, A. (1999) Intercropping *Acacia albida* with maize (*Zea mays*) and green gram (*Phaseolus aureus*) at Mtwapa, Cost Province. *Kenya Agro-forestry System* 14-193-205.

Khare, A.K.; Koshta, L.D. and Upadhyaya, S.D. (2001). Potential for turmeric grown under shaded perennial of agro-forestry system. *National Seminar on commercial processing and marketing of medicinal and aromatic plants*, Nov. 27-29, JNKV, Jabalpur pp 121-122

Koshta, L.D.; Khare, A.K. and Upadhyaya, S.D. 2001. Differential response of planting geometry of ginger grown under agri-horticultural system. *National Seminar on commercial processing and marketing of medicinal & aromatic plants*, Nov. 27-29, JNKVV, pp. 121

Manjunath, B.L. and Itnal, C.J. (2003). Integrated farming system in enhancing the productivity of marginal rice holdings in Goa. *Indian Jounal of Agronomy* 48 (1): 1-3.

Project Report, (2003). Efficiency of honey bees (*Apis mellifera*) on honey production and crop yields, JNKVV, Jabalpur

Solanki, K.R. and Ram Newaj (1999). Agro-forestry; An alternate land use system for dryland agriculture in 50 years of dryland agricultural research in India. H.H. Singh, Y.S. Ramakrishna, K.L. Sharma and B. Venkateswarlu (Eds.). Central Research Institute for Dryland Agricultrure, Hyderabad. Chapter 37 PP 463-474.